Cost Summary Explanati

Mission Data

Name: Popular name (post launch name if there is a name change)

Prime Contractor: Organization primarily responsible for systems engineering and overall project management

Funding Organization: Organization or government responsible for majority of funding. Many of the missions had funding, components, or services contributed by other organizations.

Mission: Principal mission. Most missions had multiple payloads and objectives

Mass: Spacecraft (including payload) dry mass, excluding upper stages, attachment mechanisms, and propellant, but including on-board propulsion systems

Launch Date: Parentheses used for estimated launch dates for spacecraft to be launched after 1/1/96.

Page Ref: Location in the text of the main case study discussion. See index for additional references.

Estimated Bus Cost

These estimates are for the spacecraft bus only. Additional estimated cost data is given on the page facing the first page of each case study. Explanations of the derivation of these estimates are given on page 348.

SMAD: Cost estimate based on Chap. 20 of *Space Mission Analysis and Design*, 2nd ed. [Larson and Wertz, 1992]. This is an empirical, mass-based estimate of what the spacecraft bus would have cost in a traditional program.

SSCM: Cost estimate based on the Aerospace Corp. Small Satellite Cost Model described in Sec. 8.2 of this text. This is an estimate of the spacecraft bus cost based on the mass-based portion of an empirical model of modern, small spacecraft.

Actual Cost

All costs in the table are in FY95$M. Inflation factors are approximate based on middle of the development period. Currency exchange rates are approximate values at the time of development. Costs include non-recurring development cost (or proportional share of non-recurring cost for multiple spacecraft programs), plus recurring cost for building the first flight unit, plus launch cost, plus ground segment cost (if a new ground segment is built), plus operations cost for 1 year. Explanations of the cost for each specific mission are contained in the respective case study discussions.

Spc Bus: Cost of the spacecraft bus, including the cost for systems and mission engineering, and spacecraft integration and test where these are separately identified. Includes payload cost for simple store-and-forward communications satellites and small test satellites.

Payload: Cost of the instruments where separately identified. Includes estimates of the value of instruments provided by outside organizations. (Where noted in the table, payload costs are included in the spacecraft bus cost.)

Launch: Cost of the launch and launch operations.

Ground Segment: Cost of the ground station facilities and equipment if acquired separately for the program.

Operations and Maintenance: Cost for first year of operations or for entire program if less than a year.

Average Cost/kg

All costs/kg are in FY95$K/kg. Bus and payload costs are based on dry mass of the portion of the spacecraft included in the respective cost entry. Total system cost/kg is based on the total system cost and the total spacecraft dry mass.

REDUCING SPACE MISSION COST

THE SPACE TECHNOLOGY LIBRARY
Published jointly by Microcosm Press and Kluwer Academic Publishers

The Space Technology Library Editorial Board

Other Volumes in the DoD/NASA Space Technology Series

Reducing
Space Mission Cost

Edited by

James R. Wertz
Microcosm, Inc.

and

Wiley J. Larson
United States Air Force Academy

This book is published as part of the Space Technology Series, a cooperative activity of the
United States Department of Defense and National Aeronautics and Space Administration.

Space Technology Library

Published Jointly by

Microcosm Press
El Segundo, California

Kluwer Academic Publishers
Dordrecht / Boston / London

Library of Congress Cataloging-in-Publication Data

A C.I.P. Catalogue record for this book is available from
the Library of Congress

ISBN 1-881883-05-1 (pb) (acid-free paper)
ISBN 0-7923-4021-3 (hb) (acid-free paper)

Published jointly by
Microcosm Press
401 Coral Circle, El Segundo, CA 90245-4622 USA
and
Kluwer Academic Publishers,
P.O. Box 17, 3300 AA Dordrecht, The Netherlands.

Kluwer Academic Publishers incorporates
the publishing programmes of
D. Reidel, Martinus Nijhoff, Dr. W. Junk and MTP Press.

Sold and distributed in the USA and Canada
by Microcosm
401 Coral Circle, El Segundo, CA 90245-4622 USA
and Kluwer Academic Publishers,
101 Philip Drive, Norwell, MA 02061 USA

In all other countries, sold and distributed
by Kluwer Academic Publishers Group,
P.O. Box 322, 3300 AA Dordrecht, The Netherlands.

Printed on acid-free paper

Table of Contents

List of Authors

David A. Bearden. Section Manager, The Aerospace Corporation, El Segundo, CA. M.S. (Aerospace Engineering), University of Southern California; B.S. (Mechanical Engineering), University of Utah. Chapter 8—*Cost Modeling*.

Bruce D. Berkowitz. Adjunct Professor, Carnegie Mellon University, Pittsburgh, PA. Ph.D., M.A. (Government), University of Rochester; B.A. (Government), Stetson University. Section 12.1—*Clementine*.

Richard Boudreault. Director General, Centre Technologique en Aérospatiale, Quebec, Canada. M. Eng. (Aerospace), Cornell University; B.Sc. (Physics Honours), University of Montreal. Chapter 8—*Cost Modeling*.

Gregg E. Burgess. Senior Principle Engineer, Advanced Systems Group, Orbital Sciences Corporation, Dulles, VA. M.S., S.B. (Aeronautical and Astronautical Engineering), Massachusetts Institute of Technology. Section 13.2—*ORBCOMM*.

Richard S. Caputo. Pluto Preproject Team, Jet Propulsion Laboratory, Pasadena, CA. M.S. (Mechnical Engineering), Carnegie Institute of Technology; B.S. (Mechnical Engineering), Manhattan College. Section 12.2—*Pluto Express*.

Gilberto Colón. Instrument Systems Manager, Goddard Space Flight Center, Greenbelt, MD. B.S. (Chemical Engineering), University of Puerto Rico. Section 11.3—*SAMPEX*.

Robert J. Diersing. Department Head, Accounting and Computer Information Systems, Texas A & M University, Kingsville, TX. Ph.D., (Computer Science), Texas A & M University; M.B.A., Corpus Christi State University; M.S. (Computer Science), B.B.A. (Electronic Data Processing), Texas A & I University. Section 13.3, *AMSAT*.

Orlando Figueroa. Project Manager, Explorer Program, Goddard Space Flight Center, Greenbelt, MD. Graduate Studies, University of Maryland; B.S. (Mechanical Engineering), University of Puerto Rico. Section 11.3—*SAMPEX*.

Rick Fleeter. President, AeroAstro, Inc., Herndon, VA. Ph.D. (Thermodynamics), Brown University; M.Sc. (Aeronautics and Astronautics), Stanford University; A.B. (Engineering and Economics), Brown University. Chapter 5—*Reducing Spacecraft Cost*.

Sven Grahn. General Manager, Science Systems Division, Swedish Space Corporation, Solna, Sweden. M.Sc. (Engineering Physics), Royal Institute of Technology, Stockholm, Sweden. Section 11.2—*Freja*.

Herbert Hecht. President, SoHaR, Inc., Beverly Hills, CA. Ph.D. (Engineering), University of California at Los Angeles; M.S. (Electrical Engineering), Polytechnic University of New York; B.S. (Electrical Engineering), City College, New York. Chapter 9—*Reliability Considerations*.

Donald M. Horan. Chief Scientist and Director of Operations, Clementine Project, Naval Research Laboratory, Washington, D.C. Ph.D. (Physics), The Catholic University of America; M.S. (Physics), University of Maryland; B.S. (Physics), Canisius College. Section 12.1—*Clementine.*

Edward J. Jorgensen. Member of Technical Staff, Jet Propulsion Laboratory, Pasadena, CA. Ph.D., M.A. (Economics); M.S., (Engineering); B.S. (Mechanical Engineering), University of Arizona. Chapter 7—*Design-To-Cost for Space Missions.*

John A. Landshof. Senior Staff, Mission Operations and Ground Systems Group, The Johns Hopkins University Applied Physics Laboratory, Laurel, MD. S.B. (Aeronautical and Astronautical Engineering), Massachusetts Institute of Technology. Chapter 6—*Reducing Mission Operations Cost.*

Wiley J. Larson. Professor, International Space University, Strasbourg, France. Managing Editor, Space Technology Series, United States Air Force Academy, Colorado Springs, CO. D.E. (Spacecraft Design), Texas A & M University; M.S. (Electrical Engineering), University of Wyoming; B.S. (Electrical Engineering), University of Michigan. Editor, Section 2.1—*The Government Perspective on Reducing Cost.*

Kim Leschly. Member of Technical Staff, Jet Propulsion Laboratory, Pasadena, CA. M.S. (Mechnical Engineering), Technical University of Denmark. Section 11.1— *Ørsted.*

John R. London, III. Deputy Missile Defense Architect, Ballistic Missile Defense Organization, Washington, D.C., Lt. Col. U. S. Air Force. M.S., Florida Institute of Technology; B.S., Clemson University. Chapter 4—*Reducing Launch Cost.*

Madeleine H. Marshall. Supervisor, Mission Operations and Ground Systems Group, The Johns Hopkins University Applied Physics Laboratory, Laurel, MD. M.S. (Computer Science), Johns Hopkins University; B.A. (Mathematics), Smith College. Chapter 6—*Reducing Mission Operations Cost.*

François Martel. Vice President, Advanced Programs, AeroAstro, Inc., Herndon, VA. Doctorate (Geophysics and Space Science), University Paul Sabatier, Toulouse, France. Chapter 11.4—*HETE.*

Ed Milton. General Manager, Surrey Satellite Technology Limited, Surrey, England. M. Eng. (Electrical and Electronic Engineering), University of Surrey, United Kingdom; D.M.S. (Management), University of Portsmouth, United Kingdom. Section 3.1—*Cost Effective Hardware and Technology.*

Harold E. Price. President, BekTek, Inc., Bethel Park, PA. B.S. (Computer Science), Indiana University of Pennsylvania. Section. 13.4—*PoSAT-1.*

George Sebestyen. President, Defense Systems, Inc., McLean, VA. D.Sc., M.S., S.B. (Electronics Engineering), Massachusetts Institute of Technology; Section 13.1— *RADCAL.*

Jerry Jon Sellers. Spacecraft Research Engineer, AFIT/University of Surrey, U.K. M.S. (Physical Science), University of Houston; M.S. (Aeronautical and Astronautical Engineering), Stanford University; B.S. (Human Factors Engineering), United States Air Force Academy. Section 3.1—*Cost-Effective Hardware and Technology*.

Robert Shishko. Senior Economist, Jet Propulsion Laboratory, Pasadena, CA. Ph.D., M. Phil. (Economics), Yale University; S.B. (Economics), S.B. (Political Science), Massachusetts Institute of Technology. Chapter 7—*Design-To-Cost for Space Missions*.

Robert L. Staehle. Pluto Preproject Manager, Jet Propulsion Laboratory, Pasadena, CA. B.S. (Aeronautical and Astronautical Engineering), Purdue University. Section 12.2—*Pluto Express*.

Jozef C. van der Ha. Ground Systems Manager, European Space Agency/ESOC, Darmstadt, Germany. Ph.D. (Mechanical Engineering), University of British Columbia, Vancouver, Canada; Masters (Applied Mathematics & Mechanics), Technical University, Eindhoven, The Netherlands; Bachelors (Mathematics), Technical University, Eindhoven, The Netherlands. Chapter 6—*Reducing Mission Operations Cost*.

Jeffrey W. Ward. Technical Director, Surrey Satellite Technology Ltd., Surrey, U.K. Ph.D. (Satellite Engineering), University of Surrey, United Kingdom; B.Sc. (Electrical/Computer Engineering), University of Michigan. Section 13.4—*PoSAT-1*.

James R. Wertz. President, Microcosm, Inc., Torrance, CA. Ph.D. (Relativity & Cosmology), University of Texas at Austin; M.S. (Administration of Science and Technology), George Washington University; S.B. (Physics), Massachusetts Institute of Technology. Editor, Chapter 1—*Introduction*; Section 2.2—*Radical Cost Reduction Methods*; Section 3.2—*Software*; Chapter 8—*Cost Modeling*; Chapter 10—*Implementation Strategies and Problems*.

Preface

Nearly all of us agree that we need to reduce the cost of space missions. Clearly we don't agree on how to do this or we would already have done it and there would be no need for this book. Nonetheless, substantial evidence shows that major cost reductions are possible. The purposes of this book are to summarize that evidence, to present data on methods that have worked on prior programs and that have been suggested for future programs, and to provide a common framework within which we can all work at the goal of making space exploration affordable.

Evidence that cost reduction is possible comes from several directions. For 30 years, the Soviet Union launched 1 to 2 satellites per week on a total budget approximately comparable to, and probably less than, that of the United States. The capability of these satellites was limited and, consequently, the cost per unit of performance may have been comparable to those launched in the West. However, the number of spacecraft launched and the mass of material launched far exceeded the totals launched by the West. These facts imply a cost per kilogram far lower than the United States and Western Europe have achieved.

A second body of evidence comes from the LightSat community in both the U.S. and Europe. For many years, this community has developed ever more capable satellites for costs on the order of one tenth that of traditional programs. In addition, both the Soviet launchers and western LightSats have had equal or better reliability than large traditional programs. The case studies in this book cost 50% to less than 10% of their expected cost based on a traditional cost model.

Traditional cost reduction methods are those that have or easily could be applied within the formal framework of a normal government or commercial satellite procurement. These methods, such as Design-to-Cost and Concurrent Engineering, represent a formalized approach to changing how we design space systems. They can contain cost growth and potentially drive costs down by 10% to 50%.

In contrast to the traditional approaches, *radical* cost reduction methods offer the possibility of reducing costs by a factor of 2 to 10 or more by making radical departures in how satellites are bought, built, and operated. These methods seek to change the paradigm by which satellite acquisition works and may at times be inconsistent with traditional methods and with each other. In almost all cases, radical cost reduction requires changing the rules of the game. The most obvious of these changes is to allow trading on requirements—that is, meeting the overall mission objectives by creating a compromise between what we want and what is available at low cost.

Neither traditional nor radical cost reduction methods are inherently better. Which approaches to apply, or even whether any major attempts at cost reduction are appropriate, will depend on the nature of the individual program. The current process of space mission design and development was not created in a vacuum or by individuals unconcerned with cost. Almost any attempt to reduce cost introduces some elements of compromise. We take up these issues explicitly in Chapters 1 and 10 and discuss how to set cost vs. performance objectives.

We do not claim that this book brings "new" ideas to the community or definitive answers as to what is right for your particular program. Rather we have attempted to collect, synthesize, and articulate what is known today and what people and organizations have suggested about cost reduction methodologies. We have tried to provide both motivation and methods to tackle the problem of cost: a critical first step toward truly expanding and extending the spaceflight revolution much as the aircraft revolution expanded early in the twentieth century. We **can** reduce costs dramatically and we **must** do so to ensure the continued growth and utilization of space.

Projects of this sort involve many people who get little credit for a great deal of work. The Air Force Academy provided the most essential ingredient—the funding needed to make the work happen. Sandy Welsh at OAO provided excellent contract support and Eugene deGeus at Kluwer provided coordination and advice.

Perry Luckett provided English and grammar editing for the entire volume. Although frequently painful for the authors, Perry did a truly amazing job of transforming "techno-speak" into English. Frank Redd of Utah State University, Chris Elliot of Smith System Engineering Limited, Rex Ridenoure of JPL, and Bryant Cramer of GSFC provided substantial reviewing assistance. At Microcosm, Simon Dawson provided much of the analysis for the cost summaries and many of the figures. Simon, Hans Koenigsmann, and John Collins prepared the Appendix and reviewed and indexed the entire book. Leslie Sakaguchi led a Microcosm team that provided typing, reviewing, proofing, and indexing. And it was fascinating to have the book worked on by three generations of Klungles—Donna, Jeanine, and Taylor Michelle providing the moral encouragement. Jeanine did an outstanding job creating and recreating figures and graphics, as well as page layout and overall editing and fixing. Finally, as the authors and editors well know, the book would not have been possible without Donna's technical, management, and diplomatic skills. In part, because this book needed to be very different than anything most of us have ever written, creating the book was exceptionally painful task for all of the authors. Donna managed to coax the material out of the authors and transformed frequently incoherent input into a finished book. Thank you Donna. It was very much appreciated.

January 1996

Wiley J. Larson
Department of Astronautics
United States Air Force Academy
Colorado Springs, CO 80840-6224
FAX: (719) 333-3723

James R. Wertz
Microcosm, Inc.
401 Coral Circle
El Segundo, CA 90245-4622
FAX: (310) 726-4110
E-mail: jwertz@smad.com

Part I
Process Changes to Reduce Cost

Chapter 1

Introduction

James R. Wertz, *Microcosm, Inc.*

The purpose of this book is to speed the process of reducing space mission cost by:

- Examining the reasons for high cost
- Providing a clear recipe for reducing cost
- Describing what works and doesn't work in practice
- Providing both data and sources of data appropriate to specific problems
- Providing specific examples

It should be clear that there is no simple answer to the problem of reducing mission cost. If, for example, we could reduce the cost of all missions by simply changing the solar array technology or the materials spacecraft are made of, all prime contractors would have done this long ago and there would be no need for this book.

The methods proposed throughout this book—and summarized in Sec. 1.3, Chap. 2, and Chap. 10—are simple and straightforward, yet subtle to implement. Still, they can be dramatically effective. By most measures, the programs described in the case studies at the end of this book and on the inside front cover have reduced cost by three to more than ten times below what would have been the case using more traditional approaches to space mission design. In addition, the historical evidence seems to be that low cost comes without decreasing reliability. Although no statistical studies are available, anecdotal evidence suggests that small, low-cost spacecraft are equally or more reliable than their more traditional, larger counterparts. This may be in part because they are simpler spacecraft with larger tolerances and fewer components, which tends to strongly increase reliability.

This book is organized into two parts. First, we discuss the overall process of reducing cost, ending with a summary of implementation procedures intended to provide a practical recipe for how to go about doing this for both new and ongoing missions. Part II then presents 10 case studies which illustrate how this is carried out in practice with both the results which can be achieved and the potential pitfalls that may be encountered.

We begin with several basic definitions. Historically, *cost models* have been used to provide purely empirical estimates of what it costs to build and fly spacecraft. In these models, the cost is represented as a simple function of spacecraft parameters such as mass or power, or the lines of code for ground systems. Cost models ignore the subtleties of how we manage and run programs to answer the basic question: What would it cost if we built it the way we built it last time? As described in Sec. 8.2, there are a number of cost models based on substantial historical samples which can provide a good estimate of mission cost for spacecraft built using the traditional rules of the game. For comparison purposes throughout this book, we'll use the cost model presented in Chap. 20 of SMAD[*] because it's widely available and easy to use and apply. We define a *traditional mission* as one which more or less follows the design rules and cost estimates presented in SMAD.

In contrast to the traditional mission, a *reduced-cost* mission is one for which the life-cycle cost is substantially lower than that predicted by traditional cost models for that mission or spacecraft. For example, Pluto Express, with an estimated cost of $150 million, might be regarded as expensive, but it clearly qualifies as a reduced-cost mission by the standards of traditional, interplanetary spacecraft.

We define a *low-cost* mission as one for which the life-cycle cost is substantially less than the average cost of a space mission of that type. This definition implies that we need to have in mind some idea of what we mean by the average cost. Of course this will depend on the type of mission, how it is built, and the time frame in which it is built. For concreteness, Table 1-1 provides a working definition of what we mean by low-cost space missions.

TABLE 1-1. Empirical definition of a low-cost mission. As a working definition we refer to a mission as "low-cost" if it is three or more times less expensive than typical missions in that class.

Class	Typical Cost[1]	Low Cost	Typical Mission	Low-Cost Mission
Low-Earth orbit	$150M to $2,000+M	<$50M	DMSP, GRO	ALEXIS, Freja
Geosynchronous	$250M to $2,500+M	<$75M	Intelsat, TDRS	Ball GEO comsat[2]
Interplanetary	$1,500M to $3,000+M	<$500M	Galileo, Cassini	Clementine

[1]Life-cycle cost including development, spacecraft, launch, and operations
[2]Proposed, but not built

1.1 Range of Cost Options

As illustrated in Fig. 1-1, we can design space missions to fall in any range from minimum cost to maximum performance. The underlying curve of performance versus cost is simply an offset exponential. That is, even getting a brick into orbit costs something. Beyond that, the cost of getting more performance is proportional to the performance that we have. This simple model explains a number of characteristics that are common to space missions as well as many other development activities. At

---

[*] "SMAD" is used throughout this book for *Space Mission Analysis and Design,* 2nd edition. [Larson and Wertz, 1992]

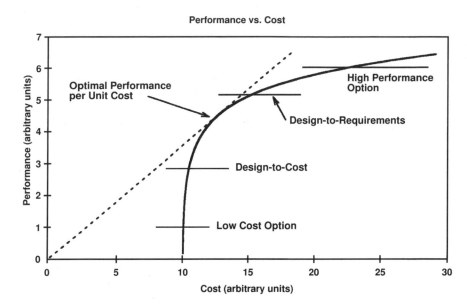

Fig. 1-1. The Range of Cost Options. For the underlying curve, we assume that there is a minimum cost to launch anything into orbit. Beyond that, the cost of additional performance is proportional to the performance already achieved. See text for discussion.

the very low end, a little more money will substantially improve performance. At the high end, additional performance is extremely expensive.

None of the cost options is necessarily better or worse than any of the others. It depends on the objectives that we're trying to achieve. The key is to decide what our cost objectives are and design the mission, the spacecraft, and the operations to be consistent with these objectives. This is simply another way of saying that we need balance in our engineering solutions. On a $1 billion science mission, it makes no sense to save $100,000 by using a lower-quality star sensor. Similarly, on a $2 million Earth-observing platform, it would be foolish to spend $1 million on a high-accuracy star sensor.

The five options illustrated in Fig. 1-1 are briefly described below.

High Performance Option

The purpose here is to obtain the best available performance. Missions of this type are used in some military and science missions, for which the objective is to push the state-of-the-art, open new ways to understand the universe, or provide critical knowledge for national defense. On the science side, examples would be Space Telescope, GRO, or AXAF. On the military side, these would include MilStar and various reconnaissance satellites. Typically, these programs are very expensive. The last increment of performance costs a great deal. Nonetheless, they also advance the state-of-the-art and create new knowledge, both about the subject being studied and about high-performance, high-precision spacecraft.

Design-to-Requirements

This is the traditional approach used for most military and science missions, such as TDRS, DMSP, or ISEE. We can also apply design-to-requirements to the very expensive, high-performance missions. The basic objective is to meet a set of pre-defined requirements at minimum cost and risk. Because the requirements are fixed, the cost is variable and is typically high and not easily controlled. We can estimate the costs in advance, but in most cases we won't know them until after we've built the spacecraft. Typically, this type of spacecraft will meet its operational requirements, but may or may not use financial resources effectively.

Optimized Performance per Unit Cost

The third category is to obtain the highest performance per dollar. This is used in commercial programs, such as long-term communications systems. What matters is not so much the total amount spent on each spacecraft, but rather the cost per unit of performance, so we can get the best return-on-investment.

In Fig. 1-1, these spacecraft are represented by a straight line passing through the origin, tangent to the Performance vs. Cost curve. This straight line represents the maximum performance per unit cost and, therefore, is the optimal satellite size. On the whole, optimized performance per unit of cost will still mean expensive satellites. It typically requires a large spacecraft with substantial payback to recover the burden of the launch cost. Examples here include most of the commercial communications satellites which are launched into geosynchronous orbit, such as the IntelSat series. These satellites are also frequently built on the basis of design-to-requirements. However, many such satellites have been built, so that the requirements can be tailored to come very close to the optimal knee in the curve. The purpose of using design-to-requirements is to provide for competitive bids such that the spacecraft will be built by the lowest bidder who can meet the proposed requirements.

Design-to-Cost

In this category, the objective is to obtain the best performance available within a fixed cost. Here costs are moderate usually less than for the optimal system, and are pre-defined such that the program is threatened with termination if it exceeds cost. This type of mission is becoming more popular, particularly for science missions, in which variable requirements may be more acceptable. Examples would include the Discovery missions and Pluto Express. The cost ceiling simply becomes a way to control cost and the design process tries to get the best performance within the fixed cost constraint. Chap. 7 describes this process in more detail.

Low-Cost Mission

These are the very cheapest space missions for which the goal is to obtain a modest level of performance at very low cost. It's used for SmallSats and some space experiments. (See Sec. 1.2.) However, there are also potential applications in multi-satellite systems in which the cost is multiplied by having a large number of satellites. With ever more constrained budgets, applications for the low-cost options are continuing to grow. Examples here would include ALEXIS, GLOMR, and the UoSATs. Usually, we choose this option because no more money is available for the project. If performance within this budget is acceptable, the mission will fly. If not, it will simply be dropped.

One of the most interesting characteristics of the very low-cost missions is that, like personal computers, the available performance is continuing to increase. Many of the case study missions in the second half of this book are examples of the low-cost option. In most cases, the organizations building them are improving performance while still keeping the cost extremely low. Even the least expensive amateur radio satellites are gaining significantly in their complexity and performance. Thus, the "low-cost" option is rapidly pushing up performance and, in due course, may be replaced by even smaller and lower-cost *MicroSats*, such as those AeroAstro is offering.

1.2 Small vs. Low-Cost Missions

Our goal is to reduce the cost of all missions, but much of our conversation will focus on small satellites. For this book, we define a *small spacecraft* (also called a *SmallSat* or *LightSat*) as any spacecraft weighing less than 400 kg including propellant. Small spacecraft aren't necessarily low cost, and low-cost spacecraft aren't necessarily small. But the two categories overlap a lot. We can also identify small spacecraft far more easily than reduced-cost or low-cost ones. The weight of nearly all spacecraft is well known, whereas cost may or may not be known for a particular spacecraft and the meaning of the cost data may be very ambiguous. Consequently, we will often use small spacecraft as somewhat synonymous with low-cost ones. Although we recognize that this is not strictly true, it gives us a way to assess technical parameters and characteristics much more easily than we otherwise could.

Figure 1-2 summarizes the satellite launches distributed by mass for 1990 to 1994. It is clear from this figure that the concept of a small satellite has an empirical definition. This was the reason for our choice of 400 kg as a number which covered all satellites up to the beginning of what might be called traditional large satellites represented by the data to the right in the figure. It is clear from the figure that there is a genuinely different class of satellites, apart from the traditional large ones. By definition these are the small satellites; in addition they are typically, but not necessarily, low-cost as well.

Also note that small satellites aren't new; they've been with us since the beginning of the space program. Figure 1-3 shows a number of perspectives on the history of small satellites. Early in the space program, particularly for the U.S., most spacecraft were small because the U.S. had only limited capacity to launch larger ones. A marked decline in the American space program occurred at the time of the Apollo program in the late 1960s. It also represented a shift from predominantly small spacecraft to a greater percentage of what we would now call larger, more traditional spacecraft. But small spacecraft still accounted for about 20% of the launches in the United States, the Soviet Union, and the rest of the world. Also, we can see the effect of the *Challenger* disaster in the late 1980s, the Soviet Union's breakup in the early 1990s, and the continuing rise in space activity by the rest of the world.

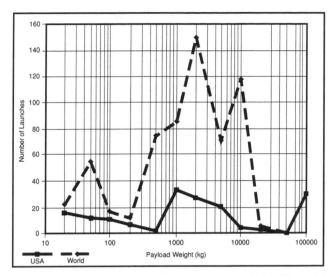

Fig. 1-2. Total U.S. and Worldwide Space Launches, 1990–1994, Plotted by Spacecraft Mass. Note the division into two distinct mass groups. This data shows we can empirically define a small spacecraft as one whose mass is less than 400 kg.

The appendix summarizes the main characteristics of many small spacecraft. The data on these spacecraft is from the KISS Database. One of the major messages from the large number of small and low-cost spacecraft is that we can reduce costs dramatically in nearly all types of missions. Examining the results of these missions shows we can do good science, good communications, good testing, and good operational missions with small, low-cost satellites. And their reliability is much the same as that of more expensive spacecraft. We don't necessarily do science or operations the same way in small, low-cost missions, but it can still be good science.

1.3 Lessons Learned from the Case Study Missions

Table 1-2 lists the case study missions presented in Part II. (Also see the inside front and rear covers for cost data summaries.) Like many of the rest of the tables, it's unusual for a technical book in that it lists the cost for each of the missions. This illustrates what we might call the first rule of reducing mission cost:

Reducing space mission cost is hard if we know what the costs are and virtually impossible if we don't.

There is a very strong tendency, within both the government and private industry, not to want to talk about real cost. Cost and price data is extremely sensitive and much of it is proprietary. Cost data is exceptionally hard to compare across program lines and components. Data for any specific program may include or exclude launch, ground stations, operations, infrastructure costs, support personnel, and so forth. The cost of individual components depends far more on how we buy them rather than what

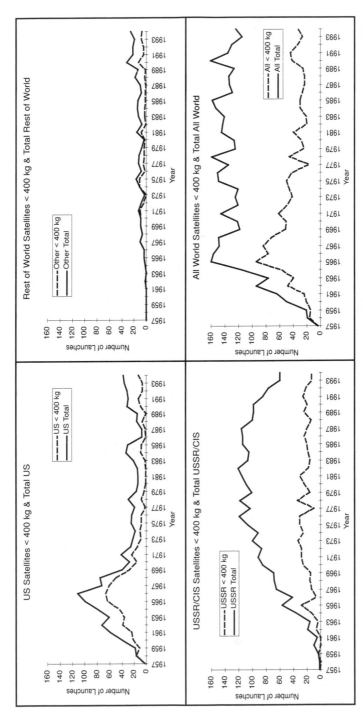

Fig. 1-3. Comparison of Annual Worldwide Launch of Small Spacecraft with that of All Spacecraft Since the Beginning of the Space Program. These curves are likely to shift in the future as low-Earth orbit communications constellations begin to be launched and as the impact of the drive toward reduced cost missions is reflected in actual launches. As discussed in the text, several historical trends are also evident.

TABLE 1-2. Case Studies Used in the Reducing Space Mission Cost Project. For more cost and technical summaries, see the inside front and rear covers, the introduction to the Case Studies at the beginning of Part II, and the page facing each case study. Cost is the total life-cycle cost in FY95$M.

Mission	Launch Wt (kg)	Cost (FY95$M)	Comments
Ørsted	60	$18.4M	Danish magnetic field mission
Freja	256	$24.5M	Swedish magnetosphere mission
SAMPEX	160	$72.6M	First of the SMEX series
HETE	125	$29.1M	High-energy science mission
Clementine	1693	$85.0M	Failed part way into mission
Pluto Express	103	$284M	Projected cost for FY95 version
RADCAL	92	$16.6M	Excludes operations and maintenance
ORBCOMM	47	$15.7M	First two test satellites
AO-13	140	$1.24M	Amateur radio satellite
AO-16	9	$0.20M	Amateur radio microsatellite
PoSAT-1	49	$2.10M	Portuguese technology transfer satellite

we buy. The number of documents, meetings, analyses, tests, and reports determines the cost.

Cost data is frequently hidden to make costs appear lower. Costs can be buried in infrastructure costs, as when a government center or company division doesn't include the cost of managers or engineers who are supporting the program. Costs may or may not include non-recurring components, such as refurbishing buildings or maintaining and supporting equipment. As an example, no one knows what a Space Shuttle launch actually costs. Estimates range from $200 million to more than $500 million per launch. Perhaps most important, revealing cost may make it harder to get a program funded. If we emphasize a program's total life-cycle cost Congress, or the system buyer, may decide that it isn't worth that much.

Still, if we're going to successfully reduce cost, we have to talk about real cost and reveal cost data for virtually all space programs. This is one of the ways the space business must change if we are to succeed in dramatically reducing space mission cost.

We began work on this book hoping that we could sort out a few common characteristics among low-cost missions and then show that low cost would be straightforward if we follow a few simple rules. Unfortunately, that hasn't proven to be true. We haven't found a common denominator among the low-cost missions. Some principles and approaches are common, but there are also major differences in how cost reduction is achieved. Thus, how we do something is more important than what we do. Somehow, our approach to engineering and managing determines whether a mission will be high-cost or low-cost. For nearly all of the "rules" this book gives, we can find counter-examples of low-cost missions which don't follow them —as well as missions for which these same rules increase rather than decrease cost. This implies that we need to change how we think about missions and how we undertake them in order to strongly reduce cost. Although this idea seems nebulous, the case studies show we can get dramatic positive results. We'll return to this issue of changing the paradigm of doing space business at the end of the chapter.

Keeping in mind the above caveat about applying rules with care and common sense, Table 1-3 summarizes the main ways to reduce mission cost as proposed in the process chapters and implemented in the case studies. Almost all of the methods have advantages and disadvantages that depend mainly on how they're applied. A few of the rules, such as greater use of microprocessors or more autonomous systems, deal with applying advanced technology. However, most of the rules—such as trading on requirements, reducing the cost of failure, or providing larger margins—deal more with the process or methodology by which we design and build space systems. Experience has shown that how we tackle the mission design, not where or how we obtain the components, is what really matters in reducing cost. As mentioned earlier, dramatic cost reduction occurs predominantly, although not entirely, in small spacecraft. Small spacecraft tend to be inherently simpler and lower cost. We ordinarily have fewer payloads and fewer requirements and constraints. Small programs are usually more willing to accept risk and to settle for what can be achieved at moderate cost rather than pushing to fulfill a specific mission requirement. This reinforces our tendency to think of small spacecraft as low in cost, although we need to keep in mind that the two are certainly not synonymous. Large spacecraft with much lower cost include the proposed Industrial Space Facility by Westinghouse or the NEAR Asteroid Flyby Mission built by APL. Similarly, very small spacecraft which remain small by pushing the bounds of technology will almost certainly not be low-cost (except for the launch), particularly if we consider the non-recurring cost for developing the technology.

Looking at the case studies and the missions listed in the Appendix clearly shows that dramatic cost reduction can occur in government, universities, or industry. No one seems to have a monopoly on finding ways to reduce cost. As discussed in Sec. 2.2.2, low-cost missions occur mainly in smaller organizations, such as Aero-Astro, DSI, the University of Surrey, Utah State, or the Swedish National Space Board. But some of the more well-known examples have occurred within very large organizations, such as the Ballistic Missile Defense Organization, Naval Research Laboratory, JPL, or the Lockheed Skunk Works for Aircraft. We believe there are good reasons why small organizations have at least some advantage in reducing cost. First, dramatically reducing cost isn't easy and typically involves choosing from bad options. Thus it occurs most often whenever money simply isn't available, so hard choices **must** be made. This occurs most commonly in small programs and small organizations. They've learned to make these hard choices because they do so for nearly all of their decisions. (See Sec. 2.2.2 for further discussion of why small organizations have an advantage.)

There is some evidence within the case studies that small teams within large organizations may have some of the best of both worlds. The space group within NRL, which built Clementine, is a good example. Spacecraft aren't the U.S. Navy's main line of work. The Naval Center for Space Technology can operate as a largely independent organization, while having at least some protection from disappearing by being an important segment of the Navy's organization. Similarly, the Pluto project office at JPL, which produced the Pluto Express design, has been able to do so in part by setting itself aside from the JPL's and NASA's traditional project organizations. By convincing their parent organizations there was no other choice, they've been able to define new rules of business that have allowed them to define a mission at dramatically lower cost than was previously possible. All of this illustrates that it isn't

TABLE 1-3. Summary of Cost-Reduction Methods. As discussed in the text, how we apply the rules is the key to reducing cost

Method	Mechanism	Comments
Trading on requirements	Eliminates non-critical requirements; permits use of low-cost technology	Makes traditional competition difficult
Concurrent engineering	Allows schedule compression; reduces mistakes; increases feedback between engineering and manufacturing	High non-recurring cost relative to lowest-cost programs; can achieve optimal design
Design-to-cost	Adjusts requirements and approach until cost goal has been achieved; makes cost paramount	Spacecraft have rarely used it
Schedule compression	Reduces overhead of standing army; forcing program to move rapidly does drive down cost	Often results in a poor design due to lack of up-front mission engineering; must reduce work required to be consistent with schedule
Reducing the cost of failure	Allows both ambitious goals and calculated risk in order to make major progress	Fear of failure feeds the cost-growth spiral; major break-throughs require accepting the possibility of failure—particularly in test
Using microprocessors	Minimizes weight; provides high capability in a small package; allows **on-orbit** reprogramming	Problem of single-event upsets; high costs of flight software; very difficult to manage and control software development
Large margins	Reduces testing; better flexibility; reduces cost of engineering, manufacturing, and operations	Margins traditionally kept small for best performance—drives up development cost
Using non-space equipment	Takes advantage of existing designs and potential for mass production	Typically not optimal; must be space qualified
Autonomous systems	Reduces operations costs	Can increase non-recurring cost
Standardized components and interfaces	Reduces cost and risk by **reusing** hardware; standardization is a major requirement for other types of manufacturing	Has been remarkably unsuccessful in space; **sub-optimal** in terms of weight and power

really the organizational size, but the mentality and decision-making process within the group that seems to be most important. If we ignore the management and business aspects and simply ask "Where can we buy low-cost solar arrays?" we will not in the end produce spacecraft at a significantly lower cost.

1.3.1 Are Cost Reductions Real?

Most space programs normally are run so as to create the lowest-cost design consistent with the mission requirements and constraints. Very few programs try to drive up cost. Thus, systems and missions must be fundamentally different in some respects to dramatically reduce cost. We won't be able to reproduce the TDRS or

DMSP spacecraft, with all of the testing that has gone into them, for much less than they cost originally. So the real question isn't, "Can we build the same spacecraft for less?" Rather, it is, "Can we meet the overall mission objectives for much less money." The best evidence for this is to look at the actual cost data from the case study missions. This data is summarized on the inside front and rear covers and is shown in more detail in Fig. 1-4 for four representative case studies. (See the separate case studies for more on the cost of each mission.) The column for comparison, labeled "Expected Cost," is based on long-term, empirical data reflected in the SMAD cost model. In addition, comparison cost data is also given from the Small Satellite Cost Model developed by The Aerospace Corporation and described in detail in Sec. 8.2. If we look over the cost data and review what the missions achieve, a "preponderance of the evidence" suggests we can reduce cost two to ten times without affecting the reliability. But, it's also true that low-cost missions aren't identical to their more traditional counterparts, either in what is built or how the program is run. Developing low-cost or reduced-cost missions requires truly changing the way we do business. With these changes it appears possible to reduce cost in essentially all aspects of the mission: non-recurring, production, launch, and operations.

The AMSAT Lesson

It's particularly important to point out the lesson of the Amateur Radio Satellite Corporation (AMSAT). They are, without question, the world leader in producing very low-cost, highly reliable spacecraft. Most engineering organizations tend to discard AMSAT as irrelevant because it's a group of amateurs with all-volunteer labor. But if we look carefully at their data and accomplishments (Sec. 13.3), we can't easily dismiss the important lessons to be learned.

AMSAT has no contracts, no formal management structure, no QA organization, and no penalty clauses or other ways to compel people to act in a particular way. They are an international group with components being built worldwide by people who, in many cases, have never met. Their available budget is remarkably small for each mission. Yet, AMSAT has never delivered a satellite late for launch. Depending slightly on how one counts, they have launched more than 40 satellites to date, of which only one, a French amateur satellite, failed shortly after launch. The rest have been successful. The current AMSAT spacecraft have 3-axis-stabilized attitude control systems and propulsion systems which perform major orbit maneuvers. The AMSAT now under construction will be approximately 400 kg and 300 W and only marginally falls into our small-satellite category. The total budget of $3 to $4 million dollars is far more expensive than any prior AMSAT, but still dramatically cheaper than many satellites with less capability.

AMSAT is a collection of talented and highly motivated people. It is clear, as pointed out in the AMSAT case study, that much of the AMSAT success comes from this high level of motivation which cannot be achieved by penalty clauses and QA representatives. In the end, substituting bureaucratic requirements and regulations for motivation and skill is a very poor trade. Many people in the United States and throughout the world want to explore and exploit space. They're highly motivated, highly skilled, and have innovative ideas on how to do it. This is the group we need to find and use. Ultimately, what matters most in getting good results at low-cost is not equipment and capital assets, the size of the organization, or the way the spec is

Pluto Express – Two 103 kg interplanetary probes:

	Expected Cost*	Small Spacecraft Model**	Actual Cost‡
Spacecraft Bus	$1,140M	$12M	$226M
Payload†	$50M	$1M	$9M
Launch	$565M	$332M	$43M
Ground Segment††	N/A	—	Incl. in spc cost
Ops. + Main. (annual)	$40M	—	$6M
Total (through launch + 1 yr)	**$1,795M**	—	**$284M**

* Initial expected cost is based on the buy of 2 additional, unmodified CRAF/Cassini spacecraft.
** The Small Spacecraft Model does not include interplanetary spacecraft and is not intended for this application.
† Does not include cost of the Russian "Drop Zond" to be deployed to the surface of Pluto.
†† Ground segment development cost was incorporated into spacecraft cost to facilitate cost trades during mission definition.
‡ An inflation factor of 1.000 has been used to inflate to FY1995$ [SMAD, Table 20-1].

RADCAL – 92 kg military test satellite:

	Expected Cost	Small Spacecraft Model	Actual Cost*
Spacecraft Bus	$35.8M	$7.8M	$4.4M†
Payload†	$18.0M	$0.7M	Incl. in spc cost
Launch	$16.6M	$7.1M	$12.2M‡
Ground Segment	$37.8M	—	Incl. in spc cost
Ops. + Main. (annual)	$4.5M	—	N/A
Total (through launch + 1 yr)	**$112.7M**	—	**$16.6M + O&M**

* An inflation factor of 1.106 has been used to inflate to FY1995$ [SMAD, Table 20-1].
† Includes cost of 2 ground stations
‡ Estimate based on Scout launch vehicle cost.

AMSAT-OSCAR-13 (AO-13) – 140 kg amateur radio satellite:

	Expected Cost	Small Spacecraft Model	Actual Cost*
Spacecraft Bus	$34.4M	$7.0M	$0.96M
Payload	$18.0M	$0.7M	Incl. in spc cost
Launch	$66.4M	$9.8M	$0.28M
Ground Segment	$7.6M	—	N/A
Ops. + Main. (annual)	$1.1M	—	N/A
Total (through launch + 1 yr)	**$127.4M**	—	**$1.24M**

* An inflation factor of 1.377 has been used to inflate to FY1995$ [SMAD, Table 20-1].

Freja – 256 kg magnetospheric and auroral research satellite:

	Expected Cost	Small Spacecraft Model	Actual Cost*
Spacecraft Bus	$44.4M	$12.8M	$12.5M
Payload	$32.4M	$6.0M	$6.0M
Launch	$66.4M	$6.6M	$4.8M
Ground Segment	$118.2M	—	$0.8M
Ops. + Main. (annual)	$9.8M	—	$0.4M
Total (through launch + 1 yr)	**$271.2M**	—	**$24.5M**

* An inflation factor of 1.000 has been used to inflate to FY1995$ [SMAD, Table 20-1].

Fig. 1-4. Cost Data for Representative Case Studies. The "expected cost" is based on the SMAD mission model, as explained on page 348. It estimates what the mission would have cost had it been built using traditional spacecraft rules and methods. The costs for the Small Spacecraft Model are those that would be expected based on an empirical model of modern, small spacecraft. Sec. 8.2 explains it fully.

written. What matters is the people who do the work, their motivation, their skill, and their freedom to innovate.

Pluto Express is an interesting example of these lessons applied to a major space program. Key people creating Pluto Express have pointed out that they have drawn many of their rules for lower cost from AMSAT and have tried to apply these rules to a large, interplanetary program. This is a major accomplishment because JPL is neither a small nor a traditionally low-cost organization. The cuts they've achieved in the Pluto Express design may or may not be enough to allow the Pluto mission to be funded. Yet, this case study can be exceptionally valuable for larger programs who need to apply the lessons of small missions to inherently more complex and expensive ones. The fastest, most economical, and most reliable way to reduce cost is to learn from those who have done it and to adjust these lessons to the needs of your program.

1.4 The Need for a New Paradigm

The low-cost and small-satellite communities have been very successful throughout the history of the space program, both in the U.S. and internationally. Satellites have flown with high reliability and at far lower cost. It's clear that there are techniques for doing business more economically in space. Whether or not these are better solutions will depend on the objectives of each specific mission.

Perhaps the principal message of the case studies in this book is that the solution to high cost typically doesn't lie in higher tech, ever-more optimized spacecraft, but in the application of balance, common sense, and simple, good engineering to mission and spacecraft design. For example, autonomy exists in both the cheapest and most expensive satellites. Doing things at low-cost means creating a simple satellite for which autonomous operations are easy. The University of Surrey is currently operating seven of their low-cost satellites with an operations crew of one person. As another example, miniaturizing can greatly reduce launch cost but can also drive up the cost of the satellite if it requires inventing new equipment and new technology for each spacecraft. Reducing margins (structural, power, throughput) gives us the best design and maximizes the performance; it can also dramatically increase the cost of design, manufacture, test, integration, and operations. Compressing the schedule can reduce cost, but only if the program can relieve requirements and make prompt decisions.

The right answer is a combination of both **what** we do and **how** we do it. Higher margin will increase cost if it means pushing spacecraft components to operate at temperatures well beyond what they will ever see on orbit. But it can strongly lower cost if it means allowing increased structural weight to reduce the uncertainty, manufacturing tolerances, analysis time, and testing requirements. There is no single, simple solution to reducing cost; but applying common sense and being willing to work problems, rather than forcing the application of rules and policies, provide an excellent beginning. This is one of the reasons it is very difficult for the government to reduce cost. Rules and policies can't force using common sense to solve problems— they typically constrain actions and increase cost. It requires motivated, knowledgeable, and skilled engineers to achieve the results we want.

By looking at the tables on the inside front and rear covers, or the list of missions in the Appendix, we can conclude that dramatic cost reduction is possible for virtually all types of missions, all mission aspects, and all categories of sponsoring organizations:

- Spacecraft, launch, ground segment, and operations
- Science, communications, military, test, and observation systems
- Government, commercial, and academic programs

We don't have hard evidence on whether these rules are applicable to manned missions. However, we anticipate that many of them would be applicable with much of the same results—lower cost, higher reliability, and less-than-optimal performance.

Even modest cost-reduction methods should be able to reduce the total cost of space missions by several billion dollars per year. Far greater savings are possible. In almost all countries far more needs to be done in space than there is money to do it. Consequently, we hope a lot of these savings can fund doing more and reaching out faster and farther than we have before. If other transportation systems are an indication, reducing the cost of reaching space will greatly increase rather than decrease the total amount of business done there.

To strongly reduce cost, we need to change the rules by which we do business in space. We need to:

- Trade on requirements and create compromises between what we want and what we can afford
- Be more open about cost and treat it as an engineering parameter, as we do in other businesses and in our personal lives
- Learn from what others have accomplished in SmallSats and the many historically low-cost spacecraft
- Encourage innovative individuals and small groups who can think and perform "low-cost" and allow them the freedom of an occasional failure in order to achieve dramatic success

The goal of this book is to understand how to do this. The next several chapters summarize cost-reducing processes, technologies, and methods for spacecraft, launch, and operations. Chap. 10 then summarizes rules for implementing a cost reduction program, both in new missions and on-going ones. In that chapter we'll return to the issue of changing the paradigm by which we do business. Finally, the case studies show how these techniques have been implemented in real programs and what the results, successes, and problems have been. Our intention is to provide a practical road map for reducing space mission cost. We hope that it is successful for your program.

Chapter 2

Process Changes to Reduce Cost

2.1 The Government Perspective on Reducing Cost

Wiley J. Larson[*], *United States Air Force Academy*

This chapter reflects my perception of how the U.S. Government is trying to reduce the cost of space systems and doesn't represent any official position within the government. I have about 20 years of experience in government, but my biases and limited experiences may color the following comments. In addition, many of the comments and ideas presented here come from the Reducing Space Mission Cost Workshop held in March 1995 [COSA, 1995].

The government is a collection of **people**, organizations, rules, and regulations. Government people are people just like everyone else. For the most part, they care about the work they do and dedicate themselves to doing a good job. At the same time, the government is regulation-based, so people succeed by following and enforcing the rules and regulations. After many years of doing so they must overcome some "inertia" when they're asked to be more inventive. Many government people are innovative—we just need to encourage them to use their talents and reward them when they do!

At the highest levels, the U.S. Government consists of the executive, judicial, and legislative branches. The people in these positions, including committees and staffs, drive space policy **or the lack of it**. As shown in Fig. 2.1-1, the DoD and NASA are the two key government players in space. Others exist, such as the National Oceanic and Atmospheric Administration (NOAA) and the Departments of Transportation or Commerce, but I'll talk only about DoD and NASA. Within DoD, space is a relatively minor concern. In fact, the rules and regulations about acquiring and

---

[*] Much thanks to Capt. George Moore, USN, for his reviews, additions, and list of references.

operating systems are developed abstractly enough to fit tanks, airplanes, commodes, and many other obscure items, and, oh by the way, space. The acquisition process is established to handle very general procurements. Acquisition officers within DoD space must carry out the policies and procedures in the regulations of the Secretary of Defense and others.

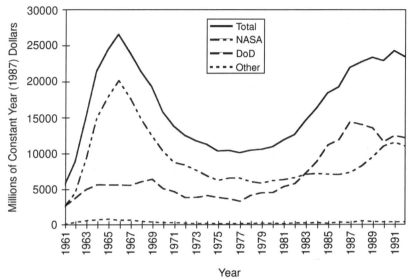

Fig. 2.1-1. U.S. Space Expenditures (Constant year dollars). Upward trend in the 1980s may be reversed as a result of major aerospace decline in the mid-1990s. "Other" refers to total outlay by Departments of Energy, Commerce, Interior, and Agriculture, and the National Science Foundation. (Data from Aerospace Industries Association [1994])

Managing the DoD's and NASA's large space programs is like trying to win a potato-sack race—with five people each having one leg in the sack while running in different directions, and 10 to 30 coaches and helpers are on the sidelines directing the people in the potato sack. The race is so long that we substitute players and coaches along the way, and they change the rules, as well as the dynamics of the team. These days, it's truly amazing that the United States can create such wonderful systems, but we **do** develop the most capable space systems in the world. Keep this in mind while reading the rest of this book.

Remember, the government is big, and although many great ideas for reducing cost arise, the government doesn't operate as one unit—it isn't monolithic. A very difficult part of reducing space mission cost is getting so many organizations to work together, and our biggest challenge is to trim the number of organizations and to streamline their structures. Minimizing the number of organizational interfaces within a program is key to reducing cost, as many of the case studies show.

Finally, understand that reducing staff, organizational size, and the cost of space systems is counter to the social values ingrained in the people charged with doing so. For many decades we've rewarded people for building large organizations with lots of people, responsible for lots of money—that's status and success. Because of our success in space activities, we have built incredibly large infrastructures and capabilities that we can't support with fewer dollars. Now, we're asking government people to reverse themselves—and to place their jobs at risk. They can do it, and we must motivate them well to do it, but how do you motivate people to eliminate their own jobs?

How can U.S. space policy help reduce cost?

I think the answer to this question is another question. What space policy? A more descriptive term is space meandering. For many years, study groups and committees have spent countless hours developing roadways and approaches that could get certain aspects of our space industry going in the right direction. But our political leaders haven't adopted them, so we've wandered in different directions, with no coherent voice or focus. Please don't misinterpret this comment—we have accomplished a lot in the last decade, but in some cases it's resulted from serendipity rather than policy.

Careful examination of how our civilian space program spends almost $15 billion a year shows that very little directly supports such policy goals as: "promoting the reduction in the cost of current space transportation systems while improving their reliability, operability, responsiveness and safety…foster international competitiveness of the U.S. commercial space transportation industry, actively considering commercial needs and factoring them into decisions on improvements in launch facilities and launch vehicles…the Departments of Transportation and Commerce will be responsible for identifying and promoting innovative types of arrangements between the U.S. Government and the private sector, as well as state and local governments, that may be used to implement applicable portions of the (National Space Transportation and Technology Policy) policy." [OSTP, 1994]

People who administrate a space project are necessarily concerned with "pleasing" those who control their future well-being. NASA's "controllers" are people who sit on the congressional authorization and appropriations committees which determine their future funding. "Space," however, as both a place and a field of activity, offers relatively little return to members of Congress who support it. Beyond the jobs attached to keeping the space-shuttle and communications-satellite industries operating, few "space-related" jobs translate into constituent support. In the end, the public's view of space activity usually determines how much one project will generate support for future ones. And the public's view depends on how well the hardware seems to work and, therefore, how well it exhibits American technological know-how [Johnson-Freese and Moore, 1993]. The Executive Branch of our government easily writes policy, but finding the dollars to carry it out is much more difficult, especially when jobs are at stake in several congressional districts.

From a national security standpoint, with the crumbling of the perceived Soviet threat, the DoD can no longer justify to Congress the need for their expensive systems. In fact, many congressional staffs believe the DoD has only **wants** for space systems—not firm requirements. Within DoD, many groups need to develop capability in space—the armed services, classified programs, unclassified programs, laboratories, and others. Because these groups are competing for funding, they've been known to stab each other in the back to justify parochial interests.

NASA is no different. Individual NASA centers are fiefdoms. They work with NASA's headquarters, but they have a life of their own. NASA's centers compete with one another and have also sabotaged each other's plans.

The point is that, with so many different, parochial voices talking to Congress, the space community is unable to convince Congress that it knows what it's doing. Interviews with congressional staffs, leaders in DoD and NASA, and people on the street suggest that our military and civilian space programs lack credibility and relevance. As a result, the space community is an easy target for anyone who would use its inconsistencies against it. Without credibility and relevance, we fall prey to the scrutiny of Congress and the American people, and perhaps rightly so.

Approaches NASA and DoD can take to reduce cost are summarized in Table 2.1-1. One of the best approaches to reduce the cost of space systems and enhance credibility is **to develop and use integrated plans and architectures** that support their future needs and requirements. We should spend time selling the plans and architectures to Congress and the American people, rather than selling individual projects. Once we have a strategic road map that provides a common frame of reference for DoD and NASA, we **can** build credibility. Of course, I'm assuming the entire process is rational, which could be a very bad assumption. The point is that the infighting within DoD and NASA only detracts from their efforts and causes everyone to suffer. In order to reduce cost and survive, the community must have strategic plans and architectures for its efforts in space.

TABLE 2.1-1. Advantages and Disadvantages of Government Cost Reduction Approaches. Each of the items listed can either increase or decrease cost, depending on how it is implemented. See text for discussion. See also Sec. 10.4.2.

Cost Reduction Approach	Advantages	Disadvantages
NASA and DoD develop and implement integrated plans and architectures	• Provides organized and coordinated approach to programs • Focuses decision making in a central location • Potential to reduce redundant efforts	• Increases the price of failure which inhibits cost reduction • Largely cuts out "small" players which are the major source of innovative, low-cost programs
Prioritize programs and projects	• Provides adequate funding to highest priority programs by eliminating low priority ones	• Innovative, high-risk programs that have the greatest potential for major cost reduction may be the first eliminated
Convergence— merge similar programs	• Can eliminate duplicate programs in other agencies	• May result in many mutually exclusive requirements • Makes requirement trades difficult • Historically expensive, hard to make work
Purchase services, not systems	• Saves government time and money associated with developing and launching space systems • Allows contractor freedom to develop system apart from government specifications and oversight	• If government is the only customer, cost of data will exceed the cost of the system • Technical expertise could be lost in the government • Only good for well-defined services

This approach may be risky. Even if our country has a plan, we can still make bad decisions. An example comes from the late 1970s and early 1980s when the government decided to abandon development of all expendable launch vehicles so we could invest in a much cheaper option—the Space Shuttle. Having a Space Shuttle is probably important in the long run, but it's not as cost-effective as it was proposed to be. The country has since recovered but this example highlights a very important concept—killing off alternatives in order to support a single option may increase cost rather than decrease it. The U.S.'s funds are limited, and funding profiles in DoD and NASA are going down, so the government must integrate plans and architectures, rank them, and make some hard decisions about which direction this country will take. In the words of Yogi Berra, "If you don't know where you're going, you'll probably get there." Enough editorializing.

The Federal Acquisition Streamlining Act of 1994 newly defines commercial items, raises the minimum value at which acquisition contracts require detailed data, makes it more difficult for the government to claim proprietary rights to jointly developed technology, and removes some of the strict procedures for cost and price accounting associated with commercially procured products. The DoD White Paper, *Acquisition Reform: A Mandate for Change*, published in February, 1994, clearly envisions increased commercial purchases by the DoD, greater use of commercial specifications and standards, reduced administrative burdens on vendors, and DoD procedures that more closely resemble commercial ones.

Also, reaction has been extremely positive to new DoD guidance which directs using military standards and specifications only if commercial standards or specifications aren't adequate. The guidance also states that any required military specification or standard must be in terms of desired performance rather than the physical characteristics or ways of producing an item. This change is part of the government's recent emphasis on *Civil-Military Integration*. The Congressional Office of Technology Assessment defines Civil-Military Integration as "the process of uniting the Defense Technology and Industrial Base" and the larger Commercial Technology and Industrial Base. Under Civil-Military Integration, common technologies, processes, labor, equipment, material, and labor would be focused on both defense and commercial needs [Office of Technology Assessment, 1994].

The second potentially significant area that offers hope for near-term cost savings is the success of many small, low-cost spacecraft, such as those described in the case studies. Small, low-cost spacecraft have been built since the very beginning of the space program. However, they are now receiving more attention and stature as money becomes scarcer. Although "better" may be largely in the eye of the beholder, these programs do produce results far faster and cheaper than the more traditional programs for large spacecraft.

Most of the low-cost programs use commercially developed technology, rather than equipment designed and manufactured under costly military specifications. They deliver valuable scientific data in a remarkably short time. AMSAT, the lowest-cost provider, has launched more than 40 spacecraft and has never delivered a spacecraft late to the launch site.

Whether we can apply these models to larger and more complex missions is uncertain, although the Pluto Express mission (Sec. 12.2) is attempting to apply these rules to develop a low-cost mission to the very edge of the solar system. Other programs are engaged in science observations, military testing, communications, and

Earth observation. The key is that these missions are occurring and provide real examples of space operations at much lower cost.

Another very effective, but unpopular, approach to reducing cost is **convergence**—merging selected NASA and DoD needs and developing one system instead of two to meet the needs. One example is weather. DoD and the National Oceanic and Atmospheric Administration (with NASA) have each developed capable weather systems—built by the same company but procured and operated by different government organizations (to different requirements and specifications). Granted, data requirements differ, and some other needs and requirements are different, but weather is weather, so why have two separate systems? These thoughts are occurring in many arenas but merging systems is fought every step of the way. When systems merge, bureaucracies decline, people lose their jobs, and organizations lose turf—and funding decreases. People who don't favor such an approach will point out that in many cases, it has resulted in disaster. An example is the DoD's attempt to merge Air Force and Navy fighter developments. Convergence can be a very effective way to reduce cost and still provide great service, but we must use it intelligently.

One potential ill-effect of convergence is that we put all of our eggs in one basket, which may cause costs to spiral upward. Using weather as an example, if the Government now makes NOAA and DoD depend on the same space system for weather data, the system becomes too important to take risks, so we create another large, all-important, space system—probably at very high cost. The key here is to remember that we can obtain worldwide weather data using several very large systems or many small systems. The latter approach would help avoid putting all of our eggs in one basket.

The concept of **dual-use** can help build support for programs, share system costs among government organizations and industry, and even reduce the cost of systems and data. Generating data and products needed for government and commerce is in vogue. The idea is to show that the government is doing something useful (relevant) for industry and the American people. Doing this effectively will result in funding support and stability that can help reduce system costs.

If Congress would fund DoD and NASA over several years, instead of changing the budgets for programs every year, we could save between 5% and 20% of the cost of doing business, according to government and industry managers. I know of no formal studies that have tackled this issue, but take building a house as an example. If every month you tell your contractor to wait for further funding, and, oh by the way, you're not sure you'll have the amount of money you promised next month, or you may not finish the house at all, you'll see several effects. Any contractor waiting to find out if you're going to have money, won't risk sending people to work, so the work will slow. In addition, a contractor who doesn't have more work for employees or subcontractors, will lose them to other work—or pay them anyway to keep them, which drives up overhead cost. These insidious effects are much the same in our space industry. The experts agree that, even though these end-of-year dynamics cost taxpayers money, Congress won't provide across-the-board, multi-year funding because individual legislators will give up too much control, along with many chances to get themselves in the public eye. Even so, if DoD and NASA can find ways to stabilize funding for selected programs, we can reduce the cost of doing business on those programs.

Another key philosophical change occurring in policy and acquisition circles is that of buying **services not systems**. For example, in remote sensing, people are asking if they should develop a space system to provide the data they need or just buy the data from some other source and let someone else worry about maintaining the systems that provide the information. This idea can drive costs up or down. Early on, in order to convince the government to buy the service, contractors would probably advertise a very competitive price. Later, after the government has abandoned all other data sources, I suspect the price would increase. The question is, which approach is more cost-effective? The answer remains to be seen.

The last concept in this section is probably the most efficient way to reduce the cost of space systems in the entire book—Just say NO! Nancy Reagan first used this phrase regarding drug use in America. But Dr. Hans Mark, for example, recommends using the same concept to reduce the cost of space systems. If we can apply it to all levels of government, we can save an incredible amount of money. Just say no to inflated and gold-plated requirements. Just say no to projects of limited utility. Just say no to reducing budgets of all projects by 10% and select projects that will and won't be cut. Just say no to selecting low performers as members of teams doing joint government and industry projects. Just say no to cutting all NASA centers by 15% to 20% and, instead, close an entire center to allow the others to flourish. You get the point, right? This recommendation actually involves decision making. Reducing cost requires active decisions that reflect the needs of the mission and funding agency, rather than the much easier passive decisions. We must make these hard decisions, trim overhead, be willing to "give" on requirements, be flexible in our approach, and get on with the business of developing highly useful space missions at low cost.

What changes to the acquisition process can help reduce costs of space missions?

The acquisition process within DoD and NASA is thought to be a bureaucratic nightmare that slows schedules and drives up the cost of space systems. There is some truth to this, but the facts speak for themselves. The U.S. develops and acquires the best and most capable systems in the world. If you're feeling bad about your acquisition process, just go to another country (any country) and compare notes with your contemporaries. You'll be very thankful to be doing business in the U.S.! Having said that, we all know we can improve.

Most people who have worked on a government procurement in any capacity will say that the acquisition process is too complex, cumbersome, and lengthy. They are right. Procurement has become more complex and burdensome over the last 20 years. Much of the difficulty comes from contractor demands for open and fair competition, affirmative action, small-business set-asides, and many other regulations required to protect the rights of all contractors.

The three approaches to making acquisition less expensive involve reducing schedules and the complexity of process and documentation, as well as using commercial practices more. (See Table 2.1-2.) Other approaches, such as improving funding and procurement stability, aren't not likely candidates for reducing cost because they require Congress and their staffs to give up a chance to vote on where and how funds are spent. One very positive approach is to raise the threshold for non-competitive contracts from $25K to $100K—hopefully higher. A very high percentage above (75%) of all contracting actions fall below the $100K threshold. Contracting officials and specialists can save much time on smaller purchases and hopefully spend more time on the larger, more difficult government purchases.

TABLE 2.1-2. **Cost Reduction Approaches in the Acquisition Process.** The objective is to return to a simpler process characteristic of the early space program. Nonetheless, each of the regulations was introduced to resolve specific defects.

Cost Reduction Approach	Advantages	Disadvantages
Decrease complexity of process • Increase dollar limits on contracts requiring the full procurement process • Minimize number of signatures required to proceed	• Eliminates paperwork • Gives contracting officers more time to devote to larger efforts	• Decreases control and oversight • Fewer fully competitive procurements
Reduce procurement schedules • Government can help by turning required paperwork around more quickly	• Reducing schedule saves time. If the government reduces its timeline, money is usually saved by the contractor	• Reduces time available for oversight and control
Reduce required contractual documentation and standards • Decrease size of Contract Data Requirements Lists • Rely more on contractor documentation and commercial standards	• Focuses attention on more useful information • Eliminates redundant information • Forces contractor and government to use the same information	• Could eliminate necessary information • Reducing documentation may reduce reliability, traceability, and maintainability

Two potentially significant areas offer hope for recognizable cost savings in the near future for space missions. The first centers on actions recently taken in Congress and given structure in the Department of Defense. DoD is clearly in favor of raising the maximum threshold for simplified acquisition contracts to $500K. According to a recent report from the Department of Defense Acquisition Law Advisory Plan, "one of the most expensive and disruptive procurement requirements involves mandatory adherence to cost principles and accounting standards enumerated in statute, in the FAR, and by the Cost Accounting Standards Board" [Law Advisory Panel, 1993]. These procedures affect all tiers of commercial vendors and require unique accounting procedures and databases which don't apply to normal, commercial business practices. Studies estimate that the cost increases directly attributable to the government's acquisition rules are 20% to 60% [Office of Technology Assessment, 1994.]. Many commercial firms simply choose not to do business with the government rather than endure the overhead and headaches associated with the FAR.

Another fruitful area in acquisition for specific missions is scrubbing requirements:

• Reducing government-required documentation

• Eliminating unnecessary government standards

• Establishing an organizational structure and acquisition process to facilitate trading on requirements

Most NASA and DoD organizations are requiring fewer items on the contract data requirements list—less complexity and documentation. Before writing RFPs, some government agencies like NASA's GSFC are holding zero-base documentation reviews: starting with no required documentation on the contract and reviewing and justifying each deliverable. Many program offices are willing to accept the contractor's own financial and planning documents instead of specifying formats. Electronically available information eliminates much of the cost for paperwork and distribution.

Our experience in space over the last 20 years has produced a staggering volume of military and civil standards and specifications. Often, we've uncovered problems and tried to avoid them downstream by creating these standards or specifications— they reflect the lessons we've learned in space. Unfortunately, the government spends little time and talent maintaining them, so they're full of guidance that may be outdated or inappropriate for a particular procurement. To solve this problem, we must scrub the standards and specifications and contract only for those that are absolutely necessary. The current approach is to eliminate as many DoD and NASA standards and specifications from the contract as possible and rely on the best practices of the commercial organizations doing the work. This can be very successful if the company truly knows how to do the work—and disastrous otherwise. If you look hard, you'll find that the best practices of many successful companies often include DoD and NASA specifications and standards.

Finally, the requirements we place on contracts probably have the most profound affect on cost of anything the government does. No matter how well we scrub them, some overly optimistic or bad requirements will arise. To avoid incredibly high costs, we must find a way—early in the design process—to trade on requirements. See Chap. 10 for a discussion.

How can the requirements process help reduce the cost of space missions?

Requirements are the key to cost and performance. Some requirements drive cost and complexity to the limit and beyond. Others are necessary, but don't push cost and performance much. The focus of the requirements process that we refer to is to plan, develop, negotiate, and satisfy needs (not wants!) for the least amount of money.

In today's political and fiscal environment, Congress and many staff groups are scrutinizing requirements for space systems in DoD and NASA. In DoD, for example, it's becoming increasingly difficult to convince Congress and the American people that there is truly a military threat in the world that warrants spending $10 to $20 billion per year to combat.

If we combine this environment with the fact that at least five or six large organizations within DoD develop and use space systems—we have a problem. In the past, these organizations have strongly supported their own parochial views, which is what they should do. But, in the process, they've often promoted their own views to Congress while sabotaging or diminishing other organizations. Thus, Congressional members and their staffs have easily pitted these organizations against one another and focused on inconsistencies. The result is that DoD has a very difficult time defending its approaches and requirements.

DoD has recognized it must be credible to Congress and the American people, so it's creating a DoD-wide space architect whose organization will, among other things, produce a road map for developing and operating space systems. The plan is that a small group, working with the space architect, will integrate approaches and systems to provide the necessary coverage and represent users, operators, developers, and funders within DoD. The space architect will oversee much of the requirements process with the goal of meeting DoD's needs at reduced the cost [*Space News*, March 1995]. Methods for reducing cost are summarized in Table 2.1-3 and discussed below.

TABLE 2.1-3.　Cost Reduction Approaches in Requirements Definition.

Cost Reduction Approach	Advantages	Disadvantages
Generate fiscally responsible requirements • Understand cost implications of requirements	• Potentially reduces the number of "impossible" requirements	• May overly constrain what we develop • May develop products that don't meet the needs
Bundle requirements to facilitate robust and affordable systems	• Allows requirements to be combined in different ways to facilitate smaller, more distributed systems • May lead to fewer large, centralized space systems	• May lead to fewer, larger, more complex systems
Establish a timely process to trade on requirements, early in design and development	• Allows flexibility in the conceptual design stage • Forces issues to be addressed up-front	• Stretches timeline • Program may appear unstable which inhibits funding

Generate fiscally responsible requirements—While generating requirements, we must consider that cost is a real constraint. When I was in a position to state operational requirements, I was told: "if it is a requirement, it must be levied. Don't worry about the cost." This may have been true 20 years ago, but today the government and its contractors must consider cost as a design variable. This means that we should try to meet requirements at the lowest possible cost and challenge requirements that have an unnecessarily high price tag. For example, in the early days of the Moon missions, NASA levied a requirement to develop a pen that orbiting astronauts could use to write in any g-environment and orientation. That pen cost several hundred thousand dollars to develop; you can buy one today for about $10. During the same time, the Soviets needed a writing instrument for their cosmonauts—they used a pencil! The point is that we must look carefully at our requirements to make sure they're truly needed. Then we must meet them in the most cost-effective way possible.

Bundle requirements to facilitate robust and affordable systems—Several philosophies are used to allocate requirements to different space systems. One philosophy is to capture all requirements in a particular application area—such as communications—and try to meet them with one very large system involving one or more large space platforms. This approach has worked for decades! But as systems get more and more complex, and as the number of requirements increases we eventually must ask if it's wise to try to meet all of them with one system or space platform. Should we put all of our eggs in one very good basket or should we distribute the requirements to other smaller, less complicated platforms or systems? Cellular communications is a good example. We can provide cellular communications from space using several very large, capable spacecraft in geosynchronous orbits, or we can use a larger number of smaller spacecraft in low-Earth orbit. The cost of the overall system may be similar, but the latter could better survive anomalies and failures.

Establish a timely process to trade on requirements—Improperly stated requirements can generate solutions the writer never intended. If we write, "develop a pen that…," most people will research and develop a pen. If the requirement is written as "develop a writing instrument that…" some people may develop a pen and others may choose a pencil or some other solution. But if we say, "provide the capability to record data for later use…" a developer has many options, including using an available pencil or pen, if appropriate. The key to stating requirements that will reduce cost is to describe **what** you need, not **how** to provide it. Let the developer create (or use) the most cost-effective approach. A good valid requirement will be verifiable, but even properly stated requirements often carry with them a particular method for verification, which essentially mandates a specific approach. We should also state a range of performance rather than specify a number. Often requirements are inter-related, and some combination of them meets the needs of the user and operator. A range of potential values for these requirements can provide the developer with design flexibility that can reduce cost. Finally, if we include reasons for a requirement, a developer can better understand its meaning and intent. This improves communication and may produce an alternative that would decrease cost and increase performance.

In general, requirements should be highly negotiable early in the development of a concept that supports the mission objectives. They should be easy to discuss and adjust as more information emerges, so **trading on requirements is** easier. Later in the development, when the system is approaching production, launch, and operations, changes aren't acceptable, except in the most extreme cases. In general, any change costs money—even changes to reduce cost will cost some money. So manage changes properly, or they'll manage you.

2.1.1 Designing Space-Mission Concepts

Developing cost-effective concepts for space missions is usually a challenge, but today's political, organizational, and requirements dynamics make this challenge almost impossible. As sponsors of a potential new program start warming up to it, they begin to weave the mission concept by promising certain roles to people and organizations. These promises can cause the cost of the program or project to increase dramatically, unless they are well thought out, rational, and integrated into a bigger picture. (Please refer to Chap. 1 and SMAD for discussion.)

Before making too many political promises or establishing organizational turf, we can use some rational strategies to reduce the mission's life-cycle cost:

* Select and carry out the proper class of mission
* Develop innovative approaches to the mission but use them conservatively
* Look for integrated mission concepts that blend resources from land, sea, air, and space
* Do inter-element trades early—they're the best chance to reduce cost

Advantages and disadvantages are given in Table 2.1-4. When we begin a new mission, one of the first things we should do is to recognize we can do it in many ways, using combinations of payloads, spacecraft, orbits, launch systems, and supporting infrastructures. We may be dealing with large or small missions, but we must recognize we can carry out the large mission using several large assets or many smaller ones. Examples include ESA's Cluster mission, NASA's Earth Observing System, and Motorola's Iridium Program. In each case, teams chose to use several smaller assets versus one big asset.

TABLE 2.1-4. Cost Reduction in Conceptual Mission Design. The greatest impact on cost occurs during the conceptual design and definition phase. (See Sec. 10.2.)

Approach	Advantages	Disadvantages
Select the class of mission and implement accordingly • Trying to implement a low-cost system in a traditionally high cost environment is nearly impossible	• Matching the class of mission (low-cost, high performance, government or commercial) with a similar implementation plan may facilitate more missions	• Reducing program office size could eliminate much of the knowledge gained at great cost in prior space programs
Look for integrated mission concepts that blend resources from land, sea, air and space	• Space provides added value • Promotes internal synergism among components	• Missions may become more complex and more expensive
Do inter-element trades early (See SMAD, Chaps. 1 and 2)	• Do the big trade-off early before organizations estab-lish "turf" and politics take over	• Stretches timeline and can add cost • May lead to early infighting

The key here is to pick a class of mission and payload—small or big, simple or complex, brief or extended—and select an approach that allows us to complete the project as needed. Many small, fairly inexpensive missions grow to be too expensive and heavy because the political and organizational climate interferes. Trying to do smaller and simpler projects in a large-project organization is very difficult because established procedures and biases tend to get in the way and force the project to look like its predecessors. Some organizations, such as the Goddard Space Flight Center, have found ways to foster smaller and less expensive missions by establishing small groups that operate semi-autonomously. (See the SAMPEX case study in Sec. 11.3).

Second, we must recognize that many of the people in government and industry have been in the space industry for a long time and tend to focus on the old approaches to missions—methods that have worked. We must listen to and create more innovative approaches to the missions we develop, and then use approaches and techniques that will ensure our success. According to George Sebestyen "Most important cost savings occurs while deciding how to meet operational requirements, not how to implement a set of technical specifications." [COSA, 1995]

Further, we must recognize that our industry tends to focus on "all-space" solutions, partly because that arena is what we know best. We often try to satisfy space-mission objectives and perform most missions using very large, complex, elegant space resources. Sometimes this is necessary, but it can get expensive. We recommend that, in the beginning of a new program or project, we develop at least some mission concepts that blend space, air, land, and sea assets to do the mission.

Finally, the people establishing the mission concept and space-mission architecture, be they government or industry, should look for trades among the mission elements. Example trades include space versus ground processing, centralized versus distributed control, spacecraft versus launch-system propulsion, and many others. [SMAD, Sec. E.2]

TABLE 2.1-5. Managing Projects to Reduce Cost. The key issue is to provide the appropriate management for each specific project.

Approach	Advantage	Disadvantage
Select appropriate approach for government interaction with contractors	Avoid mismatch of expectations (and $) of a large government program office on a small contractor's program office or vice versa	• Excuse for "business as usual" on large programs
Design-to-Cost	• Sets definite cost constraints, facilitates more affordable systems • Asks the question "how much capability can I get for this much money?"	• Answer may be "not enough" • May be highly inefficient to meet an arbitrary cost assignment ("penny wise and pound foolish")
Compress the schedule, when appropriate	Time costs money. Reducing time it takes to field a system saves money	• Shortening schedules too much can result in cost increases due to overtime and mistakes or oversights • May minimize up-front systems engineering
Manage the requirements and design change process with an iron fist	• Allowing requirements to be flexible early in the program can facilitate trading on requirements which has great potential to cut cost of systems • Disallowing changes late in the acquisition process can usually save money as well	• Heavy-handed management kills innovation and prevents technical knowledge from being used wisely
Perform zero-base documentation review • Use contractor documentation instead of government specific documents • Reuse selected, stable documents	• Reduces total amount of documentation and saves time and money	• Could eliminate some necessary documents and have to add them later • Government review may be harder to perform or impossible • Hampers standardization; may decrease maintainability
Look for the best combination of cost regulation approaches—integrate your approach to reducing cost	• All of the above • Most programs are looking for a combination of cost-reduction methods that work for them. Experiment with different combinations	• All of the above • May stifle flexibility and innovation • Best approach may be to apply "common sense" as done at Skunk Works (See Sec. 2.2.7.)

2.1.2 Managing Projects to Reduce Cost

Many excellent books, references, and consultants provide solid advice and techniques to improve management efficiency. We recommend you read as many of these as possible. Many of the concepts in this section amplify solid management approaches that successful projects have used for years. The following approaches summarized in Table 2.1-5, are drawn from the case studies in the back of this book and from a national workshop on reducing space-mission costs [COSA, 1995]. Most ideas in the case studies are from smaller, lower-cost programs. The ideas from the workshop focus more on larger projects from around the country. No single approach will give us the cost reductions we need, but I believe **the right combination of approaches selected from this book can strongly decrease the cost of space projects.**

One key is to begin the project with the **right project leader**. The selected leader should have all necessary skills. Some people handle large, complex, bureaucratic organizations well, whereas others are better with smaller teams. Some people can deal with lots of technical ambiguity, but others prefer to let someone else handle the technical details. Some great managers are constantly involved—they're hands-on; other great managers stand back and manage from afar. The point is to pick the person best suited to your type of project. Large and small projects require different skills.

Another key to reduced cost is **compatible government and industry managers**—they should be able to work together and complement one another's abilities. Some projects have actually merged government and industry into one team, headed by one person, with government and industry team members working side-by-side.

We also must **select the appropriate approach for government interaction with contractors**. Matching a large government team (civil servants and consultants) against a small contractor organization is a formula for driving up cost. Members of the government's team will try to do their jobs well. They'll poke their noses into many things, generate lots of questions, and identify potential problems, which will quickly overwhelm the small contractor's organization. The result will be chaos, or contractors will increase the size of their teams to answer the questions. The type of contract between government and industry is also important. If the contract is cost plus, the contractor will probably try to increase the number of people; if the contract is fixed price, the contractor will probably dig in and challenge the interactions as unnecessary and bothersome. The point here is to carefully decide how much government and contractor interaction is necessary. It will vary from no interaction—which is probably a formula for disaster—to living with the contractor—which can be very expensive and time consuming.

One axiom seems to hold for all projects—establish the project guidelines and scope and give the project managers sole responsibility and accountability. Let them make and carry out decisions without a lot of "help." Committees are great for gathering sage advice and wisdom, but don't let them decide anything—when things go wrong, no one seems to remember being at the meeting. **Strive for accountability** with all people. Our industry uses matrix management a lot, especially for large programs, but wisdom suggests that, if you want to reduce cost, avoid matrix management and increase individual responsibility and accountability.

One way to hold project managers and teams accountable is to require that they work the entire project from concept through operations. This gives them a long-term perspective and, hopefully, the incentive to reduce overall life-cycle cost. Many of the case studies for short-duration projects amplify this concept.

As project manager remember that the **two things that drive cost of space systems most are the organizational structure and the acquisition process**. So, if you manage the organization and the process well, you can reduce cost.

The next key topic is **design-to-cost**. We have seen many projects use this concept to drive down the cost of a particular mission, but the approach has problems. Usually, design-to-cost mainly emphasizes cost—then technical performance and schedule. We must recognize that the members of the management triad—cost, schedule, and technical performance—are strongly related. There is some truth to the saying, "you can meet any two of the three." In design-to-cost, we adjust the space mission concept architecture, the performance requirements, and the overall approach to meet an overall cost goal—usually life-cycle cost. This is like going to a house builder in Washington, D.C. with $100,000 and asking what size and type of house you can get. You may not get exactly what you need, but hopefully you'll be able to live there. Many trade-offs exist while using design-to-cost. NASA is using this approach in many new projects, and lots of universities and small companies use it because they have very little money. The case studies amplify on this approach.

Schedule plays a key role as well. Usually, the longer the project, the more it will cost, so projects commonly **compress the schedule.** This is a double-edged sword— you can compress schedule to reduce cost, but if you compress it too much, you'll increase the cost. A tight schedule is a wonderful excuse to speed up procurement. Many contracting organizations respond very well when given a challenge; others require a lot of patience. The axiom here is to **track cost and schedule in near-real time and don't accept schedule slips**.

As discussed in the section on conceptual design of space missions, it's important to keep your mind open for new concepts and approaches to do the mission—this is the area that can significantly reduce cost. Once an acceptable concept, mission architecture, and set of requirements have been created, you must **manage requirements and design changes with an iron fist**. On one hand you must respond to trading on requirements and listen for cost-saving ideas. On the other hand, you must be leery of changes in later phases of development because changes cost money! Early on in the project, when ideas abound, it's easy and inexpensive to change your mind. As the project proceeds, and many things are committed to paper, computer, and designs, changing becomes more expensive. Once the hardware and software have been developed, change is incredibly expensive. So remember, **changes cost money**. For example, you'll find that many of the activities aimed at reducing the Space Station's cost actually increased it!

Funding cuts are changes that can increase rather than decrease the cost of a project. Often funding cuts come in decrements of, say, 10% per year to organizations. Organizations deal with these cuts differently. Some managers choose to spread the cut across the whole organization, so every project takes a 10% hit. The idea being that this is fair and everyone can withstand that small reduction. In addition, this approach is easy to defend. Other managers tend to identify projects that will take little or no reduction and those that will bear the brunt of the funding cut. Passing the 10% cut to every project is the easy way—it changes all projects and equally drives up the costs. But it probably makes more sense to leave some projects alone and cut others entirely. Remember, we expect cuts for the next several years. Ten percent per year for 4 or 5 years is a lot of cuts to pass on to all projects. Make the tough decisions! You might apply this concept to NASA as a whole. Does it make more sense to have every center take a 20% cut in people or to close a redundant center altogether and allow the others to flourish? To truly reduce space mission cost, we must make hard decisions.

Another key to reducing cost through project management is to reduce and simplify the
- Number of organizational interfaces
- Number of hardware and software interfaces
- Amount of documentation and review

for individual projects. The cost of a project will increase dramatically with the addition of each new government or contractor organization. Maintaining these interfaces takes an incredible amount of time and money. Similarly, integrating hardware and software interfaces requires a lot of coordination and increases the chance for error. So keep these interfaces to a minimum.

Useful documentation costs money. Unnecessary documentation costs money. The key is to be able to tell the difference and eliminate the latter. One way to determine which documentation a project truly needs is to do a **zero-base documentation review** justifying every document request before putting it in the contract. This is a painful exercise, but it can save many dollars. Another approach is to **use appropriate contractor documentation** instead of government documents. Finally, some documents from ongoing projects can be reused with minor changes, again reducing cost. We recommend combining these approaches to save money on documentation.

TABLE 2.1-6. Full-time Equivalent People Required to Operate and Maintain Geosynchronous Communications Spacecraft. The low-cost example contrasts with typical systems. The savings in people mainly results from the approach and philosophy, not necessarily the low-Earth orbit.

Organization	Number of People per Spacecraft	Comments
DoD	27 to 30	The commander of this organization maintains that these numbers are low by 25%.
NASA	25 to 29	
ESA	22 to 25	
Commercial	12 to 14	Requires half the people compared to government operations.
University of Surrey—Low-Cost Developer and Operator	7 spacecraft per person	The spacecraft are fairly simple with standard interfaces and protocols for mission operations.

2.1.3 Mission Operations and Ground Infrastructure

The mission operations and ground infrastructures for space in this country have grown significantly over the last 25 years, due mostly to developing and operating many successful programs. Current funding doesn't support these massive NASA and DoD infrastructures, so our leaders are trying to find ways to reduce them and operate missions more cost effectively.

Two key costs for mission operations are people and facilities. The question is "How many people does it take to operate a spacecraft?" The answer is, "It depends." It depends on the type and class of mission, philosophy of the operating organization, number of spacecraft being operated, and much more. To better understand this problem, I studied geosynchronous communications spacecraft operated by DoD, NASA, ESA, and a commercial organization. This study included all people (full-time equivalents) associated with operations—console operators, engineers, software developers and maintainers, data clerks, managers, and administrators. Table 2.1-6 shows the results.

Reducing the cost of the mission operations and ground infrastructures requires

- Using standard interfaces, protocols, and procedures
- Reusing people, procedures and software, plus having mission operators participate in mission and spacecraft design early in the process
- Using selected autonomy on the spacecraft
- Automating (carefully) on the ground
- Working data flow issues early
- Providing adequate resource margins on the spacecraft
- Accounting directly for the cost of mission operations

The government is considering many of these ideas. In addition, we can operate a spacecraft in several ways. Lately, the government-owned, contractor-operated approach has received much attention. In this approach, the government would build and launch a spacecraft, then, a commercial organization would operate it under contract to the government. The advantage of this approach is that it reduces the number of civil servants doing operations. It **may** reduce operating costs, but that remains to be proven.

Another option is for NASA and DoD to research, develop and, perhaps, launch spacecraft, but leave operations to DoD. This allows NASA to get out of the operations business altogether and focus on what it does best—research and development.

2.1.4 Other Key Issues for Reducing Space Mission Cost

We need to reduce cost by focusing on

- Launch systems and operations
- Spacecraft design, development, and test
- New technologies

These areas are discussed throughout the book. Chapter 4 provides most of the information on launch systems, but doesn't discuss saving money by reducing the size and mass of a particular spacecraft so it fits on a smaller class of launch vehicle. This is one way to limit the spacecraft's size and mass. The concern here is that as we stuff more and more capability on a smaller spacecraft we tend to miniaturize systems and add more mechanisms to the spacecraft. Both of these actions can increase cost and risk and may not necessarily drive down the cost of the spacecraft. The true savings comes from lower cost for launch vehicles, as presented in detail in Chap. 4.

Technologies to reduce the cost of space missions are discussed throughout the book. But the government particularly emphasizes technologies that reduce the cost of:

- Launch operations, including onboard health monitoring, GPS trajectory monitoring, and hybrid rocket propulsion.

- Mission operations, including autonomous orbit determination, guidance and control, onboard data processing and health monitoring, and standardized communication interfaces.

- Space missions, including those that improve up-front development of mission concepts, operations planning and systems engineering, miniaturized electronics, solar-electric power generation, electric propulsion, and autonomous navigation of spacecraft.

2.2 Radical Cost Reduction Methods

James R. Wertz, *Microcosm, Inc.*

Traditional approaches to reducing space mission cost are those that fall within the confines of the normal procurement and development process. These methods reduce cost by improved engineering or management, or perhaps by bending the rules but not breaking them. More-or-less traditional procurement policies and engineering standards are applied. Traditional methods include, for example, concurrent engineering, integrated product development teams, or standardization. Depending on how they are applied and the baseline for comparison, traditional approaches might reasonably be expected to reduce cost by 5% to 30%.

In contrast, what we call *radical* cost reduction methods are those that require genuinely changing the rules of the game, i.e., shifting the paradigm by which spacecraft are bought or built. This includes techniques such as using non-space equipment, conducting trades on the fundamental mission requirements, or allowing a higher element of risk. While these changes may seem straightforward, they are not. It's difficult to change the rules we work by. Nonetheless, experience has shown that the possible rewards can be substantial. Radical cost reduction methods have the potential to reduce costs by a factor of 2 to more than 10 and, strangely, with little or no loss in system reliability. Table 2.2-1 compares the two approaches.

The key problem we must address in making major cost reductions is breaking two cycles:

- High cost ➙ not allowed to fail ➙ more high cost
- High cost/lb ➙ high level of optimization ➙ specialized parts ➙ no standardization ➙ high cost/lb

Individually each step is appropriate. Collectively they lead to a space program which we cannot afford. In order to break the cycle, we need to find ways to change the rules.

Essentially all of the radical departures which have successfully brought about dramatic cost reduction, as described elsewhere in this book, call for simultaneously changing the process and the technology. (See Table 2.2-2.) Changing the process but not the technology implies, for example, finding cheaper methods for building the Space Shuttle. Ultimately, although we may reduce costs somewhat, building the Space Shuttle at low cost is probably not possible. It is a complex vehicle with a huge number of precision parts, most of which are hand built. On the other hand, using even simple technology with the full burden of the government procurement processes, paper work, and regulations results in the $500 hammer or $1,000 toilet seat. We need to change the design to allow for building spacecraft with an ordinary screwdriver and, at the same time, change the procurement regulations to allow buying that screwdriver at the local hardware.

TABLE 2.2-1. **Traditional vs. Radical Approaches to Reducing Space Mission Cost.** The two approaches differ dramatically and are largely incompatible with each other. Which approach is preferred depends principally on the philosophical orientation of the organizations involved.

Characteristic	Traditional Approach	Radical Approach
Philosophical orientation	Fits within the structure and processes of traditional development and procurement	Requires a significant shift in the engineering and development paradigm
Typical non-space examples	• Improve manufacturing process • Redesign with fewer parts	• Development of Model T • Development of personal computer
Space examples	• Concurrent engineering • Integrated Product Development (IPD) teams • Design-to-Cost • Standardization	• Trading on requirements • Reducing the cost of failure • Using large margins • Using non-space equipment
Typical savings	5% to 30%	50% to 95%
Advantages	• Methods and results well understood in advance • Minimal risk	• Potentially far greater cost reduction • Needed for breakthroughs to occur
Disadvantages	• Can't achieve breakthroughs	• Changing the rules is painful • Results are uncertain • Potentially large risk

TABLE 2.2-2. **Radical Cost Reduction Methods Typically Require Changing Both the Technology and Process Simultaneously.** See text for discussion.

Technology Changes	Process Changes
• Non-space equipment	• Trading on requirements
• Large margins	• Devalue optimization
• Autonomy	• Eliminate weight criteria
• Miniaturization	• Avoid "critical national need" syndrome
	• Schedule compression

Typically, introducing radical cost reduction methods is neither easy nor desirable if it can be avoided. It is driven principally by simply not having enough money to do it any other way. If our options are to either dramatically reduce cost or not fly the mission, then we should look seriously at radical cost reduction methods. Applying these methods means doing things we otherwise wouldn't choose to do—backing off on requirements, providing a less-than-optimal return on investment, or settling for what can be achieved at moderate cost.

Strongly reducing cost is becoming increasingly important because almost all current programs are cost constrained. However, even if we have the money available, we can learn a great deal from the radical approaches to cost reduction. Overall, there are more space projects that need to be done than budget available to do them.

If we are truly going to explore new frontiers, we will have to do so with constrained budgets, shorter schedules, and a genuinely new approach to exploring space.

2.2.1 Changing the Paradigm

Changing the paradigm means making basic changes in the philosophical approach to building spacecraft. Described below are four philosophical elements of particular importance to reducing cost.

Reducing the Cost of Failure

High technology programs cannot both reduce cost and guarantee success at all stages. It is particularly important to allow failures early in a program. It is the only way to gain confidence later on and to be sure we understand failure mechanisms and how to overcome them. For example, if we build a test engine that is dramatically expensive, we want to be sure the test engine never fails. This, in turn, means we won't learn the real failure mechanisms or fully understand what causes failures to occur. The most common result is an overly expensive design and a **higher** risk of failure because we haven't fully wrung out the failure modes.

A number of programs have recognized this problem and begun by stating a willingness to take increased risk. Unfortunately, this willingness to accept risks plummets when the program gets closer to launch. This has occurred on several programs I am familiar with, both in government and industry. Uncertain future funding drives this change. If funding for subsequent flights is not fully assured, the very first flights, tests, or products must be successful. This, in turn, means we can't allow the initial products to fail, and drives up cost. Robert Parkinson has provided the illustrative example in Fig. 2.2-1 of how reducing the cost of failure has the potential to greatly reduce mission cost.

Hubble Space Telescope

- It also took 25 years to arrive, in which time Earthbound telescopes had much improved.

- Had Hubble cost $150 million, it would have flown earlier, provided "hands-on experience," and been replaced by something better in a reasonably short time.

- If it had cost $150 million, the odd fault would not have been a major disaster, but something correctable on Hubble II.

- By now we would have been on Hubble IV or Hubble V, and be getting better performance than the current telescope for less total cost.

Fig. 2.2-1. Alternative Approach to Developing a Space Telescope. Robert Parkinson of British Aerospace has proposed this approach as a general model for lowering cost by reducing the cost of failure. [Parkinson, 1993]

Devalue Optimization

The objective in an engineering-oriented program should be to get where we need to go, rather than to optimize the process of getting there. In engineering terms: "Better is the enemy of good enough." Unfortunately, the very high cost of space products leads to a strong economic demand for optimization, which in turn leads to specialization of components and extremely high cost.

Constrain the Tyranny of Weight Optimization

A particularly strong example of the problem of optimization is trying to achieve minimum spacecraft weight. This is perhaps the single most economically destructive force in spacecraft design. Specifically, weight optimization leads to:

- Uniquely designed structures

- Minimal use of standard components

- Materials and processes unique to space

- Inadequate margin in all elements, including software

Traditional, weight-based cost models make this situation more difficult because they assume minimizing weight will result in the lowest cost. In practice, we need to balance the legitimate need for limiting weight because launch is very expensive against the excess goal of optimizing weight. The strong tendency is to argue that if geosynchronous spacecraft are worth $50,000/lb on orbit, we can legitimately spend $40,000 to provide another pound of on-orbit spacecraft. The problem, of course, is that repeating this process many times causes the spacecraft to cost $50,000/lb in the first place.

Avoid "Critical for National Needs" Syndrome

Many programs start out presenting themselves as critical to the national defense or to our role as a world leader in science or technology. Unfortunately, they frequently must do so to sell the program in the first place because it otherwise won't get funded. Still, this mindset is a death knell for cost savings. In fact, there are remarkably few space programs that the world **must** have. The United States existed for nearly 200 years without space assets and would probably survive for a reasonable period without them in the future. Exploring and using space **is** important, but not so important that it must be achieved at any cost.

Finally, in preparing this book, we have found that no single set of rules can make the cost reduction process work for all missions. Finding an effective balance is the main key to ensuring success, but it's also the basic difficulty with trying to achieve low cost through policies and procedures. It is very difficult to develop a procedure that says, in effect, "use common sense."

Many of the cost reduction recommendations are essentially contradictory. Optimizing weight is a good example. Trying to optimize the spacecraft weight ultimately drives up cost by creating a design totally lacking in flexibility and margin. On the other hand, we clearly can't afford to disregard weight entirely. The weight of the spacecraft is a key characteristic for spacecraft design and in determining launch costs. Lighter spacecraft cost less to launch. Programs that have been successful in making major cost reductions have found techniques to achieve a balance—keeping the weight low and scaling back the objectives of the mission to provide reasonable margin and flexibility within that low weight. The rest of this section describes how to achieve that "reasonable balance" while dramatically reducing cost.

2.2.2 Creating a Small-Team Mentality

Empirically, major breakthroughs in cost seem to have come largely from small organizations and small teams. The leading players in the small and reduced-cost space communities, aren't the major aerospace organizations we normally think of as leaders in space exploration. Rather they are:

- Companies such as AeroAstro, DSI, Spectrum, and Swedish Space Corp.
- Universities such as Utah State University, University of Surrey, and University of Bremen
- U. S. Government organizations such as NRL and Los Alamos
- Foreign space programs in countries such as Denmark and Sweden

These organizations are not necessarily small in terms of budget, personnel, or power, but "think small" in terms of space activity. For example, the U.S. Navy is not a small organization in any sense. Like any large organization, it has its full share of bureaucratic policies and procedures. However, space exploration is not the Navy's main line of work. The Naval Research Laboratory (NRL) engages in a wide variety of research and development, of which space activities represent a small part. Consequently, the NRL Center for Space Technology can behave, in many respects, like a small organization. This small-team mentality was at least partly responsible for NRL successfully producing and launching Clementine and other low-cost spacecraft, as discussed in Sec. 12.1.

Perhaps the principal characteristic that connects the small organizations is the absolute demand to achieve something by working smarter, rather than by spending more money. **If they can't play at low cost, they have to get out of the game**. In the end this represents both an enormous advantage and a major limitation of small organizations. They must be continuously inventive and creative, or find some other line of work. As a consequence, many of the very good small organizations won't survive the vicissitudes of the space marketplace, in which individual contracts can represent most or all of a small organization's business. This suggests that one of the best places to develop low-cost spacecraft on a continuing basis is in a small group within a larger organization which provides some level of protection from imminent disaster. NRL, Los Alamos, or the Lockheed Skunk Works may be examples of this good, long-term approach.

Another key characteristic of the small group environment is ease of communication. When the entire activity is being done by fewer than 30 people, they don't need a huge number of formal memos, policies, and procedures. Everyone is busy, people understand their responsibilities, and the entire organization knows when problems or issues need to be resolved. These are the subjects of lunchtime discussions, and of engineers working till midnight and falling asleep at their desks. Problems are solved in similarly human ways. Everyone in the group helps look for a solution, and a few give up some of their own time to help out. It is far easier to create a coherent, unified team of 20 or 30 people than of 2,000 or 3,000.

A potential counterexample to the cohesive small team is AMSAT, perhaps the most successful of the very low-cost satellite builders. As described in Sec. 13.3, AMSAT has been exceptionally successful in launching large numbers of very low-cost satellites. Nonetheless, these satellites have been designed and built by people who, in some cases, have never met each other and who often communicate entirely or nearly entirely by radio. Yet, components are built, brought together, and assembled to create successful spacecraft. AMSAT has never been late in delivering a satellite to launch and has had only one failure in 41 spacecraft.

What is the small-team mindset that allows major cost reductions to occur? One characteristic is a point of view that will find a way to make it work or change the requirements until it does. A parallel is the everyday example of a family buying a car or house on limited budget. We all know how to do this. We look at our budget and estimate what we can afford. We look at the marketplace and see what is available within our price range and probably a bit above and below it. We compare and evaluate until we find a solution that meets our needs within a price we can afford. Unfortunately, we usually leave this mentality at home when we go to work.

We tend to make more far reaching trades in buying a house than in buying spacecraft. In house hunting, we're systems oriented from the outset, so we typically do less jockeying on budgeting for individual elements. Problems still exist—one family member wants a swimming pool, another wants a large recreation room; but individuals recognize the need to keep this in check and find a balance that satisfies the collective objective. There is tension and bickering, but everyone feels responsible for success. There is a programmatic willingness to change the rules, accept the achieved level of success, and, in some cases, accept the possibility that it won't work and we may need to try again. While no analogy is perfect, I believe the example of buying a house illustrates how small organizations with a common purpose can, in some respects, achieve objectives that are much more difficult for large, diverse organizations. The strength of large organizations is their diversity and their capacity to commit enormous resources for multibillion dollar programs. Their forte is not the small, low-cost program. Nonetheless, the case studies clearly show that what matters most is not the size of the organization but the mentality which they bring to bear. Reducing space mission cost doesn't require small companies to do the work, it only requires elements of a small company's mindset.

2.2.3 Trading On Requirements

As discussed in many of the case studies, perhaps the most important single technique for radical cost reduction is *trading on requirements*. This means comparing alternative numerical requirements in terms of their capacity to meet the broad, typically non-quantitative, mission objectives. For example, our broad mission objective might be to map the surface of Mars on a regular basis, looking for seasonal variations. Ultimately, the system will achieve some ground resolution and some revisit frequency. However, it is not clear what precise numerical values are needed to achieve the mission objectives. We will trade on these numerical requirements in order to meet our objective of looking for changes on Mars, while maintaining the budget. In a cost constrained program, we will ultimately look at what we believe can be achieved for the assigned cost and ask whether it is worth flying the mission. If so, we proceed. If not, the mission is curtailed until the budget can be increased or the cost lowered.

Table 2.2-3 summarizes trading on requirements using the example of geolocation accuracy for a sensor on a spacecraft in low-Earth orbit. To achieve our science objectives, we might, for example, choose an initial goal of providing geolocation to within 500 m directly below the spacecraft. This gives us the general range we wish to achieve. Next, we look at the system design and develop a mapping budget for the mission. SMAD Sec. 5.4 details how to create this mapping budget. The components of the mapping budget will fall into three broad categories as shown in Table 2.2-4.

TABLE 2.2-3. Process for Trading on Requirements. See Sec. 10.2 and SMAD for more detailed discussion. For illustrative purposes, the example is oversimplified. Most requirements trades have multiple effects on both cost and performance.

Step	Example	How Traditionally Done
1. Document **reason** for requirements	Geolocation accuracy for science data reduction (co-ordination with area maps)	Not done
2. State functional rather than technical requirements	Provide geolocation to within 500 m at nadir	Not done (i.e., required attitude determination accuracy = 0.05 deg, 3 σ)
3. Identify system drivers for the requirement	Attitude sensor accuracy	Not done
4. Identify main options	(a) $500K Earth sensor @ 0.10 deg (b) $2,000K star sensor @ 0.01 deg	(a) $500K Earth sensor @ 0.10 deg (b) $2,000K star sensor @ 0.01 deg
5. Determine consequences in terms of mission performance	(a) geolocation to 1 km (b) geolocation to 100 m	(a) Won't work (b) Meets requirement with margin
6. Select preferred option	Discuss cost with customers and users. Is it worth $1.5M more per spacecraft to achieve geolocation of 100 m vs. 1 km?	Option (b) is only acceptable choice

TABLE 2.2-4. Elements of Requirements Budgets Normally Fall into One of the Three Categories Shown Below. Category 3 is key to the process of trading on requirements. For a detailed discussion of the development of requirements budgets, see SMAD Sec. 5.4.

Category	Example for Geolocation Budget
1. Elements which easily meet the allocation	Timing error
2. Elements for which the accuracy is not easily changed	Spacecraft ephemeris error based on either ground tracking or GPS
3. Elements for which increased accuracy is available at increased cost	Spacecraft attitude error using: 1. Magnetometer 2. Earth or Sun sensors 3. Star sensors

For trading on requirements, the critical components are those for which additional accuracy is available at additional cost. For geolocation, this would typically be the attitude determination or attitude control elements of the mapping budget. For example, we might be able to achieve 0.1 deg accuracy for $500,000/spacecraft using Earth sensors or 0.01 deg for $2 million using star sensors. When we apply these two options to our problem of geolocation, they could result, for example, in a 1 km geolocation error for the Earth sensor system and 100 m geolocation if we use star sensors. The critical part is then obtaining feedback and making reasonable and balanced judgments. Does using star sensors justify the additional cost? From a purely numerical set of requirements, we have no way of knowing. We need to go back to our basic objectives, assess what we are trying to achieve and how well we can achieve it, and then judge the value of the contribution of the higher accuracy component. If our basic mission costs $200 million and geolocation precision is the primary goal, the added cost for star sensors will mean relatively little so it's almost certainly worthwhile. If the geolocation problem is one part of a low cost, $10 million science survey, then the added cost for the star sensors is almost certainly not worthwhile, if it contributes only to geolocation. Trading on requirements means finding a balance between what we want and what we can afford. It is a key element of what we might call intelligent cost reduction.

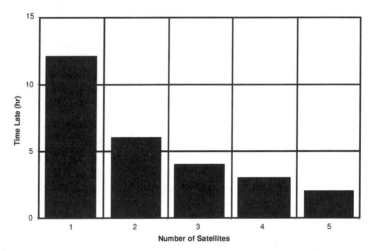

Fig. 2.2-2. Coverage Data for a Hypothetical Constellation. The initial requirement is assumed to be a time late of less than 5 hours. When trading on requirements, the key question is whether the combined cost and performance goals are best met by two satellites or by three [adapted from SMAD, Sec. 3.4].

A second example is illustrated in Fig. 2.2-2, which shows hypothetical coverage data for a sequence of satellites. The vertical component is the *time late*—that is, how long after an event occurs is the information reported to the user. Having timely data is typically valuable, but it requires a larger number of satellites and, therefore, a larger investment. In our hypothetical example, we assume the requirement was to ensure a time late of less than 5 hours, which calls for a three satellite constellation. Should we build three satellites or build only two, reduce the cost, and settle for a time late of about 6 hours? We have no way of answering a priori. We need to go back to the user and understand the purpose of the data. What impact will time late have on the use of the data and the utility of the satellite system? The user may be able to decide what is needed by looking at Fig. 2.2-2 or may need more information. What percentage of the time is the data going to be 6 hours late? What is the average time late? the 95th percentile? the 99th percentile? What is the mean time that it takes to get information? It might even be reasonable to build and launch one satellite, look at the utility of the data, and then decide whether to assemble and launch one or two additional satellites. If the times are longer than we would like, can we combine the space system with some other approach to achieve our objective? Can we get the data to the user sooner by sending it directly to the ground and then using phone or E-mail communications? Typically, radical cost reduction methods require being creative in our solutions.

Effectively trading on requirements demands several other ancillary changes in the way we do business in space:

- Setting functional, rather than technical requirements (i.e., stating **what** we want done, not **how**)
- Documenting the source of requirements, so we know **why** they are needed
- Trading explicitly on the principal driving requirements
- Being willing to give up some level of performance

The *driving requirements* are those that are primarily responsible for performance, cost, risk, and schedule of the system as a whole, so we want to look at them for making trades. Do some requirements have a strong impact on cost but perhaps not as much impact on performance? (SMAD Sec. 3.1 shows how to identify the driving requirements.) Finally, the most important element is being willing to give up performance in order to achieve our mission within the appropriate cost constraints. We may not be able to survey Mars with the resolution we would have liked, but it may be better than the alternative of no Mars mission at all.

2.2.4 Compressing the Schedule

A compressed schedule can be a major tool for cost reduction. For most programs, cost is dominated by the labor cost associated with the group of people designing, developing, building, testing, launching, and operating the spacecraft. A relatively small fraction goes into materials, purchased goods, and other capital items. Consequently, with a compressed schedule, there is simply not enough time to spend lots of money. Decisions are made more quickly and, therefore, less expensively simply because there isn't enough time to drag issues out.

However, it should be clear that schedule compression by itself is not enough. We have to couple it with some level of release from other requirements. Otherwise, as illustrated in Fig. 2.2-3, we're simply asking people to do the same job in less time, which isn't productive.

Key elements which allow schedule compression include:

The Ability to Trade on Requirements

This is frequently critical to a compressed schedule because of the need to go with what works or what exists, rather than taking the time and expense to build something new and unique to the program. (See Sec. 2.2.3.)

Rapid and Accessible Decision Making

Almost nothing slows a program more than long and involved decision making. In a quick-turnaround program the decision making must be responsive. It must also be reasonably accessible to engineers, so we can avoid poor technical decisions and the engineers who are designing the system can understand what motivates the decision makers so they can predict the likely future direction of the program.

" ... DAMN CAT DIED! ... "

Fig. 2.2-3. Schedule Compression Without Changing the Rules or Requirements is a Lot Like Getting a Cat to Move a Piano Up the Stairs. The whip doesn't help much. [Cartoon by Jack Friedenthal, created for the TDRS program.]

A Small Team

As discussed in Sec. 2.2.2, a small team greatly increases efficiency and, therefore, allows substantial schedule compression. Small teams don't necessarily work faster, but they have far more efficient communication, which allows the program to move much more rapidly.

Concurrent Activity

There is little potential for schedule compression if all actions have to be done serially, i.e., if each step can begin only after the preceding step has been completed, reviewed, and approved. Concurrent activities, such as writing software while hardware is still being built, has the obvious risk that changes in one area may affect another. However, this is also a strength of concurrent activity, in that it allows the design to mature in parallel and effects between subsystems and components can, at times, be more easily taken into account.

An excellent example of effective schedule compression is the Lockheed Skunk Works described in Sec. 2.2.7. The rapid schedule appears to have been one of the key reasons that the Skunk Works was successful in reducing the cost of many major programs.

Schedule compression does have a number of dangers and pitfalls. One of the most serious impediments to schedule compression is incremental funding. As clearly indicated by Kelly Johnson's rules for the Skunk Works, incremental funding is a schedule and programmatic disaster. Very few things are more apt to make the individual engineer want to see a program fail than continually being pressed to work faster and harder, then being told to stop and sit for 2 weeks (which ultimately becomes 3 months), while the customer contemplates the state-of-the-world and potential future funding.

Lastly don't use schedule compression to avoid up-front analysis and design. Not knowing where you're going is rarely a good shortcut to getting there efficiently. You may also end up getting quickly to some place you weren't going, which is typically of limited value.

2.2.5 Increasing Design Margins

The goal of a good, low-cost design should be to make the system manufacturable. It should also be reasonably forgiving of errors or changes in either the mission or the system. Designing the entire system with limited margin is a major cause of redesign and results in more failures than otherwise would be the case. To reduce weight, a major program that I worked on used titanium bolts to hold antennas on the end of a boom and drilled a hole in the head of each bolt to further lighten it. Unfortunately, the holes were drilled a bit too deep, the design margin was inadequate, and during testing one of the antennas fell off. People who build spacecraft are only human. A reasonable design margin provides forgiveness for small mistakes.

As an example of how minimal margins can increase cost, consider what would happen if automobiles were built with the level of optimization typically done for spacecraft. We ask the dealer for a different radio, which costs an additional $100. However, we need a new dashboard panel to accommodate the new radio. Because the speakers are higher power, we need a new battery and a new power system to charge the battery. The radio weighs more, so we need to beef up the suspension to support the added weight in the front. These changes are absurd in any normal manufacturing environment but are all too common for spacecraft.

TABLE 2.2-5. **Main Reasons that Increased Margin Reduces Cost.** See the text for more discussion.

• Less testing required
• Normal manufacturing tolerances acceptable
• Fewer rejects and reworks
• Less failures in both test and operations
• More robust design leads to less redesign
• Potential for standardized components
• Higher level of component and design reuse
• Can use more commercial grade components
• Can accept less certainty about the environment
• Reduces operations cost for planning and analysis

Table 2.2-5 summarizes the main reasons that increased margin reduces cost. Unfortunately, increased margin keeps us from optimizing the design, which is the normal practice for spacecraft. Many people have tried vigorously to bring more standardization to the space community as a means of reducing cost. On the whole, standardization has been remarkably unsuccessful. There have been NASA standard spacecraft, NASA standard subsystems, and NASA standard components, most of which have been used for one or two spacecraft. Unfortunately, standardization implies design margin. To use the same equipment on many missions, it must have more capability than we require for any individual mission. This means it will weigh a bit more and take a bit more power than a component designed exclusively for that

mission. The result is that standardization simply doesn't sell in the hardware arena. If we allow spacecraft margins to increase, there is a far greater potential for successful standardization, which in turn can reduce the cost and risk of many components and subsystems.

One of the less obvious effects of increased design margin is to reduce operations costs. As described in Sec. 6.4.2, a minimal margin design means we must plan and analyze each operations activity in detail to ensure that none of the constraints or conditions is violated. Spacecraft with increased design margins have far fewer operational constraints and, therefore, are more economical to operate and more flexible in their capacity to respond to changing needs or on-orbit problems.

The impact of increased design margin depends strongly on how it is implemented. Making structural members thicker, using a larger resistor than necessary, or providing additional RAM adds robustness, flexibility, and margin with no significant adverse impact other than the added weight for a non-optimal design. Pushing the qual temperature 30 deg higher than the spacecraft will ever see or pushing the vibration spec to beyond reasonable limits will only add cost and may create otherwise avoidable failures. Testing a readily available commercial component to see what environment it can withstand makes a great deal of sense. Forcing long-lead special parts to pass tests beyond the environment they will ever encounter does not. We need to implement with common sense.

Finally, the use of increased margin must be balanced and reasonable. Launching spacecraft is still remarkably expensive and the cost mainly depends on weight. Still the case studies clearly show that companies which have been very successful in building spacecraft at dramatically lower cost have done so in part by increasing margins and taking full advantage of these margins to reduce their cost.

2.2.6 Using Non-Space and COTS Hardware and Software

It is clear from the case studies that a principal cost reduction technique is to use commercial-off-the-shelf (COTS) software and non-space hardware. In general, very low-cost space programs simply don't use normal space equipment. The cost of almost any standard space component would dwarf the budget of the very low-cost programs.

Non-space hardware and software can be either commercial spin-offs of products previously created for space, or products created for everyday business or personal use. In either case, the cost savings can be dramatic. The *Hubble Space Telescope Guide Star Catalog* contains approximately 19 million stars and non-stellar objects and was created at considerable cost. However, it is of use not only to spacecraft and professional astronomers, but to amateur astronomers as well. Consequently, the Hubble Guide Star Catalog with access and graphing software is available from amateur astronomical organizations on two CD ROMs for $69.95. If you want to sort, analyze, view star catalogs, or map out specific regions of the sky, this can be done by excellent amateur observing programs for either Macintosh or PCs that are available for a few hundred dollars. These programs, several of which are listed in Sec. 3.2.4, use a number of different star catalogs which may be appropriate for space applications. When the Hubble Guide Star Catalog first became available, the developers simply modified their existing programs to use it as one of their alternative sources of data.

Ground-based computer applications have far surpassed space applications in terms of processing capability and requirements. Consequently, the low-cost space community has long since made the collective decision to use standard 80X86 computers for space. This is driven in part by the desire to develop software on normal office PCs and to use commercial compilers, debuggers, and software tools readily available for a few hundred dollars. Not only are the costs dramatically lower, but the product quality is typically much higher, in part because the commercial components have been tested by the community millions of times in widely varying applications. In contrast, special purpose space software using new and unique software tools typically results in debugging the compiler and software tools at the same time the application is being written.

The key issue in nearly all applications of non-space components is the willingness to give up optimization in order to achieve dramatic cost reduction. The commercial version of an astronomical observing or image processing program will almost certainly not exactly fit the specifications for a particular space application. On the other hand, it can cost hundreds of dollars, rather than hundreds of thousands or million of dollars, and can typically be on your desk within a week. Once we overcome the psychological trauma of using a commercial component, it can be a very compelling solution.

As with any other commercial activity, none of the components will meet your specifications exactly. Consequently, the methodology for using commercial components is the same we use in buying office equipment or cars—find the product that comes closest to meeting your needs and use it as is or modify it. One possibility is to have the software developer or hardware manufacturer who created the commercial product modify it to meet the specific needs of your program. Although this will be substantially more expensive than using the commercial product directly, it will ordinarily still be far cheaper than creating something from scratch. For example, Iida [1993] has recommended using automotive parts for space applications, in part because of their very low cost and high reliability.

If the appropriate product does not exist in the commercial market, many of the low-cost space contractors simply create their own, using commercially available materials, parts, and supplies. While this is more expensive than buying components already built, it is nonetheless much cheaper than a typical space procurement.

The most important requirement in using non-space equipment for space applications is knowing what is required for a component to work in space. The gravity, vacuum, radiation, and thermal environment are, of course, substantially different than most Earth-based applications. Therefore, we need to evaluate each proposed component in terms of its capacity to survive and function in space and to survive the launch environment. A great deal of information is available on the space environment. (See, for example, the references in SMAD Chap. 8 or the summary by Tascione [1994].) Unfortunately, not as much information is available as would be desirable on how to evaluate non-space components for use in space. An excellent and relatively complete set of background data on the behavior of systems in the space environment is provided by Dewitt, Duston, and Hyder [1993]. An exceptionally productive area of research for either government laboratories or universities interested in assisting the process of dramatically reducing space cost would be to examine in more detail and document the issues and requirements for using commercial equipment in the space environment.

2.2.7 The Lockheed Skunk Works

The Lockheed Advanced Development Projects division, far better known as the Skunk Works,[†] was created in 1943 by Clarence "Kelly" Johnson. [Johnson, 1985; Rich, 1994; Burrows, 1994] The Skunk Works has become almost synonymous with the process of doing business differently. It became renowned for high-tech aircraft that were at least occasionally brought in ahead of schedule and under budget. Among the more famous systems developed at the Skunk Works were:

- The F-104, the world's first Mach 2 fighter
- The U-2 high altitude spy plane
- The Mach 3.2 SR-71 Blackbird reconnaissance plane
- The Air Force F-117A Nighthawk stealth fighter

In some respects, an equally famous Skunk Works project was the 1950s XFV-1 vertical takeoff and landing airplane. After substantial experimentation and development work, Johnson told the Navy that the project was impractical and should be cancelled. It was.

Most of the technical work undertaken by the Skunk Works is classified, so that the results are made public only substantially after the fact, if at all. Of course, some of the more famous results have become well known, such as the fact that Lockheed returned $2 million of the $20 million development provided for the U-2 and under ran the F-117A development budget by $30 million.

Johnson's definition of the Skunk Works was:

> *"Skunk Works is a concentration of a few good people solving problems far in advance—and at a fraction of the cost—of other groups in the aircraft industry by applying the simplest, most straightforward methods possible to develop and produce new projects. All it is really is the application of common sense to some pretty tough problems."* [Johnson, 1985]

The rules by which the Skunk Works operated have also become well known. Kelly's methods are embodied in a set of 14 points listed in Fig. 2.2-4. Equally important, but less often quoted, is Johnson's commitment to Skunk Works' employees:

> *"I owe you a challenging, worthwhile job, providing stable employment, fair pay, a chance to advance, and an opportunity to contribute to our nation's first line of defense. I owe you good management and sound projects to work on, good equipment to work with and good work areas..."* [Johnson, 1985]

The Lockheed Skunk Works was able to reduce cost in at least some high-technology aircraft programs. The question for us is whether this experience is applicable to reducing space mission cost. I believe the answer is both yes and no. Many of the Skunk Works rules are applicable to space missions. However, some key

---

[†] The nickname "Skunk Works" came from the "Skonk Works" in Al Capp's Li'l Abner comic strip. Somewhere in the woods outside of Dogpatch, the Skonk Works made Kickapoo Joy Juice from old shoes, dead skunks, and other aromatic ingredients.

elements of the Skunk Works environment were significantly different than they are in the modern space industry.

The basic tenet of the Skunk Works was that they tried hard to **not** follow the traditional rules. Much of what was articulated in Johnson's 14 point plan is very similar to the recommendations here and carried out in the case study examples. They maintained an extremely small organization, both within the Skunk Works and with the customer. They had local authority with rapid decision making. They minimized paper work. They provided timely, continuous funding. They intentionally did not follow all of the standard MIL-SPECs. These are key issues, equally applicable to space programs, both large and small.

However, there are also differences which are important to keep in mind. Perhaps the biggest difference is that there is no evidence that the Skunk Works was ever strongly funding-constrained over the period of time when most of the major break-throughs were made. Indeed, the Skunk Works was an extremely well funded organization that worried much more about finding high-tech solutions than where the next contract was coming from or how to avoid laying off employees. It is far easier to return 10% of your budget at the end of the program if you got everything you asked for in the first place and didn't have to try to achieve the impossible for 50% less than what you believe it will likely cost. A key statement in Johnson's charge to his employees, is that he owed them "stable employment." The Skunk Works asked its customer to cancel at least some programs which did not look promising. In all probability, this was done without having to be overly concerned about having to lay off Skunk Works' employees. This is a dramatic and important difference between the environment at the time of the major Skunk Works' breakthroughs and the environment that exists in today's space community

A second major difference is that "build and test" cycles are much more workable in the aircraft environment, both during development and operations. A test flight was not only reasonable, many would be conducted before a final plane was produced. While the SR-71 may not have been produced in large quantities, far more than one was built. This is very different than today's space environment in which, except for communications constellations, most spacecraft are built one at a time, there is no test satellite launched prior to on-orbit operation, and the first satellite launched must work from the outset. These are technical challenges which can and have been overcome, but they do make the task, in some respects, significantly more difficult than the Skunk Works encountered.

Finally, the objective of the Skunk Works was not to build low-cost airplanes, but to design special purpose, ultra-high-tech planes, quickly and more cheaply than other organizations. Lockheed itself did not apply the Skunk Works approach to commercial airliners or other planes where cost rather than performance was the dominant consideration. It's not at all clear that a low-cost aircraft company could survive the Skunk Works' high rates. This suggests that the Skunk Works' approach might be most applicable in the space arena to reducing the cost of Space Station, the National Aerospace Plane, or Brilliant Eyes, more than to programs whose scale is more modest. However, even in the arena of large, high-tech space programs, it is difficult to envision a program having the funding certainty and continuity that many of the Skunk Works projects enjoyed. Nonetheless, the Skunk Works experience base provides an excellent lesson for us in finding ways to do business differently in order to create unique, high-tech products and drive down the cost of access to space.

Basic Operating Rules of the Lockheed Skunk Works

1. The Skunk Works manager must be delegated practically complete control of his program in all aspects. He should report to a division president or higher.

2. Strong **but small** project offices must be provided both by the military and industry.

3. The number of people having any connection with the project must be restricted in an almost vicious manner. Use a small number of good people (10% to 25% compared to the so-called normal systems).

4. A very simple drawing and drawing release system with great flexibility for making changes must be provided.

5. There must be a minimum number of reports required, but **important** work must be recorded thoroughly.

6. There must be a monthly cost review covering not only what has been spent and committed but also projected costs to the conclusion of the program. Don't have the books 90 days late and don't surprise the customer with sudden overruns.

7. The contractor must be delegated and must assume more than **normal** responsibility to get good vendor bids for subcontract on the project. Commercial bid procedures are very often better than military ones.

8. The inspection system as currently used by ADP [ed. note: ADP=Advanced Development Projects="Skunk Works"], which has been approved by both the Air Force and Navy, meets the intent of existing military requirements and should be used on new projects. Push more basic inspection responsibility back to subcontractors and vendors. Don't duplicate so much inspection.

9. The contractor **must** be delegated the authority to test his final product in flight. He can and must test it in the initial stages. If he doesn't, he rapidly loses his competency to design other vehicles.

10. The specifications applying to the hardware must be agreed to in **advance** of contracting. The ADP practice of having a specification section stating clearly which important military specification items will not knowingly be complied with and reasons therefore is highly recommended.

11. Funding a program must be **timely** so that the contractor doesn't have to keep running to the bank to support government projects.

12. There must be mutual trust between the military project organization and the contractor with very close cooperation and liaison on a day-to-day basis. This cuts down misunderstanding and correspondence to an absolute minimum.

13. Access by outsiders to the project and its personnel must be strictly controlled by appropriate security measures.

14. Because only a few people will be used in engineering and most other areas, ways must be provided to reward good performance by **pay not based on the number of personnel supervised**.

Fig. 2.2-4. Kelly Johnson's Rules for the Lockheed Skunk Works. The Skunk Works is one of the most famous examples of "doing business differently." Some of the rules are applicable to reducing space mission cost. See text for discussion. [Johnson, 1985].

2.2.8 Impact of Very Low Cost on Performance and Reliability

A key technique used throughout the case studies and other low-cost programs, is to trade performance and risk for dramatically reduced cost. By most measures, the risk is almost certainly larger for dramatically reduced cost spacecraft. As discussed below, this does **not** mean that the reliability is less. It means that we are less certain of being able to achieve all of our performance objectives within the dramatically lower cost and schedule goals which we have set. With a spacecraft being developed by 20 people, an automobile accident can set the systems engineering group back by 2 months. The more aggressive our goals, the less certain we are of achieving all of them. Of course, if our goal was to reduce cost by a factor of 5 and we have a 50% cost overrun, we are still a factor of 3 below where we started.

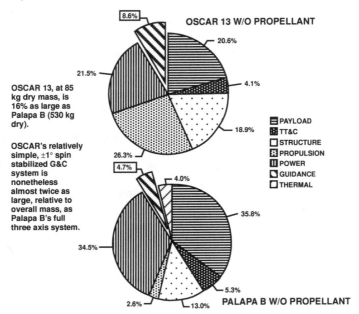

Fig. 2.2-5. Comparison of Satellite G&C Efficiency. Comparison of OSCAR 13 with Palapa B illustrates the difficulty in scaling spacecraft G&C systems for small satellites. [from Fleeter, SMAD Sec. 22.4.]

Where are the disadvantages of dramatically reduced cost systems? What are we giving up in order to achieve the benefits of reduced cost? The answer, illustrated in Fig. 2.2-5, is that we are largely giving up performance [Fleeter, SMAD Sec. 22.4]. The figure compares the mass distribution by subsystem for OSCAR 13, a relatively simple and very low-cost AMSAT communication spacecraft, and Palapa B, a much more traditional, 3-axis stabilized communication satellite. Even though OSCAR 13 was a relatively simple, spin-stabilized spacecraft, the attitude control system accounted for 8.6% of the spacecraft's mass. The much more capable, 3-axis control system on Palapa B, represented only 4.7% of the spacecraft mass. Similar effects apply to most of the other bus subsystems as well. Of course, if the bus subsystems all consume a higher percentage of the satellite's mass and resources, then something must have less. That something is the spacecraft payload. For OSCAR 13, the payload is only 20.6% of the spacecraft mass, whereas for Palapa B, it is 35.8%. In a traditional spacecraft design, the payload will be not only a higher percentage of the mass, but typically will be a more optimized design as well, such that the performance will be substantially greater than for the more inexpensive satellite. Of course, this is simply another way of saying what we have seen in Sec. 1.1—for the very low-cost missions, the performance is substantially less per unit cost than for the more traditional, high-cost missions.

One might expect a similar decrease in reliability associated with using low-cost construction techniques, simple non-redundant spacecraft, and non-space equipment. However, this does not appear to be the case. Although some data is provided in Chap. 9, there have been no definitive studies on the relative reliability between small, low-cost satellites, such as those covered by the case studies, and more traditional large satellites. Nonetheless the anecdotal evidence suggests that small, low-cost spacecraft are equally or more reliable, than their larger, more complex counterparts. The Amateur Radio Satellite Organization has launched 41 satellites, only one of which has failed to work on-orbit. (See Sec. 13.3.) This is an enviable success record for almost any space organization. Other organizations which have built a large number of small satellites—such as DSI, University of Surrey, or APL—also have excellent success records. Even those companies which have built only one or two small satellites, such as University of Bremen, CRI, or Weber State University, have typically been successful in their first small satellite venture.

Both small and large satellites have problems and total failures on-orbit. Among the more recent large programs, the Space Telescope mirror problem and the stuck antenna on Galileo have received considerable public attention. In the small satellite arena the MSTI-1 control system failed, a solar panel broke off ALEXIS during launch, and Clementine was effectively lost by an on-orbit software glitch. A major difference between the two types of programs is that the small programs cost a great deal less and usually do not have redundant equipment or extensive backup modes. Nonetheless, and perhaps surprisingly, the small spacecraft appear to have been able to recover in ways that are approximately comparable to those used in the larger programs. Space Telescope's mirror problem was very successfully corrected by a subsequent launch. Galileo is transmitting pictures at an extremely low data rate, and although it will have degraded performance, substantial data will be available. MSTI-1 spun up because of a control failure, but was still able to take photographs to validate the equipment that it was intended to test. Similarly, Clementine obtained good lunar photographs and was able to test most of it's equipment prior to the accident. Subsequent to the accident, the spacecraft has been re-contacted and appears to be functioning. The broken solar array on ALEXIS, which also destroyed the magnetometer used for attitude determination, caused substantial problems with reconstructing science data. However, considerable ground processing has allowed recovery of a good deal of that data and ALEXIS will make strong, substantial scientific contributions, in spite of the failure of a major part of the attitude control and power systems.

I believe there are a number of reasons why small spacecraft tend to have substantially higher reliability than what we would expect given their dramatically low cost. First, small, low-cost spacecraft are much simpler. This means they are easier to design and build and have far fewer components. Basically, there is less that can go wrong on a small spacecraft. When we add physical redundancy in a large spacecraft, we also add the interconnects and logic for selecting the components to be used. All of this increases the complexity and, therefore, reduces somewhat the reliability enhancement that was intended. Second, small spacecraft frequently use functional redundancy, rather than physical redundancy. This means that there are other ways to get the job done, that have either been designed as a part of the engineering process, or been invented once an on-orbit failure occurs, simply because the spacecraft design was reasonably robust. Thus, ALEXIS was able to use the Sun sensor and Earth sensor to achieve sufficient attitude information to reduce the scientific data and to control the spacecraft within acceptable limits. In a process sense, this is similar to Galileo using the low-gain antenna to send data back when the high-gain antenna became no longer available.

Finally, I believe another key reason for the apparently high reliability of low-cost spacecraft is the large design margins that are frequently used in constructing them. Since the designers are not driven by an overriding demand to optimize performance, they will try hard to use equipment which they feel certain will work. I remember well one of my first encounters with a very small satellite, which had a structure of aluminum plate. If they wanted to move a box from one side to the other in order to balance the spacecraft, they simply unbolted it, drilled new holes, tapped them, and remounted the box—a process that could easily be done by one person in half a day. In contrast, moving a component late in spacecraft integration on a large program would have a dramatic cost impact and almost certainly delay the schedule substantially as well, with an associated additional cost. Not only was the welded aluminum structure easy to work with, it was also highly reliable. That spacecraft was not going to come apart on anything short of re-entry. This situation is analogous in many respects to the difference between a sports car which is optimized for performance, but may or may not make it to the end of the race, and a well built pickup truck, which does not have anywhere near the performance of the sports car, but is capable of withstanding both bad and unpredictable circumstances. In many respects, small, low-cost spacecraft are the pick-up trucks of space. Both pick-up trucks and sports cars are needed for a strong space program.

References

Burrows, William E. April/May 1994. "How the Skunk Works Works." *Air & Space,* vol. 9, no. 1, pp. 30–40.

Colorado Office of Space Advocacy (COSA) and USAF Academy. March 1995. Reducing Space Mission Cost Workshop notes.

The Department of Defense Law Advisory Panel, Streamlining Defense Acquisition Laws. March 1993. *Executive Summary: Report of the DoD Acquisition Law Advisory Panel.* Defense Systems Management College. Footnote 1, pp. 13–14.

Dewitt, Robert N., Dwight Duston, and Anthony K. Hyder, eds. 1993. *The Behavior of Systems in the Space Environment.* Dordrecht: Kluwer Academic Publishers.

Iida, Takashi, et al. 12–14 November 1993. *Application of Car Electronic Parts to Small Satellites.* Japan–U.S. Cooperation in Space Project Workshop. Maui, Hawaii.

Johnson, Clarence L. "Kelly", with Maggie Smith. 1985. *Kelly: More Than My Share of It All.* Washington, D.C.: Smithsonian Institute Press.

Johnson-Freeze, Joan and George, M. Moore. 2 December 1993. "Clash of the Titans of Space Policy," *Nature*, vol. 366, p. 400.

Office of Technology and Assessment, U. S. Congress. September 1994. *Assessing the Potential for Civil-Military Integration: Technologies, Process and Practices.* OTA-ISS-611, p. 5. Washington, D.C.: U.S. Government Printing Office.

Office of Science and Technology Policy, The White House. August 5, 1994. *National Space Transportation Policy*, Press Release, p. 2.

Parkinson, Robert. 1993. "Introduction and Methodology of Space Cost Engineering." AIAA Short Course, April 28–30, 1993.

Rich, Ben and Janos Leo. 1994. *Skunk Works*. New York and Boston: Little, Brown and Company.

Rich, Ben. October 1994. "Inside the Skunk Works." *Popular Science*, pp. 52–61.

SMAD = Larson, Wiley J. and James R. Wertz, eds. 1992. *Space Mission Analysis and Design,* 2nd edition. Torrance: Microcosm Press and Dordrecht: Kluwer Academic Publishers.

Tascione, Thomas F. 1994. *Introduction to the Space Environment,* 2nd edition. Malibar: Krieger Publishing Company.

Chapter 3

Technology for Reduced Cost Missions

3.1 Cost-Effective Hardware and Technology

Jerry Jon Sellers, *United States Air Force*
Ed Milton, *Surrey Satellite Technology, Ltd.*

Identifying, selecting, and qualifying hardware to fly in space can be as much an exercise in philosophy and judgment as a rigorous engineering discipline. Certainly, over the years, companies providing "space-qualified" components have evolved rigorous standards and procedures meant to prevent their component from failing in the harsh environment of space. Although this type of hardware is plentiful, it's also expensive. Our challenge is to find alternatives to traditional "space-qualified" hardware that can fulfill the same mission at a lower cost. Conceptually, this challenge is no different from the one we face every time we go shopping. We all want the best-quality product at the lowest possible price. Unfortunately, in the world of space engineering, it is rare to find "close-out sales" that offer top-quality products at a discount. Thus, just as we all must choose between steak and hot dogs at the supermarket, we must often trade off the quality and price of space hardware to cost-effectively meet our requirements.

In this chapter, we'll highlight this trade-off between quality and price as it applies to space hardware. Our purpose is to provide a useful context for understanding how to select and fly low-cost space hardware. To that end, we'll identify hardware costs that drive each phase of a mission and discuss ways to reduce costs incrementally at each step. Notice this is a **process,** not a magic formula. No single hardware vendor or satellite builder has **the answer** on low-cost hardware. We can consider strategies for reducing hardware costs only within the context of a particular mission.

The fundamental question we face in this chapter is "why is space hardware so expensive?" Some people, such as space economist Chris Elliot [Sellers, 94], argue that space missions (and their associated hardware) are so expensive simply because we expect them to be. As Fig. 3.1-1 illustrates, when we assume missions are high-cost, decisions occur that make them high-cost. To break out of this vicious circle, you must first accept that missions **can** be inexpensive. Only then can you make decisions that allow them to be.

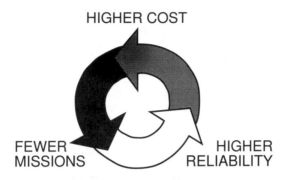

HIGHER COST

FEWER MISSIONS

HIGHER RELIABILITY

Fig. 3.1-1. Vicious Circle of a High-Cost Mission. By assuming that space is expensive, we plan fewer missions and demand greater reliability. This becomes a self-fulfilling prophecy which creates higher costs.

So, assuming we have made this leap of faith and accept that low-cost spacecraft are achievable, how do we get the necessary low-cost hardware? To simplify our discussion, we'll arbitrarily divide the life cycle of a mission, as it pertains to hardware cost, into three phases—mission definition, mission design, and acquisition.

During each of these mission phases, decisions and actions take place that directly influence the bottom-line price for hardware. In the first part of the chapter, we'll examine these mission phases in detail to identify hardware cost drivers. By understanding the cost sensitivity of these decisions and actions, you can better capitalize on low-cost options for hardware technology. As we will show, producing or procuring low-cost hardware most often requires an innovative definition of requirements tempered by a sound understanding of mission drivers. In the last section of the chapter, we'll present several anecdotes from the University of Surrey's satellite-engineering group. Their experience shows how these principles have reduced the cost of spacecraft hardware and components.

Table 3.1-1 offers a "road map" to our approach. Although we'll present the approach linearly, this process requires feedback, iteration, and imagination. Our purpose is to provide you with a comprehensive but flexible approach that covers the critical aspects of selecting and qualifying low-cost hardware. Along the way, we'll provide the context you need to make informed judgments about the infinite number of trade-offs you'll face.

3.1.1 Mission Definition

Despite the best cost-cutting attention of any engineer or program manager, certain harsh realities of planning and operating space missions can't simply be budgeted away. During mission definition, we address the "who, what, where, why, and when" of the mission. The answers to these questions determine the entire mission environment.

Among the biggest cost drivers for space hardware are the launch and space environment and the political environment.

TABLE 3.1-1.   Cost Drivers for Space Hardware During Each Phase of the Mission Life Cycle.

Mission Phase	Cost Driver	Technical Impact	Cost Implications
Mission definition	Launch and Space Environment— what the hardware will undergo	Hardware performance requirements e.g., radiation hardness Materials and practices to use or avoid	Space-environment requirements may preclude options to use cheap, terrestrial material e.g., within the Van Allen belts you may not be able to use conventional semi-conductors
	Political Environment— what hardware decisions are possible	Specified quality level of components Acceptable risk vs. allowable risk of a given technical approach, e.g., what you can technically accept vs. what you can politically allow	If the customer or political force requires space-qualified components, this will drive up the cost May not be able to use technologies that cost less but carry more risk
Mission design	Performance Requirements and Margins—what the hardware must do	Subsystem performance specifications Engineering margins to be followed during component manufacturing and testing	Higher performance leads to higher cost Specified performance may preclude the use of some technology solutions Overly constraining margins increase manufacturing, analysis, and testing costs
	Quality Level—how you will manage performance uncertainty	Specific components, materials and processes you must meet Requirements for acceptance tests	"Space-qualified" components cost more than their industrial-grade counterparts Quality verification during acceptance testing increases manpower and documentation costs
	System Architecture— how the hardware will be put together	Type, number, and deployment of components Redundancy philosophy	Some system architectures may lead to a simpler, lower-cost design with little or no performance degradation True functional redundancy is expensive
Hardware acquisition	Hardware Source— the decision to buy, build, or "borrow" specific hardware	Component-manufacturing process Insight into technical aspects of hardware	Type and level of in-house expertise and facilities
	Procurement Process— how you will work with suppliers	Risk sharing Acceptance testing requirements and procedures	Level of supplier oversight—risk sharing, documentation, and meetings
	Space Qualifying—how do you ensure it will work in space	Qualification requirements and procedures Number of dedicated components needed for qualification	Type and level of in-house expertise and facilities In-house hardware and expertise

Launch and Space Environment

During mission definition, we identify the launcher and the mission orbit. Of course, the launch price is a big part of the mission costs, but the actual launcher and orbit also directly affect available hardware options and their associated cost. We have to screen and test potential hardware to ensure it will survive the punishing ride into space and work in the harsh space environment. Because of this inflexible constraint, some promising, low-cost, terrestrial hardware must be rejected because materials or other characteristics are incompatible with launch requirements or space itself. In this way, the launch and space environment strongly drives our technical approach.

Before arriving in orbit, a spacecraft undergoes the massive accelerations and vibrations of launch, which normally demands most from the structural design. Launch agencies provide their customers with a detailed description of the launch environment, including specific loads and energy spectrum, as well as any mechanical requirements and procedures which customers must strictly follow to integrate their spacecraft smoothly into the launch vehicle. Normally, these procedures and requirements are in the *Interface Control Document* (ICD) (or DCI for Ariane [DCI]). Isakowitz (1995) also describes launch environments for various launchers. However, the final authority is the launch agency itself.

The primary purpose of the ICD (or equivalent) is to define technically and control **all** interfaces between the spacecraft and the launcher, including the formal requirements for the service the launch agency is to provide. An example would be to deploy the spacecraft into a particular orbit with some specified orientation. The ICD also establishes the operational requirements for the launch campaign. Examples of information spelled out in the ICD include [Arianespace, 1991]

- Spacecraft mass, alignment, and inertia characteristics including reference axes
- Spacecraft configuration inside the fairing
- Induced environment the spacecraft must be designed to tolerate
  - Static and dynamic accelerations such as longitudinal-acceleration profile, vibration level (e.g., ≤ 1.25 g's 5–100 Hz longitudinal, ≤ 0.8 g's 5–18 Hz and ≤ 0.6 g's 18–100 Hz lateral), acoustic vibrations, and shocks
  - The spacecraft's frequency requirements, such as fundamental frequency >50 Hz in thrust axis, > 45 Hz in lateral axes
  - Thermal characteristics, such as power input < 500 W/m^2 under fairing, 360 W dissipated power from primary spacecraft during countdown and boost, and external temperature limits for your spacecraft
  - Decompression; all enclosed volumes must be vented

In addition to the launch environment requirements listed above, the ICD defines [Arianespace, 1991]

- Contamination and cleanliness
  - Organic contamination on the spacecraft, such as clean-room specifications and outgassing from primary payload and interstage
  - Air cleanliness inside the fairing on the payload platform
  - Design and test requirements for spacecraft (e.g., outgassing criteria: total mass change $\leq 1\%$, volatile condensable material $\leq 0.1\%$ measured according to ESA PSS-01-702)
- Ground environment before launch (e.g., mechanical and thermal loads during launch processing)
- Thermal analysis requirements
- Characteristics and compatibility of radio systems
  - Transmission plan

- Radioelectric compatibility (e.g., analysis requirements, spurious radiation limited to -55 dBW (EIRP) on discrete frequencies in the frequency bands used for telemetry, limits on radiated energy in the launch vehicle's telecommand frequencies)

- Electromagnetic compatibility

- Mechanical interfaces and constraints (e.g., separation spring with $\Delta V = 0.5$ to 2 m/s with 1 deg half-cone angle)

- Electrical interfaces and constraints (e.g., battery-charging umbilical, ground reference points)

- Pyrotechnic orders for separation (e.g., voltage, current, pulse width)

- Launch campaign services (e.g., facilities, equipment, communications, transportation, technical support)

Along with the purely technical or mechanical requirements spelled out in the ICD and described above, other requirements are specified. Meeting these requirements is especially important when low-cost satellites ride "piggyback" on the primary spacecraft, whose customer bears most of the launch costs. In these situations especially, the launch agency won't tolerate low-cost, corner-cutting options which could endanger the primary payload. Any designer of low-cost spacecraft should factor these launch-agency requirements into the design from the very beginning of the program to avoid potentially costly changes later on. For example, WRR 127-1 *Range Safety Requirements* [1993] govern propulsion systems intended for launch operations out of Vandenberg AFB in California. Military Standards also dictate requirements for specific components; for example, MIL-STD-1522A covers "Standard General Requirements for Safe Design and Operation of Pressurized Missile and Space Systems."

Once your hardware has survived the launch, it next faces the unpleasant environment of space. See Sellers [1994] for a relatively painless introduction to this topic, SMAD for detailed design requirements, and the *Handbook of Geophysics and the Space Environment* [Jursa, 1985] for a physicist's comprehensive view. Table 3.1-2 summarizes key aspects that affect hardware selection.

Free Fall

The free-fall environment opens up many potential applications for space manufacturing, but, it also has many drawbacks. To begin with, free fall takes away convenient methods of convective heat transfer, as discussed below. However, the most important impact is on fluid management. Fluids, such as propellants, are much more difficult to handle when no preferred "down" direction tells them where to settle. As a result, you must plan some passive or active means of ensuring propellants can safely flow to the engine, avoiding "burps" of pressurant gas which can lead to potentially dangerous engine instability. Passive means include settling burns, spin stabilization, or propellant-management devices which rely on the liquids' surface tension. More complex (and expensive) active means include bladders, pistons, or pumps. For a complete description of spacecraft tanking options, refer to Humble [1995].

TABLE 3.1-2. Summary of Effects and Recommendations Associated with the Space Environment.

Environment	Effects	Things to Do	Things to Avoid
Launch	Vibration Dynamic and static loads Acoustic energy Shocks Decompression	Determine loads, energy spectrum and minimum critical frequencies from launch agency In general, design and procure all components to meet the launch agency's requirements Do spacecraft vibration tests to ensure compliance Do spacecraft thermal and vacuum test to ensure outgassing compliance	Unrepeatable manufacturing processes Large surface areas Overly constraining margins Unsuitable vacuum materials
Free fall	Fluid, primarily propellant, management	Provide for passive or active means of propellant management, e.g., spin stabilization, bladders	Uncontrolled fluids
Vacuum or neutral atmosphere	Outgassing Cold-welding Atomic oxygen	Check all spacecraft materials against outgassing data	Materials with known outgassing properties Bare metals in moveable mechanisms in direct contact
Thermal	Conduction is the primary means of moving heat within the spacecraft Radiation is the only way to remove heat from the spacecraft	Ensure enough conduction paths are established to move heat away from hot components Ensure there is enough area to radiate waste heat into space	Extremities on the outside of the spacecraft Heat-dissipating components in the center of PCBs Materials that "age" with thermal cycling
Radiation	Charged particles cause single-event phenomenon, which disrupts operation of electronic components Build-up of charged particles on spacecraft surfaces can lead to uneven charging and damaging discharges	Provide layered redundancy for suspect components Provide bit-error-correction code in spacecraft software Test new components to determine rad-hardness Use rad-hard components Ground the spacecraft's exterior surfaces	Rad-soft components in high-radiation orbits Exposed components electrically isolated from the rest of the spacecraft Certain types of plastic components that break down under exposure to radiation

Vacuum or Neutral Atmosphere

The most important implication of the vacuum environment has to do with the outgassing of materials. During manufacture in Earth's atmosphere, minute quantities of gas become absorbed in porous material. When this material encounters a vacuum, the gas releases. In addition, some materials, such as PVC or cadmium, slowly sublime in a vacuum. These processes, known as *outgassing,* are similar to that observed when you open a soda can: gas dissolved in the liquid under pressure releases when the pressure decreases. Although the quantities released for most materials is very small, they can damage sensors and solar-cell covers, coating them with a light film and thereby blurring them or decreasing their sensitivity and efficiency. Vapor can cause discharge around areas of high field strength (antennas or waveguide, for example) and create paths that leak current on printed circuits. The most likely cause of the failure of UoSAT-4 in 1990, for example, is a filter-dielectric breakdown due to insufficient venting of outgassing materials.

To prevent these problems, you should screen all materials early on in designing or selecting a component to determine their potential for outgassing. Outgassing criteria are normally expressed in terms of total mass change ($\pm 1\%$ normally acceptable) and volatile condensable material ($\pm 0.1\%$ normally acceptable). Fortunately, most materials you'll find in typical ground components have already been tested, so you can reference their outgassing properties in Campbell [1993] or Jollet [1986]. Some materials not recommended for use in the space vacuum are:

- Many types of plastic (e.g., PVC which contains chlorine)
- Solvent-based adhesives or paints
- Cadmium (which vaporizes and forms conductive whiskers)

In addition, designers must be wary of creating sealed volumes containing air at ground level which will expand and burst in vacuum. All enclosed spaces must have adequate venting holes or be designed to withstand one atmosphere pressure.

Another consideration in the vacuum environment is *cold welding*. Two metals in contact with each other in a vacuum can cold weld together over time. This has serious implications for joints and mechanisms to be used only very late in a mission. Although specially adapted, space-compatible lubricants have been developed to alleviate this problem, these lubricants may degrade when stored on Earth for long periods before launch (for example, due to launch delays).

Although we must consider effects from the vacuum of space, we must also recognize it isn't a complete vacuum. One constituent in particular, atomic oxygen, can cause problems with some spacecraft materials. Atomic oxygen is especially troublesome in low-Earth orbit. Bombardment of the upper atmosphere by charged particles from the Sun and galaxy causes oxygen molecules to split apart. In the denser part of the atmosphere they quickly recombine to form O_2. However, in the upper atmosphere these atoms are sparse, so we can find highly reactive atomic oxygen. Atomic oxygen is known to attack spacecraft surfaces—especially carbon, silver, and Teflon—breaking them down and degrading them. Long-term effects of atomic oxygen are still being investigated.

Thermal

From basic physics, we know thermal energy can move from place to place in three ways: convection, conduction, and radiation. Convection is based on the flow of a liquid or gas (such as air) past a surface, carrying heat away. It is perhaps the most important mechanism for heat transfer in our daily lives. Unfortunately, in the vacuum and free-fall environment of space, convection can't take place, so we're left with conduction and radiation.

In terms of spacecraft design, we need to control heat internally generated and heat radiated from nearby hot bodies (like the Sun or the Earth). Inside a spacecraft, the main way to control heat is by conduction to a cooler part of the spacecraft and, ultimately, radiation to outer space. This may be through the mounting chassis or by heat-pipes. Unfortunately, good conductors are usually dense, which adds to the mass budget. Compared to terrestrial designs, printed circuit boards in vacuum are prone to very localized heating, causing hot-spots and often requiring analysis for each component.

We must also control the heat radiated towards the spacecraft and dissipate heat generated from within. The spacecraft surfaces that aren't solar arrays or optics are usually helping with thermal control via surface coatings or multilayer thermal insulation. Note that extreme cold temperatures are just as dangerous to a spacecraft as high temperatures, so heaters are sometimes necessary, especially for propellants such as hydrazine.

We must also consider thermal expansion when joining two metallic items. Items of the same material, with the same coefficient of thermal expansion, can be bolted together or bonded with silicone or epoxy adhesive. However, dissimilar metals will expand and contract at different rates, causing epoxy bonds to fracture and creating a potential for bolts to loosen with time.

Without convective heat transfer to rely on, designers must carefully examine paths for heat conduction. For large heat-transfer requirements, heat pipes may be needed. These devices use the capillary action of a working fluid for passive pumping thus creating an efficient heat convection path. However, heat pipes are relatively complex and expensive compared to passive means.

Radiation

The "space-radiation environment" refers to charged particles in the form of electrons, protons, and atomic nuclei (neutrons quickly decay into protons and electrons) which stream out from the Sun as part of the solar wind. Another important source of these particles is actually outside the solar system. These so-called *galactic cosmic rays* consist of charged particles (and sometimes heavy atomic nuclei) which are the residue from dying stars. The interaction of these charged particles with spacecraft surfaces and materials produces potentially damaging ionizing radiation.

Traditionally, spacecraft have used radiation-hardened (or "rad-hard") electronics derived from military programs to provide (relatively simple) data-handling under these harsh conditions. These specially designed versions of their terrestrial counterparts can continue to operate even after accumulating very large life-time dosages. However, since the 1970s, the sales of military integrated circuits has steadily declined as a percentage of the total market for integrated circuits. In the 1990s, many military programs are disappearing, budgets are shrinking, and economic pressure on the manufacturers of rad-hard components is increasing as the market declines, leading to reduced availability of new, rad-hard, integrated circuits. These conditions are creating a growing demand for interest in commercial alternatives. The key phrase for the future is *dual-use technology:* aligning military technologies with the commercial semiconductor market. The commercial market benefits from huge investments intended to achieve higher performance, lower power, and lower cost per function.

Yet, ground-based tests indicate many reasons why it may be **un**-desirable to fly commercial electronics in the hostile radiation environment of space near the Earth. A main reason is the commercial devices' relative susceptibility to so-called *single-event phenomena,* which occur when a single particle (e.g., proton, or heavy-ion) passes through the device. These phenomena have several forms (e.g., *single-event upset, multiple-bit upset, single-event latch-up,* and *single-event hard-error*). All of these can be readily produced on the ground using particle-accelerators to expose devices to beams of high-energy protons or heavy ions. Ironically, such effects seem to increase as devices become more advanced and geometries reduce to the micron and submicron scale.

Total-dose damage—damage caused by long-term exposure to ionizing radiation—is a further concern. The extended operational lifetimes required of modern spacecraft can give rise to very large (10s–100s of krad (Si)*) accumulated doses, although the rates of these doses are relatively low compared to other ionizing environments, such as the interior of nuclear reactors.

Ground-based testing using X-ray and gamma-ray sources (often delivering radiation at high dose-rates) reveals that many commercial devices fail after a few tens of krad (Si). But questions remain about the veracity of these tests, and, in any case, given appropriate shielding, a survival of a few tens of krad (Si) is quite adequate for many space missions.

Thus, we have a dilemma. On one hand, to be sure a space system will work perfectly, we seem to require expensive, yet virtually obsolete, technology. Although this technology is almost guaranteed to survive, it handles data relatively poorly. On the other hand, we'd like to take advantage of the latest advances in commercial technology in order to construct highly capable data-handling systems, and yet ground tests urge caution. Unfortunately, because ground tests aren't a perfect analog of the space environment, they may raise questions: Do these effects really happen in space as much as we'd predict on the basis of these kinds of tests? Are we confident enough in our testing techniques to write off what could be very attractive technology for space use?

We can best answer these questions by actually flying commercial technologies in space to see how they really behave there. Ideally, we should measure the radiation environment at the same time and ground test identical devices to verify the whole prediction process.

All circuit designers must be aware of potential problems when designing for the radiation environment. When it comes to selecting electronic components for a mission, you can avoid the problem by flying only rad-hard components with a proven flight history or take the calculated risk of flying terrestrial electronics that aren't rad-hard. In either case, add software bit-error correction to protect against the likelihood of single-event upset. If you choose to fly a device that isn't rad-hard, or a new device without flight history, you may plan the spacecraft's computer architecture to include some functional redundancy, with a less capable but proven processor as a backup. The University of Surrey has successfully used this approach.

The abundance of charged particles in the space environment also creates the potential for unevenly charging spacecraft surfaces. Exposed, electrically isolated structures, such as antennas, can pick up a differential charge relative to the rest of the spacecraft that can lead to arcing and associated damage of surfaces. Ground exposed structures lightly to the rest of the spacecraft structure to avoid this problem.

Finally, certain materials can be damaged by exposure to radiation, including most plastics and ordinary glass, which gradually clouds over. (Solar cell covers use specially treated glass.)

The Political Environment

In addition to the well-defined launch and space environment, there is an equally constraining but less well-defined political environment, which is just as hard to ignore. The space environment provides us with definite requirements for hardware

---

* krad is a measure of energy deposition in a material ("dose"), equal to 1,000 ergs/g. The material should always be in brackets, for example, (Si) = Silicon. The SI unit of dose is the gray (= 1 J kg^{-1}) and is equivalent to 100 rads. Although the gray is the official scientific unit, the rad is still predominant in engineering literature.

selection; yet, this political environment's influence can be both definite and subtle. Whatever cost-cutting techniques we apply to a space mission—whether it's using some piece of terrestrial hardware or eliminating a particular documentation require-ment—the overriding concern, at least for commercial endeavors, is "will your customer go for it?"

Mission politics strongly influence a basic principle for selecting components: *acceptable risk*. What is acceptable risk? This is perhaps the most important, but least politically correct, question we can answer for any mission. Risk (or its inverse—reliability) is something to be budgeted along with power, mass, and every other precious commodity on a spacecraft. In the traditional vicious circle of high-cost missions, we've had to minimize risk throughout the entire system. Yet, trying to operate inside a very tight risk margin is very expensive.

Obviously, if designers were willing to accept high risk, we could achieve cheaper missions. However, even a low-cost mission is only low-cost in relative terms. Although NASA may view a $1 million microsatellite mission as very cheap, it's very expensive in the eyes of a university budget. So, the trick in low-cost engineering is to effectively spend the risk budget in order to achieve the best return. One way to do this is to look for areas in which you're paying for the same risk reduction more than once. And to best control these areas you should center most risk on the prime con-tractor or rationally share it among subcontractors.

Unfortunately, the best engineering solution is often irrelevant in the face of political demands. We can discuss options for buying or producing low-cost system hardware only within this context. Although the purely technical environment deter-mines the **acceptable risk** of an engineering solution, the political dimension deter-mines **allowable risk**.

In some cases, the customers' requirements may be more subtle than simply specifying the exclusive use of MIL-STD components. Customers may dictate a level of documentation or analysis which, from previous mission experience, you know will only drive up the cost without enhancing the spacecraft's quality. When this happens, try to educate customers to understand they must accept **all** the managerial, procedural, and engineering practices which allow for "faster, better, cheaper" satel-lites. It is a "package deal" that aims to achieve the most cost-effective solution over-all with one string attached—higher (but, hopefully acceptable) risk. Because it is a package deal, you can't pick and choose which practices to follow or ignore and still expect to get shorter schedules and lower cost.

Regardless of how cleverly you can manage to cut corners on a particular engineering solution and how confident you are it will work, if you can't convince customers or the launch authority the risk is acceptable, it won't be allowed. Fortunately, for some small missions, especially those university researchers under-take, the customer and designer are one and the same. When you're your own cus-tomer, you are more likely to fully appreciate all the mission trade-offs and to raise the threshold of acceptable risk.

In addition to internal issues, external agencies also drive hardware costs. Parts bought abroad often require the supplier to get government export licenses, which can be very time-consuming, especially for new technologies. The cost of this adminis-tration will add to the price of the goods. Also, remember that getting a license may take many months or may not be granted at all. These delays and risks aren't accept-able to projects using short schedules to contain costs.

3.1.2 Mission Design

Once the mission is defined, mission design can begin in earnest from the top down. During this phase, many key designs and philosophies may drive up costs for spacecraft hardware. Among the most critical of these are fundamental performance requirements (with corresponding margins), required levels of quality for components, and the basic definition of the system architecture. During this phase, we have the greatest opportunity to design low cost (or high cost) into each component or system.

Much has been written on the science of cost engineering. Dean and Unal [1991] provide an excellent overview of this topic as applied to the aerospace industry. As they point out, too often satellite designers focus on reducing mass in order to reduce cost—a misdirected approach. Parametric analysis consistently reveals that we can best reduce costs by reducing complexity. Reducing mass alone can only drive up costs because of the associated increase in production complexity. In other words, the best advice is **keep it simple.** For mechanical components, this means reducing specification difficulty, parts count, and tolerance for assembly and finish.

Let's explore each of these areas to determine their effect on hardware costs.

Performance Requirements with Margins

Mission performance requirements strongly determine your choice of hardware for a mission. For example, mission ΔV requirements, coupled with mass constraints, can limit you to consider only those propulsion options having a high enough specific impulse. Unfortunately, requirements cost money. The entire design team must recognize that, for most systems, there is an exponential relationship between performance (whether measured in reliability, specific impulse, or bits/s) and cost. In some cases, you may be able to relax a single requirement by 10% and save many times that from the system budget.

Another point to consider is the performance margin. *Margin* is the engineering tolerance given to a specification. For example, the specification for pressure at an engine inlet may be 250 psi ±1 psi. The smaller the margin, the more it will cost to ensure compliance, so over-specifying the margin can needlessly drive up the cost. This simple fact has been verified both in practice and through parametric study [Dean, 1991]. Spacecraft engineering for low cost focuses on increasing such margins by controlling requirements, which can lead to a wider range of possible solutions as well as less testing and analysis. By changing the margin for inlet pressure in the above example to ±10 psi, little is lost in overall performance. Yet, it's much easier and cheaper to ensure compliance, and inlet pressure becomes less sensitive to small changes in configuration or performance of related support systems.

In other cases, such as the thermal tolerance on a particular sensor, a tighter margin (20 °C ±10° vs. ±30°) would be easier to achieve, possibly enabling you to use a less expensive, terrestrial sensor. Regardless of your perspective, when specifying the margin for a performance requirement, keep in mind two important points:

- How difficult is this margin to achieve?
- How will you verify the margin has been met?

Normally, we test a design before committing to manufacture the flight model. However, in systems that depend on precision items operating with very small margins, the breadboards and demonstrator models tend to be as costly and complex as flight units. For this reason, modeling and analysis tools simulate the precise components that would be too costly to make for a demonstrator model. Tools and models can also lead to proof of concept in less time than breadboards, which may involve complex machining or components needing long lead times.

Models also allow us to test the equipment under a wider range of environmental conditions than is possible with physical units—a common practice to observe failure mechanisms.

But, we often need as much effort to validate the model as we did to create it in the first place. This drives the models to be more complex, which in turn increases costs. Unless we understand the risks and limitations of analysis, it can easily become more expensive than building flight-model hardware. Of course, the lowest cost comes from a design that has sufficient margins to make modeling or demonstrators unnecessary.

Unfortunately, for small, fast missions, engineers can sometimes be uncertain of how wide their design margins are until a failure occurs. Rapid mission turnover, in some cases less than one year from blank paper to launch, can be a double-edged sword. On one hand, the short schedule dictates only the most crucial tests, which saves both time and money. On the other hand, this means engineers and operators can't "live" with their hardware long enough to fully explore all possible failure modes and know how the system will work under abnormal conditions. Again, there is a classic cost vs. risk trade-off. But in a world of low-cost, small satellites, failures are part of doing business. Only by accepting failures and being willing to learn from them for future missions can we succeed.

Quality Level

Space Qualified

Unfortunately, there is no universally accepted definition of what makes a particular component "space qualified." The most obvious definition would be a component that had operated in space for a considerable time under similar operating conditions without failure. In some cases, however, even this argument doesn't hold up. (See Surrey Satellite Technology Ltd. diode example below.)

As any engineer knows, no amount of testing or analysis can guarantee that a particular component will **never** fail. But trying to manage this uncertainty is what the whole "space-qualified" process is about.

A simple bolt is a classic example of how the most basic of components (but still a potential cause of failure) can attract expensive precautions. Probably, a bolt bought at the local hardware store would do fine, but one for a space mission would typically have to meet specifications requiring particular titanium alloys, traceability of raw materials, precision threads, and high-torque heads. Such a bolt may be hundreds of times more expensive than the hardware-store equivalent. Plus we must add the administrative cost of writing and maintaining the specifications, ensuring the high-quality parts remain traceable, and using specialized tools and suppliers to handle the high-torque heads. Even so, this effort and expense are often considered necessary to manage the uncertainty of mission success: one rogue bolt may make all the difference.

To understand how we manage the uncertainty of hardware items, let's look at a quality spectrum for components. *Quality spectrum* means the increasing confidence or reliability from commercial- and industrial-grade components at one end to MIL-SPEC, HighREL, and finally, Space Qualified at the top end. We've defined these grades below:

Commercial- and industrial-grade components: These components are basically the same as you would get from the hardware store or mail-order catalog. Such components come "as is" without any extra documentation or traceability. Quality checks, if done at all, would be by statistical process control (checks done on random samples, perhaps 1 per 10,000).

MIL-SPEC/MIL-STD: Components built to exacting military specifications. Military applications often require rugged components that can work reliably in much harsher environments and under more demanding conditions than those typically expected for industrial-grade items. As a result, military-procurement requirements, or MIL-SPECs, typically dictate such things as materials, manufacturing processes, packaging, transport, and possibly the number per batch to be sampled. Such specifications exist for virtually every item the military buys from screwdrivers to jet engines.

Integrated circuits (IC) illustrate the difference in these quality grades. Industrial-grade ICs may operate over more temperatures than commercial devices. Both would be injection molded to encapsulate the chip. In contrast, the MIL-SPEC devices would be mounted in a ceramic shell enabling inspection during assembly, imparting less stress during manufacture and giving greater protection against moisture.

It is important to note, however, that the U.S. Department of Defense is now phasing out most MIL-SPEC requirements. In the past, U.S. military organizations needed a waiver to procure any item, whether it was an aircraft carrier or a chocolate cake, without referencing the proper MIL-SPEC. In this age of dwindling military budgets, Pentagon leaders have recognized the senseless cost of requiring a MIL-SPEC for **everything** needed to feed, clothe, and arm a modern fighting force. Furthermore, with the current pace of change, especially in the area of electronics, keeping MIL-SPECs up-to-date has become impossible. Finally, officials have found that by placing the burden of design back on the contractor, rather than spelling out every detail, they can save vast amounts of money. They believe we must "focus on functional requirements—form, fit, and function—and leave the design solution to the contractors." [Kitfield, 94] A Pentagon Process Action Team found that a "tiering" effect had evolved over the years: on average, each MIL-SPEC cited in a requirement referred to another eight MIL-SPECs, and so on, exponentially. Thus, DoD has now turned around its old requirement. From now on, military organizations must get a waiver to **use** a MIL-SPEC. For all normal procurement they must use simple commercial standards, such as Federal Aviation Administration requirements for aircraft safety. Of course, they will still use MIL-SPECs for items not available commercially, such as nuclear-propulsion systems or stealth aircraft. But only those MIL-SPECs directly cited will apply to a contract. Second- and third-tier specs will be for guidance only.

HighREL: There is no standard definition of high-reliability components. In all cases, HighREL meets or exceeds MIL-SPEC requirements, usually adding individual testing of all components. HighREL may also include some or all of the criteria under SpaceQual below.

SpaceQual: No standard definition exists, but a typical "Space Qualified" or "S-Spec" component satisfies some or all of the following criteria:

- *Pedigree*: It has operated in space under comparable conditions in a similar application without failure for a significant amount of time—sometimes months or years. Space agencies and private companies sometimes document components and materials that have operated in space. NASA, for example, keeps a database of components and materials which designers can reference. [MIL-STD-975]

- *Environmental Testing:* The component or the materials used in the component have survived vacuum, temperature, radiation, and other unique aspects of the space environment with documented results. The most often quoted example of environmental testing for materials is outgassing in vacuum. Outgassing figures for a material are usually given as a percentage mass change over time at a given vacuum level as well as the amount of collected, condensable, volatile material and water-vapor content.

- *Traceability*: Certificates of traceability document the life of a component, and even of the material used to make the component, along with the processes used at every step of manufacture.

- *FMEA*: Failure Modes and Effects Analysis is a standard engineering tool for identifying and removing potential causes of failure or reducing the consequences. Simple components may have only one failure mode (e.g., a diode breaks) with a corresponding effect (e.g., current doesn't flow to circuit.) For more complex components, the FMEA could fill hundreds of pages with varying degrees of failure or degraded performance.

- *Reliability Analysis:* A formal engineering or statistical analysis attempting to quantify the likelihood of a component's failure for given operating conditions. The reliability of a component may be quoted as mean time between failure or the number of failures in a given period. A key to reliability analysis is **repeatability** of component performance. Extensive testing aims to determine the statistical mean and standard deviation of this performance. Designers then try to reduce the standard deviation so even a 3σ component will work well within the desired specifications.

- *Declared Materials List*: A detailed list of all materials used, including procurement specifications, outgassing properties, and procedures.

- *Declared Parts List:* Assemblies use parts lists that lend themselves to quality auditing. The list can include specifications, suppliers, and screening and burn-in requirements. This approach can lead to a multiplying effect: each component requires a parts list with traceability, FMEA, and reliability analysis for each subcomponent. And each subcomponent requires the same analysis and documentation for each sub-subcomponent.

- *Declared Processes List*: A specific process may apply for using each material and part. An example would be how to prepare surfaces for a particular adhesive. A formal procedure will control the process detailing step-by-step instructions, inspection criteria, skill level of operators, and tooling.

We have objective measures for "space qualified," such as formal certification through rigorous testing. At the same time, we have subjective measures: we've flown it in space before and it worked. Ironically, the political environment may demand more than subjective criteria. A customer or launch authority may require you to use a component objectively certified as "space quality," even though it hasn't flown in space, rather than allowing you to use one that is only subjectively qualified without all the necessary paperwork. For example, a commercial company required Surrey Satellite Technology, Ltd. to use all MIL-SPEC parts in building their satellite. In one example, Surrey substituted MIL-SPEC diodes for commercial ones in the power system at nearly 100 times the cost. Of these, one failed during vibration testing (something the commercial-grade diodes had never done) and one, when tested, turned out to be a transistor rather than a diode! From this example, we learn that MIL-SPEC parts have a higher **perceived** quality and reliability, but we can't take these qualities on faith. Despite their high up-front cost, you must still test them with typical rigor.

To summarize, components for space programs usually consist of parts and materials, both of which will affect quality. Strict procedures govern handling of parts and materials. Thus parts, materials, and good handling are necessary to build quality components, as illustrated in Fig. 3.1-2. The best parts can't make a good component under poor procedures. Similarly, the most strictly followed procedure can't make a good component from substandard parts.

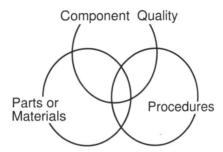

Fig. 3.1-2. The Quality of a Component is More Than the Sum of its Parts or Materials. It also depends on the quality of manufacturing procedures.

Formal Systems and Standards

Industry has various quality-assurance procedures which we can take "off-the-shelf" and apply to any company. Standards such as ISO 9000 or BS 5750 guide companies on how to track paperwork, handle invoices, and carry out other necessary procedures. One goal of these standards is to make business processes repeatable, but they're too generic to address the often unique requirements of the space industry, which has evolved its own formal systems and conventions.

Procedures for space qualification and assurance are present throughout a component's design, manufacture, and testing. During design, formal reviews and other mechanisms ensure quality is "designed in" from the start. Apparently minor changes during design can radically affect the system's eventual reliability. Procedures such as FMEA, for example, are most effective early—during design. Unfortunately, quantitative techniques often require a relatively complete design, which then makes anything other than trivial changes expensive.

Rigid procedures control manufacturing, equipment, ambient environment, and the assembly technicians' training and certification. Once the component is assembled, procedures for acceptance testing specify the type and number of tests, sampling rate, test plans, and test reports. Engineering and qualification models are universally used to qualify parts, materials, and processes by testing a flight-representative unit to qualification levels, which are more severe than flight-acceptance levels.

All these procedures and qualification testing come with a high price tag. The question then becomes, "are these procedures worth the cost?" For large satellites, the component cost may represent only a few percent of the program's overall budget, so the potential cost savings of eliminating a manufacturing procedure here, or a material requirement there, may appear to be small. However, for every formal requirement, someone must produce a part or material specification and a procedure. The organization must also provide the infrastructure to support this activity, which results in an enormous and complex (often bureaucratic) system of inspectors, technical writers, managers, and administrators. Hence, the purchase price of a material may be small compared to its assembled cost.

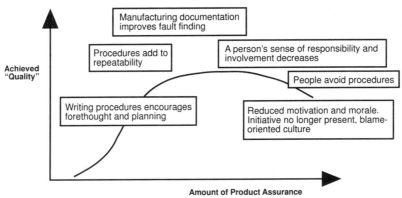

Fig. 3.1-3. How Many Space-Quality Procedures and How Much Documentation are Enough? Excessive product assurance can actually reduce achieved quality.

For smaller-scale missions, reducing these unnecessarily stringent requirements can save a lot of money. The challenge is to eliminate procedures that don't improve the component's quality or reliability. For example, consider the corporate culture created by large formal systems that require the next operator in the line to check the previous person's work at every stage. "Poor quality unless inspected otherwise" becomes the theme, and people feel more secure avoiding personal risks and initiatives rather than focusing on the most appropriate action for the task in hand. Without care, formal systems can become bureaucracies, allowing people to operate with a "warm fuzzy" feeling that each component will work or with ammunition to shift the blame in case of failure. Any serious effort to reduce cost must forego "warm fuzzies" as a needless luxury and focus more on fixing problems than fixing blame. Fig. 3.1-3 illustrates the trade-off between no rigorous quality assurance procedures and too many.

Obviously, some procedures make project planning more efficient and manufacturing more repeatable. But as these procedures and documentation requirements become too onerous, they can have a negative effect, reducing workers' motivation, fulfillment, and morale while eliminating personal initiative and responsibility. So how much documentation is enough? We can't say exactly, but we can give examples of successful spacecraft that flew with minimum documentation. Similarly, we can find examples of spacecraft that failed despite warehouses full of documentation. As a general rule, documentation is good if it's useful. You can start by asking whether you have enough documentation—for components and for the program—to:

- Completely reproduce the design and production for another unit

- Identify, track, and resolve performance anomalies or procedural problems detected during manufacture, assembly, or testing

- Quantify system performance to detect and resolve in-flight anomalies

System Architectures

System architecture is a broad term used to describe the selection, deployment, and interaction of components within a system. It is another cost driver established during the mission design phase. For example, the propulsion system architecture determines the "plumbing" layout of the tanks, valves, and engines, as well as their concept for operations. This is fundamentally a design exercise that ultimately determines the number and type of components. It identifies trade-offs between cost and performance that directly correspond to system cost.

Several important philosophies that influence the system architecture also directly drive cost. One of these is the approach to redundancy. For small, low-cost satellites, true 1:1 hardware redundancy is virtually impossible because of its prohibitive mass, power, volume, as well as the additional overhead needed to manage it. Yet, even on low-cost missions, we must deal with the inevitable single-point failures.

The philosophical approach to redundancy at SSTL—*degraded functional redundancy*—begins with circuit boards and reflects on the mission design. The aim is to reduce the effect of single-point failures by allowing continued, if somewhat degraded, performance of spacecraft systems and even the entire mission. This approach is described below.

While we understand that low-power circuits are inherently reliable, we still identify paths that allow software to reconfigure hardware and isolate potential failure modes. In many cases, "sacrificial circuits" can be turned off to protect other, more vital areas on the same board. To carry out this approach, board designs allow for a minimum-state control mode, whereby bypass channels can be configured when only rudimentary spacecraft commanding is available.

At higher levels, this degraded functional redundancy is part of selecting chips for RAM and microprocessors—choosing them from several different suppliers so a generic failure of one design leaves an acceptable amount of memory. New missions are usually built around the newest generation of microprocessor chips, with interfaces designed to use all of the latest capability. But backup microprocessors employing older, flight-proven hardware allow for degraded command and control of bus and payload functions. Also, using special-purpose microcontrollers to independently command and control the payload and other functions allows greater flexibility in case of failure. Normally, we send payload and other commands through the primary

CPU, but we can bypass it if it fails and send commands directly to subsystems or the payload. Thus, the ability to operationally reconfigure hardware and software gives the entire system more reliability without resorting to individual HighREL parts or redundant components.

Over more than 25 orbit-years of operational experience with UoSAT spacecraft, this approach to redundancy management has proven very effective. On several missions, this functionally redundant architecture has converted potentially catastrophic failures to partial or even full capability. For example, on the UoSAT-2 microsatellite, a failure took out the "data-valid" flag between a receiver and telecommand unit, causing all data to be ignored. Fortunately, designers had put in a matrix of backup data lines which allowed rerouting of the same data to the telecommand unit through the onboard computer. Thus, updating the onboard software allowed us to work around the problem.

3.1.3 Hardware Acquisition

During satellite design, engineers must identify suitable technology and then adapt it to fly in space. Although technology is everywhere, it's hard to advise designers how to find the best ideas for space. Looking for technology tends to fall into high-tech and low-tech approaches. The National Research Council [Technology for Small Spacecraft, 1994] outlines the high-tech approach, identifying new and emerging space technologies ready for use in small satellites—but in many cases at high cost.

TABLE 3.1-3. Names and Definitions of Hardware. Adapted from SMAD.

Term	Definition
Piece part or part	Individual part such as a resistor, integrated circuit, bearing, circuit board, or housing
Component	Complete functional unit, such as a control-electronics assembly, an antenna, a battery, or a power-control unit
Assembly	Functional group of parts, such as a hinge assembly, an antenna feed, or a deployment boom
Subsystem	All the components and assemblies that comprise a functionally distinct part of a spacecraft, such as the electrical-power subsystem or propulsion subsystem
Spacecraft	Complete vehicle

The other side of the coin is the low-tech approach exemplified by Takashi Iida, et al. [1993], who point out the possibility of using automotive electronics in low-cost satellites. The authors argue that, apart from the vacuum and thermal environments, modern automotive electronics are designed to operate at low-cost and with high reliability, making them prime candidates for spacecraft. Does this mean CD players or cigarette lighters in space? No, but perhaps one day soon it will be possible to take GPS receivers or control-area network chips from the family sedan and easily adapt them for a satellite. Unfortunately, as we'll see, much of the effort and expense involved with qualifying technology for space has to do with the vacuum and radiation environments—the two things that automotive technology, and terrestrial technology in general, aren't designed to deal with.

Before discussing specific methods for building or buying space hardware, let's apply a few definitions. We'll adopt Reeves' convention in Chap. 12 of SMAD to distinguish between parts, components, assemblies, subsystems, and spacecraft. (See Table 3.1-3.)

Hardware Source

You can choose different routes to getting spacecraft hardware, but the route you choose is another important cost driver. You can design, manufacture, and test your own hardware, or you can buy complete assemblies from subcontractors and other manufacturers. Each of these routes can lead to lower costs. But before discussing these two options, let's look briefly at a third—free hardware.

Free Hardware

Although there may be no such thing as a free launch, sometimes space hardware is free. By free hardware we mean hardware that already exists in production overruns, flight spares, cancelled programs, or even museums (as was the case with gravity-gradient booms used in early UoSAT missions). But you can take advantage of these one-time good deals only if you're aware of them and you can adapt your mission design to them. For example, it would be frustrating to find out you could have used a donated hydrazine/NTO engine after you've already committed the structural design to an MMH/NTO engine.

This idea of using free components has precedent. In the world of low-cost spacecraft engineering—at least for amateur and university researchers—scrounging for donated hardware is nearly an art. The OSCAR-series spacecraft are a good example. (See AMSAT, Sec. 13.3.) Many of their critical components were either donated or provided at a much reduced cost over the years.

Another option is to cooperate on research and development with a component supplier who wishes to flight test a new component. In this way, you benefit from a very low-cost or even free component, and the supplier gains potentially valuable flight experience. The FASAT-alfa spacecraft, built for the Chilean Air Force by SSTL, contains an advanced data recorder provided by Sanders, a Lockheed/Martin Company, as part of a cooperative R&D agreement. So before you build or buy, see if you can beg or borrow!

TABLE 3.1-4. Deciding to Build or Buy Hardware.

Before you decide to build or buy your hardware...
1. Review and update mission requirements.
2. Define performance specifications and margins.
3. Determine if required hardware is available for purchase (or begging or borrowing).
4. Determine if you have enough expertise and facilities to build it.
5. Weigh the advantages and disadvantages of building. (See Table 3.1-5.)

Building or Buying Hardware

To build or to buy, that is the question. We understand that we must buy some parts or components (applying the criteria used in the next section) and then build components, assemblies, or whole subsystems. So, no matter how much of the spacecraft you can build from scratch, you're going to have to buy some basic components. However, for certain hardware, such as a magnetorquer or flight batteries, the build vs. buy debate is quite important. Table 3.1-4 describes what to do when deciding to build or buy your own hardware. Also, as discussed below, you may be able to buy equivalent terrestrial or commercial-grade hardware and do your own qualification program.

Table 3.1-5 summarizes the advantages and disadvantages of building vs. buying your own hardware. As you can see, the primary advantages you gain are in increased span of control over a particular component, which creates the **potential** for cost savings but by no means guarantees it. The other main advantage stems from the opportunity to introduce new or untried hardware. Small satellites have been in the forefront of space technology, trying and perfecting new technologies or old technologies never used in space. The disadvantages flow from the hassle of having to do it all yourself. We're presenting building vs. buying as a black and white issue. But in many cases you're more likely to face a gray situation, in which you buy significant parts (or even components) and then must engineer them to match your needs. For example, you may use a terrestrial-grade component such as a GPS receiver to get low cost and high reliability and then significantly modify it for space.

TABLE 3.1-5. Advantages and Disadvantages of Building versus Buying Your Flight Hardware.

Option	Advantages	Disadvantages
Building your hardware	*YOU...* • Control the specifications and performance • Control the design and interfaces • Control the schedule • Control the cost • Can introduce new or untried technology • Spend all overhead in-house • Gain expertise that should make it even cheaper next time	*YOU...* • Carry all the risk • Need the in-house expertise to design the entire component and manufacture it • Need to acceptance test it • Need to space qualify it
Buying your hardware	*YOU...* • Share risk with supplier • Use tried and tested hardware • Have no development costs • Learn from subcontractors	*YOU...* • Have less control over specifications • Have less control over schedule and cost • Spend overhead outside your organization

Component Manufacturing

Once you've decided to build, the basic steps in manufacturing, integration, and testing differ very little from those all industries use. In Chap. 12 of SMAD, Reeves provides an excellent tutorial on spacecraft manufacturing and testing, so we'll adapt his approach to the following discussion. Table 3.1-6 summarizes the steps in this process.

TABLE 3.1-6. Steps for Manufacturing Components.

Step	Actions	Guidelines
Component design	Assemble multidisciplinary design team	Avoid unnecessary design iterations and "margin-on-top-of-margin"
	Budget design resources	Strictly limit time and resources to avoid over design and excessive iteration
	Prepare engineering data, including complete drawings, specifications, procedures, and all supporting information	Maximize margins Keep the design simple
Parts selection	Define quality level for piece parts.	"Space Qualified" or MIL-SPEC parts have higher perceived quality and reliability Commercial-grade parts have several inherent advantages
	Apply preferred parts list	Use parts you have experience with Keep list short to allow for highest quantity in each individual order
	Procure and qualify parts	Apply same basic process for assembly purchase outlined in Table 3.1-7 Space qualify parts as needed per Table 3.1-8
Material selection and processes	Apply preferred materials list	Use materials you have experience with Use inexpensive materials whenever possible, i.e., stainless steel vs. titanium Avoid "space-disqualified" materials Avoid toxic materials
	Define material-handling processes	Understand and limit yourself to in-house facilities and material-handling capabilities as much as possible Keep it easy (e.g., metal is easier to work with than composites) Limit the use of adhesives—bolts are easier and less permanent Make the most of margins—if you have mass margin, you can use stainless steel instead of titanium
Manufacture & test component	Plan and execute manufacturing process	Write detailed procedures specifying inspection criteria at each stage
	Apply procedures for acceptance testing	

Component design begins when you assemble a multidisciplinary design team. This team should contain experts from the functions responsible for the component: electrical, power, structures, and operations. By getting all these experts involved at the outset, you'll communicate much better. This can pay dividends later on because you can apply reasonable margins to the component design from the very beginning. Often, when this is not the case, each functional area tends to add margin—creating "margin-on-top-of-margin" which can overly constrain the entire design. Recall from the discussion above that margins = money. Maintaining margins that are easiest to achieve on all engineering specifications throughout the design process means less money for compliance and for verifying compliance. An integrated design team can better trade and budget margins on all critical interfaces. They can also streamline the design process by eliminating unnecessary iterations.

The design resources themselves—staffing, equipment, and time—also may drive costs up during this step. Therefore, you must carefully budget these precious design resources. Strictly monitoring and limiting the time and resources devoted to a particular design will help you avoid over design and excessive iteration. "Better is the enemy of good" should be your mantra for designing low-cost hardware.

The final activity during component design is preparing engineering data, including complete drawings, specifications, and procedures, plus all supporting information. Each project must establish its own standards for these critical documents. We can't overstate their importance. They are the primary means of communicating the design to the manufacturing team and documenting it. Attention to detail is crucial. Preparing drawings and supporting documentation requires a lot of labor, so it's expensive. Each time design changes or other program requirements require them to be redone, time and money add to the bottom line. Furthermore, each iteration, as with all human operations, risks mistakes. For these reasons, consider any changes to technical drawings or other data a critical engineering operation, not a simple documentation exercise.

The next step in manufacturing a component is selecting parts. First, define quality levels for all piece parts. Recall from the discussion above the differences between these levels of quality. "Space Qualified" or MIL-SPEC parts have the highest perceived quality and reliability. Ordinary commercial-grade parts come with varying, and often unknown, pedigrees—even within the same batch of delivered parts. Still you must weigh several advantages of commercial-grade parts:

- They are readily available, perhaps from multiple sources. Don't forget, it's just as easy to damage a highly expensive, long-lead component as it is a commercial part. In some cases, replaceability may be more important to the mission than having the highest spec.
- Development tools for commercial parts are readily available.
- Manufacturing processes may not be traceable but they are likely to be right due to very large production runs.
- Data sheets tend to be more complete.
- Commercial parts will likely go through many more design iterations to resolve anomalies that turn up during years of actual use under various conditions.
- A component built from SpaceQual parts is not necessarily space qualified, i.e., the whole can be less than the sum of its parts.

Furthermore, it's important to consider the hidden costs that go along with higher-grade, HighREL, or SpaceQual parts:

- You must write more detailed procurement specifications, costing you extra time and staffing.
- Because they are more expensive, you may not be able to scrap an individual part if you have doubts about it.
- Expensive parts imply more expensive inspection and handling procedures.

Of course, commercial parts carry several disadvantages:

- Sometimes it's difficult to buy small quantities because suppliers may be used to customers who order thousands of parts. If you want only a few, it may be difficult or impossible to get them to cooperate at a reasonable price.

- Different manufacturers may supply the same part but with subtle differences that may degrade the final assembly's performance.
- You are a slave to industry-standard requirements and trends.

Each project manager must weigh these advantages and disadvantages to determine what quality level piece parts must have for a given component. One way to avoid repeating this decision is to establish a list of preferred parts and apply it as much as possible to each new component. This list would include parts that you, or other space-component manufacturers, have used and have good experience with. Keep this list as short as possible and don't duplicate performance items to encourage low counts of individual parts for new components. See, for example, the Freja parts list in Sec. 11.2.

After selecting parts, you must carefully select materials for each component. Recall from the discussion above that many materials aren't compatible with the space environment. Here too, it's useful to establish and apply a list of preferred materials. To select for low-cost materials:

- Use materials you, or others, have had experience with
- Use inexpensive materials (stainless steel or aluminum instead of titanium or composites)
- Avoid "space-disqualified" materials (e.g., ordinary plastics)
- Avoid toxic materials (e.g., beryllium)
- Use adhesives that will cure at room temperature to reduce time and labor costs for baking. Most will accelerate in an oven, if necessary.

Along with each preferred material, establish a handling process. In general,

- Understand and limit yourself as much as possible to in-house facilities and material-handling capabilities
- Keep it easy (e.g., metal is easier to work with than composites)
- Make the most of system margins—if you have mass margin, you can use stainless steel instead of titanium
- Consider not using alochrom to prevent stress-corrosion cracking on aluminum from exposure to moisture, etc. Short design-to-launch times alone will reduce this effect. Dropping alochrom cuts the cost of all aluminum components, of which there are usually many.
- Remember that surface finishes, grades of material, and machining tolerances don't always need to be high (although they usually are!).
- Avoid bonded joints wherever possible; screws are easier for disassembly and rework. Bonded joints often require custom tooling to hold work while it's curing.

With the component designed and all parts and materials identified, you can finally build it. Here too, careful planning can avoid unnecessary costs. Engineering data should include assembly diagrams, but you should match these diagrams with procedures and inspection criteria at each assembly stage.

Finally, the fully assembled component must be tested. Volumes have been written on test procedures for new components. For our purposes, we need to separate *acceptance testing* from *space-qualification* testing. In both cases, the theory of type testing applies, as defined by Reeves in Chap. 12 of SMAD. Type-test theory assumes that if a "representative article (*type-test article*) passes a sequence of *qualification tests*, all other articles built to the same engineering data should also pass. In other words, the design is *qualified.*"

The only way to ensure each article is identical is by controlling the engineering data and manufacturing processes. Once the basic design is qualified, we can use less severe *acceptance tests* to certify proper workmanship for individual articles. Unfortunately, applying strict control to engineering data and manufacturing processes is one of the big reasons "space-qualified" parts and assemblies are so expensive!

One way around this dilemma for low-cost missions is to carefully choose mass-produced, commercial-grade parts, applying the practices for low-cost materials described above and, if possible, using parts already on an approved parts list. With this approach, the part itself is space qualified by deduction. That is, if you can't find any reason why it is space disqualified, such as radiation intolerance or out-gassing, you can assume it's qualified for space. Bypassing the formal qualification process for each part allows you to apply your resources for space-qualification testing only to assembled components or the entire spacecraft.

Procurement Process

Whether you are buying entire components or just individual parts, another important cost driver is the procurement process itself. Procurement offers several chances to affect the cost of identical quality items. Table 3.1-7 outlines the important steps for ensuring quality and low cost when buying components.

Cost drivers for hardware procurement begin when the decisions made during the mission-design phase get translated into procurement specifications that must be communicated to potential suppliers. Although the mission design should define flexible requirements, actual hardware options come to light during the hardware-acquisition phase. To take advantage of these hardware options, you must be ready to revisit earlier decisions and possibly change the system architecture or compromise on performance.

With the requirements specified, the next step is to find a supplier. Here too, decisions made earlier can drive costs. The mission's political environment (customer demands, for example) may dictate a specific supplier either by name or by default through carefully worded requirements. If this is the case, you may be stuck with the one supplier regardless of the potential cost. But if you're not so constrained, the list of potential suppliers depends on your hardware needs. ZARM [1995] has compiled a large database of hardware suppliers. The Aerospace Corporation is compiling a similar database. SMAD also lists some potential suppliers for certain subsystems. References such as these are a starting point for your search, but you shouldn't limit yourself to the ones listed. Technology and suppliers change rapidly, and no single database can hope to keep pace with these changes. Also, don't be misled by the costs quoted for a subsystem—magnetorquers, for example. The price quoted in a general reference will probably vary a lot (either up or down) from the

TABLE 3.1-7. Steps and Cost Drivers for Low-cost Procurement. This process applies to components or individual parts.

Step	Cost Driver	Guidelines
Define procurement specifications	Specified performance	Don't over-specify Maintain large margins
	Specified quality level	Recognize you can often get the "same" part with less "assured quality"
Select a supplier	Number of potential suppliers	Several data bases list suppliers of space hardware Final customer may specify your supplier
	Request for quote (RFQ)	Many commercial parts can be bought off-the-shelf, so you don't need an RFQ. Skip to next step. Don't over define your requirements and margins or tell them how to do their job. [General George S. Patton once said, "Never tell people how to do things. Tell them what to do and they'll surprise you with their ingenuity."] Specify requirements only; don't design the internals or tell them how to run their company
	Review of bids and selection of a supplier	Look for proven track record Look at previous projects to get a feel for the company's scope and approach to projects, i.e., don't ask a company to do more (or less) than they can
Establish a relationship with the supplier	Risk sharing between supplier and customer	Get involved with suppliers Penalty and performance clauses can sometimes cost more than they are worth and don't share the risk
	Acceptance test requirements	Work hands-on when possible Paperwork doesn't substitute for direct knowledge Avoid redundant tests—if the supplier tests it, you don't have to repeat it. If you're planning to do the test anyway, then they don't have to do it.
	Contract management	Define milestones and documentation requirements Limit formal meetings—use frequent informal meetings Define change procedures
Receive goods	In-house screening and burn-in procedures	Do these only as required

price you'll pay for real flight hardware once you've negotiated contract issues (such as torque, power requirements, mass, delivery, and testing requirements) with the supplier.

Once you've narrowed down the field to a short list of potential suppliers, you must select one. For commercial-grade parts, this may be a simple matter of going through some catalogs, picking out the necessary parts, and placing an order. This is probably the ideal, lowest-cost situation, with a minimum of procurement administration. However, for non-standard or more complex requirements, you may need to prepare a formal request for quote (RFQ) and send it out. If you are truly serious about taking all necessary steps to reduce the cost of your space hardware, the RFQ is your one chance to find a supplier who is willing to work with you toward this goal. The RFQ should set the tone for this cooperative venture. It should clearly state all technical requirements but also explain where you can accept new, lower-cost ideas. The RFQ shouldn't tell the supplier how to design the component, only how it will work. The University of Surrey used the following wording in a request for quotation for a low-cost propulsion system to get potential suppliers to work with us on lowering costs:

"Our experience in building other satellite systems has demonstrated that there are ways to lower component costs by relaxing performance requirements and judiciously modifying the procurement and parts production process. We know that a $50,000 component cannot simply be changed into a $20,000 component by changing the amount of paperwork that gets shuffled. We recognize that the less expensive component cannot have the same assured quality as its more expensive counterpart; however, in many cases it may be sufficient for our higher-risk, low-cost missions. We are asking **you** to tell **us** how best to reduce the cost of each component with understandable consequences."

The emphasis is on flexibility and cooperation. Recognize that your subsystem suppliers may have as much, if not more, to lose from a mission failure than you do—especially when you're trying to buy major spacecraft subsystems such as batteries, solar panels, or rocket engines. By tapping into that motivation for success, you can work together to lower cost while maintaining high quality. This means you must be prepared to view your supplier as a partner in the mission, as well as to consider their suggestions for alternate technology, hardware, or architecture that will lower your final cost. The RFQ should focus on tangible results, not philosophies on how suppliers should run their companies. An adversarial or dictatorial approach may get you good hardware in the end, but it may not be cheaper.

Once you've reviewed the bids responding to your request for quotation and selected a supplier, you'll fully define your supplier relationships. One of the most important aspects of this relationship is risk allocation. Experience at SSTL suggests that you'll get lower-cost hardware by keeping as much risk as possible with the prime contractor. SSTL has found that penalty and performance clauses can sometimes cost more than they are worth and don't share the risk. Experience at Freja (see Sec. 11.2, Freja case study) echoes this experience. Risk-taking by your supplier is costly, and this cost will pass on to you. Of course, as Grahn observes, to absorb technical risk you must have your own expertise.

By retaining risk, you can also better manage risk margins and their effects on testing requirements. Margins and testing cost money. Wider margins and fewer tests lower hardware costs. This leads to another important aspect of the supplier relationship, which involves requirements for acceptance tests. Recall the definition of acceptance tests given above. They are intended to verify that a given component is acceptable if it has already passed more rigorous qualification tests. Suppliers of space-qualified hardware who are used to dealing with traditional customers, such as NASA or DoD, will have well-established procedures for acceptance tests. Most likely, these will have evolved in response to rigidly defined performance criteria and as part of formal Quality Assurance for large programs. To reduce hardware cost, you must help suppliers re-examine these acceptance test procedures. While most of these tests may still be necessary, some may verify performance you don't require. Others, such as thermal and vacuum tests, you may be planning to repeat anyway at the subsystem or spacecraft level. By carefully working with the supplier, you may be able to eliminate some of these tests and take the savings in time and money, with an understandable increase in risk.

The final potential cost driver concerning suppliers involves managing the procurement contract itself. Milestones, documentation, formal meetings, and change procedures must all be spelled out carefully in the contract to eventually save time and money. Documentation requirements, for example, are expensive. Obviously, you must have basic engineering data, such as drawings and interface definitions, but you should keep them to a minimum. It's interesting to note that, even when you specify (and paid for) only minimum documentation, some suppliers will often provide full documentation anyway simply because their system is set up to do so! (See Sec. 11.2 Freja case study.) Formal meetings and communications can also be expensive luxuries. In this age of faxes and E-mail, you can even manage transatlantic contracts with little travel if you use proper discipline.

Even though the hardware has been bought and paid for, your own organization can still manage costs once you receive them. To do so, you'll need to control how you screen parts and burn-in the hardware.

In-house screening of low-cost hardware follows two schools of thought. The first, as articulated in the Freja study, stresses how a good screening program can detect substandard parts as they come in the door. Grahn maintains such a screening program fully justifies its small cost by identifying bad parts at the earliest point, thus preventing subsequent schedule delays. Also, in-house screening lessens the need to buy devices already screened by the manufacturer. This may reduce the need for suppliers to do acceptance testing, further lowering your costs.

SSTL follows the second philosophy. Instead of screening every part as it comes in the door, SSTL maintains an adequate inventory of flight spares for time-critical components. They only screen parts which are easy to screen, such as resistors and capacitors, as well as those which are especially prone to defects, such as crystals. SSTL tests many parts for the first time at the assembly level. If a part fails, they can rapidly replace it from flight spares. This obviates the need for a formal screening program and gains extra time in the schedule with an identifiable risk. In SSTL's experience, individual parts (especially electronic parts) rarely fail, so this philosophy has served them well.

Regardless of which philosophy you choose to adopt, it's important to choose one or the other. Because of lead times up to several months on some space hardware, especially high-reliability components, there is a tremendous risk to the program schedule if you don't screen in-coming parts and have insufficient flight spares in case of failure, however unlikely.

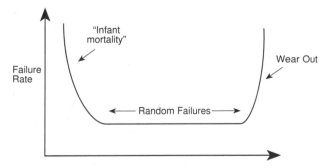

Fig. 3.1-4. "Bathtub" Curve for Reliability.

Recall from basic reliability theory that all mass-produced components, from satellites to light bulbs, conform to a "bathtub curve" of failure rate over time as shown in Fig. 3.1-4. In other words, you would expect some decreasingly high percentage to fail right out of the box, the "infant mortality phase," followed by a steady period of random failures and then dramatic increases as components finally wear out. The classic purpose of burn-in time is to weed out components that will fail initially, so later failures will be mainly random. You can do this burn-in on individual components and on the final, integrated spacecraft.

Unfortunately, we can't know exactly when components will fail initially. Furthermore, a very long burn-in time (on the order of months) is desirable to identify statistical failures but can be very expensive in terms of program staffing, equipment, and (perhaps more significant) time added to the program schedule. Thus, for low-cost hardware, complete burn-in at the component level is often not practical. Components are tested to make sure they work and to screen immediate failures, but you should reserve most of the burn-in time for the integrated spacecraft. Although this does add to program risk if a component fails later, it provides more opportunity to identify failures due to *concurrent system operations,* such as if the transmitter fails whenever the battery's charge regulator switches on. It also gives operators more time to "live" with their satellite and identify any nuances of its personality that could be valuable during on-orbit commissioning or troubleshooting.

In some cases, this burn-in process can be used as part of a program to substitute non-SpaceQual components for SpaceQual ones. However, you should recognize the opposite is **not** true—even SpaceQual parts need burn-in. Engineering organizations that work with low-cost spacecraft have found no correlation between a component's failure during burn-in and its quality level.

TABLE 3.1-8. Space-Qualification Testing for Components (or Spacecraft).

Step	Purpose	Equipment/Facilities Required	Process
Vibration and shock testing	Ensure component (or spacecraft) will survive launch Comply with launch authority's requirements	Vibration table and fixture enabling 3-axis testing Acoustic chamber	Do low-level vibration survey to set baseline Do high-level vibration test to qualification level Repeat low-level survey to look for changes
Thermal and vacuum testing	Induce and measure outgassing to ensure compliance with launch requirements Ensure component (or spacecraft) will work in a vacuum under extreme flight temperatures without convective heat dissipation Comply with launch authority's requirements	Thermal chamber (desirable) Thermal and vacuum chamber Equipment to measure outgassing: e.g., cold-finger or gas analyzer Effects on temperature extremes on electronic components can be tested less expensively in a dry-nitrogen thermal chamber before going to the more expensive thermal and vacuum chamber	Operate and characterize performance at room temperature and pressure Repeat test thermal chamber over hot-room temperature–cold-room temperature cycles Look for changes in results and compare results to models Repeat thermal test cycle in vacuum chamber Look for changes in results and compare results to models
Electromagnetic compatibility	Ensure individual components don't generate electromagnetic energy that interferes with other spacecraft components Comply with launch authority's requirements	Radiated tests: Sensitive receiver Anechoic chamber Antenna with known gain Conduction susceptibility matched "box"	Listen for emitted signals, especially at the harmonics of the clock frequencies Check for normal operation while injecting signals or power losses

Space Qualifying

With the component assembled, assuming all parts have been screened for space disqualification, you can begin space-qualification testing. Table 3.1-8 outlines the basic steps in space-qualification testing, assuming the component (or part) has already undergone acceptance testing to verify that it meets performance specifications. After each step, test the component to ensure it still works (that you didn't break it by testing it!).

The levels used in qualification testing described in Table 3.1-8 are much greater than those actually expected in flight. The higher the level of qualification tests, the more confidence in the design at lower levels. However, this artificial requirement is often a main design driver and hence profoundly affects the cost. To keep costs down, spacecraft engineers must carefully examine and set all qualification levels at an economic level.

If the technology passes the qualification tests, it's space qualified for these particular conditions providing you use exactly the same parts, materials, and processes. These limitations highlight the importance of repeatability between qualification and flight models.

You can apply these procedures for space-qualification testing to dedicated qualification hardware or to the first set of flight hardware. Or you may skip qualification testing altogether if you can prove to yourself, the customer, and the launch authority that the component is qualified by similarity.

TABLE 3.1-9. **Approaches to Space-Qualification Testing.** Adapted from Reeves, Chap. 12, of SMAD.

Approach	Actions
Use dedicated qualification hardware	Build and test a separate set of qualification components at qualification levels. Assemble this set of components or a second set of qualification components into the spacecraft and test them together at qualification levels.
Qualify the first set of flight hardware	Test the first set of flight components at qualification levels. Assemble these components into the spacecraft and test it at qualification levels. Then, launch the spacecraft. This is the *protoflight* approach.
Qualify by similarity	Demonstrate that the component and the environment are identical to previously qualified hardware.

Table 3.1-9 describes each of these options. Obviously, you would like to avoid this expensive process as much as possible. Therefore, for low-cost missions, you want to qualify as much hardware as possible by similarity. When that isn't possible (as is often the case), try the protoflight approach, which reserves full space-qualification testing for the assembled spacecraft. The protoflight approach can be much more cost-effective than qualifying each component and then separately qualifying the entire spacecraft. But it risks uncovering disqualified components only very late in the game—possibly too late to fix them before flight.

3.1.4 Examples from the University of Surrey

Since 1981, teams at the University of Surrey—and more recently its commercial arm, Surrey Satellite Technology, Ltd. (SSTL)—have built, launched, and operated

12 microsatellites. This section will highlight how they reduced component costs by focusing on the cost drivers for spacecraft hardware discussed above. Specific examples will focus on the use of adhesives, flight batteries, magnetic torquers, and ongoing research into propulsion systems.

Adhesives

Many low-cost space programs have found it best to avoid using adhesives altogether in favor of bolts and fasteners for load-bearing joints. Yet, for some applications, adhesives may be essential. Examples include conformal coating or adding into honeycomb panel tie-down points that don't bear loads.

Why Space Adhesives Are Expensive

Expensive working practices have evolved to manage the many reliability hazards that can affect the properties of a bonded joint. Table 3.1-10 steps through each stage in an adhesive's life, highlights the hazards it can encounter, and lists how to alleviate these hazards.

TABLE 3.1-10. Stages in Adhesives' Lives, the Hazards They Encounter, and Expensive Steps Needed to Alleviate These Hazards.

Stages in an Adhesive's Life Cycle	Hazards	Steps to Control Hazard
Space qualifying	Loss of mechanical properties in the space environment Reliable, repeatable, and sufficient for the task?	Test many samples of the adhesive
Manufacture	Batch-to-batch repeatability Contamination	Buyer raises procurement specifications stating the required properties of the material Buyer does batch testing on delivery to ensure compliance
Storage	Contamination due to particles or moisture Excessive storage time Shelf life depends on temperature	Improve storage specifications. Require quarantined storage in controlled environment. Staff must check temperature, shelf life.
Mixing and application to joint	Poor surface preparation (keying, contamination) Catalyst not entirely mixed with adhesive Adhesive begins to cure before being applied	Make sure procedures state how to prepare surfaces and mix adhesive. Prepare process-controlled samples and test them for compliance with specifications each time you use an adhesive.
Curing	Incomplete cure due to wrong temperature Poor mixing of catalyst. Movement of parts during cure Inadequate or excessive thickness of adhesive (the bondline)	Use calibrated ovens. Cure process-controlled samples. Use tools to hold items during cure. Inspect.

One of the biggest expenses for adhesives (and most spacecraft hardware) is the qualification program. A typical qualification program would involve preparing test pieces, applying the adhesive to them, and subjecting them to peel and shear forces at different temperatures. You would then inspect the results for sufficient strength, stiffness, and repeatability. The tests would also have to determine the sensitivity of properties to variations in the curing process and effects of bonding different materials with different surface finishes. Then, you'd test cured samples for outgassing properties and often examine microsections.

Following satisfactory testing, the test engineers write two reports. The first is the *Procurement Specification,* which covers batch testing of future consignments. The second is the *Manufacturing Procedure,* which describes how to use the adhesive, what properties to expect, how to prepare process-control samples, and criteria for test and inspection.

When materials are procured, the supplier will have to warrant that they meet the procurement specification from the qualification program. Most adhesive manufacturers will test for routine quality control. Even so, most organizations will make their own test pieces to confirm that the batch meets the specifications. Although this is expensive and can delay procurement of the material by weeks, high-profile projects may not be willing to accept the risk of using an adhesive they hadn't checked themselves.

Checking adhesives on delivery has highlighted poor batches due to some upset during production. The remedy is to reject the batch for another. Sometimes, production has drifted with time or suppliers have changed processes. In such circumstances, suppliers are unlikely to stop production, so you'll either have to accept the batch or use another supplier or material.

A common cause for adhesives failing to meet the procurement specification is slight contamination, weakness, or too much outgassing. Many designs can tolerate these problems, so engineers or project managers often accept such batches to stay on schedule (delivering and testing a replacement batch may take several weeks).

Reducing Adhesive Costs

The many hours of testing, plus specialized staff and equipment, mean that qualification is very expensive. But we can't reduce this cost much without undermining the validity of the results. For this reason, it is very unlikely that a low-cost project would try to use an unqualified material. Thus, when low-cost programs need adhesives, engineers seek out ones that have already passed a qualification program. A good place to look is on the list of accepted materials for an existing program.

If a material is partly qualified, a low-cost program can save money by extrapolating or showing its similarity to other adhesives. This requires very experienced materials people, but university teams normally have them.

We can save more money by focusing on requirements and margins. By limiting adhesives to noncritical applications and ensuring very forgiving margins, we can be more lenient with noncompliant materials. Taking this to the limit, it may be possible to avoid the qualification process altogether.

SSTL has evolved its designs to use only small quantities of adhesive, mainly in non-critical applications with considerable margins. The materials are all very readily available and mass-produced, which means little concern with availability and batch repeatability. SSTL doesn't test incoming batches. Because they use only a few different types of adhesive, it's easy to maintain enough stock and regularly replace it. This simplifies planning and scheduling.

Many adhesives need to be mixed with a catalyst just before use. This is a critical operation and is often restricted to qualified, trained staff. Some organizations mix large batches of the adhesive and put it into cold storage for later use in small portions. Before handling areas can receive the adhesive, they must be meticulously cleaned and lightly abraded (depending on the application and the materials). There are no shortcuts here for the low-cost program, although you can save a little time by allowing the experienced operator to determine whether the surface is ready rather than doing tests. (For example, the "waterbreak" test uses a drop of water to determine surface finish through capillary action overcoming surface tension. This, of course, requires a drying period.)

When the adhesive is applied to the bonding area, specialized tooling is normally used to hold the items in place. This tooling can be expensive and takes valuable time to design and manufacture—another reason why low-cost programs avoid adhesives in favor of fasteners.

Many adhesives, especially epoxies, provide a range of temperatures and durations for curing. Curing at room temperature is usually cheaper because you don't need expensive ovens and staff don't need to be in the vicinity.

Batteries

SSTL spacecraft typically operate in the relatively benign environment of low-Earth orbit, but batteries must still withstand harsh vacuum, multiple thermal cycles, and approximately 5,500 charge and discharge cycles per year at 15% to 20% depth of discharge. Table 3.1-11 compares typical commercial and space-qualified NiCd cells. As the table shows, there is nearly a 40:1 cost difference between the two types of cells. SSTL's engineers decided several years ago to buy commercial-grade cells and space-qualify them in-house.

TABLE 3.1-11. Comparison of Space-Qualified and Commercial Battery Cells.

	Space-Qualified Cell	Commercial Cell
Type	VOS7[SAFT]	GH6000F [GATES]
Capacity (rated)	7 A·hr	6 A·hr
Capacity (maximum)	9.9 A·hr	8.6 A·hr
Shape	Prismatic	Cylindrical
Mass	336 g	218 g
Energy density	36 W-hr/kg	48 W-hr/kg
Price	$4,000	$12

Table 3.1-12 summarizes SSTL's process for qualifying battery cells. By starting out with 5 to 10 times the required number of cells, the process gradually eliminates cells with defective structure or charge and discharge characteristics, until a set of flight-quality cells is left. As you can imagine, this consumes a lot of time and labor. Over the years, SSTL has developed automated test equipment to reduce these labor costs by enabling 24-hour, unattended testing and data logging. The obvious question is, "Is it worth it?" Cost analysis at SSTL indicates that the real cost for the first mission approaches $3,000 per cell (compared to $4,000 for purchased space-quality cells). However, once they've paid for the battery-selection infrastructure, the price drops to around $1,000 per cell—a 4:1 savings. How about the flight performance?

As of June 1994, SSTL's battery cells selected in this way have logged more than 2 million hours of on-orbit performance (some with more than 60,000 charge and discharge cycles) without a single failure.

TABLE 3.1-12. Steps in SSTL's Battery-Qualification Program.

Step	Procedures
Acquisition	• Purchase 5 to 10 times the required number of unsleeved cells • Spot-weld tags to positive and negative terminals • Do pre-dispatch screening

Inspection	• Look for:
	    Poor quality spot-welds
	    Damaged seal on positive facet
	    Evidence of cell-wall oxidation
	    Casing dents
	• Weigh cells; reject those outside of >0.1gm variation

Vibration testing	• Do radiography before and after vibration
	• Vibration range:
	    30–150 Hz (+6 dB/octave)
	    150–700 Hz (0.1dB g²/Hz)
	    700–2000 Hz (–3 dB/octave)

Environmental testing	• Do charge conditioning:
	    +0.3 A/40 hr
	    3.0 A/0.5 hr
	    +0.6 A/3.0 hr
	• Check overcharge: 240% OVC
	• Measure capacity: –0.6 A/1V per cell
	• Test weld: –30 A/15 s
	• Check charge retention: –0.6 A/1V per cell, 1 Ω 16 hr discharge, 24-hr open hold

Data analysis	• Reject environmental failures • Reject cells with charge anomalies • Analyze testing data to match cells with similar characteristics • Match overcharge voltages • Match capacities
Cell selection	• Assemble flight battery composed of qualified cells

Magnetic Torquers

Magnetic torquers are available commercially in various configurations at costs of up to $100,000 or more, depending on actual specifications. Looking for a way to reduce costs on this vital spacecraft component, SSTL engineers decided to build their own from scratch. To better fit it inside a microsatellite's limited volume, they decided to drop the metallic-rod core and use an "air" core.

Until recently, the magnetorquers on SSTL's microsatellites were designed directly into the spacecraft structure. SSTL's microsatellites use a basic box-like shape with solar panels on four sides. Wires were wrapped around the edges of the each of the six faces. The ±X and ±Y faces each had 200 turns of wire around an area of 0.15 m² which, at a nominal current of 0.28 amps, produces an 8.4 A-m² electrical field. The ±Z faces used the same length of wire, but being only 0.059 m² in area, they had 334 turns for an electric field of 5.5 A-m². Because the spacecraft orbit is polar, the Earth's magnetic field will vary between equatorial and polar crossings. Thus, for example, the range of available torque from the ±Z torquers is 1.188×10^{-4} to 2.376×10^{-4} Nm.

Although this design proved cheap and effective for over 10 missions, engineers found that the wire-winding operation tended to break wires. The latest generation of SSTL microsatellites employs a more robust approach, whereby the magnetorquer consists of wire windings in several layers on a printed circuit board layup. These provide effectively the same torque as the earlier model, but they're much easier to manufacture consistently. The flat, printed, circuit boards are then fixed directly to the back of each solar panel or other facet using a qualified adhesive. This eliminates the tedious and error-prone winding of 200 turns of wire and allows for more consistent performance at very low cost (hundreds of dollars each).

Propulsion Systems

The University of Surrey is researching low-cost options for propulsion systems on a 250 kg-class minisatellite. Researchers have tried to apply the cost-driver lessons discussed in this chapter at every phase of mission design. We're presenting the preliminary results as a case study in technology.

The industry's typical cost models for satellite propulsion systems, such as Smith & Horton [1984], or even modified cost models based on experience with small satellites, such as those developed by Burgess, Lao & Bearden [1995], would estimate the cost at over $1.2 million. (See also Sec. 8.2.) However, by combining system engineering and procurement practices for low-cost spacecraft with a willingness to accept higher risk, researchers have been able to significantly reduce the cost of a propulsion system that can give a minisatellite about 240 m/s ΔV. The total cost for all system hardware is projected to be ~$200,000, which is 15% of that predicted by typical cost models.

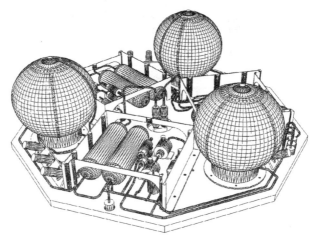

Fig. 3.1-5. Diagram of the UoSAT-12's Propulsion System Integrated onto the Attach Fitting Plate. Propellant tanks, ullage tanks, accumulators and valves are visible. The LEROS-20 engine is about the size of a wine glass and protrudes through the center of the plate.

Traditional large spacecraft use large bipropellant engines (400 + N) for major orbit corrections. These engines come with a comparably large price tag (>$200,000), as indicated above. Therefore, designers had to look for a less expensive option. They found one by adapting mission requirements to fit an engine rather than the traditional reverse approach. British Aerospace, Royal Ordnance Rocket Motors Division, has a long history of developing, testing, and manufacturing bipropellant engines. Royal Ordnance, along with several other manufacturers, make an industry-standard engine which provides 20-N of thrust while burning MMH/MON at a specific impulse of up to 305 s. Large satellites use these engines for attitude control, but using them on a small spacecraft for primary propulsion innovatively applies existing technology. It would offer a relatively inexpensive engine that the industry already supported.

Figure 3.1-5 diagrams the entire propulsion system designed around the LEROS-20 engine built by Royal Ordnance. The engine itself is about the size of a wine glass, with a total mass of 0.45 kg. Propellant mass flow rate is on the order of 7.5 gm/s. This 20-N thrust engine results from Royal Ordnance's years of research and development, which increased the engine's reliability and performance while slashing its production cost. By focusing on design and manufacturing techniques, Royal Ordnance's team has been able to develop a production engine at very low cost, as discussed in Wood [1990] and Gray [1990].

Why are these engines a low-cost option for small satellites? To begin with, these small engines are produced in much larger numbers than the 400-N version; thus, we're more likely to be able to take advantage of existing production runs. In addition, this is the same engine repeatedly tested for large, expensive programs. Thus, we have direct evidence of basic qualification without resorting to expensive, and in this case redundant, qualification testing. As a result, 20-N engines are usually 75% less expensive than their larger, 400-N thrust siblings.

On top of these basic savings, Royal Ordnance has been able to reduce the cost of these engines by nearly another 50% through careful attention to design and manufacturing issues. For any liquid engine, injection and cooling affect performance most. In the LEROS-20, these problems are handled together to achieve good, reliable performance at low cost. The engine is film and radiation cooled—a simple, cost-effective design. Furthermore, the film-cooling barrier allows us to get the best core combustion for maximum performance.

In addition, the LEROS-20 design incorporates several features which reduce manufacturing costs:

- **Low total parts count**—the entire thruster assembly has only five parts.

- **Few manufacturing operations**—the entire thruster assembly requires only three welds.

- **No special manufacturing processes requiring specialized tools to fixtures**—injector holes drilled with standard micro drills rather than electro-discharge.

- **Minimum number of hot fire tests**—extensive use of water flow to qualitatively and quantitatively evaluate injector performance.

Finally, by working with Royal Ordnance on procuring the engine, we've devised flexible requirements for acceptance testing which give us confidence in the engine's performance while reducing the final cost another 10%. Working together, we've decided to:

- **Take maximum advantage of qualification by similarity**—the LEROS-20 already has a strong development heritage, and the engine valves also have a proven flight heritage.

- **Reduce acceptance tests**—acceptance tests for thrusters and valves would include leakage and functional tests only. Use minimum hot-fire tests to verify performance over blow-down range.

- **Reduce performance requirements**—accept a nominal I_{sp} of 290 s vs. 293 s. Operate over the engine's established blow-down range of 20 to 10 bar inlet pressure (total performance loss ~7%).

Why use bipropellant rather than the more conventional monopropellant? The main reason was cooperative research between Surrey Satellite Technology, Ltd. and Royal Ordnance, but there were also compelling technical reasons. As their name implies, monopropellant rockets use a single propellant, hydrazine, which decomposes in a catalyst bed to deliver a specific impulse of around 230 s. Monopropellant systems are an industry standard and are the most commonly used in spacecraft today. Conventional wisdom within the industry assumes a monopropellant system is always simpler and cheaper than a bipropellant one. But our complete system trade between the monopropellant and bipropellant options discovered this is not necessarily the case. Although a monopropellant system is inherently somewhat simpler than the bipropellant option (only one propellant instead of two), at a systems level this advantage is not so profound. For an equivalent total impulse, the two systems are equally complex: the monopropellant system needs another tank due to its lower performance whereas the bipropellant system needs some additional valves. As a result, when you add up the price of all the components, the two systems cost roughly the same (even including the price of ground-support equipment for the additional propellant). But the bipropellant system saves more than 5 kg in total mass for a 250 kg spacecraft with 200 m/s ΔV required. Most important, it allows the system architecture to grow for missions with greater impulse requirements (such as Lunar missions), for which the bipropellant system's greater efficiency leads to an exponentially greater savings in mass and volume.

Best of all, this is a long-term solution which can be readily and economically adapted to a variety of future missions. With the LEROS-20 selected as the main engine, engineers turned their attention to reducing the cost of the rest of the system, again by focusing on the cost drivers identified earlier.

Mission Definition

During mission definition, we focused on understanding the "political" environment and how it would affect our mission design. Obviously, any technical solution would have to be acceptable to the launch authority. Because the system was intended to fly from various U.S. and foreign launchers, we didn't know all of the range-safety requirements. Therefore, we chose Vandenberg AFB's launch requirements as a baseline, assuming they would represent—and possibly exceed—those applied at other launch sites. Thus, our mission architecture had to meet two important requirements:

- At least three physical and electrical breaks between propellant tanks and the combustion chamber.

• At least two physical and electrical breaks between high-pressure gas and the outside.

Mission Design

Beginning mission design with system specifications and margins, we allowed ample mass margins for the propulsion system—based on lessons we had learned in designing low-cost spacecraft hardware. We targeted the satellite's total deployed mass to be < 250 kg with < 50 kg for the "wet" mass of the propulsion system. However, in every case, we traded mass in favor of cost. This mass flexibility is very important. Not having to count every gram had a significant knock-on effect, which ultimately lowered the system cost. Specifically, it allowed us to use:

• A "bang-bang" pressure-control scheme which uses high-pressure solenoid valves to control downstream pressure through computer-controlled feedback from a pressure transducer in place of a much more expensive regulator.

• Cold-gas thrusters and nitrogen-control valves that we could select based on availability, cost, and reliability rather than mass.

• Tanks that could be manufactured (using off-the-shelf material) to a higher-than-necessary operating pressure. This approach reduced cost, delivery time, and the additional worry of complying with tight margins for pressure.

• Standard, 37-deg, flared fittings for aerospace, instead of an all-welded construction. This also greatly simplifies system integration.

• Standard stainless-steel line (pressure rated to >5,000 psi) throughout the system instead of a lighter (but much more expensive) titanium line.

TABLE 3.1-13. Mass Breakdown for Proposed University of Surrey Propulsion System.

Component	Quantity	Unit Mass (kg)	Total Mass (kg)
Liquid engine	1	0.75	0.75
Propellant tank	2	1.00	2.00
Nitrogen tank	1	3.36	3.36
Accumulator & ullage tanks	6	0.5	3.0
"Bang-bang" valves	2	0.9	1.8
Pressure transducer	6	0.23	1.38
Flow restrictors	3	0.15	0.45
Relief valve	1	0.45	0.45
Filter	3	0.23	0.69
Fill/Drain valve	5	0.15	0.75
Cold-gas thruster	8	0.15	1.2
Ullage tank Isolation valve	2	0.9	1.8
TOTAL			**17.63**

Table 3.1-13 breaks down the system's mass. While the system engineering approach easily met the design goals, the overall system mass does exceed an estimate based on a more conventional approach by about 38%. However, this increase represents only about 2% of the deployed spacecraft's mass. As we'll see, this overall approach has saved enough money to prove its worth for our type of low-cost program.

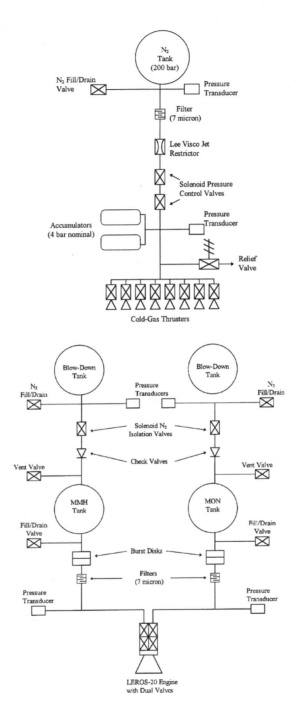

Fig. 3.1-6. Low-Cost Architecture for a Proposed University of Surrey Propulsion System.

Figure 3.1-6 shows the complete system architecture. The system includes several unique features that make it simple, safe, easy to operate, and low-cost. To begin with, the cold-gas system is separate from the liquid system (conventional systems would use a common reservoir of pressurant gas). Two "bang-bang" valves, controlled by feedback from a pressure transducer, regulate the pressure. This gives the added advantage of selectable pressure, which allows us to vary the cold-gas thruster's torque on demand.

On the liquid side, each propellant tank starts the mission at an initial blow-down pressure. The tanks need no further pressure control or operations. This architecture decouples the propellant tanks from the main N_2 supply and includes only enough nitrogen and ullage to initially pressurize the tanks to blow-down pressure. Thus, the liquid system doesn't need a pressure-regulation device.

The two ullage tanks, operating at 35 bar along with one liter of additional ullage in each tank, will give sufficient nitrogen to operate over the LEROS-20's blow-down range. Traditional designs would pressurize both tanks from a common source, requiring redundant check valves upstream. By decoupling the tanks upstream, we eliminate the possibility of propellant leaking past check valves and causing a catastrophic reaction. Simple, one-shot burst disks are used downstream of the tanks as an additional isolator before pressurizing the tanks, rather than more complex and expensive pyrotechnic valves, or even more expensive latching valves.

In addition, we designed the propulsion system keeping in mind this philosophy about redundancy:

- Include the minimum redundancy to comply with range-safety requirements.
- Keep it simple—the fewer things there are to go wrong, the less redundancy you need.
- Don't put redundant components upstream of single-point failures.
- Where possible, allow for *degraded functional redundancy,* whereby a single failure allows for continued, if somewhat degraded, mission performance.

This philosophical approach was applied throughout the propulsion system, from circuit boards to tanks. The overall mission design of the spacecraft itself also reflects this approach.

All our efforts at reducing cost in the design would be for naught if we couldn't buy the system components at a low cost. Fortunately, SSTL was able to establish a working relationship with Arde, Inc. of Norwood, NJ. Like SSTL, Arde was keenly interested in cutting through the normal red tape found in space missions and focusing on the essential elements to get a good product at low cost. Working with Arde, we've developed a procurement strategy that emphasizes

- Flexible system specifications
- Standard, proven designs for essential modules with simple interfaces
- Careful selection of components from large production runs (in some cases, these may already be in place to support larger, more formal programs)
- No formal configuration management
- No data items from suppliers (other than a certificate of compliance)
- Standard practices and procedures for suppliers

Together with Arde, SSTL has evaluated the system requirements and relaxed them whenever possible to arrive at a design solution that could be delivered on schedule. Low cost was the major driver in selecting the final approach and specific components. The system relies as much as possible on existing hardware from suppliers with a known record for good, quality products to reduce risk during ground testing and in flight.

We evaluated the systems requirements from the very beginning of the design phase so we could include key space-grade components. These components would require little or no analytical effort to determine their structural and thermal compatibility with the space environment. We used each component's history to preclude unnecessary testing and its associated cost. Experienced engineers used their judgment, based on traditional space programs, to select the components.

Simple and easily manufactured work packages characterize the design of the propulsion system. We've broken the system into subassembly kits containing all necessary tanks, valves, ullage and accumulator bottles, pressure transducers, filters, flow restrictors, burst disks and connectors. Each kit can be assembled and tested at the factory. Spacecraft integration will require adding lines and mechanical brackets to hold the components in place. This overall approach is designed to support rapid system assembly and testing, reducing cost as well as program risk.

Overall, our ongoing research indicates designers can reduce propulsion-system costs, like those of any subsystem, by focusing on the important cost drivers at every phase of the mission.

3.2 Software

James R. Wertz, *Microcosm, Inc.*

At one time space exploration was a major user and principal driver for the development of advanced computer systems and software technologies. Special-purpose processors and software were developed, maintained, and controlled, very much as S-level parts are maintained. But microprocessors and software are now part of virtually every aspect of business, science, manufacturing, and process control, so it no longer makes sense to work with special-purpose computers or software. Today, space programs should be using the same tools that business, science, and engineering use to get things done. Because of the radiation and vacuum environment, space computers will continue to be specially built; however, they will increasingly be rad-hard versions of commercial processors. Space software can and should use the low-cost, low-risk tools and processes available from business and aviation for ground and on-orbit processing.

Future spacecraft will become more software dominated than spacecraft have been in the past. This provides a new level of flexibility since onboard software can be upgraded in much the same fashion as ground-based software. In addition, autonomous spacecraft operations can substantially reduce the cost of ground operations by using *"autonomy in moderation,"*—automating repetitive tasks and leaving the higher-level logic to operators.

Because of the unique character of spacecraft and space missions, most onboard applications software is written specifically for space. In addition, although some commercial software can and should be used, much of the ground software will also need to be special-purpose or modified commercial software. In software development for space applications, the principal keys to reducing both cost and risk are:

- Provide large margins
- Use commercial tools
- Accommodate singularities and change
- Provide large margins
- Reuse existing software
- Use good software engineering practices
- Provide large margins

Each of the above principles and ways of using them are explained below.

3.2.1 Low-Cost Autonomy

In autonomous space operations, experience tends to contradict logic. The highest levels of autonomy occur in the cheapest and the most expensive spacecraft. Low-cost LightSats cannot afford expensive ground operations. For the lowest-cost systems, operations aren't staffed most of the time simply because there aren't people available to watch over them. At the other extreme, substantial autonomy exists in very large and expensive systems for technical reasons, such as the need for very rapid response in military systems or the long communication delays in interplanetary spacecraft.

The principal reasons for autonomous spacecraft are:

- Reduced operations cost
- Improved reliability
- Improved survivability
- Enable missions not otherwise feasible

Survivability may or may not be important for many low-cost missions. Missions which wouldn't work without autonomy include activities—such as direct down-link of results to end users, electric propulsion orbit transfer, or open-loop payload pointing—in which the spacecraft must make its own decisions about carrying out day-to-day activities. Typically, however, the most important reason for autonomy is to reduce both cost and risk. Fig. 3.2-1 shows how autonomous systems do this by being much less complex than ones which require extensive data collection, processing, and decision making on the ground. Simplicity and fewer actions by operators reduce cost and can decrease the requirements on the system. It is the elimination of personnel-intensive activities on the ground that is the primary driver in reducing cost.

As Table 3.2-1 shows, to get "autonomy in moderation," we want to automate elements which are repetitive or are involved in safing the spacecraft if a failure occurs. On the other hand, operators are better than software at long-term planning and solving problems.

Traditional Approach

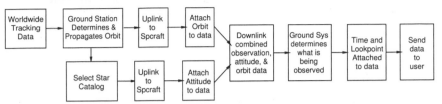

- **Operations Intensive**
- **Look Point determined after the fact**

- **Needs high accuracy to support long term orbit propagation**
- **Many opportunities for Communications or Operations Errors**

Autonomous Approach

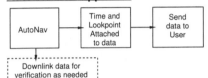

- **All intermediate steps occur in real time**
- **Accuracy requirement can be reduced to that needed for real time support**
- **"Direct to User" data flow is both feasible & economical**

Fig. 3.2-1. Comparison of Traditional and Autonomous Approaches to Satellite Navigation. Using autonomous systems can greatly reduce system complexity and, consequently, reduce both cost and risk. [from SMAD, Sec. 2.1.2]

TABLE 3.2-1. Development of Low-Cost, Autonomous Systems. The objective is to automate repetitive tasks and spacecraft safing while allowing people to identify and fix problems and create long-term plans.

Functions that should be automated	Functions that should not be automated
• Attitude determination and control • Orbit determination and control • Payload data processing • Repetitive housekeeping, e.g., battery charging, active thermal control • Anomaly recognition • Spacecraft safing	• Problem resolution • Identification and implementation of fixes • One-time activities (e.g., deployment and check-out) • Long term operations planning • Emergency handling beyond safing

A rule of thumb for autonomous operations is that automating 20% of the operations activity will handle roughly 80% of the situations that arise and that operators should handle the rest. We can further reduce cost by giving operators more real-time, visually oriented tools to provide better insight into system performance. (See Chap. 6.) Traditionally, spacecraft commanding hasn't been easy to do, so operators have erred inadvertently, partly because they couldn't recognize the consequences of a particular command sequence.

Spacecraft attitude determination and control are now autonomous on nearly all spacecraft. In addition, autonomous navigation is now available at low cost through GPS and the Microcosm Autonomous Navigation System, MANS. (See SMAD Sec. 11.7 for a discussion of autonomous navigation systems.) Work is also under way on autonomous navigation for interplanetary missions.

Orbit control should be autonomous, but so far, this has not been implemented in space. The lowest-cost spacecraft have no propulsion system and, therefore, no orbit control. This is acceptable for many small satellites and is certainly the least expensive approach. But we need orbit control when any of the following are required:

- Targeting, such as in interplanetary missions

- Overcoming secular perturbations, such as maintaining altitude against atmospheric drag

- Maintaining relative orientations, such as for constellation maintenance

SMAD Sec. 11.7 discusses the implementation of autonomous orbit control. Unquestionably the biggest problem is tradition—we haven't done it that way in the past, so we won't do it now. Still, in purely technical terms, autonomous orbit control, is far safer than autonomous attitude control, which nearly all spacecraft use. Any loss of attitude control, even for a few seconds, is a potential disaster. Pointing antennas away from the Earth may interrupt communication with the ground; pointing solar arrays away from the Sun will lose power; and pointing sensitive instruments at the Sun may destroy them. These dangers make continuous attitude control critical for most spacecraft. On the other hand, the loss of a low-thrust orbit-control system will typically not even be noticed for several orbits, unless the spacecraft itself notifies the ground. Orbital motion will continue, and the spacecraft will slowly drift relative to its assigned position. Ground personnel have ample time to react and correct problems, such as by activating back-up thrusters or providing revised control logic. In addition, the computational burden is essentially negligible for autonomous orbit control, relative to either attitude control or autonomous navigation. Although the computations are comparably complex, attitude control systems work at a frequency of 1 Hz to 10 Hz, whereas orbit control systems will work at frequencies of 10^{-4} Hz to 10^{-5} Hz. Consequently, the required throughput becomes extremely low.

Irrespective of the analytical ease of implementing autonomous orbit control, in order to be done it must be implemented in a low-cost, low-risk fashion. This is best done via what we call "*supervised autonomy*." This means putting the control logic onboard, putting the commands onboard to implement the autonomous activity, and transitioning authority to the on board system when operations personnel feel adequately comfortable. This stepped process provides a high level of comfort:

Stage 1—Require ground approval before execution

Stage 2—Allow ample time for ground override before the onboard system automatically carries out a command

Stage 3—Run autonomously, sending commands to the ground for occasional verification

Stage 4—Fully automate operations, with ground analysis only when a problem occurs

There is no technical reason for not having very low-cost, nearly fully autonomous satellites. Autonomy can reduce mission cost and risk and can be implemented in a low-cost, low-risk process. The major problem is to introduce a new way of thinking into the community.

3.2.2 Reducing Software Cost and Risk*

We can reduce software cost and risk. The key elements are to use existing software whenever possible, to employ good commercial tools and practices, and to make the onboard software reprogrammable from the ground with large margins in both memory and throughput. The lowest-cost, lowest-risk approach to software development is to maximize the use of available commercial systems and software. This implies using a commercial processor and developing most onboard software on office personal computers. As illustrated in the case studies, this approach has been very successfully adopted by the low-cost spacecraft community which uses software extensively in small, low-cost spacecraft.

Typically, 50% to 80% of the total life-cycle cost of software goes to maintenance and upgrades. The highest percentage applies to quick development efforts, which operate for a long time. The lowest percentage applies to large efforts used only once, such as an extensive software system for a one-of-a-kind spacecraft. Therefore, we should design space software for ground and onboard applications so we can maintain it. This in turn means using good software practices, modern tools and methods, and software development by individuals trained in software, rather than by engineers whose principal training is in subsystem analysis and design.

In this section we present an overview of the issues involved in reducing software development costs, and look more specifically at the implications for space software. In this context, "space software" refers mainly to spacecraft flight code. Many of the same points also apply to embedded ground segment code. The term in this context specifically excludes commercial mission analysis tools. The development process and the resulting space software products will be compared and contrasted to other production quality software. Remember, space software must meet highly demanding quality standards, so we should compare it only to other software which must attain similar levels of quality and exclude trivial, "throwaway" development efforts.

The major cost drivers for software development are:

- Complexity
- Change
- Resource constraints
- Productivity of personnel
- Project structure and management
- Development tools and techniques

Complexity

The single most significant driver of software-development cost is the system's complexity. Not surprisingly, more complex systems are more costly to develop than simpler ones. The complexity of a software system is largely driven by the functional requirements. In most cases the only option for reducing functional complexity is to trade between requirements and complexity. It is often the case that a small percentage of the functional requirements significantly increases the complexity of the system.

---

* Jeff Cloots of Microcosm, Inc. contributed much of this section.

The basic design approach and higher level system design also contribute to the complexity of the software. It is possible to design and develop different system implementations with different levels of complexity to address the same set of functions. It is therefore important that the system design not add too much to the inherent complexity of the system.

Industry studies have shown that the decisions made early in a project strongly affect the system's eventual cost. That's because early design decisions drive the cost and effort of detailed design, development, test, and support. Typical estimates are that projects commit as much as 90% to their total cost committed in the first 5% of their elapsed time.

Complexity is insidious. It feeds on itself making the system more and more complex. A complex system design leads to complex programming, which requires more testing, maintenance, and support. Also, more complex efforts usually involve a larger staff and therefore require more management. Thus, cost grows faster than complexity. A system with twice the complexity (by some arbitrary measure) of a less complex system will cost more than twice as much to develop and maintain. Therefore, reducing complexity greatly reduces cost.

Change

The second most significant cost driver is change. (On relatively small, simple projects, change can dominate more than complexity.) Some change is unavoidable when developing software, but if we can manage the number and timing of changes, we can control their effect on project cost.

For each phase of software development (requirements, top-level design, detailed design, development and test, and support), incorporating a change costs ten times as much. During requirements definition, the immediate cost of a change is the cost of including the new requirement in the documentation. The change will, of course, affect other development costs. If we need a functional change after requirements definition, costs rise tremendously. All preceding design and development work must be reviewed to determine if the change affects it. In essence, the change forces the development of at least parts of the system to back up to requirements definition and then revisit the subsequent design and development steps incorporating the change.

Although changes are inevitable, we can reduce their number and their effect on the system. The most important and productive technique is the requirement specification, which lays out the system requirements in black and white. It must be generated early in the project, before design and development efforts begin. The customer and the developers must review, understand, and approve it because it's the basic "contract" between them. Signing the requirements specification should mean they all understand and agree to what will be developed. Accepting a change should depend on its value in relation to its effect on development cost and schedule.

Good software design and programming practices can limit the effects of changes that inevitably occur throughout the project. Using good programming structure, modularity, and layering can greatly reduce the impact of changes by localizing their effects and therefore keeping rework to a minimum.

The last major element for reducing both cost and risk due to changes is to provide the tools and operational procedures to update software on orbit. On-orbit reprogrammability is one of the most important characteristics of space software. A principal reason for the extremely high cost of space missions is that we can't get at the equipment. We can't fix things that go wrong, unfurl antennas that are stuck, or change the hardware to meet evolving mission needs. We can, however, get at the software to modify, upgrade, or expand it to meet evolving requirements. Upgrading software isn't a symptom of sloppy work. Missions evolve with changing needs and new events, such as the unexpected collision of a comet with Jupiter. We can use one spacecraft for multiple missions, such as the ISEE-C spacecraft which was transformed late in its operational life into a mission to Comet Giacobinni-Zinner. All of this will be far more successful and economical if we have well established procedures available for upgrading the onboard software. Of course, we may choose not to make a particular upgrade. But to choose not to be **able** to make upgrades is simply poor mission design.

The easiest reprogramming is to adjust constants, gains, and software switches and this is commonly done. However, there is a strong operational tendency to not allow re-programming on orbit. It is comparable, in many respects, to the tendency to not fly new hardware or to not allow a change in hardware configuration once something works. Nonetheless, it is critical to have both the capability and the operational procedures in place and tested, so we can complete a needed upgrade with minimum cost, time, and risk. There are clearly operational problems with communications and verification that need to be worked out via operational procedures. This should **not** be left as a problem to work after launch.

A particular problem arises in updating active control software or other spacecraft critical elements. We cannot turn off the spacecraft control system in order to spend several orbits uploading a new one. The best solution is to design the system so we can upload revised software to the spacecraft's spare memory (remember the need for on-orbit margin) and then download and verify the new on-orbit code. Finally, we transition to the new control code, so we can return to the old code if the new software doesn't work right. This provides a robust approach to making on-orbit revisions, which would otherwise represent a high risk to the spacecraft itself.

Resource Constraints and On-Orbit Processing

Space software, especially for onboard systems, must run on low-power processors and extremely limited memory. As a result, developers have to squeeze every millisecond and byte out of their code. While this might lead to efficient code, it causes very inefficient development. Resource constraints have non-linear effects. Trying to squeeze code into half its normal space costs several times what the "normal" development would have cost. Also, each additional attempt to save resources is less productive than the last. Reducing requirements by 10% may be reasonably productive, but the next 10% will be harder and more costly.

As shown in Fig. 3.2-2, the margin at the time of System Requirements Review should be at least four times the anticipated need for throughput and memory. There are a number of reasons why this is needed. First, software requirements tend to evolve as the system design evolves. Thus we need to be able to accommodate changes as the design proceeds, and we don't know in advance the precise amount of memory or throughput we'll need. In space-based systems, exception handling is a key issue in creating robust software. However, exception handling requires a lot of available memory, even though the code may run only infrequently. In addition, the

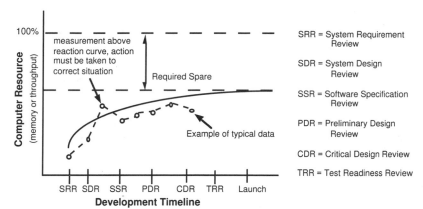

Fig. 3.2-2. Computer Resources Margin Needed to Reduce both Cost and Risk. The needed margin applies to both memory and throughput. The system must be launched with spare capacity on board to accommodate both on-orbit reprogramming and evolving needs while the system is on orbit. [from SMAD, Sec. 16.2.2]

system should launch with at least 100% margin in both throughput and memory to allow later updates and reprogramming that doesn't interfere with ongoing operations.

The need for margin is not to accommodate poor software design or coding, but comes about because of the natural evolution of the mission itself and overcoming problems that arise. I know of no group that is more careful, or does a better or more comprehensive job of onboard programming, than JPL. Still, virtually all interplanetary missions have required updates or revisions to the onboard software. This can result from hardware anomalies, changes in mission conditions, or simply the evolution of the program toward more, newer, or different tasks.

I believe most managers and software engineers are aware of the exceptionally high cost associated with putting too much software in too small a computer. Nonetheless, the need for large margins is perhaps the single most violated, well-known rule in space software. This comes about partly because spacecraft are hardware and weight driven and tend to be optimized from the very beginning. Most of us recognize that the right way to buy a computer is to decide first what the computer must do, then purchase or design the software, and only then buy the appropriate hardware to run the application. But defining software first is almost unworkable in a traditional spacecraft program. It's simply unacceptable to tell the program manager you're still sizing the software and you will determine the computer's weight and power sometime during the next mission phase. It's also very difficult—in real programs with tight budgets—to defend large margins when every other subsystem is drilling holes in equipment cases and throwing out individual components to minimize weight and power. Yet, clearly, defining the hardware first dramatically increases risk and spends hundreds of thousands of dollars in added software cost to save thousands of dollars in computer and power costs.

Software developed to reduce on-orbit data must accommodate mathematical singularities. Because of measurement noise, singular conditions will almost certainly arise. For example, locating an object relative to the background stars often involves generating the intersection points of two cones centered on reference stars. Whenever the two cones are nearly tangent, measurement noise will sometimes make the measured cones not intersect. This will result in trying to take undefined inverse trig functions, such as the arcsin of 1.0001. We can handle such singularities in two ways. A "no result" flag can be returned. Or a "best answer" can be returned if the offending operation is very close to an acceptable numerical solution, such as rounding 1.0001 down to 1 in inverse trig functions. While taking these issues into account slightly increases cost during development, it significantly reduces the cost of operations and debugging when singularities are encountered on orbit—and they will be.

Another consideration is to do the processing as near the source of the data as possible. This minimizes the cost of moving data and also reduces the potential for error. For example, an observation system to detect forest fires could well have an extremely large data stream being generated by an onboard payload sensor. If all of the data were brought to the ground, then an enormous amount of data from over the world's oceans and poles would be put through the spacecraft communications system only to be deleted in preliminary processing. Clearly, it would be much better to use a low level of pre-processing and discard useless data immediately. Similarly, the orbit and attitude information should be attached to the payload data as early in the processing cycle as possible in order to avoid errors in correlating them. Onboard systems have an advantage because they can attach the information much earlier in the data stream. It's difficult, time consuming, and extremely error prone to associate data collected from two very different sources and brought together well after the fact. Although it's important to process the data as early as possible, we must also be able to get at or reconstruct the raw data to analyze what is happening onboard the spacecraft. Thus, we need onboard processes and operational procedures to obtain the raw data for ground analysis as needed.

Productivity of Personnel

Studies have shown that there can be a 10-to-1 ratio in the ability of programmers measured by productivity, code size, execution time, or other quantitative measures of programming skills. Obviously then, the skill levels of the development staff can strongly affect a project's schedule and budget. The best source of trained developers is previous, similar projects. Further, it's important to keep the core developers together throughout the project to reduce the need to bring new people up to speed.

Project Structure and Management

A good organizational structure doesn't necessarily guarantee a good product, but it usually helps. Conversely, a poor project organization doesn't necessarily doom a project, but it is a good bet. Many project structures and management approaches have been used—successfully on some projects and unsuccessfully on others. Project structures must be compatible with the intended product, the corporate culture, and the individual skills and personalities of the staff.

Sometimes the architecture of a system reflects the project's organizational structure. This is often not a good idea because it usually won't result in the best system architecture. In other cases the parallelism is the result of organizing the project to follow the high level system design. To best accomplish this, it is first necessary for an individual (or small group) to create the top level system design. Project personnel, either individually or in teams, can then be assigned to various system components. This is a good approach to organizing a project to effectively implement a system.

Development Tools and Techniques

Using the latest and greatest development tools and techniques often appears to be the magic answer for improved productivity, product quality, and successful projects. This is overly optimistic. Although development tools may help productivity on certain activities, they won't turn average developers into superstars or overcome a poorly thought out system design. Also, many development tools are tailored to a product or development technique, or they force a certain way of thinking. If these are inappropriate for a particular project, using the tool can hinder rather than help it.

Another tempting trap is to count on using structured design, object-oriented design, rapid prototyping, or another "technique-du-jour" to ensure an efficient and effective development process and a quality product. Product quality (and project productivity) comes from the people involved much more than from the tools and techniques.

Much of the older space software was developed using unique, specially-developed tools and development environments. Dramatic cost reduction can be obtained by using commercial tools and commercially available development environments. Because of the widespread use of software, commercial tools will usually contain fewer errors and will be maintained and updated. This typically includes upgrades to support new hardware. Using special-purpose tools and compilers means the application developer may spend a lot of his resources debugging the tools and compilers, an expensive approach to applications development.

Another issue which arises frequently is the choice of development language for space applications. The government or system purchaser may mandate this choice. DoD will typically require Ada. If some development choice is available, then the first rule would be to avoid using assembly language, unless absolutely forced by the nature of the application. Assembly language applications tend to be more costly to develop and dramatically more difficult to debug and maintain. Among the higher level languages, Ada and C are the most common for space applications. Both have advantages and disadvantages. The best choice is to select the language that is already known and used by the most people who will program, monitor, and manage the development. Familiarity with the programming environment is much more important than the choice of language, so long as you use a reasonably robust language. (For further discussion, see SMAD Sec. 16.1.)

Conclusions

In many respects, the development of space software is similar to the development of any other production quality software product. This implies that many of the same techniques for improving the process for developing other types of production software can also be applied to space software. But space software has some unique features that require different approaches to controlling development cost.

Production software and space software are both usually developed through large team efforts. Thus, many management techniques apply to both, but unfortunately, they share many pitfalls, as well.

Also similar is the inherent complexity of the systems. Space software is by its nature complex because it is handles many complex functions. Therefore, we must manage complexity by wisely choosing what the software will do and what to leave to the operations staff, designing the system as simply as possible, and thoroughly documenting to help developers and others understand it, especially for maintenance.

Finally, space software and production software are similar in their need to be robust. A commercial software package must be robust because the market demands quality. A product that doesn't address the customers' needs, or that occasionally aborts, won't succeed in the marketplace. (In the case of software developed by a company for its own use, non-robust software may not fail in a marketing sense, but it will fail the company by continually requiring resources for maintenance and extensions.) Robustness in space software is also critical. The software's inability to deal with circumstances that arise at best increases the mission cost by requiring more operations support or changes to the software. In an extreme case, such a failure can end (and has ended) a mission.

The single largest difference between typical space software and other production software is their perceived lifetime. Successful production software usually results from taking a long-term view. The software is designed and built to operate for many years. Such a view justifies the "extra" expense of developing a quality product. In addition, the software will often be the baseline for future development and, potentially, for spin-off products. This mode of thinking influences developers to consider ease of maintenance, flexibility, and reusability. Typical space software, on the other hand, is for a single use. Thus, its developers often don't find these issues important, which is a major reason space software costs so much. Developers see it as unique, so they don't try to make the software reusable. As a result, when the next similar development effort comes along, no suitable software exists to serve as the baseline for the new system. Therefore, the costly and time-consuming process of custom development repeats itself.

The second major difference is the feasibility and process of software maintenance. Properly developed production-quality software is designed for ongoing maintenance. In most cases, maintenance and upgrades occur periodically throughout the life of the software. In contrast, changes to space software after initial "release" usually result from unplanned events. The need to update the software is often immediate and mission critical. However, the mechanisms to perform these updates are usually difficult and time consuming. Updates must be developed and checked out on the ground and then uploaded to the spacecraft and verified.

Finally, space software is typically more resource limited than other production software, without a means to upgrade the resources available. For a typical mission, we select hardware very early. Years may then elapse before the mission actually flies. In the meantime, requirements growth and "creeping featurism" drive up the software's resource requirements. The developers are forced to try to cram the software into the available resource. Once in operation, it's impossible (except in a few rare instances of using the Shuttle to retrieve or upgrade a spacecraft) to add more resources. Therefore, we must force on-orbit software fixes or enhancements into the remaining resources or replace other functionality. Either solution is expensive and time-consuming.

In conclusion, developing good, low-cost software consists of the following steps:

1. *Define the requirements as completely as possible*—This will reduce the number of expensive, time-consuming changes required throughout the system's life. It will also largely determine the system's complexity. Carefully determine what software should do and what people should do. Trying to do too much in software, especially handling rare occurrences, will be expensive and will make it more complex. A rule of thumb in software development is that 20% of the code handles 80% of the work, and the other 80% of the code covers 20% of the work. Based on this rule, adding "just a little more functionality" can be very expensive.

2. *Pull together a good development staff and provide them with adequate tools and a reasonable management structure*—The productivity edge of better performers, in the long run, is clearly worth any extra salary or benefits. Once you assemble a good team, make sure your organizational structure and management approach doesn't interfere with their productivity.

3. *Pay particular attention to system design at the top level*—This phase will strongly affect the system's cost, schedule, and capabilities. Decisions during design will affect the system throughout its life. A useful design technique is to identify the system components, partition functions among the components, and then define the interfaces between the components before doing the internal component design.

4. *Test, test, and test some more*—Testing should begin as soon as individual software components (modules) are available and should continue until on-orbit operations. (In reality, a lot of testing occurs after the system should have started operations. Any resulting changes at this point are orders of magnitude more expensive than testing earlier in the product cycle.) As the system development progresses from individual modules to subsystems to the final integrated system, testing should progress from unit test, to subsystem test, and finally, to integrated system test. Unfortunately, testing is often overlooked in project scheduling. The result is rushing to test and, usually, flying an incompletely tested system. The bottom line is that testing will occur somewhere along the line. You can either plan for it and do it when appropriate, or you can do it after the system is supposed to be operational. It's much easier and more cost effective to test in the lab than in the field, especially when the "field" is in space.

5. *Throughout the program, manage change*—Changes will occur. It just isn't possible to fully specify, design, and develop a system that will work completely and perfectly when flown. You must identify and incorporate changes as early as possible, design a system that is receptive to anticipated changes, thoroughly consider whether proposed changes (especially late in the project) are necessary, and provide enough computer resources to change (and, likely, expand) onboard software.

3.2.3 Commercial Software for Ground and On-Orbit Applications

The single largest step in reducing software cost and risk is the use of commercial, off-the-shelf (COTS) software. The basic role of COTS software in space is to spread the development cost over multiple programs and reduce the risk by using software that has been tested and used many times before. Because the number of purchasers of space software is extremely small, the savings will be nowhere near as large as for commercial word processors. Nonetheless, reductions in cost, schedule, and risk can be substantial. Most COTS software should be at least five times cheaper than program-unique software and is typically ten or more times less expensive. In addition, COTS software will ordinarily have much better documentation and user interfaces and will be more flexible and robust, able to support various missions and circumstances.

The use of COTS software is growing, but most large companies and government agencies still develop their own space software for several reasons. One of the best ways to develop and maintain expertise is to create your own systems and models. Thus, organizations may want to support their own software group, particularly when money is tight. Also, it's hard to overcome the perception that it costs less to incrementally upgrade one's own system than to bear the cost and uncertainty of new COTS tools. In this trade, the custom systems often don't include maintenance costs. Finally, customers often don't know what COTS tools are available. Professional aerospace software doesn't appear in normal software outlets, advertising budgets are small, and most information is word-of-mouth through people already in the community. Despite these substantial obstacles, more companies are using COTS software in response to the strong demand to reduce cost.

The government can either support or undermine the community's use of COTS software. They support its use by helping its developer and by buying and using it on government programs. At the same time, the government will frequently undermine this approach by providing free, government-developed software or by funding government agencies or Federally-Funded Research Centers to distribute software which competes directly with COTS products. With such a small market, releasing free products may make it nearly impossible for the commercial software vendor to compete. To date, the government has done some of both. It has no unified position on whether it will support or compete with commercial developers of space software.

In order to use COTS software to reduce space-system cost, we need to change the way we do business in aerospace software. We need to adapt to software being not exactly what we want, look for ways to make existing software satisfy the need, or modify COTS software to more closely match requirements. This is a normal part of doing business in other fields. Very few firms choose to write their own word processor, even though no single word processor precisely meets all the needs. Instead, they choose one that most closely matches what they want in terms of functions, support, and ease of use. We should use the same criteria for COTS space software. In addition, we need to set realistic expectations concerning what COTS software can do. Clearly, we can't expect the low prices and extensive support that buyers of globally marketed commercial software enjoy. We have to adjust our expectations to the smaller market for space-related software, which means costs will be much higher than for normal commercial products. Maintenance and upgrades will ordinarily require an ongoing maintenance contract. Within the aerospace community, a standard arrangement is for a maintenance and upgrade contract to cost 15% of the purchase price per year.

Using COTS software and reusing existing non-commercial software requires a different mindset than continuously redeveloping software. We need to understand both the strengths and weaknesses of the relatively small space commercial software industry. Because the number of copies sold is extremely small, most space software companies are cottage industries with an extremely small staff and limited resources. We shouldn't expect space-software developers to change their products at no cost to meet the needs of a unique program or organization. For example, it would be unrealistic to expect a vendor of commercial software for low-Earth orbit spacecraft to modify the software for interplanetary missions at no cost, because few groups will buy interplanetary software. On the other hand, the small size of the industry means developers are eager to satisfy the customers' needs, so most are willing to work with their customers and to accept contracts to modify their products for specific applications. This can still be far less expensive than developing software completely from scratch.

There is a hierarchy of software cost, going from using COTS software as is, to developing a new system entirely from scratch. In order of increasing cost, the main options are to

1. Use COTS software as sold

2. Use COTS software libraries

3. Modify COTS software to meet specific program needs (Modifications may be done by the mission developer, by the prime contractor, or by the software developer)

4. Reuse existing flight or ground software systems or modules

5. Develop new systems based largely on existing software components

6. Develop new systems from scratch using formal requirements and development processes

This hierarchy contains several potential traps. It may often seem that the most economical approach would be for the prime contractor or end-user to modify COTS software to meet their needs. However, it is likely that the COTS software developer is in a better position to make modifications economically and quickly. Although the end-user is more familiar with the objectives and the mission, the software developer is more familiar with the organization, structure, and code for the existing product.

Secondly, there is frequently a strong desire to reuse existing code. This will almost certainly be cheaper if the code was developed to be maintainable and those who developed it are still available. On the other hand, for program-unique code developed on an individual project, with formal specifications and no requirement for maintainability, it may be cheaper, more efficient, and less risky simply to discard the old software and begin again.

At the end of this section is a list of principal COTS software products and sources. Remember, however, that the availability of products changes rapidly, and remaining aware of the market is difficult. New programs appear frequently, and many programs either leave the market or are no longer supported.

COTS space software is typically purchased directly from the software publisher. Virtually none of the professional space products is available through normal commercial outlets. However, some of the smaller software packages are either packaged with books or sold by book publishers. These can be obtained through normal book channels.

Several of the programs listed below come from other fields, such as astronomy or image processing. These tend to be for purposes other than space mission analysis and design. Nonetheless, they are dramatically less expensive because of a much larger market and, therefore, can be useful if they can fill at least some of your program's needs. We haven't included prices in the list, because these become out-of-date too quickly. In addition, we haven't included general software applicable to space, such as compilers, spreadsheets, CAD programs, or structural analysis programs.

3.2.4 Annotated Bibliography of Commercial Space Software

The availability of commercial space software changes rapidly. Professional software is typically bought from the publisher. Many of the general astronomy products, including star catalogs, are available from Sky Publishing Corp., Cambridge, MA.

AttSim (Microcosm, Inc., El Segundo, CA, http://www.smad.com). Spacecraft ACS simulator and flight software development platform. Simulates spacecraft position, attitude, and environmental forces. Supports all phases of ACS development process and includes 5 pre-designed controllers: zero momentum (reaction wheels and torquer), zero momentum with thruster, momentum-biased (with Earth, Sun, or star sensor and gyrocompass), inertial scanner (momentum-biased), gravity-gradient torquer. Serves as hardware-in-the-loop simulator and test platform. Can analyze control system performance and related parameters, allowing selection of appropriate sensors and actuators. (PC)

AutoCon (AI Solutions, Lanham, MD, http://www.ai-solutions.com). Spacecraft mission analysis and operations support tool designed to automate routine operations for spacecraft from low-Earth to geostationary orbits. Provides capabilities for rapid orbit propagation, ephemeris generation, event calculation, maneuver planning, maneuver automation, and generation of reports and plots. Flight version of AutoCon (AutoCon-F) is also available to automate maneuver planning onboard operational spacecraft. In use on NASA EO-1 mission. (PC)

Dance of the Planets, QED edition (ARC Science Simulations, Loveland, CO, http://www.arcscience.com). Amateur astronomy program based on a gravitational model of the solar system. Shows dynamical motion of the planets, satellites, 5,600 asteroids, 1,400 comets, or a variety of spacecraft from any location in space, including onboard a spacecraft or celestial body. Advanced version has star field to 10th magnitude with 380,000 stars. (PC)

HPOP, High Precision Orbit Propagator (Microcosm, El Segundo, CA, http://www.smad.com). Advanced HPOP. High accuracy orbit propagator for Earth and interplanetary orbits. Includes perturbations up to 360 x 360 Earth geopotential model, solar radiation pressure, atmospheric drag, and solar, lunar, and planetary perturbations. (PC, Workstations)

Hubble Space Telescope Guide Star Catalog (Astronomical Society of the Pacific, San Francisco, CA, http://www.astrosociety.org). Catalog of location, brightness, and classification of 19 million objects in the Hubble catalog with access software. Read and display graphically on MAC or as text only on PC.

MicroGLOBE, Space Mission Geometry System (Microcosm, El Segundo, CA, http://www.smad.com). Creates static or dynamic plots of the entire sky as seen by the spacecraft. Distortion-free plots of sensor fields-of-view, the Earth's disk, the Sun and other celestial objects, and other parts of the same spacecraft — all in relative motion. Assists in providing rigorous solutions for mission geometry problems. (PC)

OMNI EDGE Developer Option (Autometric, Inc., Colorado Springs, CO, http://www.autometric.com). EDGE is a C++ toolkit with visualization, simulation, and analyzing libraries. It allows the user to create a complex, synthetic view to see the world as it really is from outer space to sea level. The EDO Visualization Component is the foundation of EDGE. Other components enhance this foundation by providing libraries for integrating imagery, maps, terrain, time, and weather. Visualization options consist of both 2-D and 3-D windows. (PC, Workstations)

OrbSim2w (GAO Associates, Plymouth, MA, http://www.gaoassociates.com). OrbSim2w allows user to: Visually display orbit data, generate disk files of numerical information available from the ephemeris generator, compute and display the communication coverage areas of satellites, generate satellite-to-satellite data (e.g. communication or CSAT/ASAT situations), and allow for the specification of hypothetical satellites created in simulation either by direct element generation or by simulating launch injection data. (PC)

ORBWIN (AIAA Mission Design Software that accompanies the book *Spacecraft Mission Design* by Charles Brown). Orbit design and analysis tool for orbits about any solar system body. Utilities include ephemeris generation, plane changes, Julian date, and propellant mass calculations. Includes basic perturbations, such as regression of nodes and rotation of apsides. Calculates field-of-view characteristics, including swath width, angle to the horizon, and distance to the horizon for a spacecraft at a given altitude. Groundtrack plots as a function of time for one orbit. (PC)

PRO (AIAA Propulsion Design Software that accompanies the book *Spacecraft Propulsion* by Charles Brown). Basic design/performance of spacecraft propulsion systems, appropriate for feasibility study phase of a space mission, with emphasis on sizing and preliminary performance calculations. Composed of 6 modules: Propulsion Requirements, Rocket Engine Design, Pulsing Engine Performance, Blowdown System Performance, System Weight Statement, Utilities. (PC)

SatLife, Satellite Lifetime Prediction Program (Microcosm, El Segundo, CA, http://www.smad.com). Developed to create the satellite lifetime charts for SMAD. Integrates the orbit lifetime equations for both circular and elliptical orbits without using an orbit propagator. (MAC, PC)

SatTrack v4.5 (Bester Tracking Systems, Inc., Emeryville, CA, http://www.bester.com/stsuite.htm). Satellite orbit analysis software package, consisting of five major GUI-based tools, and a number of auxiliary programs provided for retrieval, archiving and generation of orbital element sets, or for post-processing of numerical data created with SatTrack. Real-time Tracking Tool displays the location of Earth satellites on world map and sky view charts and directly controls ground station subsystems. Batch Mode Tool allows prediction of satellite passes over specified ground stations and generates a wide variety of numerical data for many different types of analyses, including orbit event and station contact schedules. Data created with Batch Mode Tool can be analyzed with the Graphics Visualization Tool, which generates XY and polar displays allowing you to view your data in many different ways. Orbit event and contact schedules are shown as Gantt charts. The Globe Display Tool is used to display non-Keplerian trajectories of ground-based vehicles, ships, airplanes, balloons or missiles in either real-time or play-back mode. Link Analysis Tool aids system engineers in design of spacecraft communications subsystems and ground stations. (UNIX Workstations, MAC)

SC Modeler 2.6 (AVM Dynamics, Toronto, Canada, http://www.avmdynamics.com/index1.htm) is an interactive modeling tool for the visualization and topology analysis of satellite networks. SC Modeler allows you to specify orbital parameters of a satellite constellation and interactively examine an instantly generated 3-D view of the constellation. All intersatellite links or links available for a selected network node (satellite or ground station) can be displayed. (PC)

Spacecraft Control System (Princeton Satellite Systems, NJ, http://www.psatellite.com). Flight-proven COTS satellite attitude control system, and a high-fidelity simulation with which to test it. User starts the design with a working control system, and changes provided modules to adapt to the requirements of specific spacecraft. Uses a C-like command language. Control system is modular in design, making it simple to replace or customize blocks. Matlab scripts are included. Base package includes a complete control system that uses momentum wheels, thrusters and magnetic torquers with a static Earth sensor for attitude determination. (PC)

Sky Map Pro 10 (Sky Publishing, Cambridge, MA, http://www.skypub.com). Detailed characteristics for 15 million stars down to magnitude 15 plus 200,000 deep sky objects. Printed catalogs also available. (PC)

Space Radiation 5.0 (Space Radiation Associates, Eugene, OR, http://www.spacerad.com). Space environment and effects modeling tool that models ionizing radiation environment in space and the atmosphere including trapped protons and electrons, solar protons, galactic cosmic radiation, and neutrons. Radiation environment models: AP-8 trapped proton models; AE-8 trapped electron models; CREME cosmic radiation models; JPL 1991 solar proton model; IGRF/DGRF magnetic field models; Total dose from electrons, protons, and heavy ions; Dose equivalent for human exposure. Basic calculations supported include: Orbit generation; Shielding of Earth's magnetic field; Spacecraft shielding, including distributions; Radiation transport through materials; Radiation effects including single event upsets, total dose, displacement damage, and biological dose equivalent, and solar cell damage. (PC)

STK, Satellite Tool Kit (Analytical Graphics, Inc., Malvern, Pennsylvania, http://www.stk.com). General purpose mission design and ground track generation software. Includes multiple map projections with day/night areas, ground stations, ground tracks for multiple satellites, swath coverage, antenna and spacecraft sensor coverage regions, and other features. Combining the functionality of STK, STK/VO, STK/

High Resolution Maps, and STK/VO Earth Imagery, users can visualize in great detail the field-of-view of their imagers at any given period and determine many aspects of the quality of coverage of a defined region over time. (PC, Workstations)

STK Modules

Astrogator: Orbit maneuver and space mission planning STK module for use by spacecraft operations and mission analysis staff. Has customized thrust models, target ephemerides, spacecraft attitude, and the ability to solve for and optimize solutions. Used in conjunction with STK/VO and a number of planetary models, STK/Astrogator generates animated, 3D images of space missions ranging from near-Earth maneuvers to interplanetary missions.

Chains module extends STK's pair-wise access determination capabilities. Allows users to develop networks of objects for visibility-related analysis. A chain is a combination of STK objects such as satellites, facilities, ships, sensors, etc., ordered to model a communications or data-transfer path. Solves multi-satellite, multi-target problems, such as simultaneous viewing of a target by two satellites or simultaneous viewing of a target and a relay satellite.

Comm module can define and analyze detailed communications systems. Can generate detailed link reports and graphs as well as 2D and 3D map graphics indicating signal quality accounting for free space, rain, atmospheric losses, and interference introduced by other satellite systems. Complete link analysis can be done over time using STK's satellite orbit propagation/geometry engine coupled with all defined receiver and transmitter properties.

Coverage module provides capabilities for analyzing the coverage performance over time of satellites, ground facilities and vehicles, missiles, aircraft, and ships. Optimization studies such as gap analysis, stereo imaging opportunities, and minimum response time can be evaluated. Allows users to define areas of interest, coverage assets (satellites, ground stations, etc.), time period, and metrics of coverage quality.

Interceptor Flight Tool (IFT) module generates interceptor flight trajectories against ballistic missiles, aircraft, and satellite targets, simulates a range of interceptor system types from low altitude theater to high altitude strategic interceptors. STK/IFT provides for a single intercept of a single threat trajectory.

Missile Flight Tool (MFT) is a high-fidelity, missile-trajectory generation module. Contains a set of missile databases that represent a complete spectrum of missile types and performance capabilities. Multiple-stage missile trajectories can be analyzed and visualized easily.

Mission Utility Systems Engineering (MUSE) (Microcosm, El Segundo, CA). Extends STK simulation environment by adding powerful analytical components and mechanisms for space mission analysis and system design, including generic Figures of Merit. Also available as a Development Environment, allowing user to broadly extend the capabilities of both STK and MUSE. Functions and algorithms written in the C programming language can quickly be integrated into the MUSE Output Manager, existing STK objects can be enhanced with new functionality. MUSE allows new objects to be added to STK.

Precision Orbit Determination System (PODS) module is designed to process space-craft tracking data and determine spacecraft orbits and related parameters. Can process tracking data from a variety of sources, including those gathered by ground antennas (e.g., angles, range, range rate), Tracking and Data Relay Satellite System (TDRSS) data (e.g., satellite-to-satellite/ satellite-to-station relay data), and Global Positioning System (GPS) position data collected by GPS receivers.

Radar provides thorough analysis and graphic displays of radar systems. The module also allows the user to model an important characteristic of radar targets, radar cross section (RCS), to calculate and display access and to generate reports and graphs of radar system performance. Can simulate both monostatic and bistatic radar systems and supports operations in Synthetic Aperture Radar (SAR) and/or Search/Track modes.

SpaceVu (Microcosm, El Segundo, CA). Space mission geometry visualization tool that provides views of the spacecraft-centered celestial sphere showing the sky as seen by the spacecraft with stars, planets, Sun and Moon, fields-of-view, the Earth disk with details as seen by the spacecraft, and the motion of other spacecraft. Also supports simultaneously the view of the sky as seen by multiple ground stations.

Visualization Option (VO): 3D visualization environment that displays all scenario information from STK. Intuitive view of complex mission and orbit geometries by displaying realistic 3D views of space, airborne and terrestrial assets, sensor projections, orbit trajectories and assorted visual cues and analysis aids. STK/Advanced VO and STK/VO Earth Imagery (1 km imagery of the entire Earth) options available.

The Sky Version 5 (Software Bisque, Golden, CO, http://www.bisque.com). Amateur astronomy program that provides very user-friendly access to planetary ephemerides, star catalogs, and other celestial objects for PC users. Includes 260,000 SAO star catalog plus 19 million entries in the Hubble Guide Star catalog. (PC)

Voyager IV (Carina Software, San Leandro, CA, http://www.carinasoft.com). Amateur astronomy program that provides very user-friendly access to planetary ephemerides, star catalogs, and other celestial objects for Macintosh users. Includes 260,000 SAO star catalog plus 19 million entries in the Hubble Guide Star catalog. (MAC)

SC Modeler 2.6 (AVM Dynamics, Toronto, Canada, http://www.avmdynamics.com/index1.htm). Tool used for the conceptual design and analysis of satellite constellations, originally created for analysis of satellite communications systems. Allows analysis of the topology of space-based networks and dynamic visualization of complex communications links. Consists of a number of modules: Constellation Design (can generate a wide variety of efficient constellation designs with circular or elliptic orbits); Coverage Analysis (allows you to determine the minimum number of satellites that will be available at any time to each point at a specified latitude.); Interactive Module (can create a static or animated 3-D view of a constellation); Constraints Analysis (can determine the minimum ground elevation angle that has to be maintained by the constellation in order to provide continuous coverage at specified latitudes); Intersatellite Links Analysis. (PC)

References

Allery, M., Sellers, J.J., Sweeting, M.N. June 1994. "Results of University of Surrey On Orbit Microsatellite Experiments." Presented at the International Symposium on Small Satellite Systems and Services, Biarritz, France.

Arianespace. March 1991. *Dossier de Controle des Interfaces (Interface Control Document)*, UoSAT-F, DCI 10/392 01, Issue 2, Rev. 0.

Burgess, E.L., Lao, N.Y., Bearden, D.A. 1995. "Small-Satellite Cost-Estimating Relationships." The Aerospace Corp. Presented at the 9th Annual AIAA/USU Small Satellite Conference, Logan, Utah, 18–21 September, 1995.

Campbell, W.A. Jr., Scialdone, J.J. September 1993. *Outgassing Data for Selected Spacecraft Materials*, NASA Reference Publication 1124, Revision 3.

Dean, E., Unal, R. 1991. "Designing for Cost." Transactions of the American Association of Cost Engineers, pp. D.4.1–D.4.6, Seattle, WA, 23–26 June, 1991.

ESA PSS 01 60, *Component Selection Procurement and Control for ESA Spacecraft and Associated Equipment.*

Grahn, Sven. 1993. "The FREJA Magnetospheric Research Satellite Design and Operations." Presented at the 7th AIAA/USU Conference on Small Satellites, Utah State University, September 11–16, 1993.

Gray, C. 1990. "Development of a 110lbf Dual Mode Liquid Apogee Engine." AIAA 90-2424, AIAA/SAE/ASME/ASEE 26th Joint Propulsion Conference, Orlando, Florida, 16–18 July 1990.

Humble, R., et al, *Space Propulsion Analysis and Design*. McGraw-Hill, Inc. College Custom Series. 1995.

Iida, T., et al. 1993. "Application of Car Electronic Parts to Small Satellites." Japan-U.S. Cooperation in Space Project Workshop, Maui, HI, 12–14 November, 1993.

Isakowitz, Steven J. 1995. *International Reference Guide to Space Launch Systems*, 2nd ed. Washington, DC: American Institute of Aeronautics and Astronautics.

Jollet, P. October 1986. *Outgassing and Thermo-Optical Data for Spacecraft Materials*. MATLAB 001, Materials Section, European Space and Technology Centre, Noordwijk, The Netherlands.

Jursa, A.S. (editor). 1985. *Handbook of Geophysics and the Space Environment*. Air Force Geophysics Laboratory, Air Force Systems Command, United States Air Force.

Kitfield, James. October 1994. "The End of the Line for MilSPEC?" *Air Force Magazine*, pp. 43–45.

Meerman, Maarten J. M., Sweeting, M.N. "A Simple Adaptive Spacecraft Structure, Six Years of Layered Satellites." Presented at CNES/ESA International Conference on Spacecraft Structures & Mechanical Testing, Paris, France, June 21-24, 1994, pp. 581-592.

MIL-STD 975, *NASA Standard Parts List.*

MIL-STD-1522A, *Standard General Requirements for Safe Design and Operation of Pressurized Missile and Space Systems.*

Sellers, J.J., Astore, W.J., Crumpton, K.S., Elliot, C., Giffen, R.B., Larson, W.J. (ed). 1994. *Understanding Space: An Introduction to Astronautics.* New York, N.Y: McGraw-Hill.

Sellers, J.J., et al. September 1995. "Investigation Into Low-Cost Propulsion Options for Small Satellites." Presented at the 9th AIAA/Utah State Conference on Small Satellites, Logan, Utah, 18–21.

Sellers, J.J., Meerman, M., Paul, M., Sweeting, M. March 1995. "A Low-Cost Propulsion Option for Small Satellites." *Journal of the British Interplanetary Society*, vol. 48, pp. 129–138.

Sellers, J.J., Sweeting, M.N., "UoSAT and Other European Activities in Small Satellite Attitude Control." Presented at the 17th AAS.

SMAD = Larson, Wiley J. and James R. Wertz, eds. 1992. *Space Mission Analysis and Design,* 2nd edition. Torrance: Microcosm Press and Dordrecht: Kluwer Academic Publishers.

Smith, P., Horton, M.A. 1984. "Advanced Propulsion Systems for Geostationary Spacecraft—Study Results." The Marconi Company, AIAA-84-1230.

Ward, Jeffrey W. 1990. "Store-and-Forward Message Relay Using Microsatellites: The UoSAT-3 PACSAT Communications Payload." Presented at the 4th annual USU/AIAA Conference on Small Satellites, Logan, UT, 27–30 September, 1990.

Technology for Small Spacecraft. 1994. Panel on Small Spacecraft Technology, Committee on Advanced Space Technology, Aeronautics and Space Engineering Board, Commission on Engineering and Technical Systems, National Research Council, National Academy Press, Washington, D.C.

Underwood, C. 1995. "Commercial Microelectronic Devices in the Space Radiation Environment." Ph.D. Thesis, University of Surrey.

Wood, R.S. 1990. "Development of a Low Cost 22N Bipropellant Thruster," AIAA 90-2056, AIAA/SAE/ASME/ASEE 26th Joint Propulsion Conference, Orlando, Florida, 16–18 July 1990.

WRR 127-1,. June 30, 1993. *Range Safety Requirements*, 30th Space Wing, United States Air Force.

ZARM, 1995. "Small Satellite Components Data Base." Bremen, Germany: University of Bremen.

Chapter 4

Reducing Launch Cost

John R. London III, *Ballistic Missile Defense Organization*

The expense of launching payloads into space today is very high. Launch vehicles and their operation—whether expendable or reusable, and from small to large—cost millions to hundreds of millions of dollars per flight. And this cost is in addition to the usually very expensive payload the launch vehicle is carrying.

The cost of space launch decreased rapidly during the first decade of orbital space operations (1958–1968), primarily due to the rapid growth in booster size during this period and the resultant effect of scale on launch prices. After this initial period, however, Western launch costs flattened out and have remained relatively constant. (See Fig. 4-1.) Despite the introduction of a number of new Western launch vehicles in the last three decades, there has been no significant change in the cost of space access [Koelle, 1995]. There are a plethora of launch systems currently operational worldwide, and they cover a wide spectrum of performance, cost, and design and operational philosophies. Table 4-1 summarizes the better known systems and some of their characteristics* [Isakowitz, 1995; Poniatowski, 1994].

---

* All cost figures in this chapter are in 1995 dollars.

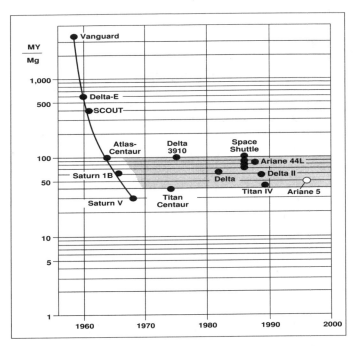

Fig. 4-1. Historical Trends in Space Launch Costs. Costs are calculated in man-years (MY) per million grams (Mg) to low-Earth orbit. Data from Koelle [1991].

TABLE 4-1. Current Launch Systems and Their Cost.

Vehicle	Liftoff Mass (kg)	Cost Range ($M)[1]	Payload Capacity to LEO ($kg)[2]	Payload Launch Efficiency ($ per kg)[3]
Pegasus	19,050	$13M–$15M	454	$30,800
LLV-1	66,225	$15M–$17M	794	$20,200
Taurus	81,650	$18M–$20M	1,451	$13,100
Titan II	155,000	$35M–$40M	1,905[4]	$19,700
Vostok SL-3	290,000	$20M–$30M	4,717[5]	$5,300
Delta II 7920	218,300	$45M–$50M	5,035	$9,400
Atlas IIA	187,700	$80M–$90M	6,760	$12,600
Ariane-44LP	420,000	$90M–$100M	8,300[6]	$11,400
Long March 2E	464,000	$40M–$50M	9,210[7]	$4,900
H-2	264,000	$150M–$200M	10,433[8]	$16,800
Titan IV	862,000	$230M–$325M	17,700	$15,700
Proton SL-13	703,000	$35M–$70M	20,000[5]	$2,600
Space Shuttle	2,040,000	$350M–$547M	23,500	$19,100

1) Per launch. Median values used for calculations. 2) Assumes the vehicle's full payload capacity. The Shuttle, in particular, usually carries much less than its full capacity. Capacities listed are for a due east launch from Kennedy Space Center / Cape Canaveral Air Station (latitude 28.5 deg) except as noted. 3) Cost divided by vehicle payload capacity to low-Earth orbit. Note these launch efficiencies are for vehicles with a wide range of lift capacities. A more meaningful use of payload launch efficiencies would be to compare different launch vehicles with the same lift capacity. 4) Polar orbit. 5) 51.6 deg latitude launch site. 6) 5.2 deg latitude launch site. 7) Similar launch latitude to Kennedy Space Center / Cape Canaveral Air Station. 8) 30.0 deg latitude launch site.

4.1 The Effects of High Launch Costs

Perhaps the most profound effect of high launch costs is on the quantity and scope of new space initiatives, be they civil, military, or commercial. In fact, these costs are the greatest limiting factor to expanded space exploitation and exploration. Small satellite initiatives are especially constrained by the high cost of space access. Assuming that civil and defense space budgets remain flat or grow only modestly in the coming years, they will be largely consumed by operational programs or programs already well under way in the development cycle. Consequently, making room for significant new space initiatives will be difficult.

A payload budget planner must allocate such a significant portion of the budget to launch services that launch cost considerations ripple powerfully through all aspects of space mission planning. Also, the cost of spacecraft has become strongly linked to the cost of launch, so reducing the cost of space systems and their missions depends to a large extent on achieving lower prices for space transportation.

Military space systems have become key components of force application planning and operations, and their utility cuts across each of the military services. Without some relief to the high cost of launch, just the expense of continuing to replenish existing systems will likely drain away significant funding (especially within the Air Force) that otherwise could be used for the acquisition of new weapon systems.

In recent years, a number of aerospace companies have succeeded in developing small, highly capable satellites at low cost (see examples in Part II, and the list on the inside rear cover of this book). They use simple design and manufacturing techniques, and they take advantage of the increasingly compact electronics and computer systems that are now available. Despite the development of these low-cost spacecraft, the cost to launch them has not decreased proportionately. In fact, the cost of small launch vehicles that are matched to the small satellite mission, based on a dollars-per-kilogram comparison, is often more than twice as much as the cost of large boosters. This is partly due to the economies of size that larger boosters enjoy. Still, the availability of low-cost small satellites has not been a driving force in reducing launch costs.

High launch costs greatly—and insidiously—increase the cost of spacecraft. A study by Hughes Aircraft [1988] said that the high cost of payloads has been driven to a large extent by the limited lift capability and restricted payload volume of current boosters. These constraints have forced payload designers into sophisticated designs which use expensive, lightweight materials, high packaging densities, and complex configurations involving a multitude of mechanisms and deployable appendages. The study also stated that the resulting designs are expensive to manufacture and require significant amounts of analysis and testing to validate. Some estimates have indicated that high launch costs are responsible for about one-half of the total cost of new satellite systems. The high cost of today's space launch systems dictate requirements such as longer satellite life, higher reliability, and multi-redundant subsystems, all of which stretch out schedules and drive up spacecraft costs. Most satellite missions (an exception being spacecraft designed for deep-space missions) do not necessarily demand these requirements be pushed to the technological limit. As we will see later in this chapter, other options may be appropriate.

The commercial launch industry in the U.S. is at risk of being increasingly diminished or even eliminated by foreign competition. Without an injection of cost-cutting leadership by U.S. government and industry in this area, future commercial boosters may be made only in other countries. U.S. spacecraft builders seeking inexpensive space access could be held hostage to foreign launch suppliers. Additionally, if the U.S. commercial launch industry continues to decline, the U.S. government may face serious problems in putting future government payloads into orbit. The government could ultimately be forced to either largely subsidize the U.S. launch industry, or to depend on foreign launch of government payloads, including those critical to U.S. national security. Finally, the failure of the U.S. commercial launch industry would represent the loss of a national technical and defense treasure, many millions of dollars in commercial revenue, and numerous jobs. It is therefore important that government and industry invest in the country's future by working together to radically drive down the cost of space launch.

Air Force Space Command specifies *four basic characteristics of any launch system*: capability, reliability, affordability, and responsiveness [Roberts, 1993]. A comprehensive analysis of current and proposed launch systems with respect to all of these areas is beyond this chapter's scope. However, improving each of these characteristics starts with the vehicle design, and the concepts discussed in this chapter will focus mainly on design issues related to reducing launch cost. Additionally, a booster that is designed to directly address affordability may provide a positive net effect on the other three characteristics.

4.2 Lowering Launch Costs Using Today's Launch Vehicles

The cost of today's launch systems can be oppressively high for prospective users. Assuming that there are no significant improvements in current launch costs and booster availability, the developer of a specific spacecraft is somewhat limited in what they can do to lower launch expenses. However, there are still some useful options available for holding down the cost of launch to the user.

4.2.1 Options for Large Spacecraft

For spacecraft users seeking to deploy large spacecraft or a large constellation of smaller spacecraft, there are several steps they can take to keep launch costs as low as possible. First of all, they should design their spacecraft to be compatible with as many launch vehicles as possible, both domestic and foreign. This will allow the spacecraft builder or user to seek opportunities for a worldwide competition among launch service providers. To be eligible to fly on a variety of foreign boosters, it may be necessary to do significant forward planning to avoid export control entanglements.

Second, the spacecraft builder should establish a significant weight and volume margin during the earliest phases of the spacecraft design, so the spacecraft does not exceed the lift or volume capacity of its chosen booster late in the development process. A collateral benefit of designing your spacecraft to fit on a number of different boosters is that if during the course of spacecraft development the satellite still manages to increase past the weight or volume design caps, you may be able to retain some booster selection flexibility and avoid a major spacecraft redesign.

Third, the spacecraft builder should carefully trade the approach of designing for the booster to provide specialized payload accommodations and services with the alternative of designing the spacecraft to be generally self-supporting (mechanically, electrically, environmentally, and operationally), allowing a standard booster interface. A *self-supporting spacecraft* will require little to no special booster accommodations and could potentially expand booster selection options. This may become increasingly important if launch providers move toward more standardized and limited payload services packages in an effort to reduce launch services costs.

Finally, the builder or user of large spacecraft or a large constellation of smaller spacecraft should consider the development of a new booster that is tailored for, and dedicated to, their specific launch service needs. The decision to procure a new launch vehicle would have to justify the nonrecurring development costs of such a capability, but would open up some possibilities for major cost savings. By designing a launch system specifically for one particular spacecraft, engineers could make the booster so synergistic with the spacecraft that it becomes an **extension** of it, and not just its means of transportation. For example, the spacecraft's guidance and control system could provide all booster guidance and control. The booster design could be focused on providing just the bare-minimum launch requirements for the unique satellite configuration. The booster design could at some point be modified to offer a more general launch capability for other payloads, thus providing a possibly lower cost alternative for space launch whose development costs have already been largely paid for. Clearly, developing a new launcher for one particular satellite system would be justifiable only in unique cases. However, the deployment of distributed satellite architectures like the 840 satellites proposed for the Teledesic communication system would warrant some consideration of this option.

4.2.2 Options for Small Spacecraft

For spacecraft users seeking to deploy smaller payloads in space, there are a number of alternatives and options available that will help minimize launch costs. (See Table 4-2.) First, all of the options already described for minimizing the launch cost of large spacecraft or large constellations of smaller spacecraft could potentially be applicable for small satellites as well. In particular, building in autonomy tends to be easier with small satellite designs, allowing flexibility of booster selection and booster integration simplicity.

Small satellite users should consider launch opportunities as secondary payloads with larger spacecraft. This option would, in most cases, not be practical unless the mission requirements of the small satellite fit into the general mission parameters of the primary, or "parent," spacecraft.[*] For example, this could be an appropriate launch approach if the small satellite's payload only needed a microgravity environment for an extended period, and it did not matter what particular orbit the satellite was in. Another way of flying as a secondary payload is to wait for a flight opportunity with a larger payload that is destined for an orbit that closely matches the desired orbit for the small satellite. This strategy would likely require the small satellite user to be very flexible concerning their launch schedule and the initiation of mission operations. Despite the inherent limitations of flying as a secondary payload, this method of space access can be extremely economical, particularly if the small sat-

---

[*] See Sec. 13.3 on AMSAT that describes some exceptions to this general rule—low-cost spacecraft with inexpensive propulsion systems capable of very large orbit maneuvers.

TABLE 4-2. **Low-Cost Alternatives to Dedicated Launches.** Many of the approaches listed can be used as alternatives to more expensive orbital flights for the achievement of systems tests or science experiments.

Option	Characteristics	Mass Limits	Principal Constraints	Approximate Cost	Sources
Balloon flights	Hours to days at ≈ 30 km altitude	Up to 70 kg for low-cost flights	Not in space, not 0-g, weather concerns	$5K to $15K	U. of Wyoming, USAFA, NSBF
Drop towers	1 to 10 sec of 0-g with immediate payload recovery	Up to 1,000 kg	Brief "flight," 5 to 50 g landing acceleration, entire experiment package dropped	≈ $10K per experiment	ZARM, JAMIC, NASA LeRC and MSFC, Vanderbilt U.
Drop tubes	1 to 5 sec of 0-g with immediate sample retrieval	<0.01 kg	Brief "flight," 20 to 50 g landing acceleration, instrumentation not dropped with sample	≈ $0.02K per experiment	ZARM, JAMIC, NASA LeRC and MSFC, Vanderbilt U.
Aircraft parabolic flights	Fair 0-g environment, repeated 0-g cycles	Effectively unlimited	Low gravity is only 10^{-2} g	$6.5K to $9K per hour	NASA LeRC and JSC, Novespace
Sounding rockets	Good 0-g environment, altitude to 1,200 km, duration of 4 to 12 minutes	Up to 600 kg	Much less than orbital velocities	$1M to $2M	NASA GSFC, NRL, ESA/ Sweden, OSC, EER, Bristol Aerosp.
GAS containers	Days to weeks of 0-g on board the Shuttle	Up to 90 kg	Very limited external interfaces	$27K for largest container	NASA GSFC
Secondary payloads	Capacity that is available in excess of primary's requirements	Up to ≈1,000 kg	Subject to primary's mission profile	<$10M	Ariane, OSC, MDA, Russia
Shared launches	Flights with other payloads having similar orbital requirements	Up to ≈ 5,000 kg	Integration challenges	Up to ≈$60M	Ariane, OSC, Russia

ellite designer focuses on making the spacecraft create only a minor impact to the overall task of integrating the payloads to the booster and flying them to orbit.

Opportunities for flights as secondary payloads are routinely available on the Ariane 4 launch vehicle. The Ariane provides an *Ariane Structure for Auxiliary Payloads (ASAP)* to carry and deploy small spacecraft [Horais, 1994]. The ASAP ring is a circular platform mounted externally to the interface between the vehicle equipment bay inner cone and the main payload adaptor. Figure 4-2 shows the placement of the usable volume for small payloads on the ASAP ring, beneath the primary payloads. The ASAP ring can accommodate individual payloads having a maximum mass of 50 kg (including interface and deployment mechanisms). The maximum aggregate mass of all secondary payloads cannot exceed 200 kg. The maximum usable volume envelope for individual payloads (including interface and deployment mechanisms) is a cube 450 mm on a side. Arianespace requires that integration, testing, operation, and deployment of secondary payloads not interfere with the primary payload's scheduled activities. Since 1990 a number of small spacecraft have gained access to space via the Ariane/ASAP [Arianespace, 1993].

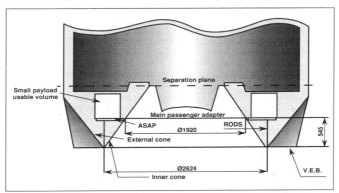

Fig. 4-2. ASAP Ring Small Payload Usable Volume Locations Within the Ariane Payload Fairing. Units are in millimeters.

The Delta II launch vehicle can accommodate secondary payloads that are attached to the side of the vehicle second stage, above the miniskirt and support truss. (See Fig. 4-3.) The integration process normally begins two years prior to launch, but the Delta II program may be able to support shorter schedules [NASA Goddard Space Flight Center, 1994]

Orbital Sciences Corporation (OSC) advertises opportunities for secondary payloads on Pegasus and Taurus launches that are already manifested with primary payloads. Categories of partial payloads on OSC boosters are: secondary, microsat, and space available. Typical secondary opportunities are in the 150 to 200 kg range for the Pegasus, and substantially greater for the Taurus [Orbital Sciences Corporation, 1995].

With the increasing number of cooperative ventures available with the Russian space program, there is a potential for many new secondary payload opportunities. The robust Russian launch capability represents frequent, large capacity access to space. One example of the new opportunities available with Russia is the Skipper program being sponsored by the Ballistic Missile Defense Organization. *Skipper* is a unique spacecraft that consists of a Russian bus and a U.S. payload designed to measure the ultraviolet emissions generated by vehicles operating in the upper atmosphere at velocities near 7 km/sec. Skipper is scheduled to be launched as a secondary payload on a Russian Molniya booster in late 1995, with the Indian IRS-1C flying as the primary payload. (See Fig. 4-4.)

Another example of the burgeoning opportunities for secondary payloads on Russian boosters is the launch of the U.S. FAISAT I and the Swedish ASTRID microsatellites on January 24, 1995. The two spacecraft were attached to the forward section of a Russian Cicada navigation satellite and launched on a Kosmos-3M rocket from the Plesetsk Cosmodrome. Also, the German SAFIR-R microsatellite was launched in November 1994 aboard a Russian Earth resources satellite, and future SAFIR launches on Russian boosters are planned [Powell, 1995].

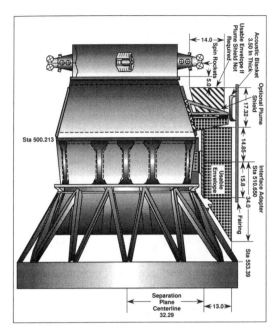

Fig. 4-3. Usable Envelope for a Separating Secondary Payload on a Delta II Launch Vehicle. Dimensions are in inches.

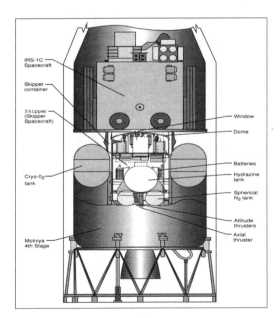

Fig. 4-4. Skipper Secondary Payload Integrated on a Russian Molniya Launch Vehicle. The Skipper spacecraft consists of a U.S. payload section attached atop a Russian bus. Skipper is integrated within the Molniya fairing beneath the IRS-1C primary payload.

The *Get Away Special* (GAS) program sponsored by NASA is heavily subsidized, but it does allow flights of small payloads on the Space Shuttle at economical prices to the user. Standard GAS payloads are not deployed, but remain in their cylindrical containers mounted inside the Orbiter's payload bay. An important advantage of the GAS program is the payload is returned to its owner or operator after several days in the low-Earth orbit environment. All GAS payloads must meet rigorous safety standards that are driven by the demands of human spaceflight. Standard GAS containers are available in two volumes: 0.142 m³ (5.0 ft³) and 0.071 m³ (2.5 ft³). Payloads of 90.7 kg can be housed in the 0.142 m³ container, and payloads of 45.4 or 27.2 kg can be supported by the 0.071 m³ container. The 0.142 m³ container has proven to be by far the most widely used by experimenters. The GAS containers are designed to be able to provide for internal pressure that can be varied from near vacuum to about one atmosphere. Each GAS container has a payload control unit that provides the equivalent of three toggle switches for the experimenter that can be operated by the Shuttle crew during the mission. However, for flight safety reasons, one of the switches must be dedicated to removing all power from the payload. A lid that opens on-orbit is available at an additional cost of 25% over the standard price. U.S. educational institutions receive a price reduction of about 60% on the standard cost of flying in GAS containers [NASA Goddard Space Flight Center, 1991].

For small payloads which need more flexible accommodations on the Shuttle, NASA offers its Hitchhiker program. *Hitchhiker* provides standard power, data, and command services for customer equipment, and is designed to provide modular and expandable accommodations. Users who want to deploy their small payloads from the Shuttle's payload bay can use Hitchhiker. The price is much higher (in the $1 to $2 million range) than what is charged for a GAS payload, but for certain payloads it could still be a very cost effective means of space access. Also, it is quite possible that an experimenter could gain NASA sponsorship for their mission, in which case NASA would not charge for flying the payload.

NASA is developing their *Space Experiment Module* program for U.S. educational institutions, which will create opportunities for even simpler and less expensive access to space. The program will subdivide a 0.142 m³ GAS container into 10 separate compartments that will be available at little or no cost to experimenters. Each compartment will have its own data system and battery power provided by NASA for individual experiments.

Users of payloads that could fly on small satellites should consider less expensive non-orbital solutions that may also allow them to meet their mission goals. Examples include drop facilities that produce a few moments of microgravity, aircraft flying parabolic trajectories that can provide longer periods of microgravity for heavy payloads, and balloon flights that offer lengthy exposure to the near-space environment.

There are a small number of drop towers available worldwide to experimenters seeking very short periods of microgravity. Drop towers typically can provide a microgravity environment of 10^{-6} to 10^{-3} g's. The *ZARM* (Center of Applied Space Technology and Microgravity) *drop tower* in Bremen, Germany has a free-fall distance of 110 m, providing a free-fall duration of 4.74 sec. In order that the free-fall of test objects not be adversely influenced by air friction, the tower is designed to operate as a vacuum chamber. The available payload envelope is 1.2 m by 0.8 m, with a maximum weight of 150 kg [de Selding, 1994; ZARM, 1990]. The *JAMIC* (Japan Microgravity Center) *drop shaft* in Hokkaido, Japan has a free-fall distance of 490 m which provides a microgravity time of 10 sec. The maximum payload dimensions that can be accommodated are 1.3 by 1.4 m, with a maximum weight of 1,000 kg [Japan Microgravity Center, 1992]. NASA's Lewis Research Center has two drop towers. The *30 m drop tower* provides a free-fall time of 2.2 sec and can support an experiment rate of five to eight drops per day. The *Zero Gravity Research Facility* has a free-fall distance of 132 m, providing a free-fall duration of 5.18 sec. The Zero Gravity Research Facility provides experimenters with a vacuum environment that can handle payloads of up to 450 kg, housed in a 2.2 by 3.4 m drop capsule. NASA's Marshall Space Flight Center has a 100 m drop tower, but it is currently not in operation [Robinson, Bayuzick, and Hofmeister, 1990].

Drop tubes differ from drop towers in that drop tower experiments experience microgravity while falling inside a capsule which normally contains peripheral equipment for data acquisition. In the case of a drop tube, only the sample falls. Low-gravity environments in the range of 10^{-2} g's are typical when a gas is present in the tube. For evacuated tubes, gravity environments as low as 10^{-11} g's are possible. There are a number of drop tubes that can support low-gravity experiments, although most are small and provide very brief periods of low gravity. The Marshall Space Flight Center operates a *105 m drop tube* that has a 0.25 m inside diameter and provides a free-fall time of 4.646 sec. It can handle sample sizes of 0.12 to 10 g's, and support up to 20 experiments per hour. Other drop tubes are located at the Jet Propulsion Laboratory (13.2 m), Vanderbilt University (30 m and 13.7 m), the Lewis Research Center (5.5 m), Grenoble, France (45 m), and the Marshall Space Flight Center (30 m, currently not operating) [Robinson, Bayuzick, and Hofmeister, 1990].

Aircraft *parabolic flights* are suitable for payloads that are heavier and require longer periods of reduced gravity than can be accommodated by drop towers. Parabolic flight profiles can even be tailored to provide a variety of planetary gravity levels. NASA's Johnson Space Center operates a *KC-135 aircraft* that can provide up to 40 periods of low gravity for 25 sec intervals each during a single flight. The KC-135's interior bay is 3 m high by 16 m long. The Lewis Research Center operates a *DC-9 aircraft* that can provide up to 40 low-gravity periods of 22 sec each. The DC-9's interior bay dimensions are 2.89 m wide, 1.98 m high, and 16 m long [NASA Research Announcement, 1994]. Novespace of Paris, France operates a Caravelle Zero G aircraft that is available commercially. Novespace plans to retire its Caravelle aircraft and replace it in 1995 with a larger Airbus A300 [de Selding, 1994]. Weaver Aerospace in Aptos, California offers low-gravity as well as lunar and Mars gravity parabolas, using a variety of privately-owned aircraft.

High altitude balloon flights can provide payloads relatively lengthy periods of access to the near-space environment. Helium-filled balloons can reach altitudes of approximately 30 km with dwell times of several hours or more, depending on payload mission requirements and weather conditions. At the end of the mission, payloads are typically cut loose by radio command and recovered by parachute. Balloon flights can support payloads up to 70 kg at relatively low cost. Balloon flights of much larger payloads are possible as well.

4.3 Analytical Techniques for Assessing Booster Efficiencies

We can use a number of analytical techniques to assess different space launchers. By relating various characteristics of a launch vehicle to its cost, we can gain some insight into the overall "efficiency" of a given booster. Different boosters can then be compared to a degree, allowing the discernment of important information that can aid in the selection of current launch vehicles and the design of new, lower cost launchers. Table 4-3 shows techniques for establishing various measures of a *vehicle's cost efficiency*. For illustrative purposes, three representative vehicles and payloads are compared using these techniques. One example of the kind of information that can be gleaned by such analytical techniques is the information on *Payload Volume Efficiency* in Table 4-3. Despite the Shuttle's high cost per kilogram to low-Earth orbit, it has a relatively low cost per cubic meter to low-Earth orbit. This indicates the Shuttle is optimized to launch very large, low-density spacecraft. Use of these analytical tools must be exercised with care, because the numerous differences between different launch vehicles can cause direct comparisons to be misleading.

TABLE 4-3. Launch Vehicle Cost Efficiencies.

Vehicle	Launch[1] Services Cost	%[2] of Life Cycle Cost	%[3] of Total Stack Cost	Payload[4] Launch Efficiency	Payload[5] Volume Efficiency
Delta II 7925/ Navstar GPS	$47.5M	21.3%[6]	53.0%	$9,400/kg	$2.01M/m³ [7]
Titan IV / DSP	$277.5M	49.1%	49.7%	$15,700/kg	Not available[8]
Space Shuttle/ TDRSS	$448.5M	45.8%	52.9%	$19,100/kg	$1.49M/m³

1) Cost range median values. 2) Percentage of a space system's total life-cycle cost that is attributable to launch costs. 3) Total stack cost is the combined cost of launch and the cost of the payload and its integration. The percentage is the amount of the stack cost attributable to launch. 4) Cost of launch services divided by vehicle payload lift capacity to low-Earth orbit. Assumes the vehicle is carrying its full payload capacity. The Shuttle, in particular, often does not. 5) Cost of launch services divided by vehicle payload volume capacity. 6) 47.2% of GPS life-cycle cost is attributable to user equipment. 7) 2.9 m diameter fairing. 8) The size of the Titan IV fairing used by DSP is classified.

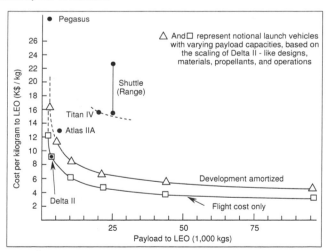

Fig. 4-5. Effect of Scale on Cost Per Kilogram to Low-Earth Orbit. Comparing raw payload launch efficiencies (cost per kilogram to low-Earth orbit) can be misleading. By projecting costs of a given vehicle (in this case, the Delta II) across a spectrum of lift capacities, we can better analyze vehicles larger and smaller than the Delta.

Payload Launch Efficiency (the cost per kilogram to low-Earth orbit) is the most frequently used method for comparing cost efficiencies of different vehicles. However, it provides the most valid comparison of values when various vehicle concepts are considered **that have the same payload lift capacity**. The efficiency ratings of various boosters with different payload lift capacities are somewhat skewed in favor of larger boosters, because of the scaling advantage that larger vehicles have over smaller vehicles. We can more equitably assess booster efficiency by comparing the cost per kilogram to orbit of various boosters to notional, standardized, Delta II-like vehicles with varying levels of performance. (See Fig. 4-5.)

There are several analytical techniques that provide a better understanding of *booster design efficiencies* by relating the weight of the booster structure and systems to payload weight, establishing the cost per kilogram of vehicle hardware, and illustrating how reusability affects vehicle weight. These techniques are identified in Table 4-4, and they compare the same three launch systems used in Table 4-3. Indicators of design efficiency historically have been expected to provide some correlation to a launch system's cost efficiency. However, despite the insight into design efficiencies these techniques can offer, they are not necessarily indicative of a booster's overall cost efficiency. As we will discuss later in this chapter, the weight of some booster designs is not as closely coupled to their cost as we would traditionally expect.

TABLE 4-4. Launch Vehicle Design Efficiencies.

Vehicle	Stack-to-payload Ratio[1]	Vehicle Dry Weight (kg)[2]	Relative Vehicle Hardware Cost[3]	Reusable Hardware Penalty[4]
Delta II 7925	46:1	20,368	$2,332/kg	N/A
Titan IV	49:1	107,957[5]	$2,570/kg	N/A
Space Shuttle	87:1	290,758[6]	$5,159/kg[7]	47.0%[8]

1) The ratio of a vehicle's gross liftoff weight to its maximum payload weight capacity. 2) Calculated with solid strap-ons unloaded and liquid propellant tanks empty. 3) Vehicle cost divided by dry weight. Actual hardware costs are somewhat lower for Delta and Titan because their calculations included the cost of propellant and launch services. 4) Percentage of hardware mass attributable to recovery systems. 5) Assumes Centaur upper stage. 6) Assumes IUS upper stage. 7) Assumes a manufacturing cost of $1.5 billion for the entire Shuttle stack. Orbiter and Solid Rocket Boosters are reusable. 8) Orbiter only.

4.4 Causes of High Launch Costs

For decades engineers have sought a solution to the high cost of space launch, with little success. To achieve a breakthrough in cutting the expense of launch, we must first have a firm understanding of some of the key causes of these high costs. Space launch vehicles typically exhibit heavily engineered complex designs that are costly to develop, manufacture, and operate, and this turns out to be the crux of the problem. Although the "standing army" at the launch base has often been cited as the chief cause of high launch costs, these personnel requirements are largely driven by the vehicle design.

4.4.1 The Heritage of Maximum Performance and Minimum Weight

Maximum performance and *minimum weight* were the overriding design drivers for U.S. ballistic missiles developed in the 1950s. These design drivers were, and still are, the norm for the aircraft manufacturing industry. Consequently, there were extraordinary efforts made to decrease structural weight and increase propulsion performance. Engineers kept design margins low in order to keep weight down. Rocket engines were configured for high combustion chamber pressures and were fed by sophisticated turbopumps. Therefore, ballistic missiles became effective carriers of long-range nuclear weapons, but they did not represent the most inexpensive designs—requirements other than minimizing cost had preeminent priority.

TABLE 4-5. Design Heritage of Launch Vehicles. Today's U.S. expendable launch fleet has a strong design legacy with the ballistic missiles of the 50s and 60s.

• Thor IRBM ➤ Delta launch vehicle (core vehicle first stage)
• Atlas E ICBM ➤ Atlas E launch vehicle
• Atlas ICBM ➤ Atlas / Centaur launch vehicle (first stage-and-a-half)
• Titan II ICBM ➤ Titan II launch vehicle
• Titan II ICBM ➤ Titan IV launch vehicle (core vehicle)

Although there have been numerous product improvements along the way, the United States' large, expendable launch vehicles are direct descendants of the liquid propellant ballistic missiles the U.S. Air Force developed in the 1950s. (See Table 4-5.) Manufacturers of the Thor, Atlas, and Titan missiles took advantage of the development costs already sunk into these systems by the U.S. government and simply derived space launch vehicles from the existing IRBM and ICBM designs. Although aerospace companies saved some up-front development expenses by using this approach, the resulting space boosters brought along the ballistic missile's maximum performance and minimum weight baggage. And even more than the ballistic missile, these launch vehicles turned out to be costly systems. Other U.S. launch systems from Saturn to the Space Shuttle, although not directly derived from ballistic missile designs, have contained the same design characteristics of maximum performance and minimum weight [Office of Technology Assessment, 1989].

4.4.2 Cost Liabilities of Expendables and Reusables

There are cost liabilities associated with both expendable and reusable launch systems. The most conspicuous liability of expendable systems is that they can only be used one time. The practice of dumping this expensive hardware in the ocean after only a few minutes of use has clearly been a major factor in the cost of expendable boosters. Consequently, reusable launch systems have been the obvious solution to reverse this practice and lower launch costs. However, the United States' single experience with a reusable launch system has not been positive economically.

The *Space Shuttle* is a partly reusable (and therefore partly expendable) launch vehicle. The *Orbiter* lands on a runway, the *Solid Rocket Boosters* parachute into the ocean and are reused (after extensive rework and recasting of the solid propellant), while the *External Tank* is expended each mission. (See Fig. 4-6.)

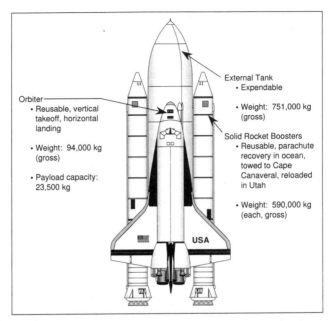

Fig. 4-6. Shuttle Vehicle Components and Their Weights. The Shuttle is a partly reusable, partly expendable launch system. The reusable components use different recovery methods.

Despite the Shuttle's poor economic track record, we must not presume that reusability automatically makes a launch system costly. In fact, it can be the key to cost-effectiveness. But the Shuttle is the only significant reusable launch vehicle that has been operated consistently over a period of years, so it is the best source of real-world data on the cost of reusable space launchers. Therefore, we must use the Shuttle as the primary case study for determining the possible cost liabilities of reusable systems. The Shuttle record highlights some key problems that must be resolved to make any future reusable launch system economically viable. Important issues with the Shuttle's reusability are listed in Table 4-6.

4.4.3 The High Cost of Human Spaceflight

There is no doubt that human presence has provided a major, if not indispensable, benefit to many of the Space Shuttle missions. Astronauts have conducted a variety of complex operations that were either impossible or impractical with automated systems. These have included on-orbit satellite retrieval and maintenance, satellite pre-deployment troubleshooting and repair, comprehensive and detailed life sciences experiments, and space manufacturing and construction activities. All of these operations depended on human interaction. However, very few required that the astronauts trained to carry out these activities had to be carried to orbit by a piloted, heavy-lift booster like the Shuttle. The reason the Shuttle was used to support many of these missions is that it was, and still is, the only operational space launch system for humans available to the U.S., and there is simply no other way to get astronauts into space.

TABLE 4-6. Shuttle Reusability Issues. The reusability methods employed by the Shuttle have a number of liabilities that affect its performance and cost.

> • Horizontal landing requirement caused design complexity; significant weight growth (e.g., wings, control surfaces, landing gear, complex hydraulic system)
>
> • Very low lift-to-drag, parachute-landed designs (e.g., capsules) are much simpler; cause less weight growth
>
> • Numerous landing sites worldwide are required to support launch and landing operations; weather a critical issue at each of these
>
> • Recovery systems are dead weight during launch and orbital operations
>
> • Launch systems are dead weight during orbital and landing operations
>
> • Reusability of Solid Rocket Boosters (SRBs) requires significant and costly logistics infrastructure
>
> • Thicker SRB cases are required because of reusability; this combined with the weight of the SRB recovery system reduces shuttle payload capacity
>
> • External Tank (at $53 million each) expended for each mission

The Space Shuttle was designed to carry both people and payloads. In retrospect, combining these two functions on a single vehicle may not have been the best idea. A piloted launch vehicle design incurs a lengthy and costly list of requirements not present in unmanned boosters. Consequently, establishing whether some future launch system will be piloted or even have the option of being made capable of carrying crews is a critical design decision that carries major cost and operational implications. Table 4-7 summarizes some of the design and operational penalties of manned launch systems.

TABLE 4-7. Design and Operational Penalties of Piloted Launch Vehicles. Although human presence on certain types of space missions can be invaluable, making launch systems capable of carrying a crew causes a number of liabilities that drive up launch costs.

> • People may be required to fly on every mission, regardless of mission characteristics or whether crew support is actually needed
>
> • Reliability requirements must be higher than the already high reliability of unmanned boosters
>
> • Requires many additional subsystems to support crew life support / operations
>
> • Requires suboptimal thrust profiles to limit acceleration loads on crew
>
> • Accidents receive much greater scrutiny, down time stretched out
>
> • Including "man-ratable" option in unmanned boosters adds the human space flight infrastructure to the development team and may complicate the design

4.4.4 Production Influences

Launch vehicles are produced in extraordinarily small quantities. Consider, for example, that one of the higher production rates in the entire launch vehicle industry is for the graphite epoxy motors used as strap-ons by the Delta booster, and the rate is only six units per month. Low production rates are usually caused by low launch rates, which, in turn, are caused by low demand. This is the case with expendable launch vehicles. The reusable Shuttle Orbiter, of course, had such low manufacturing rates (only six were ever built, including the non-spaceworthy *Enterprise*) that economy-of-scale considerations never applied. The low demand for expendable launch vehicles is caused by their own high cost as well as the high cost and low number of payloads. The high cost of payloads, as we will discuss in more detail later in this chapter, is strongly influenced by the high cost of launch vehicles. This situation has created a vicious economic circle that neither government nor the aerospace industry has found a way out of.

Another factor contributing to the low production rates (and high costs) of expendable launch vehicles is that they are not designed to accommodate high, or even modest, rate production runs. Thus, launch vehicles and spacecraft tend to be unique articles requiring a great amount of touch labor. These vehicles are subject to constant modification due to the dominant influence of design engineers throughout the manufacturing and test process. And engineers have a propensity for designing elegant solutions to problems [Keith, 1991]. Unfortunately, design engineers do not always consider the manufacturing, operational, or cost implications of their designs. For example, a Titan IV fuel torus requires more than 180 manhours of direct labor to manufacture. The process also requires many indirect, or supervisory manhours and product assurance manhours. The total manhours for the building of a single fuel torus exceeds 420, and this component is only one of a myriad required for the engine alone [Aerojet TechSystems, 1988].

4.4.5 The Hurdle of High Development Costs

Major aerospace programs usually have very large *development* (one-time non-recurring) costs for designing, developing, and prototyping the first copy of the desired system. Launch systems have historically followed this pattern, although the Space Shuttle is the only new large launch system developed by the U.S. in the last 30 years. Big development costs for new launchers create a significant early hurdle for program proponents seeking to justify their system. In the case of new weapon systems, operational necessity is typically the central issue that program supporters must address, and high development costs are not always a significant issue. However, this is usually not the case with new launch system proposals, nor should it necessarily be.

New launchers, particularly in the post-Cold War environment, must increasingly pass the cost effectiveness test—and high development costs make this difficult. These large initial outlays must be amortized somehow, and within a reasonable amount of time, so the launch system can start "paying its own way." There is a temptation to construct large, speculative, future mission models to allow projections of a high launch rate and rapid retirement of the development debt. In fact, this is one of the traps the Shuttle fell into—the Shuttle's projected large mission model and frequent launch rate never materialized (although the Shuttle's design could not have accommodated a high launch rate anyway).

The high development costs of new launch systems tend to direct decision makers away from new program starts and toward maintaining the status quo. For example, the estimated development cost in 1992 of the National Launch System was $10.5 billion, even after removing the NLS-1 heavy lifter from the initial development plans. For this money, the U.S. would have gotten by the year 2002 only the initial NLS vehicle [Gabris, Harris, and Rast, 1992]. For the same amount of dollars, a payload customer could go out today and purchase 221 launches on the Delta II 7925. This represents almost as many Delta launches as all the Delta missions flown since the program began in 1960 [Isakowitz, 1991]. The existing stable of U.S. launch vehicles is expensive and not very responsive, but they are available at known prices and their development costs have long since been paid for. In the U.S., if Congress is to appropriate dollars for a new launch system, development costs must be significantly lower or proponents must articulate sufficiently compelling justifications for large development expenditures.

One of the reasons development costs are typically very high for large aerospace systems is the considerable amount of new technology, hardware, and software development required to field the system. Further, DoD and NASA program managers have often allowed, or even used, the acquisition of major aerospace systems to serve as a mechanism to advance the state-of-the-art in key technology areas. Although technology advances may be required in many cases to achieve program objectives, these new technologies are costly and managers must minimize their employment. In the case of launch vehicles, the desire for a new booster to have maximum performance and minimum weight will demand certain technology advances and will require "repackaging" of existing systems and components to minimize weight and volume, all of which are expensive propositions.

Large, government-funded aerospace development programs generally have big budgets, are staffed with many government employees, and attract large numbers of contractor personnel not directly involved in production. This government and contractor "oversight" can make development costs spiral upward, not only from labor costs, but also from the prodigious amount of data requests and analytical studies they produce that require additional manhours for the contractor building the aerospace system to respond to. This, of course, is completely contrary to the approach taken by the early Lockheed *Skunk Works* programs, which had minimal government program office personnel and which achieved remarkable success (Johnson's rules for Skunk Works-type programs are provided in Sec. 2.2.7).

Typically, aerospace programs with big development costs will produce operational systems with big recurring costs. This has been particularly true for traditional expendable launch systems. Large development budgets, and the army of people that accompany them, usually build in costly complexity, *bells and whistles* (non-mandatory capabilities), *gold-plating* (over-optimized performance), and excessive oversight and analysis. Reusable launch systems would likely have higher development costs than those for traditional expendable vehicles, but they could have lower recurring costs (which in this case would be their operating costs). However, they would still need aggressive management to avoid the operational pitfalls normally caused by large development budgets.

4.4.6 A Zero Tolerance for Failure

The demand for increasingly greater launch reliability continues to have a major influence on space transportation costs. This pursuit of high reliability is a manifestation of a larger cultural phenomenon: a zero tolerance for failure. Launch systems and their payloads are subjected to exhaustive testing at the factory and at the launch base. Engineers also build in an extensive amount of redundancy and an array of instrumentation for remote monitoring.

Aerospace engineers who design launch vehicles have, as a matter of course, depended on the use of *redundant systems* to achieve reliability goals and enhance confidence in mission success. But redundancy carries a price. Adding redundancy increases a launch system's overall complexity, thus increasing the cost to design, build, operate, and monitor it.

Redundancy means more subsystems, more components, and more interfaces— and this means a larger work force and increased documentation to support these items throughout all phases of the launch vehicle's life cycle. Redundant systems cause a modest increase in the overall weight of a booster and a decrease in its effective payload capacity. Redundancy adds additional systems which increase the number of possible failure modes. This would not be so bad if redundant systems were treated as true backups that were not required unless the primary systems failed to function. Unfortunately, redundant systems are treated as primary systems prior to launch, and numerous launches have endured costly scrubs because a redundant system failed late in the countdown, even though the primary system was operating perfectly and was completely capable of supporting the launch.

Current launch vehicles are complicated systems that can have very complicated failure modes. When vehicle failures occur, they are usually very costly, gut-wrenching experiences. Consequently, engineers have, for years, designed extensive system monitoring instrumentation into launch vehicles for ground testing and in-flight analysis. This instrumentation and the communications equipment to get its data to the ground is the rough equivalent of the "black box" voice and data flight recorders carried on today's large aircraft. In contrast to the aircraft recorders, however, data from the booster monitoring instrumentation is captured, and in many cases analyzed, in real-time by an array of engineers and managers on the ground who closely follow each launch (in the case of the Shuttle, its entire mission). Downrange stations are necessary to provide booster tracking and communications connectivity for telemetry downlinks, and this ability for remote, instantaneous assessment of space boosters is expensive to develop, install, and staff. Even with this capability, there are no guarantees that all failure modes will be identified.

In the aftermath of a launch vehicle failure there is a strong inclination to increase the vehicle's instrumentation to enhance the probability of easier and more direct failure analysis when future failures occur. Such practices tend to put the vehicle into an instrumentation growth spiral, but the complex and fragile nature of current booster designs may not allow any other solutions.

Accountability, traceability, and quality assurance requirements have resulted in a gigantic documentation system and a commensurate number of manhours to create, update, review, and maintain it [French, 1988]. This entire system of detailed oversight is motivated by a general lack of confidence that launch vehicles will perform as planned.

Finally, range safety requirements force the inclusion of an onboard command destruct system on each launch vehicle, and a large ground infrastructure of people and equipment in case they are needed to operate it [Eastern Space and Missile Center, 1990]. Even the piloted Space Shuttle must carry a destruct system, and if the vehicle strays off course it will be destroyed.

4.4.7 Diverse Spacecraft Interface Demands

Spacecraft designs have traditionally placed great demands on the launch vehicle for various services both prior to the flight and during ascent to orbit, consequently forcing many electrical, mechanical, fluid, and structural accommodations from the booster. The Space Transportation Architecture Study, conducted by the Air Force and NASA, identified widely varying spacecraft-to-booster interface requirements as a significant contributor to high launch costs [Space Transportation Architecture Study, 1986].

The Atlas/Centaur provides an example of the kind of impact that numerous spacecraft interface requirements have had on launch vehicles. Over the years, the Atlas/Centaur has sprouted a variety of spacecraft interfaces for power, electrical, command and control, mounting bolt patterns and adapters, and environmental conditioning and ground support equipment. These interface requirements have increased the cost and complexity of payload integration an order of magnitude from what was typical of early launch programs. Integration activities for an Atlas/Centaur-class payload begin at least 36 months prior to launch [Holguin and Labbee, 1988].

4.4.8 Component and System Interface Counts

A highly redundant launch vehicle using complex technology will consist of a large number of parts, each representing a potential failure point. The higher the number of parts, the higher the number of interfaces. The more interfaces that are present (especially external interfaces), the more people that are required, both in the factory and at the launch base [Dergarabedian, 1991]. Because of the intense deliberateness and scrutiny associated with manufacturing and operating launch vehicles, the number of interfaces a booster contains has a major influence on manpower requirements and total cost. Orbital Sciences Corporation designed the Pegasus and Taurus boosters with interface minimization in mind, to allow a smaller crew for integration, checkout, and launch.

4.5 Pros and Cons of Some Key Design Alternatives

4.5.1 Manned versus Unmanned

A key decision that must be made up front about any new large launch vehicle is whether or not it should be designed to carry humans. Assuming the U.S. and other major space powers will continue to be committed to some level of human space flight, this will continue to be a critical issue when seeking to lower the cost of space launch.

Most of the types of missions the Space Shuttle is flying today can be accomplished without the Shuttle—and for much less money—onboard the international space station once it is operational. The Shuttle does provide some unique benefits for certain experiments that would not be available with the space station; for example, a limited ability to customize a mission's orbital characteristics. It also serves as an excellent logistics support platform for those rare repair missions like the one flown to the Hubble Telescope. But once NASA has an operational space station, it may not be able to afford maintaining the Shuttle merely to be a short-term orbital experiment host and occasional repair platform.

The justification for continued use of the Shuttle beyond the deployment of space station components could diminish to being little more than a very expensive and inefficient taxi for astronaut crews, since space station logistics support could be accomplished for less money through the use of existing unpiloted boosters. These prospects provide a strong rationale for developing alternate and much less expensive means for getting crews into space. One possibility is a vehicle similar to the Japanese HOPE vehicle or the HL-20 Personnel Launch System proposed at one point by NASA's Langley Research Center. Another is a capsule-like vehicle for ferrying crews. A third possibility is to design NASA's proposed X-33 reusable single-stage-to-orbit (SSTO) demonstrator to be a dedicated transporter for astronauts—but not cargo—with a minimal on-orbit dwell capability. This would provide a constrained lift and performance requirement for the X-33, and allow it to have a high degree of operational utility that would go well beyond its original purpose as an SSTO test vehicle.

At first glance it would seem obvious that any new large booster should be able to carry both people and cargo, since combining both functions on a single vehicle would mean only one new launch system would have to be developed. However, if our goal is truly to develop and operate a launch system for the lowest possible cost, such a vehicle cannot provide accommodations for astronauts. To do so would ensure higher development costs and a more complicated and costly vehicle to operate. Even if the vehicle has a modular design that would allow it to be configured for either piloted or unpiloted missions, inclusion of the option for human spaceflight could be very expensive. This option would preserve the opportunity to use the booster for crewed missions, but it may also cause the vehicle designers to become captive to the demands of the human spaceflight community. By eliminating any provisions for manned flight, unmanned launch systems could be optimized for minimum cost.

Any launch system capable of carrying humans should be optimized for flight safety, not low cost. This does not mean that a manned launcher has to be expensive, although case histories indicate it will be. Designing the manned launcher to be simple and reliable could enable the vehicle to also be safe and relatively affordable. Some added costs will come about, however, with the incorporation of systems such as ejection seats or emergency escape modules.

4.5.2 Expendable versus Reusable

The issue of expendable versus reusable launch vehicles is a hotly debated topic. Expendable boosters have been around for many years and have had a generally successful track record, albeit at very high cost. The only large reusable launch systems ever built and operated have been the U.S. Space Shuttle and the Soviet (now Russian) Buran/Energia, and these systems were only partly reusable. The Space Shuttle is an engineering marvel with significant and unique mission capabilities, but its operating costs are oppressively high. The Buran flew only once; the Russians have mothballed it for several reasons. Its operating costs were high (by Russian standards), but more significantly, almost any mission the Buran would have performed could also be handled less expensively by the existing fleet of Russian manned and unmanned boosters. Our experience to date with reusable launch systems has not yet provided hard evidence that reusable operations will be more economical than expendable launchers. Therefore, we do not yet know which solution will best reduce launch costs. (See Table 4-8.)

TABLE 4-8. Expendable versus Reusable Comparison. Both types of launch systems have a number of advantages and disadvantages.

	Expendables	Reusables
Flight hardware	Expended during one mission	Reused
Design complexity	Moderate[1]	High[2]
Performance demands	Moderate[3]	High[4]
Development cost	Lower[5]	Higher
Operating and recurring cost	Higher[6]	Lower[7]
Flight testing	Possible, but economically impractical	Yes
Sensitivity of cost to launch rate	Less	More[8]
Opportunities for manufacturing economies of scale	Significant	Minimal
Applicabilitiy for heavy lift	Good	Fair
Launch site flexibility	Low	Potentially high
Range safety requirements	Traditional	Possibly relaxed
Solid and hybrid use	Practical	Impractical

1) Potential for very simple designs. 2) High for multistage designs because of staging complexity; high for single-stage-to-orbit designs because of greater performance demands. 3) Potential to be low for simple expendable designs. 4) Very high for single-stage-to-orbit designs. 5) Based on commercial ventures such as Pegasus and Lockheed Launch Vehicle. 6) For traditional expendables derived from ballistic missiles. 7) Not yet demonstrated; likely never to be demonstrated by mostly reusable Space Shuttle. 8) Sensitivity will reduce in direct proportion to reductions in operating costs.

The debate over which approach is the most cost-effective centers on several key issues. Can a reusable system—with its expected higher development costs—amortize these costs relatively quickly through low operating costs and high flight rates? Or is it less expensive to spend fewer development dollars to field an expendable system that may have higher recurring costs? These questions capture the essence of the debate, but many other factors also come into play in determining what system truly represents the lowest cost approach.

Expendable boosters have been, and continue to be, the mainstay of the world's space launch capability, but are generally cited as a chief cause of high launch costs due to their very nature: they are thrown away after only one use. This operational mode is completely counter to that of what many view as the launch vehicle's closest corollary—the cargo-carrying aircraft. On the other hand, cruise missiles are expendable vehicles that routinely fly one-time, one-way missions, so expendability is not a concept that is unique to space launch systems. (See Table 4-9) It would be possible to design a cruise missile to fly to its destination (the target), deploy its cargo (the warhead), and then return to its launch point to be recovered and refurbished for another mission. However, it was more cost-effective to design the missile to be expended each time it was used [London, Launch Improvement and Realities, 1994].

TABLE 4-9. Examples of Expendable and Reusable Aerospace Systems. There are a large number of both expandable and reusable aerospace systems operational today. The decision to make a system expendable or reusable is based largely on cost and mission requirements.

Expendables	Reusables
• Launch vehicles • Cruise missiles • Laser-guided bombs • Ballistic missiles • Surface-to-air missiles • Air-to-air missiles • Spacecraft[3]	• Space Shuttle[1] • Transport aircraft • Strategic bomber aircraft • Tactical fighter and attack aircraft • Unmanned Aerial Vehicles (drones)[2]

1) Partly expendable. 2) A few UAV designs are expendable. 3) Very long mission life.

For an expendable system to achieve significant reductions over current launch costs, it must consist of hardware that is dramatically cheaper to manufacture and operate than the hardware making up today's expendable launchers. This imperative translates into a booster that has a vastly simpler design, uses fewer parts, and depends on the exploitation of existing technology. Such an expendable would theoretically be not only cheaper to develop, but also cheaper to build and to operate. Only when the hardware of an expendable booster becomes so inexpensive that we don't mind using it only once can an expendable achieve the kinds of low launch costs we need.

It would be possible to build an expendable single-stage-to-orbit (SSTO) booster, although before embarking on a development program we should carefully consider the potential benefits of such a vehicle. One possible benefit would be the demonstration of some key SSTO technologies and operational procedures without going straight to the more challenging development of a fully reusable SSTO. Other benefits might be the elimination of the design complexities of staging and more flexible launch basing due to relaxed range safety constraints. NASA and the Air Force demonstrated this concept to an extent when they orbited an entire Atlas B ICBM (with the exception of the booster engine section) on December 18, 1958 as part of Project SCORE [Holmes, 1962].

Reusable launch vehicles have long been heralded as the solution to high launch costs. The Space Shuttle was originally advertised to be able to place payloads into low-Earth orbit for only $660 per kilogram, but this estimate was based on an assumption of a fully reusable two-stage launch system, a rapid and low-cost vehicle turnaround, a fully-utilized Orbiter payload bay for each mission, and a very large mission model.

The lure of a completely reusable launch vehicle that could attain orbit and return without having to shed any hardware during the flight is a dream that has been beckoning designers since science fiction authors began writing about such vehicles many decades ago. One of the attractive aspects of the approach is that it emulates the operational concept of transport aircraft (i.e., take off—deliver cargo—return, having expended only propellant), and aircraft have certainly proven to be cost-effective and profitable. However, there are some technical difficulties that are common to all types of single-stage-to-orbit launch vehicles.

Assume a designer used similar technologies (i.e., the same propellants and structural fraction) and the same mission requirement to design both a single-stage-to-orbit and a two-stage-to-orbit vehicle. The two-stage-to-orbit vehicle will always have a greater payload-to-orbit capability for a given liftoff weight. We can improve the payload capacity of a single-stage rocket by decreasing its structural fraction or increasing its propulsion efficiency, but this may increase cost and reduce reliability. This is because a single-stage design engineer must seek maximum performance for key systems, and these systems will make the vehicle prone to be less reliable than a simple staged vehicle with its lower performance demands and greater design margins. Designers can increase payload capacity for an SSTO launcher by making the overall vehicle larger, but increasing vehicle size introduces other development and operational challenges. Simple physics dictates that the more stages a booster has, the less important high performance and low weight become, especially for the lower stages [Sackheim and Dergarabedian, 1995].

Although there are obvious technical challenges to any single-stage rocket program, the concept still holds the promise of ultimately becoming a cost-effective system that uses routine, "airline-like," operations to place payloads into orbit at very low cost. However, getting to this point will require a large front-end investment for what must be considered a high risk development program. Recent advances in materials technology notwithstanding, any single-stage booster design's low structural fraction demands will make it difficult, in the foreseeable future, for the vehicle to compete with simple expendable staged rockets for the launching of heavy payloads (15,000 to 75,000 kg).

The keys to making a reusable system cost effective is to make its development costs acceptably low and its operating costs extremely low. The Single Stage Rocket Technology (SSRT), or DC-X, program that was sponsored by the Pentagon's Ballistic Missile Defense Organization (BMDO) provided some important lessons in these areas. The most important aspect of the program was not its demonstration of a reusable vertical takeoff and landing launch vehicle concept. Far more significant was what the SSRT program accomplished for the allocated budget and schedule. The Strategic Defense Initiative Organization awarded a $58.9 million, two-year contract to McDonnell Douglas Space Systems Company in August 1991. The BMDO program office that managed the effort consisted of only a few people. The SSRT team designed, fabricated, and flight tested a liquid oxygen and liquid hydrogen-powered aerospace vehicle incorporating a number of design innovations—all in under two years and for a lot less than $100 million. The program succeeded in part by using very streamlined management techniques, as well as employing existing technology throughout the design and existing hardware and software whenever possible. This streamlining philosophy flowed down into the SSRT's launch operations requirements, which specified a minuscule amount of personnel and ground equipment [Worden, 1993]. These are the kinds of programmatic and operational cultural changes that will be necessary to develop any launch system that achieves a breakthrough in launch cost reductions.

Despite the current popularity of single-stage-to-orbit concepts, *partially and fully reusable multistage vehicles* still deserve serious consideration as future booster candidates. On the negative side, a multi-stage reusable vehicle incurs additional design and operational complexity due to its staging requirements. Fully reusable multistage systems that are "clean sheet" designs will require the development of at least two complete aerospace vehicles, as opposed to only one for SSTO designs. Also, the integrated configuration of multiple stages effectively represents a third unique vehicle that is markedly different from the individual stages [Sponable, 1990]. However, this additional complexity must be carefully traded against the greater demands for lightweight structures and higher efficiency propulsion imposed by SSTO designs. A staged reusable vehicle relieves many of these maximum performance and minimum weight requirements that are typical of SSTO concepts. The first "stage" of a multistage system can even be a completely air-breathing vehicle, so the aircraft-launched Pegasus booster should be considered partially reusable, and the planned X-34 vehicle would be fully reusable.

There are many architectural approaches for reusable launch systems. A fundamental decision required of any designer of reusable boosters is the launch and recovery mode, of which there are numerous options. Table 4-10 compares these options [Koelle, 1991; Sponable, 1990; Truax, 1967].

TABLE 4-10. Reusable Architecture Comparisons. A designer of a reusable launch system is faced with choosing from several vehicle configurations, each of which has unique design and operational characteristics.

	HTHL[1]	VTVL[2]	VTHL[3]	Staged, Parachute[4] Recovery
Use of atmosphere during ascent	Source of oxidizer, aerodynamic lift	No	No	No
Use of atmosphere during descent	Aerodynamic lift, energy depletion	Aerodynamic lift, energy depletion	Aerodynamic lift, energy depletion	Parachute operations
Atmospheric drag and thermal penalties	Severe during ascent, thermal severe during descent	Thermal severe during descent	Thermal severe during descent	Modest during ascent
Midair refuel compatible	Yes	No	Possible	No
Orbital refuel compatible	Yes	Yes	Yes	No
Complex external geometries	Yes	No	Yes	No
Runway dependency	Yes	No	Yes	No
Coastal launch operations required	No	No	No	Probably
Unpowered landing capable	Yes	No	Yes	Yes[5]
Horizontal-to-vertical ground translation required	No	No	Yes	Yes

1) Horizontal takeoff, horizontal landing. 2) Vertical takeoff, vertical landing. 3) Vertical takeoff, horizontal landing. 4) Or other drag inducing device (ballute, inflatable wing, etc.). 5) Parachute.

TABLE 4-11. Propellant Selection Trades. Each of the three basic families of rocket propellants offers a large set of unique characteristics for the launch vehicle designer.

	Solids	Liquids	Hybrids
Impulse density	High	Moderate to low	Moderate
Storability	Good	Limited[1]	Good[2]
Development risk	Moderate	High[3]	Very high[4]
Development cost	High	Highest[3]	Higher[4]
Design complexity	Low to moderate	Very high[3]	Moderate
Testability	None[5]	Good	Limited[5]
Inspectability	Poor	Good	Good for oxidizer tank, poor for fuel[6]
Reliability	Very high	High[7]	Probably very high[8]
Dependability	Very high	Moderate[9]	Probably high[8]
Safety concerns	Significant from factory to launch	Significant after loading on pad	Insignificant
Environmental considerations	Acid rain, ozone depletion, CO_2 increase	Ozone depletion, CO_2 increase[10]	Ozone depletion, CO_2 increase[10]
Throttleability	Predetermined thrust profile only	Good	Good
Thrust termination	Difficult	Simple	Simple
Payload environment	Relatively harsh	Relatively benign	Relatively benign
Ground handling considerations	Difficult	Moderate	Moderate to difficult
Weather considerations	Susceptible to corrosion,[11] narrow range of operating temperatures	Most susceptible to inflight wind transients, wide range of operating temperatures	Susceptible to corrosion,[11] narrow range of operating temperatures
Propellant cost	High	Moderate[12]	Moderate
Manufacturing cost	Low	High[13]	Probably moderate[8]
Engine-out applicability	Impractical	Good if vehicle properly designed	Impractical
Reusability applicability	Very poor	Good	Poor
Propellant aging concerns	Significant	None	Minimal

1) Assumes cryogenic liquids. Hypergolics have good storability, but are highly toxic and environmentally unfriendly. 2) Only for solid component of the propellant combination. 3) Only for pump-fed propulsion systems; moderate for pressure-fed systems. 4) Due to hybrid technology and cost model immaturity. 5) For actual flight system components. 6) Poor inspectability of fuel not as much of a concern due to the hybrid system's relative insensitivity to grain defects. 7) Only for pump-fed propulsion systems; very high for pressure-fed systems. 8) Projected. 9) Dependability increases with increasing system simplicity. 10) Assumes no toxic hypergolics. 11) Especially for solid booster segment field joints. 12) Assumes LOX/hydrogen. LOX/RP-1 is low; hypergolics are rated high. 13) Only for pump-fed propulsion systems; low to moderate for pressure-fed systems.

4.5.3 Solids versus Liquids versus Hybrids

The type of propellants used in designing a new booster is a key factor in the resulting development, manufacturing, and operating costs. Traditionally space launch systems have used either liquid or solid propulsion systems or some combination of the two. Lately, hybrid propellant concepts have matured sufficiently to be considered as a valid propellant alternative. A *hybrid propellant* combination typically consists of a solid and a liquid constituent—in most cases, a solid fuel and a liquid oxidizer. Table 4-11 lists some specifics of the various trades between solids, liquids, and hybrids. [Andrews and Haberman, 1991; Cook, et al., 1992; McDonald, 1992]

4.5.4 Liquid Engine Cycle Options

It is generally understood among rocket propulsion engineers that, for engines requiring high thrust and long duration operations, *turbopump feed systems* allow lower vehicle weight and higher performance when compared to pressurized gas feed systems. There are several classical pump-fed engine power cycles. The gas generator cycle (and other "open" cycles) is the least complex of the pump-fed cycles, but it also generally delivers the lowest performance. An *open cycle* denotes that the working fluid used to drive the turbine is exhausted overboard, or discharged into the engine well downstream of the nozzle throat. Examples of gas generator cycle engines are those used by the Atlas and Titan launch vehicles. The expander cycle and the staged combustion (or preburner) cycle represent pump-fed systems having a more complex "closed cycle," but they provide higher performance. A *closed cycle* denotes that the exhaust from the working fluid that drives the turbomachinery is injected into the engine combustion chamber to take advantage of its remaining energy. An example of an expander cycle engine is the Pratt & Whitney RL10, and the Space Shuttle Main Engine uses the staged combustion cycle [Sutton, 1992]. Regardless of the type of turbopump engine cycle chosen, when it comes to designing liquid boosters for maximum performance and minimum weight, there is little doubt that turbopump-fed propulsion systems are the way to go.

TABLE 4-12. Liquid Engine Cycle Comparison. Turbopump-driven engines have several classical cycles, but in this generalized comparison they are simply identified as "gas generator cycle" or "complex cycles." The pressure-fed cycle is notably different; it depends on pressure as opposed to turbomachinery to deliver propellants to the engine combustion chamber. Note use of English units.

	Pump-Fed		Pressure-Fed
	Gas Generator Cycle	Complex[1] Cycles	
Vehicle weight[2]	1.25	1.0	1.75
Chamber pressure	800 to 1,500 psi	2,000 to 4,000 psi	250 to 300 psi
I_{sp}[3]	300 (LOX / RP-1) 400 (LOX / LH$_2$)	320 (LOX / RP-1) 430 (LOX / LH$_2$)	280 (LOX / RP-1) 380 (LOX / LH$_2$)
Tank wall thickness[2]	1.0	1.0	2.0 to 2.5
Tank pressurization system required	Yes, low pressure system	Yes, low pressure system	Yes, high pressure system
Tank pressure	30 to 50 psi	30 to 50 psi	500 to 700 psi
Vehicle complexity[2]	2.5	4.0	1.0
Engine component part count	2,000 to 3,000 major component parts	5,000 to 7,000 major component parts	100 to 200 major component parts
Manufacturing complexity[2]	2.0	3.5	1.0

1) Staged combustion and expander cycles. 2) Normalized dimensionless units. 3) Average I_{sp} throughout flight, in seconds.

Alternatively, *pressure-fed propulsion systems* offer the potential for greatly simplified designs of engines and overall vehicles [Huzel and Huang, 1967]. Pressure-fed propulsion is most applicable to expendable launch vehicle designs. Pressure-fed systems are almost certainly impractical for single-stage-to-orbit concepts, but they could be used in parachute (or other drag-inducing device) recoverable strap-on booster applications or in parachute-recovered lower stages of a multistage reusable launcher. Table 4-12 compares pump-fed and pressure-fed engine cycles. Figure 4-7 shows some examples of pump-fed and pressure-fed engines.

Fig. 4-7. Examples of Pump-Fed and Pressure-Fed Engines. Clockwise from top left: the pump-fed Space Shuttle Main Engine, consisting of 70,000 piece-parts; Pratt & Whitney's standard pump-fed RL10 (left) and the pump-fed RL10A-5 used by the DC-X Single Stage Rocket Technology demonstrator; TRW's pressure-fed Lunar Module Descent Engine; and the pressure-fed thrust chamber assembly for Aerojet's Delta second-stage engine

The primary justification for the use of turbomachinery has been that turbopumps significantly increase the delivery pressure of the propellant as it is being routed to the rocket engine. The pressure of the propellant delivered to the engine combustion chamber injector is one factor that establishes combustion chamber pressure—and combustion chamber pressure is an important element in determining a rocket engine's specific impulse (the higher the pressure, the higher the specific impulse) and physical size. Engineers seeking compact, high performance engines want high chamber pressures with the attendant high performance turbomachinery. Using turbo-pumps to boost propellant pressures makes it possible to keep pressures in the large propellant tanks relatively low, allowing the tanks to have thin structural skins and be lightweight. The launch vehicle is intended to perform a task that is in direct opposi-tion to the force of gravity, so vehicle designers have traditionally sought to keep the weight of the vehicle as low as possible. The use of turbomachinery has been key to this effort.

A launch vehicle that uses tank pressurization as opposed to turbomachinery to deliver propellant at the appropriate pressure to the engine combustion chamber injector is called a pressure-fed booster. The propellant must be pressurized by a high-pressure gas source or some other mechanism to a level that exceeds the required pressure at the combustion chamber injector. Compared to pump-fed boosters, a pressure-fed design exhibits structurally stronger (usually thicker-walled) propellant tanks, as well as engines that operate at lower chamber pressures (and have a lower specific impulse). Consequently, a pressure-fed booster with the same payload ca-pacity to low-Earth orbit as a comparable pump-fed vehicle would be heavier and would need larger engines with greater thrust to compensate for the heavier dry and gross vehicle weights and the lower engine efficiency [Allman, 1987]. Typical *propellant fraction* values for pressure-fed designs can be around 0.87; structurally lighter pump-fed vehicles normally have greater propellant fraction values that can be 0.92 or higher. Table 4-13 lists propellant fractions for some existing launch systems, although these vehicles depend for the most part on solids or pump-fed liquids.

TABLE 4-13. Propellant Fraction—Current Vehicles. These vehicles depend mostly on solids or pump-fed liquids. The Atlas IIA has the highest propellant fraction of the vehi-cles listed, partially due to the unique tank design—a thin skin balloon—which it inherited from its ICBM ancestor.

Vehicle	Propellant Fraction[1]	Vehicle	Propellant Fraction[1]
Pegasus	83.7%	Space Shuttle	85.8%
Delta II 7925	91.1%	Vostok	88.9%
Atlas IIA	92.2%	Proton	89.2%
Titan IV	87.5%	Ariane	89.6%

1) Propellant fraction = propellant weight divided by gross liftoff weight (less payload).

In a mid-1980s study conducted for the Air Force Astronautics Laboratory on low-cost expendable launch vehicles, pressure-fed boosters were eliminated from further trade study consideration. The study's analysis indicated that the booster's heavier tanks and engines, and the anticipated complexity of its pressurization system, would make it more expensive than a pump-fed booster with the same payload capacity [Dyer, et al., 1987].

TABLE 4-14. Differences Between Transport Aircraft and Launch Vehicles. This table lists some of the dramatic differences between transport aircraft and expendable launch vehicles. These differences are part of the reason an expendable booster does not have to follow the aircraft design paradigm of maximum performance and minimum weight.

	Transport Aircraft	**Traditional Expendable Launch Vehicles**
Mission duration	Hours	Minutes
Flight dynamics	Accelerate to cruise, decelerate to land	Constant acceleration
Operating life requirements	Years to decades	Less than 10 minutes
Propellant	Fuel only; storable liquids	Fuel and oxidizer; cryogenics, solids, or toxic storable liquids
Primary flight mechanism	Aerodynamic lift; horizontal flight	Positive thrust-to-weight; vertical flight
Operating speeds	0 to Mach 0.8	0 to Mach 25
Operational ceilings	12 km	435 km
Takeoff mode	Horizontal—runway	Vertical—launch pad
Landing mode	Horizontal—runway	None
Crew required	Yes	No
Tare-to-load penalty ratio[1]	1:1	1:1 for uppermost stage, 48:1 for lowest stage[2]
External profile	Complex geometries, control surfaces	Simple cylinders & cones

1) The ratio represents the amount of payload (load) reduced for a given increase in gross vehicle (tare) weight. For staged vehicles, there is much less of a penalty to the payload when the weights of the lower stage increase. The penalty increases to 1:1 for the uppermost stage. For aircraft (and single-stage-to-orbit launch vehicles) the penalty is always the worst case 1:1. 2) Assumes 3-stage vehicle.

Another study conducted for the Advanced Launch System (ALS) program had similar findings. The study stated that in addition to having heavier structure and less efficient engines, pressure-fed boosters raised questions about scalability due to pressurization and combustion stability concerns, and were not amenable to having an engine-out capability. The study went on to say that "the lowest total systems cost is strongly influenced by the dry weight. Invariably, the lowest weight produced the lowest system cost by requiring less structure and propellant and smaller engines and facilities." For all of these reasons, but especially because the projected structural costs of the heavier pressure-fed design were more than twice the structural costs of competing pump-fed designs, pressure-fed engines were eliminated from consideration for the study's ALS reference design [Stofan and Isakowitz, 1989].

In both of these studies, the primary basis for determining that pressure-fed systems would be more expensive than pump-fed systems was that overall system weight is a decisive cost factor. Many cost-estimating models for aerospace systems are predicated on system dry weight being a primary cost determinate, which helps to explain the strong bias in the aerospace community toward designing minimum weight systems.

The aircraft industry has always put a premium on minimizing vehicle weight. Aerospace historian Richard Smith said that "weight is at the heart of every airplane's purpose and problems, its success, assignment of mediocrity, or condemnation to failure." The obvious goal for aircraft designers is to maximize the percentage of *useful load* (such as cargo, passengers, or expendable armament) relative to the gross takeoff weight of the aircraft. Normally this is accomplished by minimizing the weight of the aircraft structure, engines, and other components [Smith, 1986].

When some Aerospace Corporation engineers began studying launch system cost drivers in the 1960s, they arrived at a startling conclusion: contrary to the minimum weight imperative of aircraft design, the weight of a launch vehicle and its propellant was not so critical as long as your primary goal was to design the lowest cost vehicle possible [Tydon, 1970]. Table 4-14 highlights some of the reasons for this conclusion by comparing transport aircraft with traditional expendable launch vehicles.

For launch vehicles, low weight does not necessarily equal low cost. And low cost, not high performance and low weight, should be the dominant design selection criteria. As an analogy, consider the use of speed boats to haul coal up and down the Mississippi River. You obviously would have a transportation system that has high performance and low weight, but it is not the most efficient way to carry coal. Traditional aerospace cost estimating techniques assume the weight of the vehicle is a major cost driver. However, for launch vehicles, especially expendable designs, weight is less of a cost driver than most other physical parameters. Table 4-15 cites two examples that refute the notion that vehicle size or gross mass has a dominant and directly proportional influence on its cost.

TABLE 4-15. Vehicle Cost and Size Relationships. Traditional aerospace cost models assign a direct linear relationship between the size and weight of a vehicle and its cost. This table provides two examples, one historical and one current, that illustrate the opposite result from what the cost model would be expected to predict. In the case of these and many other aerospace system examples, size and weight have a secondary, if not minor, influence on overall system cost.

	Cost ($M)	Length (m)	Gross Weight (kg)
Thor IRBM	$19.0	18.30	78,000
Agena B	$21.9	7.25	6,240
Delta II 7925	$47.5	38.10	230,000
Inertial Upper Stage	$100.0	5.20	14,769

The preeminence of weight-driven cost models in the aerospace arena is a symptom of the larger problem of over-reliance on cost prediction models that are subject to very large errors. Cost predictions have a place in an aerospace system's development, but program managers must use them judiciously and with an awareness of their frailty. See Chap. 8 on cost modeling for a detailed discussion of this topic.

The initial thickness (and cost) of propellant tank raw material for a pump-fed design has a good chance of being greater than that of a comparable pressure-fed design. Often, pump-fed tankage is designed with isogrid "waffle" ribs for stiffness, so the raw tank stock must be milled down to the required lighter gage at a large additional expense. For example, the propellant tanks of the Saturn V third stage required special milling equipment during their fabrication. In an effort to reduce weight and maintain structural rigidity, the manufacturer carved out 7.5 cm^2 waffle recesses on the inside surfaces of the tanks [Bilstein, 1980].

The use of composite materials for aerospace applications has become increasingly widespread. This may be an enabling technology for the practical application of pressure-fed propulsion systems with large space launch vehicles. Composites could allow liquid propellant tanks to be of sufficient strength and light enough in weight to enable the employment of ultra-simple, pressure-fed engines [Microcosm, 1992].

If a pressure-fed vehicle is designed to be water-recoverable, its stronger propellant tank structure affords a less-than-obvious benefit: locking up the residual tank pressure prior to water impact would make the vehicle structure very strong. It would also preclude water intrusion to the vehicle interior, which would simplify the recovery and refurbishment process. This is important because salt water can be a major source of contamination [Truax, 1967].

TABLE 4-16. Liquid Propellant Comparison.

	LOX / LH$_2$	LOX / RP-1	N$_2$O$_4$ / MMH	H$_2$O$_2$ / JP-5
Performance	Excellent	Good	Good	Good
Density	Low for LH$_2$ requiring larger tanks	Moderate	Moderate	Moderate
Operational considerations	Cryogenic operations required; LH$_2$ extremely cold & prone to leakage; ground storage dewars required	Cryogenic operations and ground storage dewars required for LOX	Numerous safety precautions cause operational complexity	Numerous safety precautions cause operational complexity
Environmental considerations	Exhaust product is steam	Hydrocarbon exhaust products	Propellants are highly toxic, environmentally unfriendly	Hydrocarbon exhaust products, diluted
Safety considerations	LOX creates a fire hazard; LH$_2$ an explosion hazard	LOX creates a fire hazard	Highly toxic; vapor scrubbers or burn-off required; personnel must wear protective suits	H$_2$O$_2$ creates a fire hazard; protective suits may be required
Storability	Limited	Limited for LOX; good for RP-1	Good	Good
Cost	Moderate	Low	High	High
Reliability	Good	Good	Excellent; spontaneous ignition	Good to excellent

Launch vehicle designer Arthur Schnitt said of traditional expendable booster design practices that "we were designing every stage as if it went into space. For the top stage, which is small and extremely valuable, minimum-weight designs made sense. For the lower stages it was nonsense. Why spend millions on high-efficiency engines when you could substitute a less efficient engine and simply make it bigger?" Pressure-fed systems may be the key to these simple and low cost lower stages that could help drive launch costs down [Easterbrook, 1987].

4.5.5 Liquid Propellant Choices

For liquid propulsion systems, the choice of propellant combinations is particularly critical to the vehicle's complexity, performance, size, and cost. The employment of liquid oxygen (LOX) as an oxidizer and liquid hydrogen (LH_2) as a fuel has long been recognized as one of the highest practical energy combinations possible for chemical propulsion. But this combination also has some liabilities, and other combinations need to be considered. Table 4-16 provides a comparison between four of the better known liquid propellant combinations. Figure 4-8 compares the performance of LOX/LH_2 with LOX/RP-1, illustrating the trade between LH_2's higher performance and RP-1's higher density. Although LH_2 provides much greater specific impulse than RP-1, this higher performance is offset somewhat by LH_2's lower density and the resultant higher structural weight.

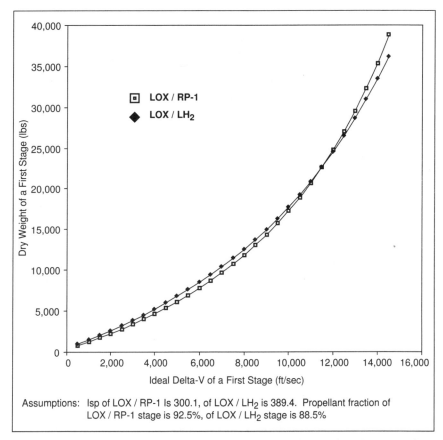

Assumptions: Isp of LOX / RP-1 Is 300.1, of LOX / LH_2 is 389.4. Propellant fraction of
LOX / RP-1 stage is 92.5%, of LOX / LH_2 stage is 88.5%

Fig. 4-8. Stage Dry Weight Versus Velocity. Most concepts for new launch system designs call for the use of liquid hydrogen (LH_2) because of the high I_{sp} it can deliver. However, the low density of LH_2 requires a larger propellant tank than other fuels, which effectively erodes much of the performance gained by its higher energy. In the figure, increases in ideal Delta-V correspond to an increase in dry weight. This is due to greater dry weight resulting in larger capacity propellant tanks. Note use of English units.

4.6 Some Possible Design Approaches for Cutting Launch Costs

4.6.1 General Considerations

Making any significant inroads toward the reduction of space launch costs is constrained by a number of factors. In the case of currently operating boosters, attempts to reduce their launch costs through means such as streamlined operations will be limited in impact since the vehicle design largely dictates its operating costs. Design modifications short of starting with a clean sheet vehicle will also have limited success in reducing cost, especially if the development costs for the modifications are accounted for and amortized properly into the recurring cost of launch.

Design considerations play a dominant role in launch vehicle costs. For example, the decisions to base the Delta booster on the Thor ballistic missile and to make the Shuttle a piloted launcher that lands like an airplane were early, top-level, design choices. Historical data indicates that 70% of the entire life-cycle costs of DoD weapon systems are fixed during the concept exploration phase of development. A NASA study conducted subsequent to initial Shuttle operations stated that a launch vehicle's configuration has a dominating influence on launch processing requirements and personnel head count, as well as on life cycle costs. The study found that simplicity was the key to reducing launch costs [NASA, 1988].

The manufacturing process for a launch vehicle is also driven by its design. Design considerations that directly affect manufacturing include the relative complexity of the design, the types of structural material and parts that will be used, and how much the design will push the state-of-the-art. The number of units produced each year strongly affects manufacturing and cost—and design choices affect component manufacturability and the amount of component commonality, which ultimately influences production rates.

The development of a launch system designed for low cost may not be possible using the aerospace industry's traditional methods of design and manufacturing. The aerospace industry must take a revolutionary approach to addressing the problem of high launch costs, and government managers must be innovative in seeking industrial sources and methods for designing and building low-cost boosters.

One method for providing a new launcher that incorporates unorthodox design approaches and could cost significantly less is to establish a vision for a rugged space truck, much like the Air Force wanted the original Advanced Launch System (ALS) to be. Program managers for the ALS spoke of cultural changes needed to develop an inexpensive booster that spanned the entire spectrum of the system's life cycle from concept development through mature and stable operations. Development of such a system will necessarily **require** a clean sheet approach. To simply incorporate existing high cost hardware and infrastructure into a vehicle design modification would void a program manager's ability to establish true changes in design, manufacturing, and operational methods. It would also ensure the perpetuation of the current development and operating practices that are at the root of high launch costs. The clean sheet effort must have an affordable development budget, and the current wisdom says this will be difficult if not impossible. The current wisdom, however, is based on previous launcher development experience, so cultural changes in how we develop launch vehicles will be essential to bringing about drastically lower design and manufacturing costs.

An example of an unorthodox new booster design that attempted to drive down the cost and complexity of access to space was Orbital Sciences Corporation's (OSC) Pegasus. OSC sought to penetrate the Scout market, and captured 100% of this market niche. Although the cost of a Pegasus flight on the basis of dollars-per-kilogram to orbit is still high, the development of the booster has provided some notable lessons for building low-cost launchers. Key to OSC's commercial development of the Pegasus air-launched booster was low development costs. These low costs were enabled by the use of off-the-shelf and relatively inexpensive subsystems and software and, most importantly, a commercial development environment that was free of government involvement.

Any clean sheet launch system design must have an affordable, nonrecurring development budget that is markedly lower than typical booster development costs of the past. The high front-end development costs that are normally part of any new launch system proposal create a fiscal roadblock that results in lots of concept false-starts and no new boosters on the launch pad. Defining a booster concept that has acceptably low development costs is critical to achieving a launch vehicle new start within the U.S. anytime in the near future, so new vehicle planners must focus on this parameter as a first order requirement.

What amount of nonrecurring cost is acceptable for a new launch system? Based on the NLS experience, $10 billion is too high. NASA is hoping to be able to stimulate development of an operational single-stage-to-orbit vehicle with a useful payload capacity for around $5 billion. Most of the funding to develop this new launcher is expected to come from the aerospace industry in the form of commercial investment money. The Air Force is budgeting around $2 billion for its Evolved Expendable Launch Vehicle program, but the resulting boosters will not be clean sheet designs. The NASA and Air Force programs have received initial concurrence from Congress to proceed, but the big annual budgets for these programs are in the future. Given the likelihood of increasing fiscal pressures, it is not known if the dollars will be appropriated to carry these programs to conclusion. Generally, proposals for clean sheet expendable launchers that have development costs budgeted for less than $1 billion are discounted as not being credible by the majority of government and aerospace industry analysts. However, some smaller aerospace companies are developing innovative and unorthodox (but not necessarily cutting-edge technology) concepts with nonrecurring costs considerably less than $1 billion; some are even less than $100 million.

There are a number of specific steps that launch vehicle designers can take to achieve dramatically lower launch costs. These steps may be viewed as somewhat unorthodox compared to conventional launch system design practices. One of the reasons these approaches may appear unconventional is that they emphasize designing for low cost rather than for high performance and low weight. Although some aspects of these approaches are applicable to any type of new launch system design, they provide the maximum utility to expendable booster designs and staged reusable system designs that depend on parachute recovery for major vehicle elements.

4.6.2 Optimize for Minimum Cost

A promising approach to achieving big reductions in launcher development costs is to move away from the design philosophy of maximum performance and minimum weight and embrace a minimum cost design approach like the one proposed by The

Aerospace Corporation in the 1960s. The fundamental premise that the Aerospace concept rested upon was that by using a clean sheet design approach, a space launch vehicle could be optimized for minimum cost, instead of maximum performance and minimum weight [Bleymaier, 1969; Schnitt and Kniss, 1968].

The key to the *design for minimum cost* methodology is to optimize vehicle cost and weight without compromising quality or reliability. The proper application of technology is very important, and the minimum cost methodology must be applied throughout the entire life cycle of the system, including both nonrecurring and recurring costs. This type of approach requires that, starting with the conceptual phase, all disciplines—design, manufacturing, launch support, quality assurance, facilities, etc.—work in close concert. Above all, the aerospace design engineer, who has been traditionally taught to be minimum-weight-oriented, must become knowledgeable of costs down to the component level [Tydon, 1970]. The concept of designing a launch vehicle for minimum cost has been studied by government agencies and the aerospace industry several times over the years, and the results have consistently indicated that major reductions in launch costs are available using this technique. For a more detailed discussion on the benefits of the design for minimum cost methodology, see London's *LEO on the Cheap* [1994].

4.6.3 Reduce the Part Count

The total number of parts contained in a given launch vehicle is a major cost driver, so minimizing the part count can go a long way toward reducing costs. Engineers assigned to the Advanced Launch System program worked hard to simplify the manufacture of key booster components and to reduce the number of parts. They understood these to be key to lowering manufacturing costs. This was not a new concept, however, but had been recognized for some time.

During the 1960s, studies were conducted for the Air Force Space and Missile Systems Center that indicated a low-cost launch vehicle could be developed with the same payload capacity (and with a heavier overall booster weight) as the Saturn V, but with nonrecurring and recurring costs that were five-and-a-half times less. These cost reductions would have been enabled by the simplified low cost vehicle's radical reductions in the cost of research and development, testing, and the required management of interfaces (since the interface count would be greatly reduced). The cost of *direct labor* (engineering, fabrication, assembling, testing, procuring, and documenting) and *burden* (overhead labor, capital equipment, facilities, and paid absences) would have been reduced tenfold. In the case of the Saturn V, the cost of materials and propellants was only three percent of the total system cost, so the higher weight of the low cost vehicle (relative to the Saturn V) would not have come close to overwhelming its cost advantages in other areas [Dergarabedian, 1991].

4.6.4 Increase Simplicity and Margin

An alternative to redundancy for achieving increased reliability is simplicity of design, coupled with more robust design margins. This is typically the kind of design solution a minimum cost design criteria would provide [TRW, 1981]. This is not the typical design philosophy for boosters today, so a change in thinking would be necessary if this approach were to enjoy widespread application by launch vehicle engineers.

Simplicity of design would reduce the number of subsystems, components, and interfaces—and the size of the work force and the amount of documentation needed to support them. The benefits of this simplification would flow down through all aspects of the launch vehicle program. Simplification, combined with large design margins, would increase confidence in the system's performance, and would allow increased use of "single-string" design practices and a reduction in testing requirements. Simplification would reduce the number of potential failure modes, and increased design margins would decrease the probability of failure in those that remained. Selective redundancy could still be incorporated into the launcher's design, but only in limited areas and in a very prudent manner.

4.6.5 Reduce Instrumentation

Current launch systems are highly instrumented machines that provide a withering amount of prelaunch and flight data to an army of technical personnel through sophisticated telemetry receiving and computational equipment. One school of thought supports this same kind of extensive vehicle instrumentation for future launch vehicles so that system downtime can be minimized if a failure occurs [Vice-President's Space Policy Advisory Board, 1992]. This approach has some merit, but program managers must carefully weigh its benefits against the added complexity, manpower overhead, and cost that such an approach requires. Designers may be able to decrease system monitoring by developing simple, forgiving vehicles with robust design margins.

Launch vehicles with simple designs and large design margins would allow a reduction of vehicle instrumentation and the ground systems and support staff required to receive, store, and analyze the data coming from it. This reduced instrumentation would also lead to savings during the vehicle's design and manufacture. These reductions are enabled by design simplifications that reduce the total number of vehicle systems and subsystems the booster has and, therefore, the total number that are available to be monitored. The larger vehicle design margins would also allow reduced instrumentation requirements by lowering the failure probability of booster systems and components, which would presumably cut back on the need to monitor them. The scaling back of vehicle instrumentation and the amount of data available for analysis would represent a major change in the way launch vehicles have been processed almost since their inception. Procedures for data analysis could be changed so that, even though all appropriate data is captured for each flight, detailed analysis would only be accomplished if there was some type of mishap or significant anomaly. This could result in huge personnel and infrastructure savings.

Range safety policies and practices that were initially developed during the 1950s could be reviewed for descoping in light of the reliability of current launch systems, expected improvements in the future, and the possibility of inland operation for fully reusable launch vehicles or SSTO expendables. At some point, we may be able to develop enough confidence in our launchers that many of these range safety practices become either autonomous to the launch vehicle, or even unnecessary.

4.6.6 Emphasize Manufacturing

Low-cost boosters that provide less expensive access to space would fuel an increase in the payload market and, thus, demand for the launchers. The change in emphasis from engineering toward manufacturing during the development process will allow the booster design to be amenable to large production runs and greater economies of scale. Although the market demand for launches may have to "catch up" before the benefits of these manufacturing-oriented design features come to full fruition, a lack of manufacturing foresight will negate the ability of a booster's production rate to expand to meet increases in market demand.

For the booster design to take full advantage of a minimum cost design criterion, the vehicle must be designed and manufactured to commercial standards, using commercial—and not government and aerospace industry—specifications, tolerances, and practices. A classic example of this approach is the tremendous cost and performance success TRW experienced when the company designed, built, and tested an ultra-simple liquid engine in the late 1960s that had been fabricated to commercial standards.

The simple design of the pressure-fed Lunar Module Descent Engine (LMDE) prompted the engine's manufacturer, TRW, to initiate a study that would apply a similar design philosophy to a larger rocket engine. TRW employed design- for-minimum-cost principles in designing a simple pressure-fed engine that would ultimately demonstrate a maximum thrust of 1,112,000 N. They contracted with a Gardena, California, commercial pipe and boiler fabricator to build the engine to "shipyard production tolerances." The manufacturing cost of the entire propulsion assembly was $33,300, and the engine was built in about two months. Ablative liners were later added for an additional $62,175. This engine was successfully tested at full thrust at the Air Force Rocket Propulsion Laboratory. (See Fig. 4-9.) Its total part count was at least two orders of magnitude lower than large pump-fed engines with similar thrust levels [Easterbrook, 1987; Elverum, 1973].

Fig. 4-9. TRW's Pressure-Fed Engine Being Test-fired at Edwards AFB in the Late 1960s.
The engine, complete with ablative liners, cost $95,475 and demonstrated 1,112,000 Newtons of thrust.

The common argument against the use of commercial standards for space boosters is that no self-respecting owner of the large satellites worth hundreds of millions of dollars that are typical today would ever let their spacecraft ride on top of a booster built to these standards. This argument implies that launch vehicles built using commercial practices would not be as reliable as current boosters. But as discussed already, the simplicity and design margin robustness of these simple boosters may be able to more than offset any possible liability incurred by not building to aerospace standards. Still, simple boosters designed for minimum cost may have to fly successfully a number of times before owners of expensive payloads develop confidence in them. In the meantime, these types of boosters may be forced to fly less expensive payloads. We will discuss at the end of this chapter, however, that availability of cheap access to space provided by such launch vehicles may enable the development of a completely new breed of inexpensive, yet highly capable, satellite systems.

For new expendable vehicle concepts, the opportunity exists to design the booster to accommodate large production runs and manufacturing economies of scale. The Russians have, either by superb planning or fortuitous necessity, designed many of their expendable boosters in a fashion that allows significant manufacturing economies. An excellent example is the *RD-107 engine*, which has enjoyed widespread use on the Vostok class of boosters for many years.

Although the four-chamber feature of the RD-107 engine caused the Vostok strap-ons to be wide at the base and heavier overall, the multiple-chamber design likely has provided an important benefit to engine production. By having to produce four smaller thrust chambers per engine instead of one large one, combined with the high production requirements for all Vostok subsystems, the Russians have enjoyed tremendous economies of scale by Western booster and manufacturing standards [Glushko, 1973; Keith, 1991].

Each Vostok booster uses four identical RD-107-powered strap-ons, as well as a sustainer core which uses a slightly modified version of the RD-107 engine called the RD-108. Consequently, each booster requires 20 identical thrust chambers [Glushko, 1973]. The Russians have launched well in excess of 1,500 Vostok-type boosters over the years, and this large number combined with the modular and high-commonality nature of the Vostok design has allowed high production rates of vehicle components. In fact, for more than 20 years, the Russians have averaged a delivery every workday of at least one turbomachinery assembly, one strap-on or sustainer core section, and four engine thrust chambers.

It is important to remember that there can be potential liabilities to multi-chamber engine designs. For a four-chamber engine, each chamber may require a reliability four times as great as a single chamber four times as big. Multiple chamber designs that use propellants which do not ignite spontaneously can have additional problems. If one of the chambers failed to ignite, it could load up with unburned propellant and be lit from an adjacent chamber, with potentially disastrous consequences.

On the other hand, thrust vector control system designs can be simplified by the availability of multiple thrust chambers. For example, we can control pitch, yaw, and roll on a four-chamber design by moving each chamber about only one radial axis. Also, a designer can provide significant control authority through the use of simple thrust vector control techniques such as variable thrust magnitudes. Therefore, we must carefully trade the benefits and liabilities of multichamber designs to determine the best overall design, manufacturing, and operational solution. Multichamber designs may be best suited for robust vehicle configurations with wide design margins.

Figure 4-10 compares the high commonality Vostok design with the U.S. Delta II and Titan IV. The *Vostok/Molniya SL-6* uses a LOX/kerosene propellant combination for its strap-ons, core first stage, second stage, and third stage. Each of the *Delta II 6925* vehicle's three stages is a completely separate design with its own unique propulsion system. There is no commonality between engine designs or propellants. The 6925 also employs a completely separate (and different) system of solid propellant strap-on boosters. If the Delta were designed today from a clean sheet of paper, it would require separate, dedicated development efforts for each of the stages and propulsion systems. The *Titan IV* with a *Centaur upper stage* has four unique stages (counting the solid strap-ons and the Centaur), and the Shuttle uses three separate and unique propulsion systems to achieve orbit. Each of the three U.S. launch vehicles uses three of the major (and different) classes of chemical propellants: cryogenics, hypergolics, and solids. The designers of these vehicles were not trying to maximize system commonality, but commonality can be an enabling technique for reducing development and manufacturing cost.

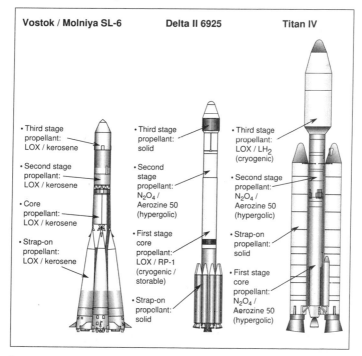

Fig. 4-10. **Commonality Comparison of Launch Vehicle Propulsion Systems.** The Vostok/ Molniya SL-6 booster has significant propellant and system commonality when compared to U.S. boosters. The SL-6 uses LOX and kerosene all the way up the stack, whereas the Delta and Titan use various combinations of solid, hypergolic, and cryogenic propellants. The SL-6 uses virtually the same engine on its core vehicle as well as on each of its four strap-ons, resulting in 20 identical thrust chambers per vehicle. System commonality like that exhibited by the Russian booster can be enabling in driving down design, manufacturing, and operating costs.

4.7 Booster and Spacecraft Cost Relationships

When confronted with the issue of high launch cost, some government and industry managers have contended that the real problem is not the excessive cost of launchers but the extreme expense of the spacecraft that fly on them. There is certainly a case to be made for this position because spacecraft sometimes exceed the cost of their launch services several times over. However, it is important to closely examine some of the root **causes** of the high cost for satellites, because the high cost of their launch has a major impact on the design, and hence the cost, of spacecraft. A fundamental influence that expensive boosters impose on a spacecraft's design is intense pressure to make every kilogram of the spacecraft pack as much capability and performance as possible.

Space-grade hardware must be very light weight, due to the price for space transportation to low-Earth orbit starting at about $10,000 per kilogram. Consequently, this light weight hardware tends to be very fragile. Yet, the Office of Technology Assessment (OTA) said that if payloads were allowed to be much heavier, manufacturers could avoid the costly exercise of removing every gram of nonessential structure. Also, heavier spacecraft designs could enable a reduction of the expensive analyses and tests normally required for assuring the adequacy of fragile spacecraft structures. *Off-the-shelf subsystems*, which are available in quantity at an economical price, could be used in place of costly *customized components* designed for minimum weight [Office of Technology Assessment, 1990].

Of course, the implementation of such cost-cutting ideas for spacecraft must be coincident with the availability of very inexpensive access to space, which does not currently exist. A family of launch vehicles that offered a large payload capacity at greatly reduced costs would provide high leverage for reducing total mission costs.

TABLE 4-17. Benefits of Spacecraft Designs With Wide Weight and Volume Margins. If launchers were available at greatly reduced cost, spacecraft could be designed with much greater weight and volume margins. Such a fundamental change in the way spacecraft are designed and built could result in very significant cost savings.

- Provide generous spacecraft design margins
  - Avoids costly, last minute exercises to reduce weight
  - Easily accommodates weight and volume growth during development
- Avoid the relentless and costly push to increasingly miniaturize spacecraft components
- Provide additional redundant systems
  - Avoids premature mission ending failures; lengthens operating life
  - This approach must be balanced against the increased cost and complexity it brings
  - More frequent replacement of less expensive, shorter operating life spacecraft may be an alternative strategy
- Standardize booster and spacecraft interfaces
  - This approach places more burden on the spacecraft but is more easily accommodated with the wider design and volume margins available; makes for much lower cost launch systems and payload integration.
- Standardize spacecraft bus designs
- Increase use of standard, off-the-shelf subsystems

Simply allowing a spacecraft design to increase in weight does not guarantee cost reductions. Normally, when a spacecraft design gets heavier, the opposite is true. This is because the weight gain is a result of added capability and/or performance. The OTA background paper cited the Milstar satellite as being the antithesis of the "grow bigger and cheaper" design philosophy. *Milstar* is a huge space vehicle, but its size is necessary to contain all of the satellite's advanced communications and other supporting subsystems. To achieve cost reductions through heavier satellite designs, the weight increases must be a result of increased design margins, design simplification, and application of previously developed subsystems that are suboptimized for weight and volume. Table 4-17 summarizes the benefits that wide weight and volume margins afford to spacecraft designs [Office of Technology Assessment, 1990]. Table 4-18 provides some specific and practical examples of different design practices that this approach enables [Hughes, 1988].

TABLE 4-18. Some Examples of the Benefits of Spacecraft Weight and Volume Growth. These are a few of the benefits that could be derived if spacecraft designers could use increased weight and volume parameters due to significantly reduced launch costs.

- Larger margins to reduce test and operations cost and increase reliability
- Larger solar arrays with simpler drive mechanisms
- Reduction in the number of complex mechanisms and appendages
- Better opportunities for standardized bus designs
- Use of heavier and less expensive structural materials
- More opportunities for using off-the-shelf components
- Easier packaging of payloads, wire harnesses, and antennas

4.8 Summary

The current high cost of launch is severely constraining new space initiatives. For launch service users who must live with the current stable of launch vehicles, options are somewhat limited for reducing their cost of space access. High launch costs have caused a profound increase in the cost of spacecraft.

To make significant reductions in launch costs, new "clean sheet" launch systems must be developed. Unfortunately, the front end nonrecurring development costs for every new launch system proposed by the U.S. government in the past decade have been unacceptably high. Also, a number of institutional barriers within government and industry have prevented major inroads in cost reduction.

The traditional design imperative of maximum performance and minimum weight has precluded minimum cost launchers, especially for expendable booster concepts. Numerous ideas for reusable launch systems have been offered up, but no clear consensus has emerged for the most cost-effective design. In defining new launch systems intended to lower the cost of space access, we must remember that what **is not** important is what the booster looks like, how many stages it has, or how many parts it does or does not throw away. What **is** important for a new low cost launch system is how much it costs to put its payload on orbit; this must be the bottom line measure of merit.

References

Aerojet TechSystems. September 23, 1988. *Rocket Engine Combustion Devices Design and Demonstration NLB8DP*. Sacramento, CA: Aerojet TechSystems.

Allman, R. M. September 25, 1987. *Minimum-Cost-Design Space Launch Vehicle*. Briefing to The Aerospace Corporation Board of Trustees AD Hoc Committee on Space Systems Cost, El Segundo, CA.

Andrews, W. Geary and E. G. Haberman. June 1991. "Solids Virtues a Solid Bet." *Aerospace America*.

Arianespace. February 1993. A.S.A.P. User's Manual, Rev. 1. Evry, France: Arianespace.

Bilstein, Roger E. 1980. *Stages to Saturn*. Washington, D.C.: National Aeronautics and Space Administration.

Bleymaier, Major General Joseph S. January–February 1969. "Future Space Booster Requirements." *Air University Review*.

Cook, Jerry R., et al. July 1992. "Hybrid Rockets: Combining the Best of Liquids and Solids." *Aerospace America*.

Dergarabedian, Paul. November 14, 1991. *Cost-Model Considerations for Launch Vehicles*. Internal study. El Segundo, CA: The Aerospace Corporation.

de Selding, Peter B. December 12–18, 1994. "ESA to Increase Parabolic Experiment Funding." *Space News*.

Dyer, J. E., et al. August 1987. *Low Cost Expendable Propulsion Study*, AFAL Report No. TR-87-020. San Diego, CA: General Dynamics Corporation

Easterbrook, Gregg. August 17, 1987. "Big Dumb Rockets." *Newsweek*.

Eastern Space and Missile Center/XR. July 24, 1990. *ESMC 2005—Functional Area Requirements and Technology Data*. Patrick Air Force Base, FL

Elverum, Gerard W. Jr. October 1973. "Scale Up to Keep Mission Costs Down." Paper presented at the 24th International Astronautical Federation Congress. Baku, USSR.

French, James R. April 1988. "Paperwork is a Launch Vehicle Roadblock." *Aerospace America*.

Gabris, Edward A., Ronald J. Harris, and Stephen A. Rast. August 31, 1992. "Progress on the National Launch System Demonstrates National Commitment." Paper presented at the World Space Congress. Washington, D.C.

Glushko, V. P. 1973. *Development of Rocketry and Space Technology in the USSR*. Moscow: Novosti Press.

Holguin, M., and M. Labbee. January 11–14, 1988. "Launch Vehicle to Payload Interface Standardization: The Quest for a Low Cost Launch System." Paper presented at the AIAA 26th Aerospace Sciences Meeting. Reno, NV.

Holmes, Jay. 1962. *America on the Moon*. Philadelphia, PA: J. B. Lippincott Company.

Horais, Brian. October 17–23, 1994. "America's Need for an ASAP Ring." *Space News*.

Hughes Aircraft Company. October 1988. *Design Guide for ALS Payloads*. El Segundo, CA: Hughes Aircraft Company.

Huzel, Dieter K. and David H. Huang. 1967. *Design of Liquid Propellant Rocket Engines*. Washington, D.C.: National Technical Information Service.

Isakowitz, Steven J. 1995. *International Reference Guide to Space Launch Systems*, 2nd edition. Washington, D.C.: American Institute of Aeronautics and Astronautics.

Japan Microgravity Center. 1992. *Japan Microgravity Center User's Guide*. Hokkaido, Japan.

Keith, Edward L. June 24–26, 1991. "Low Cost Space Transportation: Hurdles of Implementation." Paper presented at the AIAA/SAE/ASME/ASEE 27th Joint Propulsion Conference. Sacramento, CA.

Koelle, Dietrich E. March 1991. *TRANSCOST, Statistical-Analytical Model for Cost Estimation and Economic Optimization of Space Transportation Systems*. Munich, Germany: MBB Space Communications and Propulsion Systems Division, Deutsche Aerospace.

London, Lt Col John R. III. July 4–10, 1994. "Launch Improvement and Realities." *Space News*.

London, Lt Col John R. III. October 1994. *LEO on the Cheap—Methods for Achieving Drastic Reductions in Space Launch Costs*. Maxwell Air Force Base, AL: Air University Press.

McDonald, Allan J. August 28, 1992. "The Impact of Chemical Rocket Propulsion on the Earth's Environment." Paper presented at the World Space Congress. Washington, D.C.

Microcosm, Inc. August 13, 1992. "Concept for a Low Cost, High Reliability, 100% Available Space Delivery System." Presentation to USAF Space and Missile Systems Center. El Segundo, CA.

NASA Goddard Space Flight Center, Orbital Launch Services Project. 1994. *Delta Secondary Payload Planner's Guide*.

NASA Goddard Space Flight Center, Special Payloads Division. 1991. *Get Away Special (GAS) Small Self-Contained Payloads Experimenter Handbook*.

NASA Research Announcement. December 12, 1994. *Microgravity Materials Science: Research and Flight Experiment Opportunities*. NRA-94-OLMSA-06. Washington, D.C.

NASA. *Shuttle Ground Operations Efficiencies/Technology Study Final Report*, vol. 6. 1988. NAS 10-11344. Washington, D.C.

Office of Technology Assessment, Congress of the United States. February 1989. *Big Dumb Boosters—A Low-Cost Space Transportation Option?* Washington, D.C.: Government Printing Office.

Office of Technology Assessment, Congress of the United States. January 1990. *Affordable Spacecraft—Design and Launch Alternatives.* Washington, D.C.: Government Printing Office.

Orbital Sciences Corporation, Pegasus/Taurus Business Development. 1995. *Making Space Available!* Fact Sheet.

Poniatowski, Karen S. August 31, 1994. AIAA/USU Launch Vehicles Panel. Briefing presented at the AIAA/USU Small Satellites Conference. Logan, UT.

Powell, Joel W. May/June 1995. "American/Swedish Microsatellites Launched Onboard Russian Spacecraft." *Countdown.*

Roberts, Lt Col Timothy K. June 1993. "The Need for New Spacelift Vehicles," *Space Trace.*

Robinson, Michael B., Robert J. Bayuzick, and William H. Hofmeister. 1990. *Review of Drop Tube and Drop Tower Facilities and Research.* Washington, D.C.: The American Institute of Aeronautics and Astronautics.

Sackheim, R. L., and P. Dergarabedian. 1995. *Reducing the Cost of Space Transportation.* Redondo Beach, CA: TRW.

Schnitt, A. and Colonel F. W. Kniss. July 1, 1968. *Proposed Minimum Cost Space Launch Vehicle System.* Aerospace Report TOR-0158(3415-15)-1. El Segundo, CA: The Aerospace Corporation.

Smith, Richard K. Spring/March 1986. "The Weight Envelope: An Airplane's Fourth Dimension...Aviation's Bottom Line." *Aerospace Historian.*

Space Transportation Architecture Study. June 1986. GDSS-STAS-86-00. San Diego, CA: General Dynamics Space Systems Division.

Sponable, M. September 25–28, 1990. "Two Stage or Not to Stage." AIAA Space Programs and Technologies Conference. Huntsville, AL.

Stofan, Andrew J. and Steven J. Isakowitz. July 10–12, 1989. "Design Challenges for the Advanced Launch System." Paper presented at the AIAA/ASME/SAE/ASEE 25th Joint Propulsion Conference. Monterey, CA.

Sutton, George P. 1992. *Rocket Propulsion Elements.* New York: John Wiley & Sons, Inc.

Truax, Robert C. October 1967. "The Pressure-Fed Booster—Dark Horse of the Space-Race." Paper presented at the 19th International Astronautical Federation Congress.

TRW Incorporated. May 15, 1981. Low Cost Shuttle Surrogate Booster (LCSSB) Final Report. Redondo Beach, CA: TRW Incorporated.

Tydon, Walter. July 31, 1970. *Minimum Cost Design Launch Vehicle Design/Costing Study*. Aerospace Report TOR-0059(6526-01)-2, Volume I. El Segundo, CA: The Aerospace Corporation.

Vice-President's Space Policy Advisory Board. November 1992. *The Future of the U.S. Space Launch Capability*. Washington, D.C.: National Space Council.

Worden, Simon P. 7–13 June 1993. "DC-X: Trouble for the Status Quo." *Space News*.

ZARM. 1990. *Center of Applied Space Technology and Microgravity*. Bremen, Germany: University of Bremen.

Chapter 5

Reducing Spacecraft Cost

Rick Fleeter, *AeroAstro, Inc.*

In this chapter I lay out some of the strategies for lowering the cost of developing, manufacturing, and testing spacecraft. Low cost results from integrating low-cost methods into all phases of the program—even the design of the program itself. The life cycle of an ideal, low-cost, spacecraft-development program is different from more conventional programs mainly because it's shorter. Table 5-1 lists typical program phases, along with some suggestions on saving time and money along the way. As the next chapter shows, the operational phase bears much of the overall cost, largely because of the affects of satellite and mission designs. But in all phases of a space program the best way to ensure low costs is to start with the leanest possible requirements and then to stubbornly resist changing them.

The best way to master the techniques of low-cost satellite development is to work with people who have produced these devices and learn their methods. The bibliography at the end of this chapter lists references provided by veterans of this process.

5.1 Spacecraft Development Life-Cycle

5.1.1 Mission Definition

Building a satellite is like throwing a serious party—maybe a wedding. Once you've decided you must have champagne, caviar, and prime rib served by candlelight in the grand ballroom of the Ritz—with 250 of your closest family and friends, you can save only so much by careful shopping for cocktail napkins and tough negotiating with the caterer. Compared with opting for an open air ceremony

TABLE 5-1. Approaches to Designing Low-Cost Spacecraft During Various Mission Phases. See text following table for detailed discussion.

Phase	Activity	Ways to Reduce Cost
Mission definition	Define objectives and preliminary requirements Assess feasibility Design strawman Control scope	Keep mission simple Set budget objective Expose and resolve implicit assumptions Minimize procedures and standard requirements Keep schedule short Assume maximum team size of 10
Preliminary design	Design system Trade on system Define operations concept	Adjust requirements Leverage existing systems Don't linger on trade-offs—decide & move on Attend to launch and operations issues
Detailed design	Design subsystems & interfaces Design ground segment Prepare mission-operations plan	Renegotiate requirements Decide whether to build or buy Find alternatives for launch and operations
Build	Buy parts and components Assemble Test component & subsystems	Redesign to use cheapest components Review requirements if necessary to maintain budget
Integration and test	Test integrated system Modify and reassemble	Minimize test envelopes Minimize redundant testing
Pre-launch	Do first integration Simulate mission Deliver to launch site	Don't try to reduce risk at the last minute Minimize activity at launch site

capped by hot dogs, chips, and beer in the city park down the street from your house, your leverage during detailed planning and execution is weak once the overall objectives are cast in cement. Defining a mission is thus the single most important step in limiting its cost. Limit your mission scope to an absolute minimum. Ideally, stick with only one mission objective. Because political realities usually require multiple objectives to broaden constituency, rank these additional goals so you can reduce scope if costs rise.

Part of the cost is certainly the menu and venue—the objectives. But another large chunk is procedural. For your wedding party, waiters in tuxedos serving from silver platters, plus sommelier and cocktail servers, cost more than a buffet. Establish program rules that allow a minimum-cost program. These might include not adhering to MIL-SPEC procedures and components. (Of course it's politically correct to state that you'll adhere to them **as a goal**, keeping in mind this goal is like a New Year's resolution to lose 10 pounds and go to the gym at least 3 times a week.) Fight almost all reporting requirements and don't be lured by the promise of more money to do anything you don't need to do to build the system. More money means more people; more people means less team cohesion and lower efficiency. And money can't buy efficiency or cohesive work.

Only after the ground rules and objectives are extremely clear can you discover whether you can meet the program's budget objectives. If you're too high, go around the loop again—pare down objectives and spurious requirements and reexamine the budget. Be grotesquely generous in your estimation of how many people you'll need and how much needed parts and services might cost—you'll overrun by less. See Chap. 7 for other techniques to fit the program to the budget.

Once the planners are happy with the program's design, communicate it in detail to all the participants. Expect complaints and dissatisfaction—and if you don't get negative feedback, actively seek it out. People who are actually building your stuff know a lot about the problems and pitfalls you don't want to know about. Don't be an ostrich—listen to feedback, however anxiety provoking. And don't belittle the bearer of bad news—you'll shut off the supply of warnings you need to avoid trouble. One proposal was literally saved from certain overrun because an engineer volunteered that the 3-m-diameter antenna we thought was a must to get adequate link margin would cost 3 to 10 times more than the 2-m dish we had costed. We learned to live with the 2-m-diameter dish and shaved a potential 15% overrun off the parts budget. It's better to go another round now than when you're 80% through your budget and not yet half way there. Carefully consider feedback from your technical team. Even if they're wrong, they must buy into the plan, or the plan will fail.

5.1.2 Preliminary Design

Once you've set the menu and venue, you start answering certain questions in preliminary design. For your wedding party, this means how many bottles of what age of blended vs. single-malt Scotch your crowd is likely to consume, how many ounces of prime rib per serving, Perrier or Poland Spring. We're not yet ready to send out to every bakery in the state for quotes on 114 ft.^{2} of 7-layer, chocolate-rum-raspberry, wedding cake—but we are going to think about just how much 250-people-worth of desert is likely to cost and how much lead time it takes to get it designed, baked, and delivered. In AMSAT, preliminary design is where volunteers sign up for their roles and, in turn, for their budgets and schedules. They look at the requirements for the spacecraft subsystems and, based on experience, calculate back-of-the-envelope estimates of what they need to produce each subsystem. (Sec. 13.3 describes AMSAT missions.)

These mind games make you more resolved about cost and schedule, so you can opt out of hot casoullet in favor of a cooler gazpacho if you can't stand the heat. Unfortunately, most program managers and clients think that, if they change the design or decrease scope after preliminary design, lightning will strike them down, they'll be turned into a pillar of salt, or, worse, someone will cut their bonuses. Don't bother doing a preliminary design if your only objective is to squeak by the preliminary design review. Remember, descoping can be as simple as agreeing to lower the payload's operating duty cycle, or deleting a secondary payload that is complicating the mechanical design.

Conventional programs produce detailed specifications (specs) of all the necessary components to prepare for the preliminary design review. You can do this, but it's expensive and a long way around the block. Instead, whenever possible, go shopping for what exists. I don't go to the store with a cereal spec, conclude that nobody has just the right mix of ingredients in a box of the right size and color, and then call a cereal maker and tell them to custom design me a box of Fleeter's Own Cereal. I wander the isles at the local grocery store and see what works for me out of the selection available for under $5. Do you go to a car dealer and insist on 49.6 inch shoulder room, 168 horsepower, and metallic-teal trim? The manufacturer might build it for you, but they'll probably charge you about 10,000 sedans worth of dollars for it. So don't show up at the preliminary design review with a stack of purchase orders your spec-writers have crafted while sitting in front of PCs and projecting their impressions of reality into cyberspace. Just find out what you can buy that works for your design, at what price, in writing or by a telephone conversation you can record on paper (vendors seldom lie, err, or even exaggerate on costs of real, already built hardware). Then, add it all up for your reviewers. Quotes obtained on existing gear are a lot more convincing than a 50-inch-high stack of Cost Plus Fixed Fee (CPFF) ROM quotes. Remember, ROM is "**Rough** Order of Magnitude," meaning you can **easily** be off a factor of 10—or even a factor of 50—and still be within quote. When your client heads out to lunch and orders the equivalent of a burger and fries, thinking the bill will be $5, she won't be happy when the check comes in at $50 or $250, well within the calibration limits of a ROM quote.

Preliminary design ends when you have enough definition to add up your entire soiree—soup to nuts—and come up with a cost number you believe in, based on encounters with reality. For your wedding party, you'll have talked to at least one or two halls, one or two bakeries, one or two caterers for the main dishes, photographers, florists, tuxedo rental shops, and limousine services. Or, let's say you prefer building satellites to marrying off that little girl you still visualize riding the Pixie with pink streamers and training wheels. In that case, you've talked to one or two vendors of solar panels, fabricators of metal structural elements, launch providers, and radio vendors. And you've shown through systems-engineering design that your candidate components will plug and play together to meet the agreed-on objectives for mission performance. Unlike staging a wedding, your major expense isn't your vendors—it's your own organization. Even if you really know the cost of parts, your most uncertain estimate is still your own organization's labor costs.

Plan to do the cost exercise at least once before you even schedule the formal review of your preliminary design. Even better, don't have this formal review. It is enough, in a program that aspires to be cheap, fast, and limited in scope, that you and your client are comfortable with the design, its capabilities, and your estimates for carrying it out. You won't get closer to completing your mission by jetting a few dozen people away from their desks and labs, where they could have done useful work, to an arbitrarily selected airport somewhere, transporting them to hotels, boring them with an endless parade of viewgraphs, and telling them the review team must like their ties, pumps, and viewgraph style or they'll kill the program. For a streamlined program, you can replace the traditional preliminary design review with three or four people getting together on the phone; going over identical stacks of paper to review the engineering, cost numbers and schedule; and occasionally bringing a specialist in by teleconference whenever questions arise. If you can't do that, you don't have a low-cost program or your client isn't serious about low cost.

Including preliminary design reviews, or violating any of the advice in this chapter doesn't mean you've failed. I don't know of any programs that truly minimize cost. All programs minimize cost subject to political, scheduling, and organizational constraints. Still, trying to delete program requirements that have been with us since satellites were only a theory will help you more clearly map out your cost boundaries.

5.1.3 Detailed Design

This is it—you're not thinking about it, you're doing it—building a satellite. Preliminary design is really design for its own sake—looking at the thing from a distance and saying **it is good**. Or not—in which case you design it differently, maybe with renegotiated ground rules. But detailed design is the first phase of manufacturing. The management job is different now. Returning to our wedding-party analogy, you're now printing and mailing the invitations; arbitrating the location of each name tag at each table; and sending advance payments to the musicians, the florist, and the photographer. In most small-satellite programs, purchase commitments begin after the preliminary design review, particularly for items vital to developing other items. An example is procuring the flight computer (or the parts for building it). Without a flight computer, the software people are guessing their product will work. The risk of mis-specifying some part of the computer is less important than the risk of a team creating software they can't run on the actual system.

Oh, but you thought you couldn't buy anything before the critical design review? You're right that in classical programs, purchase orders aren't released until the detailed design, which means final, approved schematic diagrams, manufacturing drawings, test plans, and operational plans. But in manufacturing low-cost satellites, our designs are our parts—they're what we build the satellite out of. And if you want minimum cost and schedule, you don't pay for inventory lying around. The instant the design is done and the engineering team signs off that it's correct, the parts get ordered and the thing gets built. If it's wrong, at least you'll find that out as soon as possible.

Also, in typical, small, low-cost programs, the engineer is architect, designer, parts-procurer, assembler, tester, and sign-off authority on his or her part of the spacecraft. If you wait for critical design review to order parts, what does your engineer do during the 2 to 24 weeks before critical parts show up? You don't want to know—an idle mind is the devil's playground, according to a rather sage engineer, Ben Franklin.

Returning for a moment to our wedding party, one key to avoiding disaster is knowing the caterer's responsibilities. Does the caterer supply flowers? Whiskey? Wine? Linens? Tables and Chairs? A special runner for the bride and groom to walk on for their grand entrance? How about a special silver cake slicer for that classic scene where she stuffs the first slice into his mouth? Similarly, if you're going to have engineers off ordering parts, shooting boards, and committing to purchases, you must have an interface-control document. Specify all inputs, outputs, and mechanical environment for every board. Identify every wire and pinout. This is especially critical wherever you interface to instruments supplied by other organizations in the program. Every error in the interface-control document will bite you later, so try to get it right and update it frequently.

Building vs. buying is a very subtle issue. Most engineers believe their forte is synthesis, or building. That's an honest, professional bias which management has to allow. Most managers want to buy off the shelf—a bias the engineers have to deal with. Each build and buy decision is individual, but we commonly claim we can build a component better, cheaper, and faster ourselves. If someone expresses that opinion, talk to the vendor from whom you could buy it. Tell their sales engineer you think you can do it better, cheaper, and faster; then, let them tell you why you're wrong. Most of the time, you'll be able to decide quickly, based on a 10-minute phone conversation. If you're still on the fence, buy it. Everything takes longer and costs more than you think.

How does management differ starting with detailed design? It changes from broadly directing the preliminary design to measuring and controlling—determining daily what the project status is, where it's getting off track, and correcting virtual train wrecks before they happen. Design is art, not science. Surprises will happen. Bad news on cost and schedule as the unexpected materializes is not bad management—it's reality. Don't dig into program reserves to fix problems at this point in the program—you'll need them later. Renegotiate requirements or fall back to less ambitious baselines if cost and schedule are already plaguing you—or negotiate a larger cost and longer schedule.

5.1.4 Development

It would be very neat if programs really proceeded so linearly from preliminary design, to detailed design, to development, and so on. In development—as opposed to manufacturing the nth copy of a proven design—it's a fiction useful mainly for creating very tidy viewgraph presentations and helping organize chapters of books like this one. As things get built, we have to redesign them for many reasons, such as unavailable or substandard components, performance changes somewhere else in the system, and human fallibility. In small programs which have the advantages of a nimble workgroup, development isn't linear—it's circular. Figure 5-1 shows that in each phase of development—from conception through integration and test—you must be prepared to iterate back to the previous phase and make changes. "Build a little, test a little" is shorthand for an iterative process.

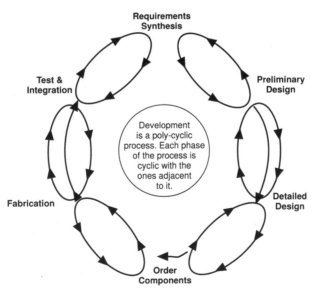

Fig. 5-1. Spacecraft Development as a Collection of Cyclic Processes. Each phase of development will require looping back to the one before it as reality affects your program and engineering plans.

But what about the big circle? Surely we don't go all the way around to integration and test only to start redeveloping requirements and a preliminary design. On a single program, we don't. But few engineers and engineering managers aim to create a single spacecraft and then retire to music or bicycle repair—or maybe they do have that goal but they fall prey to baser instincts and end up building one space system after another. I don't fervently champion lessons learned. So many of these lessons depend on each program's unique assumptions, requirements, and technologies. But I do believe in studying history—battles, business successes and failures, and solutions to problems in space-systems design. Recognizing or admitting that we're on a never-ending cycle of development is the hard part. Each program seems to be the only program. Once you recognize a learning loop exists, you can close the loop with many satisfactory solutions. AMSAT-DL, Germany's amateur-radio-satellite organization, practiced a solution I like. Into a single very large binder, about 1 m wide by 70 cm high, the group assembles every electrical schematic, every mechanical drawing used to make components, plus every useful photograph of components and their integration onto the satellite. Individual notes are sometimes recorded directly on the diagrams, as Leonardo Da Vinci did, or pages of text may accompany the graphic materials, which themselves follow the spacecraft from design through launch.

Such a book is valuable on many levels. For example, once an AMSAT spacecraft was lost in a launch-vehicle failure. It was rebuilt from memory enhanced by the book and successfully reflown in 9 months. The book is a valuable primer for a new generation of engineers and fabricators in a new program—it can answer how they solved this or that problem the last time. People signed their drawings and notes, so sometimes we'd just call them and ask for their advice and opinion. If something malfunctions on orbit, the book is there as a reference for how it looked when we last saw it.

Finally, the book tells the world just how much fine work went into that satellite. Unlike people who build cars or bicycles or computers, satellite designers never get to see their creations operate. We see pieces and we see a satellite standing on a structural support or hanging from wires. Then they're launched, and we never see them again except, very rarely, as points of light passing quickly across the evening or predawn sky. The book tangibly rewards and reinforces the team by saying "we did something remarkable."

5.1.5 Integration and Test

That the word "test" hasn't come up yet is misleading because testing drives the feedback loops discussed above. In the design stages, we test assumptions about viability, cost and component availability, and changes in the mission model. As components are built, we test them, although at times only by visual inspection or by checking fit among, say, mechanical components.

But the divergence between a low-cost program and more conventional approaches is clearest in testing. Low-cost creatures are schizophrenic about testing. We vehemently advocate testing and a devote disproportionate share of program time and money to it. But we select what we test, when, and within what envelope. It's often said (but worth repeating) that a whole microsatellite is smaller than the typical subsystem of a conventional satellite. Thus, the substantial savings in time and trouble are often worth the risk of postponing extensive environmental and operational testing until the entire satellite is integrated. By contrast, a conventional program for a larger spacecraft will devote resources to building (and testing) test fixtures and test circuits in order to exercise subsystems. These fixtures and circuits simulate for them the inputs they will receive from the rest of the unbuilt satellite. This is enormously complex, and getting the satellite's simulation to work is almost as tough as getting the satellite itself to work. We simulate—but only crudely. To test a circuit board from the spacecraft's telemetry system, we might put 5 V across its power bus—a simulation of what the satellite will do. And we might feed it a random bit stream—a partial simulation of the packetized data it will expect to see from the satellite's RF detector—to test its dynamic behavior. We might even better simulate the spacecraft's operating conditions but temper them by what is easy, quick, and cheap to do. We avoid, for instance, building more circuit boards only to simulate complex spacecraft telemetry operations and conditions.

The risk of not having simulators is, for example, that we'll have to wait around until the whole satellite is put together. Finally we'll try to talk to it across the RF link, only to find that the telemetry I/O board won't work. That risk must be balanced against the cost of building a complete simulator for satellite telemetry and testing it, which will require the RF team to grow by one or more people just for the test fixture's development. Also, several things lessen this risk. For example, in a fast program nobody is waiting around for long. The board, and all the other major subsystem boards, are likely to be done just in time or a bit after just in time to integrate them into the satellite. And even if the board is done, its creator—the person who knows how to test it—is undoubtedly busy building other parts of the satellite—so your key resource isn't just standing around. As importantly, no cheap, one-of-a-kind or low-production volume spacecraft should be built so that pulling a board takes more than an hour of labor. By virtue of its newness and its uniqueness, you should expect frequent wild goose chases into the circuitry and design accordingly.

If you plan to test only at the spacecraft level, engineers will be nervous. This is good, for they'll seek comfort in two ways. They will design the simplest possible thing—something that shouldn't screw up when expected to work for the first time, when the whole satellite is relying on it, probably in the first series of operational tests. And they'll devise clever, clandestine tests which they can carry out with little effect on the program budget or flow to make sure it works. An example would be a simple, bootlegged simulation circuit with two outboard power supplies and a little board using Actel (programmable gate array chip) logic which they can build, verify, and use in a day or two. Good engineers don't let something out of their hands without reasonable certainty it will work; what we're talking about here is the level of that verification and the means of achieving it.

Test standards are another subtle cost driver. If thermal analysis shows the CPU stays between 25 °C and –5 °C, over what range do you test? The books might tell you +80 °C to –40 °C. It's comforting to just do as the book says, but it's also suicide. You'll wash out perfectly good boards. You'll disqualify nifty components, thus creating the need to substitute complex boards filled with less elegant solutions. And you'll spend a lot more time in test. Life is filled with risk, and so is your program. Testing to extremes makes everybody feel good, but it's an illusion. Surviving an 80 °C soak might drastically shorten the life of otherwise fine components. Yet, it won't find the software bug that turns off the satellite for 18 hours the first time the ground station links up while the satellite is in the dark.

5.1.6 Pre-Launch

All programs—big, small, cheap, costly, crewed, and uncrewed—use the mission-simulation test, a dress rehearsal of the actual on-orbit mission. The difference in smaller, cheaper programs is we do it proportionately more often to make up for the limited amount of simulation and subsystem-performance testing. A way to do mission simulation more, but not pay more, is to use the ground-station hardware as the main test fixture. Do as many of the performance-verification, environmental, and mission-simulation tests as possible through the same interface you'll be using on orbit. This not only is the most effective way to uncover weak links in the system design as you'll fly it, but also gives the operations crew the best possible training on how to manage the spacecraft when they can only touch it with the RF link.

But mission simulation is more than just going through a thermal vacuum with only a serial cable and a power line attached to the satellite. Some careful thinking goes into designing a series of operations which exercise all of the spacecraft hardware and as many of the software trees' branches as possible. Include launch, separation, and initial turn-on of the spacecraft. Think of various emergencies ahead of time and develop and refine emergency procedures to deal with them if they were to happen on orbit. If you can successfully and repeatedly simulate the mission, you'll usually earn a sign-off to proceed to the launch site. There the satellite is integrated to the rocket and often operated one last time off the serial port before it's signed off for launch. Because that mission simulation is the last link in a long chain, a last line of defense against launching a satellite that doesn't work, its design requires careful thought.

5.2 General Low-Cost Philosophies

5.2.1 Keep It Simple

In the end, you can save only so much money paying yourself and your team slave wages and working long hours. Design and build something that's simple and can't help but work—you'll still find plenty of challenges. Remember the wedding-party analogy. If you want the most elaborate flowers, the best champagne, prime rib, and a 27-layer cake, hard bargaining with the caterer won't make it cheap. And trying to bring down costs late in a program, when all the subsystems are already in development, creates a lot of expensive entropy—and no savings. Remember too that later in the program, as the big day approaches, you'll inevitably tend to spend against risk. Build reserves wherever you can, start saving where you can from day one.

5.2.2 Build vs. Buy

I've treated this subject already (see Sec. 5.1.3), but here are a few more pieces of advice. If this decision were easy, they wouldn't need you to help make it. There is no pat decision-making tool here. Sometimes, though the exact part you need isn't available off the shelf, you may be able to work with the vendor to modify an existing product. So you have three options—build, buy, or modify—which demands more of your judgment. Here's some help for your list of pros and cons:

- What is the vendor's motivation? To earn money on a modification contract, or to leverage your program to justify broadening the product line?

- Are the people who designed and understand the existing product still around? Or is your vendor just building from an archival set of plans?
- By how much margin can the program plan for the vendor's proposed change overrun before you'll start to suffer?
- How well can you assess the modification's difficulty and potential problems?

And of course:

- How painful are the alternatives? These might include lowering your satellite's performance or raising its cost to use existing products. The psychology here is the loss you know vs. the loss you don't know. None of us likes to accept a loss, so we tend to go for the one we don't know as well—in this case, how much damage a vendor's flawed development program can do to your program once it's committed to using the unbuilt product. I don't like to gamble anyway, but I won't gamble at all unless I know how much I may win and how much I stand to lose.

Still thinking about developing something yourself? Ask your vendor to talk you out of it—you may be surprised at all the subtleties built into an apparently simple product. Things you wouldn't have seen until much later, likely resulting in delays and overruns.

On the other hand, it's tempting to shut down an internal team's development program and put in their place a subcontractor with less vested interest in the program's success. You may go outside because you don't want to swell your ranks or you don't have the expertise. But an alternative is to design the program so what you can't buy can be built by the team whose collective life depends on the mission's success. Your team usually is more motivated than your subcontractor's.

5.2.3 Schedule Aggressively

Let's look at the dogma that short schedules save money. This isn't true if you work your people to death—or to the point at which they mutiny and either burn the ship down to the water line or kill the captain, which may be you. It's also not true if the crew pays double or triple for everything it needs, such as express orders to make boards or overnight envelopes for every communication and shipment. Crews may also need faster computers running fancier tools to get the job done as quickly as possible. And they may just not do it right—the operation is a success but the patient, in this case your satellite in orbit, dies.

So it's with some strong reservations that I advocate the equation: Faster = Cheaper. Before rushing in where wise persons might fear to tread, here's some advice:

- *Contingency*—Have plenty of it. You're going to be advocating a fast schedule, and things are going to go wrong. **Don't** present this schedule to the client. Present it to the team as the goal and offer them rewards for attaining it and your intermediate milestones. But expect occasional, perhaps even frequent, failure, which is always part of striving for excellence.
- *Responsibility*—People can't move fast if they need permission from you for everything. Give everyone generous purchase authority, latitude to change

designs, and discretion in selecting vendors and suppliers and in techniques and venues for fabrication and testing. And this delegation must be real. If the program manager or the client starts to reel in the control line, the engineers quickly recognize they don't have the discretion to get the job done on your aggressive schedule, and they'll stop striving to attain it regardless of any goal-achievement motivators you've concocted. If your team can't handle autonomy, either you or they have to go.

- *Another kind of responsibility*—The saying about any job filling the time allotted is very true. You're going to be ripping things out of people's hands when the creators don't think they're ready. Some of these products aren't going to work and the responsibility belongs **NOT** to the engineer, but to the keeper of the maniacal schedule. You can push, but when things stop working, back off.

- *Communication*—Often, a program integrates two or more separate organizations located at different institutions. You must have a single point of contact to manage communication between organizations. Requests to many different people at the same time can be confusing and disruptive. An effective means of communications is also important. Electronic mail with nearly real-time access to each staff member's workstation is an excellent means of tying all organizations together.

Of course, ready E-mail exchanges may violate the rule about having a single point of contact; so what's a manager to do? Stay close to the team. Communicate with them—often and in satisfying detail—exactly what is going on in the program. Honesty is everything here—once they know they can't trust you, they'll build their own communication network at the expense of productivity. Let people know that, although helpful discourse program-wide by E-mail is a great thing, nobody should change designs, procedures, or other aspects of their work without confirmation from the team leader.

If not taken to extremes, E-mail has several advantages. It's relatively inexpensive and automatically documents the project's progress. Besides having this "self-documenting" feature, E-mail may also be preferable to direct contact because we can directly copy messages to the project managers and others who need to know about decisions and their background. Often the copied people will notice design decisions apparently unrelated to their work have an unexpected—perhaps very harmful—effect that they would have discovered much later if only a one-on-one conversation had led to a particular decision. When operations start in the field, say for integration and test, documents (as daily test reports) on E-mail prove invaluable. It brings all of the engineering team into contact with daily activities so they can offer information to help the systems team doing the tests or integration. E-mail is also a very effective way to broadcast minutes of technical-interchange meetings between small, specialized groups. E-mail helps to integrate the entire team into the development process.

5.2.4 Reviews

Don't rely exclusively, if at all, on conventional preliminary and critical design reviews. Schedule frequent, periodic, technical-interchange meetings to promote informal exchanges of information about the design. These meetings tend to focus on specific technical areas and involve a few of the development-team members. They'll more than make up for the increased program travel costs by preventing system-design problems and promoting a more efficient design through better collaboration among team members. Internal peer reviews are also valuable because they may be less threatening than technical-interchange meetings with customers, allowing the engineers to discuss concerns and raise potential issues that otherwise might stay submerged.

5.2.5 Documentation

Even if a program had no paper requirements at all, it would generate plenty of paper: manufacturers' specs, meeting notes, designs, test logs, E-mail conversations. When program planners say documentation, they mean paper over and above this basic, highly useful information.

I try to require only documentation that would otherwise not get done and that is undeniably necessary or helpful to the overall effort. The system block diagram with pinouts and voltages is an obvious example. A more subtle one is the interface-control documents. There may be several of these, such as one from the launch vehicle to the spacecraft and another between the spacecraft and the instruments it is housing and supporting. But the block diagram and interface-control documents are vital because they define the engineers' work packages. Finally, a systems-requirements document is worth writing mainly because it becomes a record of what everyone signed up for. It's sort of a contract—but one you can easily renegotiate. I'd advise against writing an actual requirements contract because some people would rather see the program fail than renegotiate if it turns out your goals were too ambitious.

Finally, the AMSAT style of project documentation is an attractive alternative to traditional, massive documentation which consumes thousands of engineering hours and which tends to be so arcane and detailed that no-one reads it. Yes, you run the risk of not knowing what thread cutter was used for the left-handed telescope door framish, if nobody took its picture or wrote it down or saved the purchase order or receipt, but you can get the program done on cost and budget and have a very adequate record of what was done and why.

In Table 5-2, I've recommended documentation for a low-cost mission. Note that most of the documentation is engineering materials used in developing the spacecraft.

5.2.6 Continuity of Staffing

Keeping the team largely intact throughout the program can reduce loss of information on how things go together and work. Unfortunately, it's impossible to keep everyone. To lessen turnover, keep the program short, hire plenty of people up front, and resist capability upgrades to avoid burning out the team and triggering premature departures.

But plan for none of these methods to work. People's reasons for leaving run deeper than too-hard work or a 6-month program extension. Family members get sick, an acceptance letter to a prestigious Ph.D. program with financial support comes in the mail, or a phone call offers a chance to crew in the America's Cup. Maybe your lead RF engineer doesn't like the cut of your jeans. Whatever the reason, careful planning of your program is insignificant compared with the forces pulling people from it.

TABLE 5-2. Recommended Documentation for Low-Cost Missions.

- System-requirements document
- Software-requirements document
- Hardware-requirements document
- Payload to S/C bus interface-control document
- Satellite and booster interface-control document
- Package for preliminary design review (if held)
- Package for critical design review (if held)
- Launch-loads specification (contains results from analyzing coupled loads un-less spacecraft is too small to require it)
- Report on expected thermal environment (final report being released after model correlation with results of thermal-balance test)
- Spacecraft block diagram
- Spacecraft and payload cabling diagram
- As-built schematics of the satellite bus and payload (may include markups and comments)
- Plan for environmental test
- Reports from all tests for the spacecraft bus and payload, including a log of anomalies and their resolutions
- System- and software-operations (this document should explain how to operate the satellite)
- Commented source code from the bus and payload (hope springs eternal)

To plan for the inevitable, use a buddy system. Every program performer has a partner who should ideally be involved in the same or a similar role in the program. If a person leaves, the buddy continues their work, finds a replacement, and brings the new person up to speed. The buddy is effective because buddies work closely together. They check each others' work, which is vital in its own right, and they often collaborate. Sometimes, several people together constitute a buddy, if one's work overlaps that of several others. Documentation is helpful, but careful, detailed documentation is not much more helpful. Person-to-person contact and working together on the project is the best introduction to others' work.

5.2.7 Continuous, Predictable Funding

Planning for minimum staff turnover and steady funding to successfully complete a minimum-cost program is a bit like ordering clear skies, light winds, and 78 °F weather to help create a successful outdoor wedding. If it were that simple—just passively relying on the gods—managers would be replaced by something like Shinto priests. In fact, the average program manger can't control steady, predictable funding.

Examine your goals. If you want the program to succeed (not always the goal), you should staff appropriately and commit to the capital and purchases the program needs to achieve its goals, always assuming the planned funding will be available. Your strategy is that good progress will help motivate the client to meet funding commitments. Your rationale is that, because your goal is a successful program, you can't abide curtailing necessary purchases when this action would keep the program from succeeding as planned—even with nominal funding. But if your goal is to maintain your job, keep your company solvent, and avoid layoffs of your colleagues, it's perfectly rational to lower your commitment level and recognize that the program will take a little longer and cost a little more. Although even a conservative program can bankrupt companies and layoff employees, it won't do so as quickly as deciding to go full speed ahead just before the client cancels your program.

Neither of these alternatives is right or wrong. The important thing is to face the dilemma and, as a group, decide how to proceed. Either way, consciously committing to a decision will prevent conflict as you move forward. Based on how the funding is going, it's never too late to alter your approach and start trending the program toward another outcome.

5.3   Subsystem Development

Modern society is characterized by products. Our markets sell complete, one-piece, ready-to-use things, such as clothes, radios and televisions, cars, and micro-wave-ready dinners. Stores selling thread and yarn, resistors and integrated circuits, engines, transmissions and wheels, and even many basic cooking ingredients like live chickens and flourishing basil plants, are harder to find. Professional engineers need to remind themselves that each detail, each subsystem is part of the performance of the whole.

The engineers actually do this pretty well, but managers and their clients often forget that a low-cost satellite doesn't result from mixing together inefficient, expensive subsystems. In the following sections, we look at ways to control cost and schedule in developing the key spacecraft subsystems.

5.3.1   Attitude Determination and Control

In a properly executed program, spacecraft engineers and payload engineers negotiate continually to reduce cost and complexity without unduly compromising the mission. For example, an attitude-control requirement usually has several solutions. Often, we haven't examined or thought about all of them. When a program has a tight budget, continually evaluating requirements tends to highlight the most difficult ones and forces us to reexamine them. Table 5-3 lists some of the most popular guidance options in generally ascending order of cost and complexity. But note that the cost depends more on performance requirements such as pointing accuracy, stability, and slew rate capability than on the approach to attitude control.

Once we've defined requirements and the general control technique, the attitude-control engineers can begin to search for low-cost actuators and sensors that will meet the requirements. (See Table 5-4.) Table 5-5 lists some suppliers of components.

Sensor and Actuator Highlights.

Cost, capability, and availability of various sensor and actuator options for low-cost spacecraft are discussed below.

Sun Sensors

Sun sensors range from simple incidence meters with accuracy of about 1 deg to digital devices with resolution under a milliradian. Although spacecraft designers have often developed them expressly for a mission, several firms now offer Sun-sensor sets for under $10,000.

TABLE 5-3. **Low-Cost Solutions for Guidance and Control.** Listed in approximate order of increasing cost and complexity.

Stabilization	Typical Accuracy	Advantages	Disadvantages	Cost Drivers
Unstabilized	N/A	Simple	No antenna or solar-array gain	Control spinup
Passive magnetic	±30 deg	No active components	Limited gain, single alignment, no roll control	May need damping
Passive aerodynamic	±10 deg	No active elements, unique pointing	Low altitudes only (<300 km)	Deployable drag device, dampers
Sun pointing spinner	±2 deg	Best electric power, protects payload from Sun	Active control	Magnetic systems need magnetometer + torque coils
Earth-pointing spinner	±5 deg	Sweep through nadir aids remote sensing	Active control + requires Earth sensing	Expenses for magnetometer, torque coils and Earth sensor
Earth Point (gravity-gradient)	±10 deg	Nadir staring	Weak stability easily upset, damping critical, attitude difficult to sense, no roll control	Large, deployable boom, active damping, costly sensors
Sun-pointing nonspinner	±1 deg	Best electric power, minimal angle rates on all axes	Active control, limited lifetime due to moving parts	Momentum wheel, complex control logic, more costly sensors
Earth-pointing nonspinner	±5 deg	Nadir staring	Active control, limited lifetime due to moving parts	Momentum wheel, more costly sensors
3-axis	±1 deg	Any attitude	Active control, poor solar or radio orientation	Multiple momentum wheels (4 typical), complex algorithm

TABLE 5-4. **Low-Cost Sensors and Actuators for Guidance and Control.**

Stabilization	Actuators	Sensors
Sun pointing: spin or momentum bias	Magnetic torque coils/thrusters + momentum wheel for nonspinner	Sun sensor + Earth sensor or star sensor
Earth-oriented spinning or momentum bias	Magnetic torque coils/rocket + momentum wheel for non-spinner	Earth sensor
Gravity-gradient	Gravity-gradient boom, torque coils (option) or passive damper	None or Earth sensor for active damping
3-axis	Momentum wheels (4 typical) + magnetic torque coils/rockets	One star, or at least two Earth or Sun sensors; several usually necessary for 4π sr visibility

Magnetometers

Required for effective use of magnetic torquers, magnetometers can also help us reconstruct a spacecraft's attitude. Recently developed filters claim to rebuild the attitude to less than 1.5 deg using just magnetometer data. The effectiveness of these sensors degrades when orbit height increases, and they are useless above 6,000 km. These sensors are widely available for under $10,000.

Earth Sensors

New products available from Barnes Engineering, EDO Corporation, Servo Corporation of America, and AeroAstro have significantly reduced the cost of Earth sensors with accuracy up to 0.1 deg for around $10,000.

TABLE 5-5. Some Manufacturers of Low-Cost Components for Attitude-Control Systems.

Component	Manufacturer
Sun sensors	• AeroAstro, Herndon, VA • Ithaco, Ithaca, NY • Space Sciences, White Plains, NY
Earth sensors	• Barnes Engineering (EDO Corporation), Shelton, CT • Ithaco, Ithaca, NY • Servo Corporation of America, Hicksville, NY • Space Sciences, White Plains, NY • Spiricon, Logan, UT
Magnetic torquers	• AeroAstro, Herndon, VA • Ithaco, Ithaca, NY • Fokker, Holland
Magnetometers	• Southwest Research Institute, San Antonio, TX • Israeli Aircraft Industries, Yehud, Israel • Ithaco, Ithaca, NY • KVH, Middleton, RI • Nanotesla, Annapolis, MD • Schonstedt, Reston, VA
Reaction/momentum storage wheels	• Allied Signal Aerospace Equipment Systems, Tempe, AZ • Bendix (Allied Signal Guidance & Control Division), Teterboro, NJ • CTA Space Systems, McLean, VA • Honeywell Satellite Systems Operation, Phoenix, AZ • Israeli Aircraft Industries, Yehud, Israel • Ithaco, Ithaca, NY • Satcon • Space Sciences, White Plains, NY • Teldix, Heidelberg, Germany
Angular rate sensors	• Humphrey, San Diego, CA • Litton, Woodland Hills, CA • Matra Marconi, Vélizy-Villacouvlay, Cedex, France • Robi Controls, Boulder, CO • Teldix, Heidelberg, Germany
Differential GPS	• Trimble, San Jose, CA

GPS

Onboard GPS provides time, position, and ephemeris data for satellites in low-Earth orbit. In the future, it may also offer reliable attitude information. As GPS receivers become smaller and less expensive, they will become common on virtually all missions. Units now ready for space cost from $15,000 to about $500,000.

Bus Applications of Payload Instruments

Payload instruments such as astronomical telescopes can serve double duty to provide attitude information for the bus. Although we need a more complicated interface between the spacecraft bus and payload to use this information, shared use can eliminate the need for an otherwise expensive sensor. Attitude control and instrument resolution are tightly linked in this approach, and we have to sacrifice some of the instrument's resolution because attitude control is less precise than the instrument's ability to resolve objects.

Magnetic Torquers

Magnetic torquers are the simplest of all low-cost actuators. They range from a passive bar magnet to align the spacecraft with the Earth's field, to more complicated electromagnetic torquers with air or ferromagnetic cores. The slew rate and axes are limited, but there are no expendables and prices are low. A flight set of magnetic torquers and driver electronics typically cost $25,000 to $100,000.

Momentum and Reaction Wheels

Because of demand from developers of small, low-cost spacecraft, several vendors now offer relatively small, low-cost wheels for as little as $70,000. Excepting its power draw, which can be as low as 2 W, a momentum-based solution may now be cheaper and more reliable than gravity-gradient (see below) because it has no deployable.

Gravity-Gradient Boom

To achieve gravity-gradient stabilization for nadir pointing, we need a long, deployable boom to develop the required inertia tensor. Several spacecraft have used various booms successfully. In addition to the deployable, we need an actuator to flip the spacecraft in case it stabilizes upside down, as well as a separate method of either active or passive damping.

Testing and Verification

We can't fully test attitude-control systems for spacecraft on Earth. The forces of gravity and air drag completely wash out the small effects that influence spacecraft attitude on orbit. Further, sensor inputs from the Earth's natural horizon, from stars, and from the Earth's varying magnetic field aren't readily accessible on Earth. Typically, we use a lot of computer software and complex equipment to simulate part of the space environment. Still, active and passive systems for attitude control foster concerns, and we spend resources to address those concerns. A common approach in low-cost programs has been to extensively test the actuators and sensors on their own and then test control loops in the actual flight software through simulation.

A more complicated technique, with **hardware in the loop,** simulates the space-craft's attitude dynamics on a separate computer attached to the spacecraft through the actuator-control and sensor-input lines. The analog outputs of the sensors are replaced by analog outputs from the simulation computer, and the actuators' control outputs are broken and fed to the simulation computer. This signal is taken from the spacecraft at points as close to the real actuator as possible. This ensures the simula-tion will see command signals as similar as possible to those received by the actuator. The inputs to the dynamics simulation are these actuator-control lines, and the outputs are the analog sensors' inputs. In this manner we can test the attitude-control system's mode-switching logic, how the system handles plant-modeling mistakes, and the basic control logic's robustness. This method is probably the most extensive a low-cost space program will be able to afford.

A more expensive test would be to close the loop outside of the sensors. For example, we could read the output of the actuators and use custom hardware to simulate the resulting changes in the environment observed by sensors in the attitude-control system. We can do the same thing with the actuators by using low-friction or servo-controlled suspensions, which allow the spacecraft to move under actuator power alone. This type of simulation and kinetic environment is beyond the budget of all but the largest spacecraft programs and is probably not appropriate for most low-cost satellites.

5.3.2 Power

Some designers believe they can estimate the spacecraft program's entire cost by the power budget alone. More power means larger, more expensive solar arrays, which in turn can drive up costs for the attitude-control system. Power storage using chemical batteries adds a lot of weight, and the power-regulation system's complexity grows while reliability falls as we need to control more power conditioned to suit more devices.

Low-cost satellites rely entirely on silicon and GaAs photovoltaic cells to generate power. Two references the low-cost satellite designer should have are the *Solar Cell Array Design Handbook* and *Solar Cell Radiation Handbook,* written and published by JPL. They comprehensively cover designing and constructing solar arrays for satellites.

Once designers have determined the amount of power necessary for the satellite's mission, the power-subsystem designer and systems engineer select the type and size of the solar arrays. Every solar array consists of photovoltaic cells assembled in series to produce the required end-of-life voltage. Designers build such strings into an array to generate the necessary power.

Several characteristics contribute to the end-of-life voltage. Once we agree on a few requirements, the design outcome is both predictable and rigid—with precious little wiggle room left to reduce cost. And as solar arrays get larger, structural, thermal, stability and control, and even antenna systems grow larger and more complex. Here are some things you can control to contain costs for the power system and its effects on the rest of the satellite:

- *Cut Power Required*—Your big consumers are devices that are on 100% of the time—typically the CPU, other housekeeping systems, mass memory, and possibly the command receiver and some instruments. Even if you're designing a 50 W system, a couple of watts in the CPU is very significant. Out of the 50 W, the payload may need 30 or 40, and the rest of the spacecraft has 10 or 20. A few watts of extra power at 100% duty cycle is a 15% hit.

Also, maybe you can negotiate a lowered duty cycle from the payload. Does it really need to be on all the time, or can it go into a sleep mode during uninteresting intervals? An example is passage through penumbra, when power is most "expensive" because it's produced by the batteries, which must be charged and have their own inefficiencies and costs. Just as the best way to save money on health care is not to cut physicians' salaries and quality of care but rather to stay healthy, carefully shopping for batteries and solar arrays won't save as much as needing half as many of them.

- *Point at or Near the Sun*—This simple advice stems from a simple fact—all your electric power is coming from the Sun. Batteries don't make power— they just store it. Photovoltaics don't make power—they just convert it from sunlight to electric form. The Sun is king, do what you can to worship him. It also happens to be very easy to point at the Sun—it's bright and stationary. It may prove more efficient to point at the Sun and carry six radio antennas than to point at nadir and end up with four times more photovoltaics or a complex Sun-tracking array. Which leads to...

- *Avoid Complex Sun-tracking Arrays*—Very nice ones are now available, and you do feel as though you're in the big leagues when you spec them. The satellite also looks really racy fitted with them. But deployable arrays cost more in much the same way that a twin-turbo V-8 costs more for higher insurance, the extra gas, overhauling those turbos at 60,000 miles, and the new tires you'll be buying after depositing all that rubber on the road. For deployable arrays you need logic and a logic-state machine (computer) to point them. Their motion perturbs your attitude control in some obvious and some very subtle ways. They reduce reliability because they have to move— moving parts are always a worry. Testing them takes longer and requires more equipment.

- *Avoid Deployable Arrays*—A deployable is most often a simple hinge to increase the solar-collection area on orbit—once deployed, it's stationary and therefore it's not as complex as a Sun-tracking array. But it's still a moving part to be designed, built, and tested, and it may not work. It requires several critical parts, including a latching mechanism, the hinges themselves, and some motive-force producer (usually springs) to actually deploy the panels. Because of these parts and the testing they require, deploying a couple of simple flat panels can cost $250,000 or more. It may prove cheaper to live with the fixed array size you can fit on the rocket and spend more on cells, possibly going to GaAs. Even if it's not cheaper, it will be more reliable. If you do need deployable or tracking arrays, a few potential suppliers include:

  − Starsys Research, Boulder, CO (hinges and actuators for deployment mechanisms)

  − AEC-ABLE Engineering, Goleta, CA (deployable and steerable arrays)

- Katema, El Cajon, CA
- Programmed Composites, Brea, CA
- Composite Optics, Los Angeles, CA

- *Use Commercial Batteries.* AMSAT pioneered the use of commercial, C-size NiCds. DSI/CTA has flown many gel-cell, lead-acid batteries. AeroAstro's ALEXIS used four C-cell NiCd stacks. These were built and fitted to the satellite in 1990 and lived with it through three years of testing until the 1993 launch. They have since powered the satellite through about 15,000 eclipse cycles and show no signs of degradation—in fact we haven't even conditioned them since launch. These things work well—way beyond the manufacturers' spec lifetime, assuming you select and treat them carefully. They also cost typically 5% to 10% of the price for MiL-SPEC batteries. If reliability worries you, fly four small commercial stacks instead of one large MiL-SPEC stack. You'll typically save $300,000 and gain redundancy. But keep in mind that successfully using commercial rechargables on satellites requires many tricks of the trade. If you haven't done this before, work with a manufacturer who has experience and understands the technology. Unfortunately, few companies supply commercial cells or batteries:

  - Gates Corporation, Denver, CO
  - Saft, Columbia, MD
  - Eagle-Picher Industries, Inc. Joplin, MO
  - AeroAstro corporation, Herndon, VA (builds commercial and MIL-SPEC batteries from other manufacturers' cells)

  With the focus on improving transportable, rechargeable power for laptop computing and personal mobile communications, we expect some beneficial technology transfer to small satellites in the next few years. Particularly promising are Li-ion and NiMH batteries. These bear watching. In the meantime, I recommend you read these two references on NiCd battery applications:

  - *Sealed-Cell NiCd Battery Applications Manual*, [Scott and Rusta, 1979, N80-16095]. General information on batteries and charging systems.
  - *Gates Battery Applications Manual*, a well-written manual that contains information about some of the newer types of batteries.

- *Avoid Long-lifetime Missions and Orbits with High Radiation Doses.* The operative phrase here is end-of-life. Photovoltaics and batteries degrade with time and with radiation dose, and you'll need to oversize them to maintain enough capacity on the last day of the mission. If that last day is 5,000 days after launch, you'll more than double your power-system size compared with a modest 500-day mission. Note too that commercial NiCds will not reliably survive a 5,000 day mission, so you'll have to use MiL-SPEC batteries.

- *Shop Carefully.* This is true of almost all small-satellite components, but it's truly shocking the range of prices you can find for the same installed capacity to convert solar power. For a 50 W mission, you can buy a small but comfortable summer home on Narragansett Bay for the difference in price between two identical products from two suppliers. If your program can live with minimum paper, find a vendor that isn't good with paper but makes a good solar array or battery. Suppliers of non-deployable, solar arrays include:

- Applied Solar Energy Corp. (ASEC), City of Industry, CA
- SpectroLab, Pasadena, CA
- Satellite Power Corp., Rancho Cucamonga, CA
- EEV, England
- Sharp, Japan
- CISE, Italy
- FIAR, Italy
- ASE GmbH, Germany

- *Distribute Power Regulation.* Batteries provide DC power at a single voltage. Thirty years ago, providing other voltages with good regulation was a major engineering undertaking. That's still true today, except all that juicy engineering has been subsumed into cheap, highly reliable DC regulators available from manufacturers like Interpoint, Datel, APEX, Melcher, and Pico. You will likely save time and money while increasing reliability by just piping battery voltage around the satellite and letting devices select their own converters and regulators at the site where the power is needed. A notable exception is high-voltage power supplies. Consider these not a part of your power system but a complex instrument living on your satellite, and plan accordingly.

5.3.3 Thermal

Like power systems, weddings, and almost anything you wish to imagine, a satellite's thermal system can be as simple as putting 25¢ in the parking meter or it can become the focus of your life and the lives of many others for years. Many small satellites in low-Earth orbit appear to have no thermal control at all. They are thermally a single block of metal with so many body-mounted solar panels that there is no room for thermal coatings. They work fine. But some small satellites, particularly those flown on interplanetary trajectories, have louvers, heat pipes, myriad thermal coatings, and a team of full-time thermal engineers. Having spent years on the design, these engineers remain on the payroll during mission operations, cheering the spacecraft on from Earth. Here are the leading strategies for controlling thermal-engineering costs, and what to do if you can't use them.

- *Stay Small*—Thermal conductivity is inversely proportional to length. If your satellite is a basketball-sized hunk of copper, it will be at one temperature, and your thermal model will be a zero-dimensional point (the only kind of point worthy of the name). If it's two football fields across and resembles a daddy longlegs spider, it won't and it can't. You'll need a finite-element model with thousands—maybe tens of thousands—of nodes to get even a wild guess at the temperature distribution. And that model will be worthless without a very large investment in thermal-vacuum testing to at least verify it models reality.

- *Stay Close to Earth*—Earth is a great heat source and sink when you're at 400 km. At 40,000 km it's not a significant part of the landscape. Close thermal coupling to the Earth tends to attenuate temperature fluctuations. The toughest missions are ones that start close to the Earth (all of them, so far) and end up far from both the Earth and Sun. The 40% decrease in insolation you experience going to Mars will radically alter spacecraft temperatures and require active control of heat generation, emissivity, and absorptivity.

- *Try to Live with Passive Thermal Control*—If you must do something, do something permanent, such as painting or thermal blanketing. Paint costs a few thousand dollars. Thermal blankets are tens of thousands of dollars. Active control, including louvers and spot heating, will be 10 to 20 times that.

  Easy to say, but how can you do it? Besides staying near Earth—or at least at a constant distance from the Sun, thermally tie as much of the satellite together as possible. Systems designers have to reconcile batteries that think they want to be at 15 °C with some telescope's focal-plane array that wants to be at 0 °C or colder and a CPU that can't go below –20 °C. Each of these subsystems engineers will lobby for his or her own thermal environment. You will accede to their demands at your own peril. Each thermally isolated element, having less thermal "inertia" (mass multiplied by mean heat capacity), will experience proportionately larger temperature fluctuations. Thus, you'll need more authoritative active control, mainly in the form of heaters which gobble electric power and worry everyone about whether they're on or off.

  Better to compromise among the warring factions and convince them that if they hang together you can guarantee 10 °C ± 10 °C (the ± being the fluctuations). Maybe offer them each 10% of what it would have cost you in parts, design, analysis, and assembly and test to accept the deal—you'll still be way ahead. One more thing to put on the table—but make sure everyone is exhausted and aware this is your last concession. Offer a backup electric heater **just in case**. A few simple heaters to make sure the processor boots and the batteries don't freeze won't bust your budget; if nothing else, they buy some peace of mind. Just make sure they can't get stuck in the "on" state.

- *Avoid Payloads with Special Thermal Requirements*—OK, not avoiding those payloads may be your job. But at some point payload requirements may drive your mission out of the low-cost regime. Really special requirements may drive not only the thermal design (insulating parts and so on) but also the stabilization system and the entire spacecraft layout. A part that needs constant shading from the Sun, total radiation decoupling from the spacecraft, and a 100 W thermoelectric or sterling-cycle cooler may be impossible on your budget. And not just because of the demands on the system design. Thermal design is second only to attitude control in difficulty of verification on Earth. You need at least a thermal-vacuum chamber with walls cooled well below the temperature of the coolest element on the spacecraft, plus plenty of time inside that chamber and lots of temperature sensors all over the spacecraft. Finally, you'll need the ability to turn on and off all the spacecraft's systems to look at their effects and determine if those effects track the model.

Oh, didn't I mention the model? Let's consider a typical 100 kg satellite with characteristic dimension of 1 m containing four or five **special** instruments. Along with the normal thermal concerns, such as not freezing the NiCds, understanding the satellite's thermal behavior will require several models. These will be finite-element models typically running under SINDA or NASTRAN and having 1,000 to 10,000 mesh points. Figure a couple of engineering person-years to build and verify that model, plus another person-year or so to see if it tracks the results you're getting in the vacuum chamber and to modify the parts of it that don't.

Perhaps most expensive of all is that verifying the model ties up the entire satellite—no RF range testing, no unbolting pieces that need rework—the whole thing has to be there including thermal blankets, radiators, and solar arrays.

If you *really* want a good model, your thermal-vacuum chamber will need Sun and Earth simulators. This can multiply the test costs tenfold—from about $25,000 to verify a basic thermal model to $250,000 for higher-fidelity simulation of the Sun and Earth. Plus you'll probably have to travel to find such a facility and modify your schedule to get into it when it's available. Note that the lower-cost test doesn't simulate the space environment—it puts the satellite in a vacuum environment to suppress conduction and convection and allows you to test the model, which you then use to predict on-orbit behavior. Going to the simulator chambers allows direct verification of mission performance. Most small missions live with the more economical route, and problems have rarely occurred.

- *Good Tools*—Thermal modeling requires a lot of labor if it gets beyond 10 or 20 nodes setup on a spreadsheet program. Labor is the bulk of your mission cost, and it requires management, of which less is decidedly more. You can take two main steps. Design the satellite on an integrated design tool that will either send the design directly to thermal-modeling software or will do the thermal modeling itself. Second, assuming you are going the old-fashioned way, use a basic algorithm your customer understands, such as SINDA (so they don't ask you to do it a second time on SINDA). Then, use a good pre- and post-processor and solver, such as PC-ITAS by Analytix Corp.

- *Simplify*—Thermal blankets, though expensive, may be cheaper than analyzing the complex geometry of the unblanketed spacecraft. To realize this savings, make sure you design the multilayer insulation to have simple shapes. This reduces manufacturing costs and avoids bends that would reduce insulation effectiveness while at the same time simplifying thermal analysis.

- *Thermal Tape*—It's cheaper than multilayer insulation; it's easier than paint. Very helpful on parts that must be exposed outside the blanket of multilayer insulation.

- *Avoid Requirements Based On Personal Feelings*—In thermal design, like most areas of mechanical engineering for satellites, everyone has an opinion. People don't dabble in a programmer's C code or in a processor-board designer's estimation of propagation times and collision probabilities, but everyone thinks they intuitively understand heat transfer, not to mention hinge and bolt sizing, from everyday experience. They don't.

Suppliers and references:

- Stapf Scientific, Baltimore, MD: Designing, selecting, making, and installing MLI thermal blanketing
- Sheldahl, Northfield, MN: Thermal-control materials, including tape
- *The Satellite Thermal Control Handbook* [Gilmore, 1994]: Excellent reference.

5.3.4 Telemetry

Advances in both digital electronics and low-cost, personal, mobile communications equipment have made telemetry one of the most dynamic elements of small-satellite design—perhaps second only to the onboard systems for information processing. Small satellites have historically sidestepped the costs for complex ground systems by employing their own relatively simple telemetry formats and PC-based custom ground stations. This trend has continued, but the flexibility computers are bringing to ground stations may allow even greater cost savings through shared use of highly flexible ground stations that can operate remotely and match a broad spectrum of telemetry protocols and frequencies.

Solid-state microwave amplifiers are now widely available which can provide S-band output power of tens or even hundreds of watts with efficiency comparable to travelling wave tubes. These are finding wide application in small satellites to increase data rate without driving up ground-station costs. At the same time, companies such as Telonics Corporation (Mesa, AZ) are providing low-cost tracking antenna-based ground stations. Low-cost, open-loop tracking antennas with up to 2 m aperture, including front-end electronics for RF, are now available for under $100,000.

Few low-cost telemetry systems are available off-the-shelf, so we have two choices: develop something ourselves or contact the small-spacecraft primes who provide low-cost telemetry subsystems. These include AeroAstro in Herndon, VA and Swedish Space Corporation in Solna, Sweden.

5.3.5 Information Processing

See Sec. 3.2 "Software" for an extended discussion of information-processing technology for low-cost spacecraft. I'll hit a few highlights here.

The old "command and control" system has used the enormous increase in onboard computation to become a much more important part of a satellite mission. It is almost an identifying characteristic of small satellites that they use highly capable onboard processing to eliminate power-consuming hardware and reduce costly ground operations. Modern satellites are highly autonomous, with fully automated attitude control, power management, data collection, and downlinking to an often autonomous, PC-based, ground station. Using the onboard computer to predigest mission data reduces other requirements on the system. For instance, onboard data compression can reduce the required telemetry bit rate by a factor of 10. MSTI, Clementine, HETE, ALEXIS, Freja, and Ofeq-III have all exploited highly capable onboard processing to do complex mission functions without direct commanding from the ground. This allows the satellite to carry out mission operations during most of the time when it's not in contact with any ground station.

All of these benefits come at two costs. One is that few commercially available computers are well suited to satellites. Thus, we must build our own or contact other primes to find what is already available. Secondly, software uses a lot of our spacecraft-development budget. Contrary to common lore, we can manage software costs without reverting to totally hardwired logic:

- Select equipment—including the processor, compiler, and debugger— which is proven and known to work well together. Writing software is challenging; doing so without reliable tools is virtually impossible. And impossible is expensive. This will limit your design to mainstream CPUs.

- Use the same development platform across the project. Cross-platform compatibility is often little more than a hypothesis whose verification you should leave for people who don't also have to get a satellite built quickly and cheaply.

- Agree on ground rules and stick with them. Include a style manual for how code is written and documented, adhere to a software architecture developed at the program's inception, and use a system to control versions.

- Keep it simple. The satellite has to work. But does it have to be fully commandable in real time during a four-minute pass over the ground station? Or able to compress video at 120 frames per second? Minimize requirements, particularly those which specify software performance (speed).

5.3.6 Mechanisms and Structures

We live in a world of complex mechanisms which work well. Automobiles, electrically deployed antennas, dishwashing machines, ice makers, office copiers, and one we too often take for granted—our own human bodies. Thus, it shocks us that so many satellites fail or have their missions curtailed due to faulty mechanisms much simpler than those inside an automatic teller machine. Yet, this failure is logical. It took hundreds of millions of years to develop (evolve) supposedly mundane things like the knee and the shoulder. Hundreds of millions of dollars have been spent to make automotive airbags deploy properly. Your car's "deployable" antenna has benefitted from tens of millions of automobile-years of on-the-road testing. But in space programs we must sketch a deployment mechanism for a satellite, a solar array, an antenna, or a boom and expect to build it, test it, and be willing to bet our lives on it—all within two years. That's like trying out a new Julia Child recipe for the first time when the boss is coming over for dinner in two hours and our future depends on perfection.

There are ways around this problem. Large sums of dollars and engineer-years have already been invested in many of the mechanisms you typically need. The Marmon ring used to separate spacecraft from launch vehicles is a good example. Available from Bristol Aerospace in Canada, they're not expensive and they always work.

After separation mechanisms, the next most common example is probably hinging solar panels, which have two main parts—the hinge itself and the latch. Some options for meeting this requirement were discussed in the section on solar arrays above. Deployable booms are also available. Note that the less expensive of these, an example being the Stacer boom, aren't retractable and control neither the physical length of deployment nor the rotation angle about the deployment axis. Booms having both these desirable characteristics are available from people like Spar in Canada and AEC Able Engineering in Goleta, CA.

Unfortunately, none of these options is cheap. What can be done for very low cost? AMSAT has reliably employed the "carpenter-tape" antenna. Typically, they make a coil of metal from a springy, slightly curled, metal strip. A wire holds the coil in place. When a current passes through the wire, it evaporates and the antenna deploys. Israeli Aircraft Industries (IAI) of Lod, Israel makes the wire-deployment mechanism and the antenna materials which can be deployed in this way. They have been flight-proven without failure on the Offeq series of satellites. Hinges made of similar restrained metal tapes are also reliable, but they can deploy only very small objects.

At least three low-cost actuators are available—pyrotechnic bolts, hot-wax actuators (Starsys Research, Boulder, CO), and memory metal actuators (GSH, Santa Monica, CA). The latter two can be tested and reset, which saves money and allows multiple tests to add confidence. However, the mechanisms which these actuators release are more complex, more typically tailored to a single application, and more expensive. One exception is a hinge available from Starsys Research in Boulder—it's compact and flight proven, and it costs no more than a large, Japanese, luxury sedan. IAI also manufactures hinges and other deployment mechanisms.

Some say the difference between large and small satellites is that small ones have no extrinsic structure. ALEXIS and HETE exemplify this approach. They bolt together a collection of boxes—major electronics box which houses the digital electronics, battery boxes, power-switcher box, radio boxes, and so on. The boxes, whose wall thicknesses are chosen for ease of manufacturing, are strong enough to carry launch loads and require no external structure. However, the Surrey satellites, as well as all the CTA satellites, have been built with an external shell often reinforced with stiffeners, structural plates, and hard-mount points resembling those on conventional satellites.

Like mechanisms, developing an efficient structure can take years, cost a lot of engineering time, and still fail. The major weapon here is margin. If you can avoid being squeezed for mass (and many small satellites are constrained more by volume than by mass), build everything much heavier than you believe is necessary, and you'll probably get through all the testing and the launch with no structural problems. As you weaken the structure to reduce mass, you're increasing the probability of catastrophic failure.

A lot of engineering focus is on the classical box or shell and stringer structure; unfortunately, this focus often misses the most important problems. Those large beefy structures, which we subject to all the fancy analyses our years in school trained us to do, rarely fail. What tends to fail during vibration testing and launch are much smaller, apparently less significant elements. Leading among these are connectors, wiring harnesses, and parts mounted on circuit boards. These typically aren't analyzed at all, mainly because structural engineers don't think about connectors; they think about stringers, shells, and boxes. And complex, tiny connectors are difficult to model and analyze. Luckily, the solution is usually common sense and inspection. If you can wiggle it, assume it will probably fail in test. Imagine the bumpiest roller-coaster ride of your life. Now multiply that by 10: that's what the satellite has to survive—with margin. It should be absolutely dead solid with no wiggle room at all anyplace. If wires can move, parts on a board can flex; if a connector has any load on it that hasn't been relieved, it's a potential failure point. Staking components with RTV or space-qualified epoxy works well and is inexpensive insurance. Support all connectors with wire ties or brackets. Make sure all wiring harnesses and cables are secured to the structure every inch or two. But be careful not to tie down so tightly that your small components are under undue stress all the time. This too can eventually lead to failure.

These measures, plus building a highly robust structure, will virtually guarantee a successful vibration test with minimum expense. Besides having to repair failed components, you must retest them, and retest means more test system, engineering, and technician time charged to the program. Applying foresight will save money, and in this case fixing what isn't broken (yet) is a reasonable approach.

The overwhelming choice in structural materials is still 6061T-6 aluminum. It is inexpensive, easy to machine, not prone to crack propagation, takes well to coatings, is not magnetic, and is well understood. Thick box walls made of it offer some radiation protection which you can enhance by selectively using tantalum foil. Structural modeling is easier with aluminum than with more exotic materials because it is isotropic. Aluminum honeycomb from Hexcel and other manufacturers is often used for faceplates and solar panels. Make sure your honeycomb isn't glued together with adhesives that you can't space qualify. Graphite-epoxy composite used for these same applications requires the same caution. Most terrestrial epoxy isn't usable in space. It will outgas, become brittle, and fail.

5.3.7 Propulsion

Most low-cost missions don't include propulsion. They are in low-Earth orbit, so they use magnetic or gravity-gradient stabilization. But some missions need either an orbit not available from a low-cost, dedicated, or piggyback launch or very precise orbit maintenance. Recently small satellites have begun to be built for geosynchronous and interplanetary missions where propulsion is necessary to control attitude and to modify trajectories and orbits.

By far the most frequently used propulsion alternatives for small satellites are modular solutions purchased as a **black box** or pressurized cold gas, usually nitrogen. Modular solutions include monopropellant hydrazine, storable bipropellants (usually N_2O_4 and MMH) and solid motors provided by such companies as Rocket Research; Olin; MBB division of Daimler Aerospace; Atlantic Research; and Thiokol. The primary advantage of these systems is that they provide large impulse at low mass without requiring the spacecraft team to understand the details of propulsion-system engineering. The liquid systems allow modulation of the thrust and are the only practical option for attitude control, trajectory, and orbit adjustments. Solids mainly apply for single, large impulses, which they provide reliably and efficiently. A typical solid application is a good solid nudge to move from GTO to an interplanetary trajectory, or from a circular, low-Earth, parking orbit to a highly elliptical orbit.

Costs for the propulsion modules themselves range from several hundred thousand dollars to several million dollars. There is an old rule of thumb that for every dollar spent on a subsystem, the team will spend three more dollars on its procurement, integration, and test. This may be an overestimation in this case, but not so much as you might first think. No matter how modular the propulsion system, it will make several critical demands on the spacecraft. These include structural attachment points, electrical interfaces for ignitors and valves, and perhaps most significant, mass distribution and attitude control to ensure that thrust is in the correct direction and that it doesn't upset the spacecraft. Another requirement is often onboard, autonomous determination of the trajectory or orbit, as well as upgraded attitude determination. A propulsion maneuver requires this state information to be executed correctly, and then usually results in a substantial change in the ephemerides and the spacecraft attitude. The ground station may not find the satellite right after such a maneuver, so the satellite may need to use its onboard sensors and software to recontact the ground.

Cold-gas systems, which typically use pressurized nitrogen metered through small sonic openings, are sometimes built by the spacecraft-development team. Don't make the mistake of thinking this is simple just because it is only gas phase and has no chemical reactions taking place. Typically these systems have low thrust and high pressure, requiring nozzle throats with very small diameters, which are easily clogged by foreign materials that typically reside inside pressure vessels. Valves and threaded fittings can leak, especially after attack by small solid particles suspended in a sonic gas stream. Fittings can also leak due to launch-induced vibration. Leaking can not only drain your propellant supply but also upset the spacecraft attitude and possibly kill the mission. And, as with modular systems, you'll need attachment points, electrical interfaces, and a guidance approach that will handle the torques even small thrust nozzles produce. The software and hardware that drive valves are another potential source of catastrophe. Like a teenager learning to drive, a small-satellite designer needs to respect the relatively huge amount of horsepower in even the smallest propulsion system.

One way to eliminate some of these risks is to use a prefabricated, all-welded, fluid-handling system. This makes the cold-gas option into a sort of module, which you can build and test separately and then integrate into the spacecraft almost like an electronics box or electrical batteries. Welding doesn't guarantee zero leaks, but once thoroughly cleaned and tested it should remain free of leaks and contaminants. Often a blow-away foil is attached to the nozzles at their exit plane to keep foreign materials from entering the system through the nozzle.

Alignment can be another significant engineering challenge with all these systems. Thrust forces are very large compared with naturally occurring disturbance torques, and even a small displacement between the cold-gas system and thrust axis can create torques which the attitude-control system can't overcome. For all of these reasons, carrying propulsion onboard can strongly increase cost. With planning, you can transfer many of these costs to a propulsion-system subcontractor, but the spacecraft-engineering team still has to deal with system complexity. Still, chances for low-cost piggyback launches aboard Delta, Ariane, Zenit, and Long March will continue to motivate onboard propulsion. Accepting a low-cost launch to a poor approximation of the desired orbit and then doing the final insertion with onboard propulsion may well prove more economical than paying for a dedicated launch. And, as small spacecraft are applied to more ambitious missions, they'll more likely have onboard propulsion.

AMSAT's OSCAR-13 spacecraft was launched by Ariane and used an MBB-supplied bipropellant thruster with integral helium pressurization to transition to a near-Molniya orbit, providing both orbit-raising and plane-change ΔV. The planned orbit wasn't precisely achieved because of helium leakage, but the satellite and mission were still a great success. (See Sec. 13.3 for AMSAT's experiences.) DSI (now a division of CTA) launched the seven "microsats" aboard Pegasus flight #2. Each carried a nitrogen tank and a single thrust nozzle to raise and lower orbit. The spacecraft were designed to maintain a stable constellation configuration among themselves. NEC has built several satellites, including the highly successful Planet-A for Japan's ISAS, which carried onboard liquid-propulsion systems to navigate and stabilize between planets.

5.4 Integration and Test

Simple logic: integration and test come at the end of a program. Therefore, they benefit the most from careful planning. Overspending of dollars and time throughout the program becomes most obvious during integration and test. Often, this last phase gets the blame for poor performance and sometimes undergoes corner cutting. This is wrong—and dangerous. The hallmark of a well-managed, small-spacecraft program is a well-organized, very thorough test program for the integrated system, even to the point of sacrificing some tests at the component level. Emphasis on integrated-system testing, not on component-level testing, most effectively leverages the small scale of the development job.

The interface-control document (ICD) is a tool which, if applied early in the program, will eliminate many integration problems. It should specify all electrical and mechanical interfaces in complete detail, such that an engineer can build a box or a circuit board without needing to poll every other engineer on the program—or even any other engineer, other than to check the ICD for errors. Besides ensuring that the subsystems actually will integrate (an obvious advantage at the integration phase), the ICD allows the program's engineers to begin their work without delay, and that allows you to get to the integration phase faster. Jump on the ICD early in the program and freeze it as soon as possible so the team can get to work. As soon as pieces of the design get built, take every opportunity to see if they fit together and work together. It's amazing how many little stitches in time you can sew together way ahead of the actual integration with these miniature integration tests.

While cool heads prevail—early in the program, maybe right after the ICD is finished—get everyone to agree on the test plan. Not that you'll speed up the program now, but when people get to the test phase, they become creative about all kinds of interesting new tests they'd like to try—which slow the program and add cost. Even worse, these new tests occasionally expand the envelope of required performance and environments. Thankfully, these little landmines won't be in your early plan, and you'll be able to shield your schedule against them or opt to include them as long as everyone understands the additional time and expense they'll incur. Developing this test plan will also highlight the need for particular ground support equipment (GSE) far enough in advance so you can develop or buy it before it slows the program's progress.

The key milestone test is the fully integrated satellite test or FIST. This is the end-to-end operation of the satellite, exercising all its capabilities as much as possible before launch. A well-built FIST provides a performance baseline you can test repeatedly to make sure the satellite continues to operate to spec. The FIST should include initial deployment and operation of the spacecraft. Typically the acquisition sequence and obtaining the anaerobic configuration are the most mission-threatening operations. Repeated simulation of the tasks which occur in the first few days or weeks of operation will make mission success more likely. The FIST works out the satellite and the crew which will operate it during that critical phase.

By the time the program enters test, typically 90% of the program budget is budgeted to have been spent—and, in fact, many programs have spent over 100% of the budget by this time. Thus it's good to keep a few resource-conserving steps for this phase in your back pocket.

- Do a simple modal survey in your own integration shop before going to a vibration facility. Better to find out before their clock is ticking if you have unacceptable modes.

- Similarly, do a nitrogen-atmosphere, thermal-cycling test before going to thermal vacuum. Owning a big oven and refrigerator is cheaper than a few wasted days for the crew at a thermal-vacuum chamber while a failed component gets replaced. Even though the vacuum test is a different environment, an atmospheric test over a slightly extended temperature range can anticipate 90%+ of all the thermal problems you'll find in the expensive test. Keep the following in mind:

  - Widen temperature ranges slightly to allow for convection.

  - Flush thoroughly with the dry nitrogen. Pockets of air will introduce water vapor that can become ice at low temperature and damage components.

  - Don't try to replace thermal-vacuum testing; this test pre-qualifies the satellite for the formal tests.

- The range test is typically the only chance you'll have to operate the spacecraft in a signal environment nearly as clean as you'll have on orbit. Use time at the range also to complete electromagnetic-interference testing of all spacecraft subsystems.

- For most small programs, the thermal-vacuum chamber will poorly simulate the space environment. Thus, use it mainly to validate the thermal model, not to measure the spacecraft's actual thermal behavior on orbit. The real thermal-vacuum test runs on your finite-element model once its validity is clear because it agrees with the results of the thermal-vacuum test.

Finding Things

A key part of systems engineering is leveraging what's already out there. But what is out there? This chapter contains several tables and lists of manufacturers of specific elements for small satellites. AeroAstro and New Space have created a World Wide Web site on the Internet which contains frequently updated, catalog-style data from many of the manufacturers of needed components. The site, whose address is http:// www.newspace.com/, also contains pointers to manufacturers of interest which may be listed elsewhere on the web. This is an excellent way to become acquainted with the components available and with the companies that provide them. Other databases on suppliers of small satellites and components are available from:

- ZARM, University of Bremen, Germany
- Aerospace Corporation, El Segundo, CA
- JPL, Pasadena, CA
- Microcosm, El Segundo, CA

The companion to this volume, *Directory of Space Technology Data Sources*, lists current databases and their sources, plus additional Internet sites where you can get current information. You may also want to contact some of the small-satellite primes directly to discuss how you can leverage existing products and facilities to reduce your mission cost. It's characteristic that we want to tailor every aspect of the program to our tastes, but it's easy to get bogged down redeveloping what exists. An item or approach that has flown successfully can greatly reduce cost and flight risk, while also shortening the development schedule and risk. You can contact nearly all of these primes through the newspace web site.

Bibliography

Most of the people who build small satellites seem to be too busy to write about their experiences, but I highly recommend the few references below:

Rick Fleeter, *Micro Space Craft*. An overview of the key technologies and general approaches to developing small spacecraft. The book includes plenty of advice on avoiding common pitfalls, detours, and misunderstandings along the way. Available from Edge City Press, Reston, VA; (703) 620-6650.

Martin Davidoff, *The Satellite Experimenter's Handbook*. Surveys and describes the satellites developed by the amateur (radio) satellite (AMSAT) community. AMSAT is the most prolific and successful developer of small, ultra-low-cost satellites. Available from the American Radio Relay League, Newington, Connecticut.

SMAD=Wiley J. Larson and James R. Wertz, *Space Mission Analysis and Design*, 2nd edition. See in particular the chapter by Fleeter on small-satellite design. Available from Microcosm Press, El Segundo, CA; (310) 726-4100.

Chapter 6

Reducing Mission Operations Cost

Madeleine H. Marshall, John A. Landshof
The Johns Hopkins University/APL
Jozef C. van der Ha
European Space Operations Centre

This chapter presents methods and concepts for reducing the cost of mission operations. The first section recalls the main components of a satellite mission and their relevance to the cost of mission operations. Decisions made **well before** the satellite launches largely determine this cost. Furthermore, many of these decisions are made *outside* the realm of operations, for instance in spacecraft design, without full awareness of how they'll eventually affect operations cost. On the other hand, reducing operations costs in isolation—without regarding costs for other system elements—is meaningless and, in fact, counterproductive. The actual objective should be to reduce cost over the mission's life cycle with the least effect on mission performance.

As a result, we'll strongly emphasize the environment for mission operations and how characteristics of this environment, such as spacecraft design features, determine operations cost. We can't state absolutely how much or how little operations should cost because that depends too much on the type and size of mission, on the experiences and culture of the operations center, and on the risks we're prepared to accept. We'll introduce a few examples of low-cost operations within a wide range of mission objectives and emphasize their most significant cost-saving features. Later sections describe in more detail the most important management and engineering methods for operations which have reduced, or are expected to reduce, mission life-cycle cost.

The recent motivation for reducing operations costs, particularly for NASA's space-science missions, has accelerated because funds required for mission operations were threatening to preclude new mission starts: a 1992 projection of costs for operating space-science missions and analyzing data showed they would double in real-year dollars from about $500 million in 1992 to almost $1 billion in 1999. Substantial cost-cutting has halted this trend of ever-increasing costs at a level of about $600 million per year for the rest of the decade, as shown by Ledbetter [1995].

6.1 The Environment for Mission Operations

To reduce operations costs significantly, we must understand the environment and context for mission operations. After the satellite and ground segment have been designed, developed, and tested, operators have little opportunity or incentive to reduce operations costs: the satellite must be operated within the environment defined by the space and ground systems and with the operations concept selected long before launch. Changes in the ground segment or operations concept after the satellite has launched are usually hard to justify and would be difficult and costly in the best of cases. Therefore, we **must** take operations into account while designing the system: in fact, they are a natural and integral part of system engineering for the overall mission. This concept is the most significant challenge to reducing total mission cost because it requires a reversal in well-entrenched traditions which have tended understandably to place a higher priority on pressing problems in spacecraft design rather than on considering mission life-cycle cost. To do meaningful life-cycle trades, mission funds must be centralized (under one manager's control) so they may be distributed across traditional barriers.

Mission objectives are the **raison d'être** of the satellite program: they state why someone is doing the mission. In the past, the mission objectives determined the required mission budget, but more recently budgets tend to be prescribed, and objectives must be tailored to the available funding. In the case of a *science mission*, the objectives usually aim at gathering data about a physical phenomenon by using some onboard instruments to measure it. This data is collected and preprocessed on board, transmitted to a ground station, and routed to a control center. There it undergoes further processing before being forwarded to scientists who evaluate its scientific content. For a *military mission*, the objectives may be to actively acquire and track a surveillance target, determine a state vector onboard, and hand it over to a ground-based sensor in real time. For a *commercial mission*, the objectives may be to launch and operate a satellite that provides a global communication service while generating a profit for the company. In each case, the mission objectives drive the characteristics of the system we need for effective mission operations.

Figure 6-1 shows the components of a *mission system* and their interrelationships during pre-launch design and development. A space mission system includes all elements necessary to design, develop, test, launch, and operate one (or more) spacecraft in order to achieve mission objectives, as described in SMAD, Chap. 1. All space missions include mission objectives; a mission design (which includes defining the orbit or trajectory and designing the operations concept); a space segment; a launch system; a mission operations system; and, for some missions, a data analysis system. In our terms, the *mission operations system* consists of the ground segment, the operations processes, and the mission-operations team.

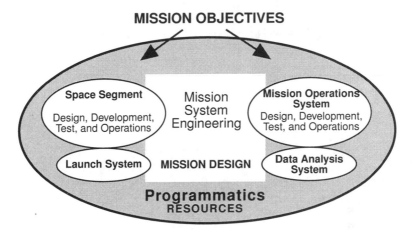

Fig. 6-1. Components of a Mission System (Pre-launch). All space missions include these components, although perhaps under different terms. Of crucial importance is the synthesized point of view for the overall system.

The context within which we develop the mission components includes such intangibles as schedule, organization, budget, risk level, and reporting requirements. Sometimes appropriately called *constraints*, considerations such as budget and schedule nonetheless are an important part of every space program, and usually are prescribed to the mission design team. We'll give the name *programmatics* to the context within which we design, develop, and operate. Programmatics significantly influence both the character of the operations concept and the cost of space missions; in fact, they tend to dominate costs.

This chapter focuses on methods which reduce the cost of the pre-launch design, development, and testing of the mission-operations system, as well as that of post-launch mission operations. **To get the most for our money, we must design all components of a mission system in concert over the mission life cycle by using system engineering.** Therefore, concepts and methods in other domains (such as spacecraft design) will be addressed in this chapter since these can significantly affect the cost of operations. In fact, ignoring system engineering almost certainly results in unnecessary complications in mission operations and increased overall mission cost.

Allocating facilities and activities to components of the mission system is a central part of system engineering. This process addresses design options and allocates resources for a space mission over the mission's life cycle and across all components, including the mission objectives, the space segment, the mission operations system, and the distribution and analysis of mission data. Given limited resources (e.g., money and time), the system engineering process must determine the most efficient use of these resources for achieving the mission objectives by looking at the mission's entire system architecture.

We represent mission operations during mission system design by developing the *operations concept* shown in Fig. 6-2. The operations concept describes **how** the mission will be conducted. Seen at the left as input to concept development are mission objectives and programmatic considerations such as schedule and budget. To save the most money, developing the operations concept must be part of system engineering, so we'll have the most room to do trades between all feasible design options. We should develop the operations concept based on an end-to-end view of the mission. That means connecting the mission objectives formed at the beginning of the program to the mission products operations must generate efficiently and cost-effectively.

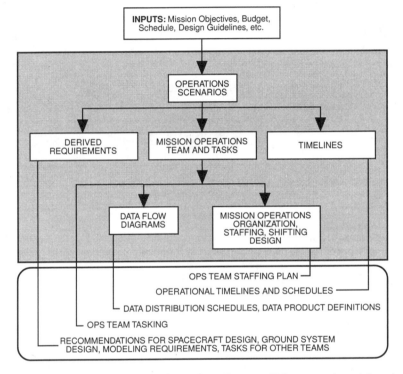

Fig. 6-2. Process for Developing the Operations Concept. This process is an integral part of system engineering for the overall mission.

The mission operations system includes the functions in Table 6-1, which must provide relevant and fault-free data and supporting services to the science or user community:

- *Mission Operations Planning*: determining the operational activities to be done over a given period of time

- *Spacecraft Operations Control*: exchanging telecommand, telemetry, and tracking data with the spacecraft

- *Mission Operations Assessment*: evaluating how well the spacecraft and ground segment achieve mission objectives

TABLE 6-1. Operations Functions and Components of the Mission Operations System.
This table summarizes the main ground-segment facilities, processes, and teams required for the different operations functions.

Operations Functions	Ground Segment Facilities	Operations Processes	Mission Operations Teams
Mission operations planning	• Workstation hardware & software • Models & simulators • Schedule management tools • Flight dynamics tools • Command translators	• Developing timelines • Developing scenarios • Sequencing & generating commands • Preparing for maneuvers • Scheduling & planning contacts	• Ground system engineers • Mission analysts • Operations engineers • Operations analysts • Flight dynamics experts
Spacecraft operations control	• Ground stations • Communications links • Operations control center • Workstations	• Uplinking commands • Downlinking telemetry • Displaying & monitoring • Processing & distributing data	• Operations experts • Mission controller • Support engineers • Data handling engineers
Mission operations assessment	• Trending software • Spacecraft, orbit & environment models • Report generators	• Analyzing trends • Identifying & investigating anomalies • Updating models	• Operations engineers • Spacecraft & ground system engineers

Table 6-1 maps these operations functions into the major elements of each component of the mission operations system: ground segment facilities, operations processes, and operations team characteristics.

We can divide operations costs into two main categories: development costs (mainly pre-launch) and operations costs (post-launch). Examples of the usually non-recurring development costs are

- Manpower to develop the ground segment and prepare for operations
- Procurement of ground-segment equipment and facilities, such as control center, workstations, ground stations
- Development and acquisition of software elements

Examples of recurring on-orbit costs are

- Operations-team manpower
- Maintaining equipment and software
- Use of facilities, such as rental of communications circuits
- Consumables, such as archive media and paper

We'll address ways to reduce cost in both categories. We won't discuss analysis of science data and interfaces with users or customers because these areas usually fall just outside of the traditional realm of mission operations. But, of course, these aspects must also be part of the engineering trades for the mission system.

For more detailed information on the elements of a ground system and on the activities and objectives of mission operations, see SMAD, Chap. 15 on "Ground System Design and Sizing" by Whitworth and Chap. 14 on "Mission Operations" by Negron and Chomas.

Although the system for a space mission includes space and ground segments, it is built and operated within the context of programmatic considerations, such as schedule, organizational structure, institutional culture, and budget. These are usually in place before a satellite program begins and aren't part of the trade space given to the mission-system engineer. A discussion of cost-reduction methods for operations can't disregard the programmatic ground rules. For example, imposed institutional reporting and documentation standards for hardware and software development can strongly drive up the cost of ground system development.

As an illustration of how programmatics affect mission operations design, we consider the programmatic requirement for distributed processing of spacecraft telemetry by a science team residing at several remote sites. The operations concept responds to the mission objectives consistent with the programmatic context in which the mission must be designed, developed, and operated. We'll therefore design a ground system that supports the required distributed processing of spacecraft telemetry. Normally, we'll do several trade studies (for example, direct downlink at remote sites versus centralized downlink and distribution on the ground). Eventually, we choose a design, perhaps taking account of other programmatic considerations (funding or time delay in receipt of data by the science team). We'll also need to evaluate the implications on other mission components and resolve them with other parties involved in the mission as part of system engineering for the mission.

6.2 Examples of Low-Cost Operations

Table 6-2 lists our selected examples of operations concepts for low-cost missions. These missions range from the $200 million NEAR mission to a $2 million UoSAT mission; this range represents all mission objectives low-cost spacecraft can now achieve. Naturally, there is a significant difference in complexity among the selected missions. NEAR is the first of NASA's Discovery science missions. It is a full-fledged interplanetary mission with epochal objectives, i.e., the first rendezvous with an asteroid. At the other extreme, UoSAT-5 is mainly a technology-demonstration mission. One or two decades ago, the objectives for any of the missions in Table 6-2 would likely have cost 5 to 10 times more to fulfill.

TABLE 6-2. Summary of Selected Missions for Small Satellites. These missions cover the full spectrum of mission objectives for present-day, low-cost spacecraft.[1]

	NEAR	Clementine	SAMPEX	ALEXIS	UoSAT-5
Agency	NASA	BMDO	NASA	DoE	Various
System responsibility	JHU/APL	NRL	GSFC	LANL	U. of Surrey
Ops center	JHU/APL	NRL	GSFC	LANL	U. of Surrey
Mass (kg)	805/400[2]	1,690/227[2]	161	113	49
Launch date	2/1996	1/1994	7/1992	4/1993	7/1991
Development phase (mos)	26	22	36	48	18
Operations phase (mos)	47	3.5	48	30	36

[1]Clementine and SAMPEX are described in Sec. 12.1 and 11.3. ALEXIS was built by AeroAstro who also designed HETE described in Sec. 11.4. UoSAT-5 was built by SSTL and is similar to PoSAT-1 described in Sec. 13.4.

[2]Beginning of life/End of life.

Table 6-3 shows the associated cost figures for development, launch, and operations. ALEXIS was actually launch-ready within 36 months after project start but had to wait 12 months because of launcher-induced delays. Launch costs range from $50 million for the 800-kg NEAR spacecraft, which a Delta II will launch into an interplanetary trajectory, to $200K for UoSAT, which Ariane 4 ASAP launched along with ESA's ERS-1 satellite into a Sun-synchronous orbit in July 1991.

TABLE 6-3. Breakdown of Costs for Selected Missions. These costs are expressed in "real-year" $ million, except for SAMPEX, which is in FY94$.

	NEAR	Clementine	SAMPEX	ALEXIS	UoSAT-5
Satellite design, I&T	122.5	48.8	45.7	6.5	1.5
Payload	(incl.)	4.8	7.2	12	(incl.)
Launcher	50	20	13	8	0.2
Ground system	(incl.)	1.4	7.6	0.5	0.1
Mission operations	17	5	4.6	1.4	0.2
Analysis of science data	19	?	8.4	1.6	(incl.)
Total, operations, analysis	36	5+	13	3	0.2
Total mission cost	208.5	80+	86.5	30	2
Mission ops cost (yr)	4.3	17.1	1.2	0.6	0.07
Mission ops staff FTE	11/27[1]	62	9	9	2
%(mission ops cost/total mission cost)	8.2%	6.3%	5.3%	4.7%	10%

[1] The NEAR mission includes a 37-month cruise phase, followed by a 12-month rendezvous phase.

The cost breakdowns in Table 6-3 for the ground segment and operations require careful interpretation. Financial protocols among the institutes result in significantly different cost calculations, which we can't take into account here. Costs for overhead and infrastructure may or may not be accounted for. Furthermore, individual "anomalies" spoil any direct comparisons. For instance, the development cost of Clementine's payload instruments isn't included in the project budget (BMDO provided them free), but the payload cost for ALEXIS amounts to about three quarters of the total system cost. Also, remember that the operations cost for ALEXIS is higher (by about 20%) than it would have been without the onboard anomaly.

In any case, Table 6-3 suggests that the total cost of ground segment, mission operations, and science data analyses runs between 10 and 25% of the total mission cost. The actual proportion depends on mission, science, and operations character-istics as well as on how long mission operations last. There's no magic "correct" ratio. To distribute costs best among spacecraft design, ground system design, and mission operations, we need to do trade studies during the design phase (system engineering) for each mission.

6.2.1 NEAR

NEAR is a low-cost mission in NASA's Discovery series intended to rendezvous with and explore a near-Earth asteroid. The spacecraft will launch on a Delta II vehicle during a 16-day launch window from February 16 to March 2, 1996. Overall mission design and operations responsibility for NEAR lies with the Johns Hopkins University Applied Physics Laboratory, APL. The APL will monitor and control the spacecraft from an on-site mission operations center, and will provide a science data center. The Deep Space Network (DSN), operated by NASA and JPL, will do spacecraft tracking, receive telemetry data, and transmit commands. The latter institute also supports spacecraft navigation, which is crucial for achieving the mission objectives during rendezvous.

The short development schedule of less than 26 months fiercely challenges the spacecraft design and operations teams because many activities which normally are sequential must run in parallel. For instance, teams must design and build the ground segment well before the spacecraft design and implementation has matured. Spacecraft design and operations teams must coordinate and cooperate efficiently to carry out such a short schedule. Figure 6-3 shows the major elements of the NEAR ground system.

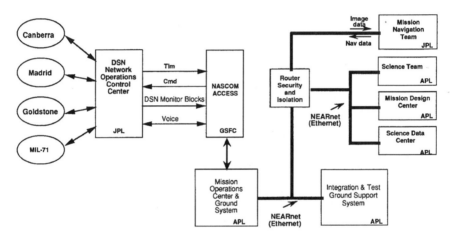

Fig. 6-3. Overview of NEAR's Ground System. The NEAR spacecraft is controlled from a Mission Operations Center located at the Johns Hopkins University Applied Physics Laboratory through the NASA Deep Space Network.

Marshall et al., [1995] and Cameron et al., [1994] describe the main characteristics of the NEAR operations concept. We've summarized them as follows:

- Mission operations will be in three distinct phases. First, there is the *launch and early operations phase*, lasting about 30 days and consisting of in-orbit checkout and calibration of the subsystems and instruments. In addition, operators may correct the trajectory if needed because of launch delivery dispersion. Second is the interplanetary *cruise phase* of approximately 35 months. Finally, the actual mission is the asteroid *rendezvous phase*, which lasts about 12 months and contains the operations related to the detailed study of EROS.

- Communications with the spacecraft during all mission phases will be through the Deep Space Network ground stations and communications network. All mission operations activities will be controlled from a dedicated mission control room at APL.

- A central part of NEAR's operations concept is employing a small number of highly trained operations specialists equipped with state-of-the-art, commercially available, software and hardware tools. The spacecraft engineering team will augment the operations team to support specific subsystems as needed.

- During the more than 3 years of cruise, an operations team of 11 people will operate the spacecraft and payload with activities such as monitoring, calibrating, and maintaining instruments, updating onboard software (if necessary), and doing trajectory maneuvers and an Earth gravity assist.

- During the rendezvous phase, an operations team of about 27 people will be available at the operations control center around the clock in four shifts; tasks will include uplinking command sequences and monitoring the spacecraft's performance. Contact with the spacecraft will be essentially continuous (three 8-hour passes per day). The team will do trajectory correction maneuvers as necessary during this period. A Science Data Center, also located at APL, will archive and distribute mission science data.

- NEAR's mission-control system may be viewed as a single system composed of the *mission operations ground segment* and the *integration and test operations ground segment*, with the latter a "prototype" of the former. A single lead engineer and a single development team handle development. The system uses the commercial EPOCH 2000 product from Integral Systems, Inc. augmented with NEAR-specific functions and interfaces.

- The concept for developing the operations system follows a so-called *evolutionary prototyping* approach to develop critical new software components. This approach offers the opportunity to incorporate system enhancements while the operations concept matures, but it also requires flexible yet disciplined management of budget, risk, and schedule.

- The mission operations team trained by supporting the spacecraft integration and test phase; the team will cross-train to handle planning, control, and assessment during the mission.

- The mission operations ground segment will provide tools to analyze and schedule spacecraft and ground operations; generate, verify, and uplink command sequences; monitor and process telemetry data; assess spacecraft performance; and investigate anomalies. Also, software will be able to do spacecraft simulations (as necessary) for telemetry predictions and command validation. The science team will receive planning information through a graphical interface.

6.2.2 Clementine

The Clementine mission combined the flight-qualification of advanced technologies for lightweight sensors with planetary exploration objectives such as lunar mapping. (See Sec. 12.1.) The design of the mission operations system was driven—like the other mission elements—by Clementine's extremely short schedule (less than 2 years). One of the consequences of this short schedule was the "cradle to grave" approach: the *same team* of engineers was responsible for designing, testing, and operating the mission. The spacecraft control system, telemetry processing, and tele-command preparations were based on experienced NRL systems with perhaps 20% mission-dedicated extensions. A dedicated high-fidelity testbed was produced, and this tool was essential for validating the onboard software and control system.

During the lunar-mapping part of the Clementine mission, contact with the space-craft was almost full-time—18 to 20 hours per day. NRL's Pomonkey ground antenna and NASA's DSN stations (primarily, the 26 m subnet) and communications network supported Clementine's data downlink of 128 kbps and uplink of 1 kbps. Operators worked around the clock with four shift teams of eight staff each, covering all of the required activities (trajectory, science planning, scheduling, control, software, space-craft systems, sensors, and data). Another 30 full-time staff were supporting various tasks during normal working hours. Finally, the mission would call on another 40 or so engineering experts when required.

An important contributor to Clementine's successful operations concept was the close (desktop-level) interaction between science representatives and operations specialists, which resulted in almost immediate adaptations of instruments settings during actual operations [Horan, 1995]. By the end of the lunar-mapping phase, new operations approaches were tested, e.g., automation of operations scheduling and autonomous onboard monitoring and control of power levels. Of particular interest was the *Spacecraft Command Language* system, which is basically a script and rule compiler for onboard stored commanding, as described by Buckley & Van Gaasbeck [1994]. This system is a prototype for standardizing high-level spacecraft control interfaces and uses identical control software onboard and on the ground. Clementine used it onboard mainly to experiment with autonomous operations; after completing the lunar-mapping phase, this system proved capable of autonomously performing all operations required for a full 5-hour mapping sequence.

6.2.3 SAMPEX

SAMPEX is the first of NASA's Small Explorer missions. It launched in July 1992 on a Scout launcher. (See Sec. 11.3.) The design of the Small Explorers has less built-in redundancy than would usually be the case for this type of science mission, but the associated somewhat higher risk of mission failure is part of the philosophy for the Small Explorer program. Reducing redundancies does require strong coordi-nation of the designs for the space and ground system in order to make the overall system robust—for instance, by enhancing autonomy and operability. Also, emphasis is on simplicity of these designs. For example, there is no means of orbit control even though operators can calculate orbit using two-way Doppler ranging measurements. One of the principal challenges for SAMPEX was the short development schedule (less than 3 years). Another was using a 386-based onboard processor which wasn't already space qualified. This 386-type processor features dynamic memory manage-ment and adds many functions to those of conventional space-qualified processors. However, this increase in functionality has many repercussions on the operations support, such as more complex management of onboard software, tables, and com-mands [Mandl et al., 1992].

Noteworthy is that SAMPEX represents NASA's first use of a solid-state memory and the *full* protocol standard for space-to-ground interfacing of telemetry and telecommand data. The international *Consultative Committee on Space Data Stand-*

ards developed this protocol standard. This committee consists of experts on transferring and processing telemetry and telecommand data—from NASA, ESA, and NASDA. Due to SAMPEX's relatively high data rate of 900 kbps and the new formats associated with the data-interface standards, mission teams had to develop new equipment for synchronizing blocks and frames. SAMPEX was also the first new mission to use Goddard's multimission core software and hardware for the *Transportable Payload Operations Control Center*. The goal of this effort was to develop a core system containing at least 80% of the functions required by each new mission [Mandl et al., 1992]. It used as much commercial off-the-shelf hardware and software as possible to make it more reusable and flexible. Industry standards were followed as much as practically feasible, for instance UNIX, Ethernet and TCP/IP. Intense coordination between the different user missions and early prototyping of the core telemetry processing functions for SAMPEX were critical in meeting the short development schedule.

An expert-system package [Hughes et al., 1994] was integrated into the control system to help isolate faults through intelligent monitoring of graphical data and rule-based processes. This system has been used to demonstrate proof-of-concept for autonomous, routine, spacecraft-control functions, including telemetry reception and monitoring and command actions during a nominal SAMPEX pass. Automated operations include verifying the pre-pass readiness by testing data flow, examining the spacecraft's event log, setting up configuration monitors, evaluating system events, and starting the daily telecommand load and the onboard data dumps.

A dedicated control room with at least three separate, dedicated, control-system "strings," each with its own software and hardware, is available for operating Small Explorer missions. The common operations team can use any of these strings to support a ground contact. They can support up to three on-orbit missions and one mission in pre-launch tests "simultaneously" (with at most two in real-time concurrent passes). During any given shift, two spacecraft analysts and one command controller are on duty to support all Small Explorer missions. At present, SAMPEX is the only Small Explorer in orbit and is operated with one 12-hour shift per day.

The flight operations team trained through heavy involvement in spacecraft tests—preparing test plans and developing *System Test and Operations Language* procedures, as well as operating the test workstation and related test facilities. An operations staff of about 20 people is needed to operate up to three or four missions at the same time.

The spacecraft autonomously monitors and controls some "Telemetry and Statistics." For instance, it uses onboard limit checking and further calculations on spacecraft housekeeping telemetry, which may then trigger autonomous actions such as shutting down one or more subsystems. In addition, associated event reports are prepared and downlinked. Another novel feature developed and validated on the SAMPEX spacecraft is the autonomous data dump, in which spacecraft data is downlinked, received on the ground, and stored in data files without human interaction.

6.2.4 ALEXIS

The experience of the extremely small satellite ALEXIS is relevant because it provides an insight into potential trends of future science missions. AeroAstro, Inc. developed the ALEXIS spacecraft bus and ground system; the Los Alamos National

Laboratory handled overall mission-system design, integration, test, and operations. ALEXIS was launched by a Pegasus launcher on April 25, 1993 and encountered severe difficulties due to mechanical problems when a solar panel broke loose during launch but remained attached by a cable bundle. All systems except for the magnetometer (which is mounted on the damaged panel) are fully functioning. Without a working magnetometer, the attitude control system couldn't point the satellite to the Sun. This problem necessitated complex recovery operations lasting until the end of July 1993, as reported by Bloch [1994]. As a result of the anomaly, the attitude corrections and control needed to take place on the ground rather than autonomously onboard as originally planned. Nevertheless, the satellite is producing state-of-the-art science results, such as a sky map in the EUV range of the spectrum and ionospheric VHF measurements with the Blackbeard instrument.

The total mission development cost for spacecraft, instruments, and integration was about $17 million up to readiness for launching in April 1992; because the actual launch took place about a year later, the mission cost another $2 million. Operations costs, including science operations, have run about $1.5 million for the first year, including the 2 people added to cover the extra ground attitude determination and control tasks resulting from the mechanical anomaly [Bloch, 1995].

ALEXIS weighs 113 kg (45 kg of which belongs to the payload) and uses a largely non-redundant design and many parts that don't meet MIL standards. The spacecraft bus processor consists of redundant 80C86-based CPUs and six onboard mass-memory boards with 96 MB total storage. But one board was mapped out after a failure shortly before launch, so a net mass memory of 78 MB is available in-orbit for science data storage between data downlinks [Priedhorsky et al., 1993a]. The rate for downlink telemetry is 750 kbps and for uplink is 9.6 kbps; the specification for attitude reconstitution is 0.25 deg, with magnetic torque coils providing attitude control. The steerable 1.8 m ground antenna is mounted at the roof of the Los Alamos Physics building and gives ALEXIS about half an hour of total contact time during four passes per day.

The procedures for acquiring data and transmitting commands had been designed for automated and unattended contacts during nights and weekends. But frequent, unpredictable ground computer crashes and spacecraft thermal problems caused by the anomaly didn't allow this approach in the beginning. Operators solved the original ground system's instability by exchanging the platform and streamlining and auto-mating many of the operations tasks. At present, telescope command loads can be uplinked to cover 5 days of operations.

The operations team consists of nine people [Roussel-Dupre, 1994], with only one person having previous satellite operations experience. The qualifications and range of specialties may well have contributed to ALEXIS's operations success. At least two team members support each pass: one from the spacecraft operations team and one duty scientist. In general, operators are on duty two weeks and off one week, whereas duty scientists are on shift duty for five days and have five days off. The operations engineers have digital pagers which summarize critical parameters of the spacecraft's status at the beginning and end of unattended contact periods. Payload operations have now become sufficiently routine that graduate students are able to do them alone. The spacecraft operations and science staffs are closely integrated, so scientists can also do operations tasks if required. ALEXIS is automating further to reduce routine operations costs for the science and attitude-control operations. The goal is to be able to manage night and weekend operations in a fully unattended manner, as originally intended.

Finally, the mission couldn't have succeeded without the spacecraft design's robustness and resource margins (in particular, the economy of the power subsystem), as well as the programmed versatility of the on-board processor which was instrumental in salvaging the mission after a serious catastrophe. The failure itself can hardly be blamed on ALEXIS's low cost and short schedule; even with unlimited resources and time, the solar panels wouldn't have been redesigned [Priedhorsky, 1993b].

6.2.5 UoSAT

UoSATs are small satellites built and operated by the University of Surrey in collaboration with a University-based research company, Surrey Satellite Technology Limited. Ten satellites have launched so far, starting with a technology demonstration mission, UoSAT-1, in 1981. These missions cost very little, typically $1 million to $2 million, and they have an extremely short development schedule of about one year. (See PoSAT-1 discussion in Sec. 13.4.)

The UoSAT program has been highly successful and must be given credit for showing that microsats aren't less robust than a typical large spacecraft: the first two missions have each operated more than 8 years, and only one (UoSAT-4 in 1990) has failed prematurely. UoSATs have handled various mission objectives and payloads using basically the same satellite platform: for instance, a specialized store-and-forward communications transponder and a medium-resolution Earth-imaging instrument were on UoSAT-5, which launched in July 1991 [Radbone & Sweeting, 1993]. Radbone & Sweeting also make a number of valuable recommendations on the management and technical aspects of small satellite missions, especially the critical integration and co-location of all team members.

The entire ground segment, including ground-station and operations-control facilities, is in a small area at the University of Surrey. Even though the ground segment contains all the main "conventional" elements, it typically has less redundant and much less sophisticated hardware and software. For example, five PCs handle all operations support, such as tracking, telemetry processing, archiving, and telecommanding.

The most significant characteristics of UoSAT operation are as follows:

- Telemetry and telecommand links are at VHF or UHF frequency bands allocated to amateur radio, so most of the equipment is commercially available.

- Communications interfaces are simple because equipment is co-located; most of the workstations are PC-compatible, so data transfer is often done on diskettes. The only external link is with NORAD for communicating the orbital elements.

- The control center is equipped with workstations dedicated to preparing and archiving telemetry, preparing telecommands, processing payload data, and simulations; in addition, off-line functions like mission planning and flight dynamics are carried out on general-purpose, office-type PCs.

- A spacecraft simulator built around the engineering model of the spacecraft electronics is used for pre-launch tests of operations procedures and for validating updated onboard software during operations.

- The ground-segment hardware and software are the same (with minor enhancements) as those used for a number of previous missions; pre-launch validation is therefore relatively straightforward.

- Telemetry reception is highly automated, and human interventions are required only in case of malfunctions, so shift work is unusual.

- Commanding is *not* automated but is done in real-time only by the person in the loop.

- Onboard autonomous orbit determination uses position information provided by an onboard GPS receiver.

- Nearly all operators were involved in the spacecraft design, and heritage is strong from one mission to the next; therefore, formal training and documentation could be minimized.

- The typical team for (day shift) operations control has only one or two people; they do all planning and control tasks. Mission planning is "by hand" and typically requires half a day for a two-week schedule.

6.3 Programmatic Methods to Reduce Operations Cost

The budgetary pressures on present-day satellite projects seriously affect all areas of spacecraft and ground-systems design, implementation, and operations. In particular, they've generated many new concepts aimed at decreasing total expenditures to achieve the particular mission objectives. Although we could apply two different operations concepts on essentially identical missions and compare effectiveness, most documented space missions are unique, one-time projects. For these missions, significant differences in objectives and programmatics prevent us from comparing two missions "apples-to-apples" for productivity. Consequently, no cost-reduction method to date has reduced operations cost by an absolute percentage. **The methods that follow, however, when used appropriately on applicable missions, will reduce the cost of space mission operations relative to conventional methods.** The methods apply differently to vastly differing mission types and institutions. For instance, a method that reduces the development cost of a ground system for a constellation of communication satellites may not be cost-effective for a unique, scientific-research satellite.

Programmatics control the context in which a space mission system is designed, developed, tested, launched, and operated. Requirements for documentation, schedule, staffing, and organization are all usually levied at the program level, so we can start reducing costs the most at this level. Are rigorous documentation trees necessary, or just done out of habit? Are schedules and budgets meaningful, or does everyone "know" that the end date will slip and cost estimates will grow? These types of questions, addressed at the top of a program, hold the key to reducing costs. Programmatics include elements of potential cost reduction which by necessity are out of the hands of the space system's developers and operators. Requirements for meetings, briefings, status reports, schedules, drawings, and technical reviews are usually defined at the top; these requirements can be very wasteful of an engineer's time, especially when teams are small. To work effectively, we must keep management and administrative tasks to an absolute minimum. We should also select team members for their competence and experience and then give them far-reaching responsibility, accountability, and empowerment on issues in which they're expert. Doing so will eliminate superfluous managerial decisions and delays, and will maximize team motivation and commitment.

Table 6-4 summarizes the major programmatic cost-reduction methods we'll address. They are what the engineering staff needs to meet objectives within the prescribed low cost and short schedule.

TABLE 6-4. Summary of Programmatic Methods for a Low-Cost System. Programmatics control the context in which a space mission is developed and operated.

Programmatic Features that Can Reduce Cost
1. *Short Development Schedule*   • Focused objectives   • Acceptance of risks   • Rapid prototyping
2. *Effective and Competent Teams*   • Co-locating teams   • Empowering staff   • Efficient team interactions
3. *Mission System Engineering*   • Concurrent engineering   • Efficient documentation   • System-level trades

6.3.1 Short Development Schedule

A *short development schedule* means 3 years at most between mission approval and launch. Normally, we can achieve such a schedule only when the payload instruments are either extremely simple or already designed. A short schedule is beneficial from many points of view. First, it enhances "mission turnaround," which benefits the user. Second, it provides significant advantages in terms of the potential use of advanced technology both onboard and on-ground. For instance, we can buy computer hardware fairly early in the cycle without fear of obsolescence by launch. A short schedule is also an effective cost reducer because time is indeed money. This simple fact is true not only in the sense that the total staff cost is roughly proportional to the length of the implementation period, but also because the work will proceed much more efficiently under a fixed minimum schedule. When a mission takes many years from design to launch, it's only natural to engage in one or more of the typical conventional development "rituals," such as extensive studies, repeated design iterations and refinements, unwarranted perfectionism, excessive documentation, and frequent reviews, which are often counterproductive to cost-effectiveness. Furthermore, it's easier to maintain staff morale and continuity of expertise over a short period.

But a short schedule powerfully drives how we define and build the spacecraft and ground segment and affects all levels of design and implementation activity. The schedule for the development program must be strongly fixed at the program level: knowledge of schedule creep will inevitably become actual creep. In case of schedule or budget problems, we may need to descope the program. Finally, a short schedule imposes a considerable burden on the engineering staff, whom we expect to work at peak level during two or more years. We must take seriously and adopt all possible precautions for the danger of staff burnout. Table 6-5 summarizes the cost-saving features of a short development schedule, which we'll discuss below.

TABLE 6-5. **Cost-Savings Potential of a Short Schedule.** A short schedule drives how we develop the mission system and affects all levels of design and implementation activity.

Method	Cost-Saving Potential
Focused objectives	• Savings due to effective implementation concept—reduces waste of effort from working on wrong issues
Acceptance of risks	• Reducing effort on marginal improvements • Reducing redundancies in spacecraft subsystems and ground facilities • Possibly reducing operations manpower
Rapid prototyping	• Savings from reducing overall development duration • Indirect savings due to more efficient interactions between teams

Focused Objectives

This term refers to the clarity of "what one needs to do." All team members must know the mission objectives, so they all can "pull in the same direction." To achieve success on a short schedule, establish well-defined and unambiguous objectives from the beginning. In particular, the mission objectives should be clear, complete, and preferably categorized into primary and secondary priorities. The principal payload instruments should be mature in terms of their functional and design characteristics. Prepare an efficient and realistic concept for carrying out the mission, with adequate cost and schedule margins for unexpected emergencies. Get agreement on this concept from all parties. Keep in reserve a reasonable potential for descoping or relaxing mission objectives or system design requirements. This reserve will allow flexibility in the face of financial, schedule, or technical problems

One element we must fix early at the mission level is the number of payload instruments. There must be relatively few (three to six) instruments, and they must be mature. This approach speeds system design by minimizing the number of simultaneous constraints the design process and operations support must take into account. Mature instruments allow early and realistic baseline models for the required payload resources in terms of power, onboard and on-ground requirements for data handling and interactions between telemetry and commands. Understanding the instruments helps us get a good start on the prototyping design concept because interfaces with spacecraft subsystems and with ground systems are relatively easy to design in the first iteration. Operations cost savings come from the relative simplicity of sequence design because operators don't have to consider as many options. Also, the design of the operations-control system and its manning will cost much less.

To carry out a short schedule and ensure effective system engineering, we must fix all system-level requirements early in the program. Focused mission objectives and mature instruments are the key enablers for fixing system-level requirements early on. The requirements definition process should be intense, however, and everyone involved in the mission should participate. After requirements have been formed and agreed on, keep them fixed throughout the design phase if possible. Fixing requirements early on enables effective design and implementation by avoiding repeated iterations that try to satisfy "floating" requirements or to marginally improve the mission. Mission operations is the only technology area left **after launch**, so it often suffers from unresolved design and development problems. "Let operations take care of that" is a common refrain when problems arise. Fixing requirements early in a short schedule helps by giving the design team for the spacecraft system and subsystems and the operations engineers the most time to work together and tackle known design problems.

Acceptance of Risks

This phrase states that we design and operate low-cost satellites fully recognizing that anomalies and failures may occur. This strategy lowers costs because trying to eliminate the last few remaining elements of risk in design and operations is extremely expensive; there certainly is a law of diminishing returns. In a cost-effective environment, we must understand that perfection isn't possible and increasing the estimated reliability from 0.95 to 0.96 likely won't be worthwhile for most unmanned missions if it costs much more money. To reduce costs over conventional approaches, we'll have to accept a somewhat higher risk of in-orbit failures induced by the enormous time pressure and by unavoidable shortcuts in the design and operations concepts.

But we must still manage risk in designing low-cost satellites. For example, we have to design the satellite to survive anomalies by providing robust safe modes. We also need to manage small teams carefully because every person is essentially indispensable on a tight schedule. At the same time, many of the typical low-cost characteristics, such as a short schedule, small and co-located teams, design simplicity, and resource margins will in fact make the system more reliable. Indeed, experience has shown that small satellites are at least as reliable as larger conventional spacecraft: for instance, only **one out of nine** UoSAT satellites launched between 1981 and 1993 has failed prematurely (UoSAT-4 in 1990).

Rapid Prototyping

Rapid prototyping means using a rough, incomplete model and then iteratively refining and completing it based on maturing requirements, an evolving system-design baseline, and the results of early standalone and interface tests. Such a method can be much faster and cheaper because various phases of a development cycle interact more efficiently. Rapid prototyping allows for a more natural, incremental, and iterated evolution in the requirements and design definitions resulting from continual interaction between users and engineers. We may also integrate the (usually incomplete) elements of a prototype and subject them to system validation tests in order to identify major system level problems at the earliest possible time. Naturally, these faster approaches carry considerable risks, but there is no alternative when we're faced with a 3-year implementation schedule. We have to tolerate higher risks to achieve the avowed goals of "faster, better, and cheaper."

A traditional mission timeline usually allows a year or more for conceptual design (Phase A studies), at least a year each for preliminary and critical design (Phase B and C), and about 2 years for fabrication, integration, testing, and launch preparations (Phase D)—a total of 5 or more years. A short-schedule mission should take no more than 3 years from start to launch, with many missions needing even less time. For some current spacecraft with relatively ambitious mission objectives, the period between mission approval and launch is not more than about 2 years (22 months for Clementine and 26 months for the first Discovery mission, NEAR). Conventional "waterfall" approaches to system development have been slow, cumbersome, and costly. Booch [1991] points out that they follow a "sacred, immutable process" consisting of rigid and sequential requirements definition, specification development, preliminary design, detailed design, fabrication, and test phases. Each of these phases is usually driven by milestones based on deliveries from one team to another, which enforces the "us against them" attitude. Furthermore, the products of each phase are "written in granite" and serve as a costly-to-change input to the next phase. For effective short-schedule and low-cost systems, these conventional approaches aren't appropriate. We must discard them and adopt modern methods of system development, such as rapid prototyping.

6.3.2 Effective and Competent Teams

Effective and competent teams ensure that the effort concentrates on the right type of activities and capably carries out these activities—a crucial condition for successful, cost-effective mission design and operations. Recruit knowledgeable and high-performance staff, preferably with previous project experience, to work effectively with minimum training. A good strategy is to build up a dedicated, in-house, core team and to add expertise by hiring consultants with the proper background whenever and for as long as needed. Competent and motivated staff are cost-effective: they're able to assign the proper priorities and urgencies to the right activities with little management supervision. Table 6-6 summarizes the cost-savings potential offered by effective and competent teams.

TABLE 6-6. Cost-Savings Potential Offered by Effective and Competent Teams. Assembling the right team in the right environment is crucial for cost-effective mission design and operations.

Method	Cost Saving Potential
Co-locating teams	• Savings from reducing coordination and interface management • Indirect savings due to effective information flow
Empowering staff	• Reduced reporting and "micro-management" • Indirect savings due to increased motivation
Efficient team interactions	• Reduced coordination and management overhead • Indirect savings due to better communications

Co-locating Teams

For mission system engineering to be most efficient, the design and operations teams (and preferably also representatives of the science or users team) must be integrated in one team on a single site. This condition is satisfied for the extremely small missions, such as ALEXIS and UoSAT, and essentially also for the NEAR mission at APL and for the SMEX missions at GSFC. Co-locating and integrating staff are crucial to generating the synergy required for cost-effective activities. The traditional management structure for large projects, which consists of separated project management, prime contractor, and many subcontractors interacting very formally, is not appropriate for low-cost systems.

Replacing strictly formalized interfaces with direct open communications leads to a faster and more effective flow of information and represents a real cost savings because it greatly reduces coordination and interface management. This also increases mutual understanding and improves the project atmosphere; everyone "pulls in the same direction" and avoids traditional "us against them" animosities. Inefficiencies caused by communications delays decrease or disappear, and teams can take advantage of informal lunch and corridor discussions, which considerably enhance their mutual commitment to the project. Internal E-mail or project meetings can quickly distribute essential project information. Co-location also offers natural, cost-effective training for operators through participation in design and test activities. Finally, it's extremely profitable to have the original design engineers on the operations site after launch when contingencies develop, even if they are working on other projects at that time.

Empowering Staff

Empower team members as much as possible by delegating decision authority to the lowest possible management or engineering level. Also, minimize the layers of management to reduce overhead costs and to avoid time delays and misunderstandings in communications between layers. Typically, the project's lead engineers must have the authority to make all decisions in their areas, as is the case for the NEAR project; naturally, the project holds them fully accountable for any decisions made. In this manner, we eliminate the diffusion of responsibilities and accountabilities that is fairly common in large organizations.

Mission operations teams on many past and present programs have full-time people whose only jobs are to fulfill the reporting requirements of upper management. This practice is expensive and unnecessary. Because the mission operations team usually communicates with many teams, unrestrained status reporting can consume a lot of staff hours. Although project and program management do need adequate insight into the development and operations processes, excessive requirements can be costly and counterproductive. Empowering teams saves money by eliminating superfluous reporting and continuous "micro-management." Indirect cost savings result from more efficient working conditions, under which all are responsible and accountable for their work: this results in a higher staff motivation and pride and therefore in more productivity. The Clementine experience, which centered on the concept of team competence and empowerment, has confirmed the importance of this aspect, as pointed out by Regeon et al. [1995].

Efficient Team Interactions

The number and size of teams to be coordinated largely determines a system's complexity, especially if the teams are distributed over different organizations at various sites. Making the organizational concept less complex in terms of the number and frequency of managerial and technical contacts will benefit the system cost and reliability. To streamline a project organization, we must remove or reduce all organizational barriers, in particular by integrating and co-locating teams. A simple, straightforward management structure requires a simpler hierarchy and fewer reporting levels. Communication is most effective when teams talk to teams, not when teams talk to management, who communicates with teams. Naturally, for this set-up to be effective, a true **project atmosphere** must exist—one which invites frank and open discussions as emphasized by Radbone and Sweeting [1992]. The conventional "us against them" mentality within large projects concentrates on not failing regardless of the risk to others; it harms the project and will in fact increase the chance of all failing.

Efficient interactions among the teams involved in system engineering are of crucial importance. Timely and efficient face-to-face communications between the various parties are essential; they're much more effective than written instructions and specifications when it comes to achieving a full understanding of the technical issues. Regular progress and coordination meetings also stimulate the necessary team interactions and inform all team members of the project's status and on-going issues. Make the most current requirements and design baselines available to all team members at all times. Keep relevant project memoranda and documents on a central and easily accessible database or file server; send updates of these documents to all project team members by E-mail.

6.3.3 Mission System Engineering

Perhaps the single most effective measure a project can take to lower the total system cost of a space mission is to strongly endorse mission system engineering. Mission system engineering can't start at the grass roots level: it can only be effective with the full support of the project leadership. Someone "at the top" must commit to allocating resources between space mission elements based on overall mission effectiveness and efficiency. Mission system engineering is extremely valuable for reducing life-cycle costs because it tries to find the best **overall** solution for the system design, while including all technical aspects of spacecraft and instrument design, as well as considering operations, life-cycle cost, and schedule.

TABLE 6-7. Cost-Savings Potential Offered by Mission System Engineering. Mission system engineering tries to find the best overall solution for the mission system design.

Method	Cost-Savings Potential
Concurrent engineering	• Indirect savings from enhanced system quality and reliability, thereby reducing probability of failures
Efficient documentation	• Reduces staff efforts for superfluous writing and document reviewing
System-level trades	• Significant savings from using the most cost-effective design for the system's life cycle

Usually, because mission operations tends to inherit all design inadequacies after launch, operations costs go down when the design and development team properly consider operational requirements [Ledbetter, 1995]. A spacecraft designed to be operable will be more reliable and will require fewer ground resources. Prefer "effective" design options over optimal ones. Best solutions are usually more time-consuming to define and more complex to implement—often, for a marginal extra return. An example is designing payload sequences under stringent resource constraints. "Better is the enemy of good enough" is a good maxim for designing low-cost mission operations. Table 6-7 summarizes the cost-savings potential offered by a system engineering approach.

Concurrent Engineering

This term refers to simultaneously and interactively designing and developing two or more parts of a system. Communications between the teams must be continuous and intensive to ensure meaningful concurrent engineering and proper attention to all points of view. A single system engineer oversees the mission system's design by initiating and resolving the relevant trades. A few experienced operators should participate in the spacecraft design full-time, so operational capabilities, constraints, and operability become an integral and natural element of design trade-offs for the spacecraft and ground systems. In the case of NEAR development, two of eight operations engineers work directly with the spacecraft designers mainly to ensure spacecraft operability and to document the system interfaces between the spacecraft and the ground [Landshof, Harvey & Marshall, 1994]. Another approach would be to have a few of the design engineers continue on to the operations phase, as Clementine did [Horan, 1995]. In practice, however, design engineers usually prefer to do design work, so this approach would work only for a relatively short mission. In a cost-effective environment, operators should do validation testing of the spacecraft. This approach reduces total staff, provides a natural training opportunity for operations engineers, and allows operators to validate and exercise operational sequences. Teams should re-use databases containing spacecraft design information throughout the design, test and operations phases. Concurrent engineering indirectly creates cost savings by reducing system complexity and increasing system reliability.

Efficient Documentation

Documentation tends to become an end in itself rather than a means to an end; what counts in the end is that the system itself is in good shape rather than the documentation. No doubt the growth in documentation is strongly related to the increasing management complexity of large space missions. Every management layer and every interface between different teams typically levies its documentation requirements, often as "self-protection" in case problems arise. We strongly believe that **documentation should be written only to support the design activities while requiring a minimum of "engineering time."** Program managers should leave the appropriateness of engineering documentation to the engineering teams and its leaders.

Documentation is useful for recording and maintaining requirements and the design baselines. As such it should **follow** the design process. In more traditional environments, with strictly sequential development phases, documentation often **drives** this process. Make documents as concise as possible: viewgraph-style turns out to be very effective and serves more than one purpose. Keep design documents under configuration control after they are reasonably mature—for instance, after a design review. Make sure operators—while consulting with design engineers—write

spacecraft-operations manuals and document interfaces between the space and ground segments. This extremely valuable training exercise leads to a better product because design engineers are usually not aware of and not interested in the operators' point of view. We can also shorten the schedule considerably if we abandon the formalities (e.g., the hierarchy and succession of requirements and design phases) of a documentation structure and adopt rapid prototyping.

TABLE 6-8. System Design Features that Can Reduce Life-cycle Cost. The designs of the spacecraft, the ground system, and the operations concept drive operations costs.

System Design Features that Can Reduce Cost

1. *Spacecraft System Design*
  • Robust design
  • Resource margins
  • Design for operability
  • Onboard autonomy

2. *Ground System Design:*
  • Re-using facilities
  • Commercial off-the-shelf components
  • Adhering to standards
  • Test & operations commonalities

3. *Operations Concept Design*
  • Small & skilled team
  • Multimission infrastructure
  • Using advanced-technology tools
  • Automating operations

System-Level Trades

Design trade-offs must be at the system-level—consisting of spacecraft, payload, ground system, and operations—rather than within one or two particular elements [Ondrus, 1992]. Trade system design and life-cycle costs, risks, reliability, and operability as early as feasible during mission planning. All team representatives should participate in management-coordination meetings and design reviews, so they can identify and do the relevant trades. A good project atmosphere is essential for effective design trades: "hidden agendas" and self-serving "us-against-them" arguments destroy the possibility of meaningful system trades. System-level trades may save a lot of money because they help us select the most cost-effective system design rather than just the best "parochial" one. Table 6-8 summarizes the three main elements of a system design which are relevant to operations costs: they form the environment for trade studies during the system-engineering process. We'll discuss each of these areas in more detail in the following subsections.

6.4 Designing Spacecraft Systems to Reduce Operations Cost

The spacecraft-system design critically drives operating costs, so system engineers must scrutinize operational capabilities and constraints. Reduce the complexity of controlling spacecraft operations as much as possible by decreasing the number of telecommands, states, conditional relationships, and flight rules, as well as the frequency and lengths of station contacts. Keep the frequency and complexity of "one-of-a-kind" operations—such as commissioning, initializing, and calibrating— as low as possible. Relax as far as possible instrument-pointing control, stability, and calibration requirements and reduce the density of payload activities in terms of number of switchings and sequences for different modes. Consider providing redundancies in functions rather than in hardware components. An example is the design of NEAR, in which the payload's imager (with plenty of ground support) can do what the single star camera can do. A good spacecraft-system design saves money because its operations concept can be simpler, which reduces the ground system and the post-launch operations staff. Table 6-9 summarizes how an effective spacecraft system design can save money.

TABLE 6-9. Cost-Saving Potential Offered by Good Spacecraft-System Design. A Spacecraft System that is robust, rich in resources, and easy to operate or even self-sufficient will cost less to operate.

Method	Cost-Saving Potential
Robust design	• Reduced staffing from simpler operations concept • Indirect savings from lower probability of failure
Resource margins	• Savings from simpler mission planning and operations • Indirect savings from higher system reliability
Design for operability	• Reduced staffing from simpler operations concept • Indirect savings from higher system reliability
Onboard autonomy	• Savings from significant decrease in monitoring and control tasks

6.4.1 Robust Spacecraft Design

Robustness of a spacecraft design refers to its inherent flexibility, versatility, or resilience. Although robustness is largely intangible, it's extremely important, especially after in-orbit failures. Its absence or presence often determines mission failure or success. For instance, the rescue of ALEXIS after its solar paddle broke loose is largely attributed to its design robustness; in his description of ALEXIS's experiences, Priedhorsky [1993b] refers to the role played by the onboard intelligence and flexibility and emphasizes the "value of robustness."

Design robustness is strongly related to spacecraft operations: typically, robustness would help us avoid complicated and time-driven control procedures, which could become critical under minor anomalies. Sometimes, relatively small modifications in spacecraft or instrument design can significantly simplify operations and thereby save on costs (usually, over and over again). For example, robustness of the thermal and power design may eliminate the need for complex analysis of every maneuver sequence, saving time and money in developing sequence uploads. A mission-level system engineer should have the authority and responsibility to do such trades at a high level. Savings can be significant due to the possibility of a simpler and more reliable operations concept, which may lower staffing after launch.

6.4.2 Resource Margins

Spacecraft *resource margins* are reserves in the on-board resources during operations. This includes onboard power and thermal capabilities, data-handling throughput, RF signal strength, as well as processor memory and speed. Enough resources should be available at all times to avoid introducing complex constraints in the design of operations sequences, which invariably decrease system reliability. Therefore, a reliable and efficient operations concept crucially depends on these margins. During the design phase, adequate spacecraft resource margins are essential to provide "room" for doing design trades. Furthermore, resource margins contribute to the design's robustness.

Providing enough resource margins significantly simplifies the mission-planning tasks, so it may allow us to manage spacecraft resources autonomously or not to manage them at all. Large onboard memories and processor speeds, for example, will enable us to store and use command sequences onboard and to automate instrument sequences for data collection. Margins always enhance the flexibility of our operations concept and ease the design of operations sequences. They also strongly reduce the probability of an anomaly and may limit damage if one occurs. As a result, the mission often lasts much longer, which naturally represents an enormous cost advantage compared to developing and launching another spacecraft. This point was dramatically demonstrated in the recovery and redesign of ESA's HIPPARCOS astrometry satellite after its Apogee Boost Motor failed to fire in August 1989 [van der Ha, 1992].

6.4.3 Design for Operability

Operability refers to a spacecraft's "ease" of operation, which is closely related to resource margins and design robustness but addresses different aspects related to the overall design concept. We must design a system for operability from the outset and not as an "afterthought," as often happens. We can normally fulfill operability requirements at little or no extra cost and usually with straightforward and sensible design choices. Too often, however, in order to reduce spacecraft development cost or to avoid annoying complications, operability issues are neglected, postponed, or relegated to the ground-system design or to the mission-operations team. While this approach is understandable (complexity versus reliability trade-offs in the spacecraft favor simplicity), it may not be best or most cost-effective overall. Projects should address operability issues as early in the design process as possible. In fact, this should come naturally as part of concurrent engineering if everyone takes the system-engineering concept seriously. After the spacecraft has launched, modifications to the ground system and onboard software can do little to improve operability.

To enhance operability, we should design the system so we can:

- Avoid time-critical and time-driven operations sequences whenever possible, recognizing that we usually can't eliminate them under contingency situations during initial deployment operations.

- Equip the spacecraft with one or more simple and reliable safe modes which are triggered autonomously onboard in case of serious malfunctions. These safe modes must allow the spacecraft to survive for a relatively long period (determined by mission characteristics) without ground support.

- Construct survey tables of telemetry monitor data onboard for automatic downlink during predetermined station passes. This would allow ground-support staff to assess the spacecraft's health immediately after establishing contact. NEAR, SAMPEX, and ALEXIS use this technique.

6.4.4 Onboard Autonomy

When a spacecraft is able to operate on its own, without direct support from the ground, it has *onboard autonomy*. One of the prime objectives of mission system engineering is to design and develop spacecraft systems that require minimal operations support. Perhaps the most obvious way to reduce operations cost is to build a spacecraft that doesn't require control from the ground. (But we must trade this feature against corresponding complications in spacecraft design and their costs). More autonomy means less mission operations support. Yet, the prevailing view is often the opposite: the more the ground does, the less the spacecraft needs to do.

System engineering trades enable us to partition requirements and capabilities between the ground and flight systems. (See Sec. 3.2.) If we work to minimize overall program costs (as we should), operations costs will usually decrease as well. Even if a low-cost program were not our goal, considering mission-operations issues while designing the spacecraft would normally reduce operations cost—for example, by improving operability. Cost savings from onboard autonomy may be substantial: for instance, if it eliminates shift work, we could save 3 or 4 staff years of effort per year for **each staff position**.

Spacecraft autonomy features which would simplify operations include:

- *Autonomous telemetry monitoring and alarming*, which reduces the workload on ground personnel, especially if the operations system is designed to immediately communicate spacecraft-generated alarms to operators. In the extreme, there is the so-called *demand access* concept, whereby the spacecraft carries out its mission basically on its own: it decides when to call on ground support, for instance when it has detected an anomaly. Autonomous monitoring minimizes the need for ground-system monitoring and reduces the number of operators required as well as the frequency or duration of contacts. During missions with long cruise phases and infrequent contacts, onboard alarming, coupled with storing alarm status in memory, can enable operators to instantaneously assess the state of spacecraft health since the last contact. This reduces the contact time required, the operations load, and thus the total cost to the program.

- *Autonomous anomaly detection, correction, and reporting*, which affects operations much as onboard telemetry monitoring and alarming do. The potential reduction in operations workload and the increase in intervals between contacts means fewer operators.

- *Automating memory management*, which allows us to use less accurate ground models of onboard processors, thereby reducing development costs. In addition, fewer commands are required to manage processor memory, reducing the costs of testing those commands as well as simplifying operations.

- *Autonomous data handling*, which means the spacecraft processes, stores, and retrieves data by instrument or subsystem without detailed operator intervention. This allows the operations team to use contact time more efficiently and send fewer commands, reducing the workload and cost of operations.

- *Autonomous orbit determination and control*, which, for example, uses position information provided by a relatively inexpensive GPS receiver and onboard software. This can significantly reduce support costs for flight dynamics.

- *Multilevel safe modes*, which allow the spacecraft to assume intermediate modes of operation between fully operational and "cocoon" mode (minimal activity, awaiting ground command). For example, a failure in the data-handling system may cause the spacecraft to shut down this system, point the antenna at the Earth (assuming guidance, navigation and control is working), and await instructions. Allowing the good subsystems to remain operational means faster handling of the anomaly. This allows for longer intervals between contacts, which reduces operations loads and costs, as well as the time spent and the assets used in recovering from a failure. Of course, complex safe modes may complicate operations to the point that operations engineers won't use them. Therefore, it's crucial to design these modes with full concurrence of the operations staff.

6.5 Designing Ground Systems to Reduce Operations Costs

Planned costs for designing and developing ground systems usually build up late in the pre-launch program. If a program gets into budget problems late in the space-craft-development phase, the budgetary ax-wielder often turns to costs for developing mission operations. Saving money in development costs at the expense of repetitive costs in mission operations after launch isn't cost-effective over the mission life-cycle; yet, this trade often occurs. Here, we discuss several approaches to saving costs in ground-system development which don't damage mission capability or add to total life-cycle cost. As we've stated, to meet the tight scheduling requirements of a short development program, we must use rapid prototyping for both hardware and soft-ware. We assume here that mission system engineering will establish requirements for operations support while aiming at the best life-cycle cost. Therefore, we shouldn't further compromise these final requirements on the ground-system design. Nevertheless, the following major drivers of ground-segment costs at the *requirement level* deserve special mention:

- *Contact frequency and duration* significantly drive spacecraft, ground-segment, and operations costs: spacecraft design cost increases for fewer and shorter contact periods, whereas ground segment and operations costs increase with more frequent and longer periods. Obviously, we can establish the best cost trade-off only through system engineering.

- *Accuracies* of on-ground reconstitution of attitude, navigation, and orbit have traditionally driven up ground system costs, so relaxing these requirements (if warranted by the mission objectives) would save money. We need to keep ac-curacy requirements for attitude pointing compatible with proven guidance and control capabilities in order to avoid costly new developments and validations. Nowadays, onboard processors are sufficiently powerful to do sophisticated estimation onboard. GPS receivers give us a very cost-effective way to do real-time orbit determination onboard, with limited or no support from the ground.

* *Requirements for telemetry data recovery* must stay relaxed: typically, demands above 90 or 95% of full-time are expensive to fulfill and force more quality and redundancy in the ground system design. A promising new development which would eliminate this problem (at the cost of reducing the rate of data throughput) is part of ALEXIS's working system design: it uses "handshaking" protocols typical of data communications on the ground, whereby any onboard data may be retransmitted until the ground computer acknowledges receipt.

The following sections describe important concepts for building ground systems that will reduce operations costs. Naturally, each operations environment has its own "pet methods" and its own traditional constraints on the degree of experimentation, so not all methods will be equally relevant to each organization. Table 6-10 summarizes potential cost savings resulting from a cost-effective design for a ground station.

TABLE 6-10. Cost-Savings Potential Offered by Efficient Ground-System Design. A ground system that builds on existing facilities, components, interface standards, or common architectures, will be cost-effective.

Method	Cost-Savings Potential
Re-using facilities	• Significant reduction in development and test efforts • Indirect savings due to increased system reliability
Off-the-shelf components	• Reduction in development and test efforts • Indirect savings due to Increased system reliability
Adhering to standards	• Reduction in development and test efforts • Indirect savings due to increased system reliability
Test and operations commonalities	• Significant savings due to combined development • Reduction in operations training cost

6.5.1 Re-using Facilities

To save the cost of developing and validating new mission-control systems, it is obviously advantageous (for cost, reliability, and schedule) to re-use software and hardware from previous projects. If an existing voice-communications system or ground-station network will work for your mission, why reinvent the wheel? In addition, a number of nice-to-have features may often come along with the existing infrastructure at little or no extra cost. However, an existing infrastructure is **not necessarily** cost-effective: maintenance and personnel costs for outdated and inefficient systems may more than negate their advantage. Evaluate each element for cost-effectiveness. Analyze the existing element in detail to ensure it can do what the mission requires. Make sure the interfaces with other system elements are (or can be easily made) compatible and the tailoring and integration will cost less than dedicated new development. Usually, your main implementation challenge will be in tailoring and integrating the existing elements from different heritages into a working system and in validating this integrated system. NASA Goddard has created *"Renaissance"*—an important initiative that intends to systematically identify and develop reusable "building blocks" within mission-operations and ground data systems [Perkins et al., 1994].

6.5.2 Commercial Off-The-Shelf Components

Commercial Off-The-Shelf (COTS) hardware and software elements are those we may purchase in the commercial market from an established vendor. (See Sec 3.2.4 for a list of COTS software products.) Always examine this option to see if it would be cost-effective for a new mission. In fact, it may be worthwhile to reformulate your support requirements to adapt them to the COTS capabilities. Or you may adapt or tailor COTS elements to your requirements for less money than for new developments. In recent years, COTS systems have shown a tremendous growth in capability: low-cost programs in particular can get a lot of "bang for the buck" from COTS systems when compared to customized systems. In the case of NEAR's ground system, the core control system—containing generic telemetry and telecommand functions and interfaces—is off-the-shelf [Marshall et al., 1995].

But we must also take into account a few major shortcomings of COTS systems: (1) COTS elements for space missions aren't the shrink-wrapped products we've come to expect in the truly commercial (in particular, PC) marketplace; they lack the smooth polish of a mass market product (e.g., documentation, on-line technical support) and must frequently be tailored and customized for each application. Therefore, make sure you consider the efforts for these modifications in the total cost of a COTS system; (2) COTS systems don't offer many functions needed to operate a complex and unique space mission. Straightforward *Telemetry, Tracking, and Control* (TT&C) operations for a commercial satellite (usually, a communications satellite) are significantly different from operations for a planetary-exploration mission, requiring more complex planning and command-sequence development. COTS products tend to be stronger in meeting the needs of commercial users than planners of scientific missions.

6.5.3 Adhering to Standards

Standards refer to commonly accepted "means of implementation." Using standards reduces costs by avoiding superfluous duplication of development efforts and by reducing the complexity of interfaces. Therefore, standards are cost-effective, especially in a multimission environment, and have found widespread application in all major space agencies. Adopt standards whenever they're more cost-effective or whenever you may need support from other agencies. A prime example would be the international interface standards for packet telemetry and telecommand data, developed by the *Consultative Committee on Space Data Standards* and used in most of NASA's and ESA's recent missions.

Study with care any potential use of standards for a particular project as they may be incompatible with existing or commercial off-the-shelf items. Standardizing onboard subsystems and interfaces (for instance, microprocessors and data-interface buses) has also proven to save money on spacecraft and ground-segment development. There is considerable interest in standardizing spacecraft-control functions, as exemplified by the *Spacecraft Command Language* system in the Clementine mission [Buckley & Van Gaasbeck, 1994]. This system uses a common control language on the ground and onboard, which helps migrate support functions between the ground and spacecraft. NASA stimulates the adoption of standard mission-control and data-acquisition architectures, for instance by means of the *Mission Operations Control Architecture*, as described by Ondrus et al. [1994].

6.5.4 Test and Operations Commonalities

It is advantageous to build systems that achieve simplicity by using common architectures. Many spacecraft Integration and Test (I&T) functions are duplicated in the mission-operations system and vice versa. Why should we develop these capabilities twice? Using a common system design for mission operations and I&T saves money not only in designing and developing the ground system but also in training and staffing during test, launch, and mission operations. Typically we build two systems for simultaneous use in I&T and operations preparations before launch; after launch, the I&T system may come in handy as a backup after some minor tailoring. Spacecraft databases can readily be designed for re-use during the operations phase. Operations staff should run, or at least actively participate in, the spacecraft-integration tests for familiarity and training. In the case of the NEAR mission, the control system developed for spacecraft integration and testing is the kernel for the operations-control system required after launch, thereby eliminating duplication in development effort. After launch, the I&T system will be reconfigured as an operational backup system. Figure 6-4 shows the relevant commonalities between I&T and mission-operations functions.

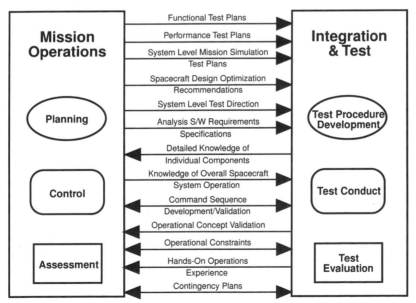

Fig. 6-4. Commonalities Between Integration and Test and Mission Operations. The arrows show the exchange of knowledge between the two activities. [Harvey, 1994]

6.6 Improving Operations Concepts to Reduce Operations Cost

This chapter has so far mainly dealt with "non-operations issues" because the largest potential savings for operations come from decisions on the system and spacecraft design. The main way to ensure that we take operations aspects into account during the mission design is by developing a meaningful and cost-effective operations concept within the system-engineering process. Document this concept in skeleton form at the earliest possible stage in the program and keep it as a "living document" as the operations concept evolves throughout the development phase.

Let's now discuss cost-saving features within operations which don't affect the rest of the system. Distributing operations costs between pre- and post-launch depends on your mission. Pre-launch development of ground-segment facilities and operations processes, team buildup and training, and system testing can be significant cost items. If the mission operations phase is short (or if it can be staffed at a very low level), pre-launch costs will be a large part of overall operations costs. If the mission is long, complex, or both, post-launch costs tend to drive overall costs. We must analyze carefully and trade off operations support options while developing our concept for end-to-end operations within the framework of the system-engineering process. In the following sections, we'll emphasize a few of the most important ways to reduce operations costs after launch. Table 6-11 summarizes the cost savings offered by a cost-effective operations concept.

TABLE 6-11. **Cost-Savings Potential Offered by a Cost-Effective Operations Concept.** An operations concept that uses fewer people with very capable tools, facilities that can support multiple missions, or automated ground systems, will cost less.

Method	Cost Saving Potential
Small & skilled team	• Savings due to increased team productivity
Multimission infrastructure	• Savings in development and test effort • Savings due to more effective use of facilities and staff
Using advanced-technology tools	• Savings due to reduction in "manual" spacecraft analysis activities and enhancement of staff effectiveness
Automating control	• Significant savings due to reduction of operations staff, in particular outside normal work hours • Minimize operations errors

6.6.1 Small and Skilled Team

The major post-launch cost for most missions is people. Therefore, cost-effectiveness demands that we reduce the number of staff needed to operate the spacecraft during post-launch operations. For example, we can build spacecraft and ground systems that require minimum support. We can also reduce staff merely by paying attention to their types and capabilities and matching them to the changes in skills needed during different mission phases. As teams become smaller, the competence and scope of their members becomes more important. **Small teams can't afford to have members with specialized or limited skills; every team member must contribute significantly to the team's overall productivity for operations to be cost-effective.** Operators should be inclined toward several disciplines at the system level. For maximum effectiveness, they may cross-train to do all operations activities, such as planning, scheduling, generating commands, real-time monitoring, and post-event analyses, as well as any archival and administrative chores. Mission operators should do various related tasks, to keep them challenged and motivated and to prepare them to move beyond their direct responsibilities, if circumstances require it.

Also note that the skills required during design and development of the mission operations system are different from those required after launch. Thus, we must add people whenever we need their skills and remove them whenever their skills are no longer compatible with the needs of the program. This may conflict with the policies of some organizations, but it's essential to controlling operations costs. Large institutions often use matrix-management techniques, which can be advantageous because they allow the project to draw from a broad mix of skilled people, paying only for the actual net time spent on the project.

6.6.2 Multimission Infrastructure

The repeated use of a standard mission-control infrastructure for several missions is an obvious and common approach for lowering ground-system costs. We can reuse facilities "vertically" within one project (e.g., re-using test software in operations), or "horizontally" across projects [Garner & Ross, 1992]. Operating with available facilities and staffing saves money and offers significant advantages in terms of system-implementation effort, testing, and operator training. ESA, for instance, has systematically developed and maintained a multimission operations infrastructure; in fact, ESA has used only three different control systems in supporting operations for about 30 satellites over the last 20 years: MSSS, SCOS, and SCOS-II [Head, 1995]. They've incorporated mission-specific support requirements by tailoring the core system. A multimission environment usually provides a readily available, common, back-up control system a number of missions may share. Also, sequential and even parallel projects may re-use staff efficiently, as NASA Goddard does to operate the SMEX missions. Multimission infrastructures also allow us to use common control rooms for more than one mission. We can do this in parallel when, for example, contact periods are exclusive, as is the case for the SOHO and ACE missions operated at NASA Goddard. Or we may work in series, as ESA/ESOC does in their mission-dedicated control rooms.

Of course, the initial development cost and schedule of a multimission control system are usually less favorable than those for a mission-dedicated control system because designers have to consider many generic and anticipated requirements. Therefore, enough missions (within a multimission program, for instance) should be available to amortize the initial investments.

6.6.3 Using Advanced-Technology Tools

Advanced-technology tools refer to the highly evolved capabilities of the hardware and software operators use for planning, control, and assessment. We should use advanced tools whenever they apply to increase the staff's overall effectiveness. In fact, NASA has mandated the application of advanced technology throughout Discovery-class missions. Powerful advanced-technology tools can reduce operations costs by enhancing productivity, allowing fewer people to do more work using fewer resources. Present state-of-the-art software tools can immediately interpret large, complex data sets and "visualize" spacecraft and other properties by means of graphics or mimic displays. Spacecraft analysts can use sophisticated software tools to prepare high-level commands. Off-the-shelf database tools are extremely effective in handling large telemetry files and spacecraft databases. Real-time animation software to visualize the spacecraft's attitude motion have proven to be extremely valuable during spacecraft design as well as in preparing maneuvers during operations. Advanced-technology tools for operations may save substantial money because they reduce operations staff, but you must trade these savings against development, validation, and training costs for installing and using the advanced tools.

We can use advanced technology to enhance productivity in two ways: (1) applying higher-level interfaces to gain access and insight into data and processes; (2) helping us make decisions. Applying advanced graphical techniques to gain insight into complex data sets is called *visualization*; using software to help make decisions falls in the category of *expert systems*.

Visualization

Everyone has seen global maps with projected spacecraft ground traces, coverage circles of ground receiving sites, and perhaps time ticks indicating when a spacecraft will (or did) pass over a particular spot. These displays were a staple of the highly publicized manned space missions of the 1960s. They are a prime example of visualization to provide immediate insight into a complex data set. This complex data included the orbital ephemeris of the spacecraft, the locations and views of each of the ground network's tracking stations, and the time the spacecraft will be available for contact at each of the ground stations. Humans excel at quickly assimilating visual information. The recent trend in returning to traditional watches and clocks from the digital variety is evidence of this phenomenon. People easily interpret the time of day from the angles of clock hands, whereas comprehending a digital clock requires assimilation and interpretation.

Computer graphics are a powerful tool for taking advantage of this characteristic of the human brain to ease operations tasks. The trend in operations-control systems is away from alphanumeric screens with numbers and cryptic mnemonics towards graphical displays, including analog dials, graphs, and trees of color-coded boxes representing spacecraft systems and subsystems. Aircraft cockpits with modern flat-panel displays use representations of analog dials and "tape" gauges for the same reasons operations systems do: these displays intuitively present more information to the user more quickly than alphanumeric displays, thus allowing fewer people to monitor and control a complex system more efficiently, more thoroughly—and with fewer errors.

Expert Systems

More advanced than visualization is using expert systems to help make operational decisions. Operations centers already use expert systems although sparingly and almost exclusively in an "assistant" role. Rule-based expert systems now help operators monitor telemetry and displays. Rule-based systems may also be used soon to help diagnose spacecraft anomalies—again, based on interpreting spacecraft telemetry. In artificial-intelligence circles, however, rule-based systems have fallen out of favor because of their inherent lack of robustness; these systems can only apply pre-programmed rules to a known data set and can be very difficult to adapt rapidly to changing conditions. For complex systems, the rule sets can get very large and difficult to manage. Also, rule-based systems require all rules to be programmed before we can use them reliably.

Model-based systems are being investigated for spacecraft operations because they address these problems. *Model-based reasoning* methods use models of systems and subsystems to estimate system states. These methods allow incremental growth in capability as models are added, refined, and updated. They can also provide qualitative and quantitative answers. We can use the methods to diagnose problems based on spacecraft telemetry, but can also use the models to support analysis for sequence generation. Model-based reasoning likely will reduce operations costs in two ways. First, it may allow us to develop a single set of spacecraft models to do planning, analysis, and assessment, thereby reducing system-development costs compared to a traditional operations design. Second, it may allow fewer analysts to generate very complex spacecraft sequences with greater confidence, thereby reducing personnel requirements while enhancing capability. Model-based reasoning may be a suitable alternative to building costly, hardware-based spacecraft simulators traditionally used to test command sequences.

We must weigh the use of advanced but experimental or immature technologies against schedule and risk impact. For example, NEAR's operations concept uses many spacecraft and environmental models during operations planning, control, and assessment. However, when the development team was defining NEAR's ground system, they rejected model-based reasoning tools for the baseline control system because these tools were experimental and NEAR's development schedule was very short. It is however possible that experiments with model-based expert systems for certain tasks may be developed and implemented after launch.

6.6.4 Automating Ground Systems

Automating the ground system means automatic and unattended performance of operations functions by on-ground hardware and software. Automation of many routine and tedious planning, monitoring, and control functions is expected to be a promising way to reduce operations costs. Once this method has been demonstrated to be robust and reliable, we should be able to save a lot of money by reducing or eliminating operations staff, in particular outside normal working hours. At present, the processing and display of telemetry data, as well as monitoring and alarming during routine-mode operations are usually automated to a large extent. High-level command languages may further reduce efforts in operations control, as will integrated databases, graphical user interfaces, and automatically generating and distributing reports. NASA's SAMPEX spacecraft has demonstrated autonomous preparation, control, and evaluation of nominal spacecraft operations during complete station passes. This includes executing the "blind" unattended downlink of onboard housekeeping and science data [Hughes et al., 1994].

The next logical step in automating operations will consist of systems that autonomously receive, process, interpret, and **respond** to spacecraft telemetry. Although *totally* automated operations are not yet feasible for scientific missions, we can already automate many functions. Automated monitoring of telemetry may not only alert an operator to an out-of-bounds condition but also spawn a complicated process to advise the operator what to do (e.g., retrieve a contingency plan from a database). It may even act itself, depending on the nature and severity of the anomaly. Trending and analysis of spacecraft data can be highly automated, generating formatted reports and delivering them electronically to the correct parties at the appropriate times (e.g., at shift changes or on Monday mornings). Clearly, automation can effectively reduce the staffing otherwise needed to perform these tasks.

References

Bloch, J., et al. 1994. "The ALEXIS Mission Recovery," *Advances in the Astronautical Sciences: Guidance and Control 1994*," Edited by R. D. Culp and R. D. Rausch, vol. 86, Paper AAS 94-062, pp. 505–520.

Bloch, J. 1995. Personal Communication, August 18.

Booch, G. 1991. *Object Oriented Design with Applications, First Edition*. Redwood City, CA: The Benjamin/Cummings Publishing Company, Inc.

Buckley, B. and J. Van Gaasbeck. 1994. "SCL: An Off-The-Shelf System for Spacecraft Control." *Third International Symposium on Space Mission Operations and Data Systems, NASA Conference Proceedings 3281*, Greenbelt, MD, November 15–18, 1994, pp. 559–568.

Cameron, G. E., J. A. Landshof, and G. W. Whitworth. 1994. "Cost Efficient Operations for Discovery Class Missions."*Third International Symposium on Space Mission Operations and Ground Data Systems, NASA Conference Proceedings 3281*, Greenbelt, MD, November 15–18, 1994, pp. 809–816.

Garner, J. T. and A. Ross. 1992. "Satellite Control Throughout the Complete Life Cycle." *ESA Bulletin 72*, pp. 107–109.

Harvey, R. J. 1994. "The Role of Mission Operations in Spacecraft Integration and Test."*Third International Symposium on Space Mission Operations and Ground Data Systems, NASA Conference Proceedings 3281*, Greenbelt, MD, November 15–18, 1994, pp. 361–369.

Head, N. 1995. "SCOS II: ESA's New Generation of Control Systems." *Acta Astronautica*, vol. 35, pp. 515–524.

Horan, D. 1995. Personal Communication, March 8.

Hughes, P. M., G. W. Shirah, and E. C. Luczak. 1994. "Using Graphics and Expert System Technologies to Support Satellite Monitoring at the NASA Goddard Space Flight Center." *Third International Symposium on Space Mission Operations and Data Systems, NASA Conference Proceedings 3281*, Greenbelt, MD, November 15–18, 1994, pp. 707–712.

Landshof, J.A., R. J. Harvey, and M. H. Marshall. 1994. Concurrent Engineering: Spacecraft and Mission Operations System Design. *Third International Symposium on Space Mission Operations and Data Systems, NASA Conference Proceedings 3281*, Greenbelt, MD, Nov. 15–18, 1994, pp. 1391–1397.

Ledbetter, K. W. 1995. "Mission Operations Costs for Scientific Spacecraft: The Revolution That is Needed." *Acta Astronautica*, vol. 35, pp. 465–473.

Mandl, D., J. Koslosky, R. Mahmot, M. Rackley, and J. Lauderdale. 1992. "SAMPEX Payload Operations Control Center Implementation."*Second International Symposium on Space Mission Operations and Data Systems, Proceedings JPL 93-5*, November 16–20, 1992, pp. 63–68.

Marshall, M. H., G. E. Cameron and J. A. Landshof. 1995. "The NEAR Mission Operations System." *Acta Astronautica*, vol. 35, pp. 501–506.

Negron, Jr., D. and A. Chomas. 1992. "Mission Operations" (chapter 14), *Space Mission Analysis and Design*, 2nd edition, Edited by W. J. Larson and J. R. Wertz, Microcosm, Inc. and Kluwer Academic Publishers.

Ondrus, P. J., R. D. Carper, and A. J. Jeffries. 1994. "The NASA Mission Operations and Control Architecture Program." *Third International Symposium on Space Mission Operations and Data Systems, NASA Conference Proceedings 3281*, Greenbelt, MD, November 15–18, 1994, pp. 1297–1303.

Ondrus, P. and M. Fatig. 1992. "Mission Engineering." *Second International Symposium on Space Mission Operations and Data System, Proceedings JPL 93-5,* November 16–20, 1992, pp. 313–318.

Perkins, D. C. and L. B. Zeigenfuss. 1994. "Renaissance Architecture for Ground Data Systems." *Third International Symposium on Space Mission Operations and Data Systems, NASA Conference Proceedings 3281*, Greenbelt, MD, November 15–18, 1994, pp. 1305–1316.

Priedhorsky, W. C., et al. 1993a. "The ALEXIS Small Satellite Project: Better, Faster, Cheaper Faces Reality." *IEEE Transactions on Nuclear Science*, vol. 40, no. 4, August, pp. 863–873.

———. 1993b. "The ALEXIS Small Satellite Project: Initial Flight Results." *AIAA Space Programs and Technology Conference and Exhibit, Paper AIAA 93-4188*, Huntsville, AL, September 21–23, 1993.

Radbone, J. M. and M. N. Sweeting. 1992. "UoSAT: A Decade of Experience Pioneering Microsatellites." *Proceedings of Small Satellites Systems and Services Conference (CNES)*, Arcachon, France, June, 1992, pp. 477–487.

Regeon, P. A, R. J. Chapman, and R. Baugh. 1995. "Clementine: The Deep Space Program Science Experiment." *Acta Astronautica*, vol. 35, pp. 307–321.

Roussel-Dupre, D., et al. 1994. "On-Orbit Science in a Small Package: Managing the ALEXIS Satellite and Experiments." *SPIE Proceedings of Conference on Advanced Microdevices and Space Science Sensors*, San Diego, CA, July 28–29, vol. 2267, pp.76–89.

SMAD = Larson, Wiley J. and James R. Wertz, eds. 1992. *Space Mission Analysis and Design,* 2nd edition. Torrance: Microcosm Press and Dordrecht: Kluwer Academic Publishers.

van der Ha, J. C. 1992. "Implementation of the Revised HIPPARCOS Mission at ESOC." *ESA Bulletin No. 69*, February, pp. 9–15.

Whitworth, G. G. 1992. "Ground System Design and Sizing" (chapter 15), *Space Mission Analysis and Design*, 2nd edition, Edited by W. J. Larson and J. R. Wertz, Microcosm, Inc. and Kluwer Academic Publishers.

Chapter 7

Design-to-Cost for Space Missions

Robert Shishko, Edward J. Jorgensen
Caltech/NASA Jet Propulsion Laboratory

Design-to-Cost (DTC) was originally a term used in DoD programs to denote a concern for a system's production cost during the definition, design, and development phases of a program or project. This concern was appropriate because DoD programs usually contemplated production runs in the hundreds or thousands of end items. The term later evolved into Design-to-Life-Cycle-Cost (DTC/LCC) when concern for operations and support was added. Operations and support costs, when viewed over the entire life of a weapon system, often comprised the majority of the life-cycle costs. Today, either term is a moniker for an approach to life-cycle cost management. Life-cycle cost management is the complete integration of life-cycle cost considerations into the systems engineering and design process—that is, life-cycle cost is treated as a system attribute and is managed accordingly. This is the way we use the term *DTC* in this chapter.

The DoD directive on DTC dates back to the 1980s [DoD, 1983]. It requires that "flyaway" or unit production cost goals and thresholds be established and presented for each major weapon system before final commitment to full-scale development. Establishment of additional cost parameters for operations and support is suggested, but not mandatory. The purpose of this approach to DTC is to control life-cycle cost by requiring designers and acquisition managers to work within a minimum performance floor and maximum affordable cost ceiling. Once managers establish these goals and thresholds, they are supposed to strive to meet them, and to report periodically on the current estimates for their system's DTC parameters.*

---

* The extent of compliance with the DoD DTC directive today is unknown, but an internal 1987 study found that of 35 DoD programs, only about two-thirds established flyaway or unit production cost goals and thresholds, and only 10% had operations and support DTC parameters [DoD,1987]. None of these 35 programs were space missions. While the DoD approach is well-intentioned, we feel that the process could be strengthened.

NASA space missions typically involve only a very small number of end items; typically one or two. The argument for doing DTC in missions, then, may not seem very compelling. However, cost fundamentally limits nearly all space missions. Given the new realities of NASA post-Cold War budgets, future missions face a budgetary environment in which lower-cost projects are the norm and a program management environment in which cost overruns can jeopardize a project. By current guidelines, overruns of as little as 15% on development or life-cycle costs may result in mission cancellation [NASA, 1993]. With hard cost constraints like that, DTC is a necessity. Even if cost is not a hard constraint, DTC provides greater cost-effectiveness and economic efficiency. In a commercial space venture, this translates into greater profit, and perhaps market share.

In our view, DTC is not just an acquisition management technique or a technical process within systems engineering. Managing cost as a system-level parameter requires both the commitment of project management as well as the technical expertise of mission and system engineers and cost estimators. Further, a space mission must have a strong DTC focus from its early conceptual studies in order for DTC to work. This is because so much—70% by some estimates—of a mission's life-cycle cost is determined during the conceptual studies and mission definition phases of a project.[*]

NASA's commitment to DTC is relatively new. The governing document for major system acquisition states that projects must maintain the capability to estimate, assess, and control life-cycle cost throughout the project cycle [NASA, 1993]. In *preliminary analysis* and *definition* (Phases A and B, respectively), this capability focuses on the effects on life-cycle cost of varying mission design and mission effectiveness parameters. Such high-level trade studies help identify whether slightly relaxing performance requirements could produce a much cheaper system or whether slightly increasing cost could produce a much more effective one. Before a project is approved for *design* and *development* (Phases C and D, respectively), it must establish targets for development and life-cycle cost. These are known as the *development cost commitment* and the *project cost commitment*, respectively. These commitments can be renegotiated, if for example, external conditions change.

During Phases C and D, DTC activity focuses on assessing the effects on life-cycle cost of refinements in the system design, operations concept, and associated downstream processes (such as fabrication, verification, operations and support, and disposal). For major changes in these areas, effects on life-cycle cost must be estimated and submitted as a part of any formal change control requests. Change requests are examined for consistency with the development cost and project cost commitments. If the final projected development or life-cycle cost exceeds the cost commitments by more than 15%, the project is subject to a *cancellation review*.

In this chapter, we describe our approach to implementing NASA's DTC mandate at the JPL. So far, our approach has been applied only to projects that are in the early conceptual studies phase (also known as pre-Phase A) or *preliminary analysis* (Phase A). These projects tend to be smaller—a few tens of millions to a few hundreds of millions of dollars—than some of the planetary projects JPL has undertaken in the past, and they are highly cost-constrained. The DTC approach could, however, be applied to both large and small projects, and to both manned and unmanned missions.

---

[*] This often-quoted figure can be found in the DoD's Systems Engineering Management Guide [DSMC, 1990], though we have also seen a Boeing Company report cited as the source.

7.1 Design-to-Cost in Systems Engineering and Management

Design-to-cost is a way of pursuing maximum mission returns **while keeping costs within a specified amount**. For organizations accustomed to working in a highly cost-constrained environment, the systems engineering process is likely to have evolved in a way that naturally supports DTC; in other organizations, DTC requires a shift to a new approach to systems engineering. The difference is illustrated in Fig. 7-1, which contrasts the requirements-driven systems engineering process with the cost-driven one. In the latter, mission goals and hard cost constraints replace hard requirements for mission performance. The resulting mission implementation reflects capabilities adjusted to meet the cost constraints.

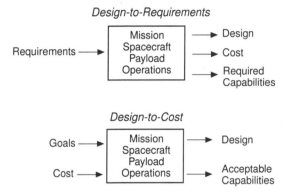

Fig. 7-1. Comparative Design Methodologies. In Design-to-Cost (DTC), requirements are replaced by goals and hard cost constraints. Capabilities adjust to meet the cost constraints.

In a requirements-driven process, project-level requirements are allocated to various systems (spacecraft, mission operations, mission design, and launch vehicle), and system-level requirements are allocated to subsystems (power or attitude control). These allocated requirements are ultimately translated into a feasible mission implementation, which is then costed. (In the past, operations were typically costed separately, and only after the mission and spacecraft design were well-understood.)

This approach has led to several problems. Subsystem designers often do not fully account for the cost burden their designs place on the whole project. This interface problem results in costs being thrown "over the fence." A particularly common example is when spacecraft designers offload costs into the verification or operations phase. Subsystem designers also tend to optimize the performance of their own subsystem rather than optimize the performance of the project as a whole. They may guard information about costs and technical margins in order to avoid risk in their own subsystems. Often they prematurely focus on a preferred technical design, resulting in little exploration of alternatives and little room for descoping.

In a cost-driven DTC process, the basic elements of good systems engineering are still present. These include:

- Flowdown of guidance in the form of mission goals, cost constraints (by year or phase, if necessary), risk and margin policies, and engineering and management plans.

- *Concurrent engineering*—that is, the simultaneous consideration of all downstream processes such as fabrication, verification, operations, and disposal.

- Rigorous, consistent evaluation of alternative mission implementations considering cost, schedule, performance, and risk.

- Successive refinement of the mission implementation through architecture, preliminary design, and detailed design, iterating as needed in response to new ideas and trade studies. In this process, mission requirements "float" until the costs are understood well enough to meet cost constraints with high probability.

- Tracking of technical progress, and DTC thresholds in particular, through the project cycle to insure that what is delivered is capable of performing a useful mission within the cost constraints.

Over the longer run, a DTC process also requires:

- Maintenance of, and improvement, in design tools, models, and engineering and cost databases.

- Continuing education of system engineers in DTC techniques and use of the design tools and models that support DTC.

These process elements would usually be performed by the system contractor with the buyer (e.g., the government) being cognizant of the approach, tools, and models. In the case of NASA missions, NASA normally performs the preliminary analysis (Phase A). During definition (Phase B), NASA organizations often conduct parallel in-house analyses to validate the contractor's efforts and to remain an informed buyer. However, the system contractor may have special insights and knowledge about costs not available to the buyer, and may view the risks differently than the buyer.

7.1.1 Impediments to Design-to-Cost

We can easily finds reasons why DTC is not always successful. One example can be found in the political process for obtaining a new project start. When management focuses primarily on selling a project, DTC may even be viewed as a threat. The DTC requirement, expressed in NASA management directives, that major projects be able to estimate and control life-cycle cost throughout the project cycle, means that cost information must be available. If a project's life-cycle cost is much larger than the development cost alone, those who question the project's affordability may use this information to kill it. Such a programmatic risk must be balanced against the value of full cost information in making better design choices.

Another impediment to DTC is technical. When DTC techniques, tools, and models are either not very well understood by project teams, or not implemented coherently, DTC efforts will be ineffective. Our experience suggests that just having the DTC tools (e.g., cost models and databases) in place is not sufficient. The lack of training and experience in what to do and how to do it means that project teams will not use DTC tools. The chief technical impediment, however, may be that the techniques and models for cost estimation are simply inadequate. Traditional statistical Cost Estimating Relationships (CERs) can be inappropriate for several reasons. The uncertainty bands on estimates may be much larger than, and in fact overwhelm, differences in the point estimates that CERs provide. Is it then reasonable to base decisions on such differences when they have a substantial probability of being illusory? Secondly, costs and cost sensitivities from CERs may be untrue. The available CERs may be weight-based when performance-based ones are needed, or, they may be based on historical data that include several decades of designs during which technology has evolved significantly. In that case, they may not produce costs and cost sensitivities that reflect the current state of technology. Lastly, there may not be any models or experts capable of estimating costs as a function of performance parameters and design attributes.

A third impediment to DTC is organizational. DTC is effective only when project teams become convinced that it is part of their job, not a separate exercise. Management at all levels must communicate that DTC practices are imperative. When communication across technical disciplines is difficult, or fails to be open or timely, DTC has a low probability of success. The same can be said for communication across project phases, when the make-up of project teams may change. Lastly, some may distrust DTC's emphasis on quantitative analysis and decision making and prefer purely instinctive engineering judgment.

7.1.2 Design-to-Cost at the Jet Propulsion Laboratory

The Jet Propulsion Laboratory's space missions have traditionally been requirements-driven, but future missions are very likely to be cost-driven, smaller, and completed in less time than in the past. To accomplish this, we have moved toward the concepts of small, *integrated product development teams*, concurrent engineering, and DTC. All three are facilitated by a new Project Design Center, which enables project teams to include life-cycle cost as a direct part of each mission implementation decision. In the design center environment, project teams concurrently conduct the DTC iterations throughout the project cycle. The center is designed to speed up the decision-making process: if a proposed mission implementation results in an estimated project cost exceeding the cost constraints, design center tools can help project teams reduce the cycle time needed to resolve the issues. These tools include multiple computer workstations that support and link cost databases, project archives, models specific to each technical design discipline, and a project-level model for DTC. The design center can connect with similar facilities across the U.S. through computer and video links.

Our DTC approach involves much greater use of trade studies based on the project-level model for DTC.[*] The DTC model helps project teams make key design decisions with better information on how technical performance parameters and design attributes affect cost. The trade studies are far more comprehensive than in the past, because the DTC model takes account of the interactions across spacecraft subsystems and between the spacecraft and the other parts of the project. There is more visibility and consistency in the assumptions being made and in estimates of project-

---

[*] The idea of developing and using a DTC model for space missions has been independently implemented by Rockwell's Space System Division, the Lockheed Space and Missiles Company, and by the Aerospace Corporation [Bell and Hsu, 1994] for use at the USAF Space and Missiles System Center. All of these implementations have a common intellectual core.

level technical and life-cycle cost implications. Using computers means there is faster turnaround when "what if" questions are posed, and less back-of-the-envelope calculation. Documenting and archiving of trade studies, which provides traceability of decisions, occurs naturally when the DTC model is used. DTC becomes a continual process—not a once or twice a year exercise.

DTC is also related to a project's risk management activities in at least three ways. First, DTC cost estimation and modeling can be performed stochastically. A Monte Carlo simulation of the total cost results in a cumulative probability distribution that can be used to calculate project cost reserves as a function of the project's desired level of confidence that actual costs will not exceed the cost cap. Secondly, the project team can use the DTC model to develop descoping options. This is best done early in the project cycle when various mission architectures are being evaluated. The DTC model provides an estimate of how much life-cycle cost can be reduced by varying technical performance parameters and design attributes. These descoping options should be archived so that they can be rapidly accessed if needed. Thirdly, the DTC model provides a means of automatically calculating a number of high-level project metrics such as estimated mission effectiveness and mass and power margins. In cost-constrained projects, estimated life-cycle cost (and its components) can be tracked so that they can be compared to their threshold values at any time during the project.

7.2 Some Basic Concepts

At this point we introduce some basic concepts to help provide a foundation for the technical portion of our approach to DTC; some readers may recognize these as familiar concepts from mission utility analysis described in SMAD Chap. 3, and may wish to proceed directly to Sec. 7.3. We present these basic concepts here in an abstract manner, but their application to an actual space mission is far from abstract. It requires a quantitative understanding of the underlying physical, engineering, and cost relationships for that mission.

We view the DTC problem as one of choosing a set, perhaps from among many such sets, of technical performance parameters and design attributes that represents a wholly feasible alternative means of accomplishing the system's intended mission within a particular cost and schedule. For a given mission these technical performance parameters and design attributes describe the relevant *design space*. We intend for the concept of a design space to include all things that describe a full mission implementation.[*] With a given state of technology, only a subset of the design space points (quantitative values for these attributes) are feasible. For example, only so much power output can be obtained from a given solar array area with currently available designs.

Each point in the design space maps into a measure of mission effectiveness (sometimes called a mission *figure of merit*). The *measure of effectiveness* expresses quantitatively the degree to which the system's objectives are achieved. For example, launch vehicle effectiveness depends on the probability of successfully injecting a

---

[*] This includes such elements as mission design, spacecraft system architecture and design, and operations concept. Combined with other elements, these describe the *space mission architecture* as it is called in SMAD [p. 10].

payload onto a usable trajectory. Mission effectiveness, in turn, depends on the system's performance parameters and attributes. For a launch vehicle system, some associated performance parameters and attributes include the mass that can be put into a specified nominal orbit, the trade between injected mass and launch velocity, and launch availability.

Simultaneously, each point in the design space maps into a quantitative measure of cost. It is also possible then to think of each point in design space as a point in the trade space between effectiveness and cost. A graph plotting the highest achievable effectiveness of designs currently feasible as a function of cost would in general yield a curved line like the one shown in Fig. 7-2. In the figure, the ordinate represents all the dimensions of effectiveness and the abscissa represents all dimensions of cost. The curved line represents the envelope of currently available technology in terms of cost effectiveness. The feasible region contains the curve and area below it.

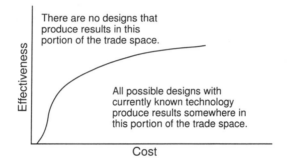

Fig. 7-2. Cost-Effectiveness Trade Surface. This surface reflects the set of non-dominated design solutions. Along the surface, an improvement in mission effectiveness can only be achieved at increased cost, holding schedule and risk constant. Sec. 1.1 discusses why the curve is shaped as shown and what its implications are for mission evaluation.

Uncertainty complicates this simple picture because exactly what cost effectiveness outcome will be realized by a particular design cannot be known in advance with certainty. For example, even the most robust design has some chance of a randomly occurring failure. A design's projected cost and effectiveness are better described by a joint probability distribution than a point. The curved line in Fig. 7-2 can be thought of as representing the envelope at some fixed confidence level. For further discussion, see Shishko [1995] pp. 4–6.

From this perspective, DTC becomes a matter of selecting the alternative that solves a constrained optimization problem. We can formally state this problem as:

$$
\begin{aligned}
&\text{max.} && E(x_1, x_2, ..., x_n) \\
&\{x\} && \\
&\text{subject to:} && C(x_1, x_2, ..., x_n) \le \mathbf{C} \\
&\text{and} && x_1, x_2, ..., x_n \text{ feasible.}
\end{aligned}
\tag{7-1}
$$

where $E(x_1, x_2, ..., x_n)$ is the function describing the effectiveness of the system design with technical performance and design attributes $x_1, x_2, ..., x_n$, and $C(x_1, x_2, ..., x_n)$ is the function describing its cost. $\mathbf{C}$ is the fixed cost that cannot be exceeded. Equation

(7-1) states that we seek to maximize effectiveness subject to a cost constraint. The first-order conditions for a solution to Eq. (7-1) have an intuitive appeal; when all the functions are well-behaved, the marginal effectiveness of a change in any x_i is proportional to its marginal cost at the optimal values.

In real space missions, it is not practical (or possible) to write down the functions in Eq. (7-1) in closed form, so it is perhaps best to think of them as just the reduced form. Our examples from actual flight projects in Sec. 7.4 will demonstrate how involved the relationships can be. We wish to emphasize, however, exactly what role Eq. (7-1) plays in DTC. DTC is **not** a mathematical problem that can be solved with enough computing power. Instead, project teams must creatively uncover the feasible region by synthesizing alternative design solutions relevant to that particular mission. DTC then leads the project team through a focused search of the feasible alternatives in order to find the most highly valued ones. This search is, in fact, carried out by performing trades using the DTC model that is built for this purpose early in the DTC process.

Another basic concept that we use is life-cycle cost, which is the most comprehensive measure of a system's cost. A system's life-cycle cost is the total cost of its acquisition, ownership, and disposal over its entire life span. It should be estimated and used in Eq. (7-1) in the evaluation of system alternatives during trade studies.* Two views of life-cycle cost are shown in Fig. 7-3. The two views basically reflect institutional and mission differences rather than any substantive ones. The important point is that in life-cycle cost, operations costs (and disposal costs) are treated concurrently with acquisition costs. DTC requires that trades between these should receive attention.

7.3 Building the Design-to-Cost Model

The idea of linking technical design tools and models together is not new; engineers working in the same technical discipline commonly exchange model results. The DTC model's strength is in the ability to link dissimilar technical disciplines through the effects on mission effectiveness and life-cycle cost. This section describes what we have done to implement the technical portions of the DTC process. Primarily, this involves building and verifying the DTC model, and using it to evaluate alternative mission implementations. In early conceptual and system definition studies there are a number of recurring questions or issues that the project team must address. Some examples include:

- What are the *system drivers* driving the design or mission costs?
- What is the best balance of subsystem technical performance levels?
- What potential payoffs do various "new" technologies offer?
- What trades of spacecraft capabilities and costs against mission operations capabilities and costs make sense?

---

* Technically, the cost function in Eq. (7-1) should be the *present value* (also known as *present discounted value*) of life-cycle cost. In the DTC models described in Sec. 7.3 and Sec. 7.4, we allow the model user to set the discount rate. The higher this rate becomes, the lower operations costs are "weighted" relative to development costs.

- How much margin of finite resource x can be gained by lowering the margin of finite resource y?
- What are the project's best descoping options?

Addressing these kinds of issues is typically done in trade studies. These trade studies play an important role in determining technical performance and design requirements for the various systems and subsystems that make up the project. The ability to address the above issues in a systematic, quantitative, and rigorous manner requires a variety of models and tools. These include cost models, subsystem performance models, system-level effectiveness models, reliability models, and decision analysis tools. When these are integrated in a particular way, the result is a DTC model. Our approach to building a DTC model focuses on the top-level metrics of interest to the project such as life-cycle cost and mission effectiveness. We describe a space mission by a set of inter-related equations that lead to the calculation of these, and other, important technical performance parameters and design variables. These equations represent a description of the space mission in a design space that incorporates cost, schedule, and performance. If the right variables are treated stochastically, we are able to represent risk as well.

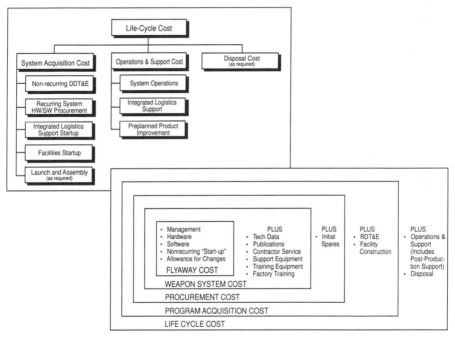

Fig. 7-3. NASA (top) and DoD (below) Views of Life-Cycle Cost. Life-cycle cost is the most comprehensive measure of the cost of a system. In some space missions, costs may be incurred over many years, which implies the need to use present value in comparing alternative implementations.

In our current DTC models for Jet Propulsion Laboratory projects, we use Excel™ 5.0, though our approach to DTC is not tied to that particular software package. Commercial spreadsheets are now capable of handling the model sizes and complexities that we are developing and using. Naturally, some software problems occur from time to time, but the advantages of using commercial spreadsheet programs in cost, portability, and extensibility are substantial. We have also found ways to link our spreadsheets with Unix programs that run on workstations. Two types of such programs include mission design and trajectory analysis programs and computer-aided design programs for solid modeling of the spacecraft. Using these links, we can change a parameter in one of these, and the information is passed to the spreadsheets where the "ripple" effects on the other subsystems and on the top-level metrics are calculated.

Our role as DTC engineers is actually in integrating the models and data that the cognizant system and subsystem design engineers already possess in one form or another. Our efforts do not duplicate what they do. Instead, we hold one-on-one meetings with each design engineer during which we acquire the mathematical equations for that system or subsystem. All of the information for the DTC model comes from the project team, and as a result, the project team owns (and eventually uses) the DTC model. The mathematical equations are organized into a set of spreadsheets, generally one for each spacecraft subsystem and separate spreadsheets for other parts of the mission. (A look ahead to Fig. 7-4 may be helpful in visualizing how the equations are typically organized.)

The construction of the DTC spreadsheet model does not happen all at once in a "big bang." The scope and detail of these equations depend on the project's phase. During early conceptual studies, we focus on the issues the project team feels is driving the design solution. For example, in the Space Infrared Telescope Facility project, the two key design challenges were attaining a telescope life of at least 2.5 years while fitting it within the fairing dimensions of the Delta II launch vehicle. A rigorous thermal analysis of various physical configurations was essential. The project team had three thermal models to choose from, and selected the one recommended by the thermal subsystem engineer. The DTC model was built around this set of equations.

The DTC model is first filled with equations and parameter values describing the project team's baseline mission implementation. For projects that are in the conceptual studies phase, this baseline may have only the barest description such as a target body, launch vehicle, launch date and feasible mission trajectory, and a spacecraft mass bogey. At this point, the equations in the DTC model might describe how the trajectory varies with changes in spacecraft mass, how acquisition costs change with that mass, and how operations costs change with the trajectory. The model is not very useful at this point because, while we might have a relationship between life-cycle cost and mass, we do not know how much spacecraft mass is truly needed in order to deliver the payload of instruments and return the data collected. In order to gain usefulness (as well as confidence in the cost and technical performance estimates), the model needs to calculate a mission effectiveness metric, and have more subsystem and project cost detail. This detail is generally required in order to perform trades both within the spacecraft and across the spacecraft and mission operations systems. The subsystem detail includes basic subsystem design information (mass, power consumption, hardware cost, and reliability estimates all tied to an equipment list) and technical performance calculations. These are then aggregated to the spacecraft level and tied to top-level metrics such as mission effectiveness. Project cost detail— including full acquisition and operations cost—is estimated and presented in a top-level project *work breakdown structure*.

The DTC model can represent mission effectiveness in several ways. For the Space Infrared Telescope Facility project, the project team chose to measure effectiveness by the telescope's expected lifetime. This metric is primarily a function of the size of the liquid He dewar that cools the sensors, as well as the reliability of the spacecraft bus and telescope. Another mission effectiveness metric we have used is the probability of getting at least x gigabits of returned science data. On missions to the outer planets, for example, this metric is highly reliability-dependent and is significantly affected by whether the spacecraft has single- or dual-string subsystems. These decisions, in turn, affect mass, volume, complexity, and cost.

We next tie the subsystem's technical performance equations to each other and to the cost equations. For most of the technical performance variables, this is fairly straightforward. The relationships among antenna size, radiated power, and data return rates are established, for example, through the standard link budget. Naming these technical performance variables in our spreadsheets and using the names in the equations ensure that changes propagate correctly throughout the spreadsheets. Naming variables also makes it easier to verify the model.

Tying in costs presents a special challenge. Spacecraft acquisition cost **must** be represented by equations that reflect its relationship to the subsystem technical performance and design attributes. The equations must be structured so that the correct *cost gradients*—how the cost changes when the performance attributes are changed—are applied. If appropriate CERs are available, then we use them. More often, we construct the appropriate CER for each subsystem using its "grass-roots" (i.e., bottom-up engineering) cost estimate and what we call the *method of standard increments*. First, for each subsystem, we separate the acquisition grass-roots cost estimate into five categories:

- Management
- Design and development
- Flight hardware
- Integration and test
- Software development.

For each technical performance attribute, we define a standard increment (say, for example a 10% increase), and we require that the subsystem design engineer then re-estimate the grass-roots costs in each of the five categories for this higher level of performance. Although this is an increase in the amount of work the subsystem design engineer must perform, the added burden is generally accepted because of the perceived value of the completed DTC model. These cost gradients are used to form the relevant local acquisition cost equations. Cost spreading functions can be used to distribute these costs (by fiscal year) over the design and development period.

The DTC model must also include the mission operations and data analysis costs. Equations representing these costs must also show cost gradients with respect to the mission design and operations concept parameters, and to the spacecraft technical performance and design attributes. Usually the operations cost equations are highly specialized to the type of mission. We currently estimate these costs by fiscal year using a model developed at the Jet Propulsion Laboratory (1992 to 1994) [Carraway, 1995]. This model, also on spreadsheets, is completely integrated into the DTC model. (In 1996, we will upgrade this model to provide a more complete set of cost gradients.)

The DTC model at this point describes the baseline mission implementation, and is now ready to use in trade studies. If the baseline mission implementation exceeds the development cost commitment or project cost commitment, i.e., the programmatic cost targets, then the baseline mission implementation is descoped sufficiently to meet them. A typical scenario starts with a proposal to insert a new technology or change an item in the equipment list with a less powerful (and usually less expensive) substitute. Working between regular project team meetings, a subsystem design engineer calls up the model on his or her computer and makes the necessary changes in parameters, equations, and databases for that subsystem. (Larger changes in the design may require that several engineers make changes in their respective parts of the model.) The system-level results, which may show an increase or decrease in life-cycle cost and mission effectiveness, are reviewed at the project team's regular face-to-face meeting. An increase in costs that violates a cost constraint causes the project team either to reject the proposed change or to start searching for a compensating change in another subsystem. (Note how the mission objectives and technical performance requirements float until the full cost implications are understood!) When the project team accepts a change, the new mission implementation becomes the baseline. The DTC process iterates this way as new ideas are suggested and tested using the model. Each decision is automatically archived by the project team by saving the DTC model used in that trade study.[*]

In later phases of a project, the DTC issues move from fundamental ones to more detailed ones. For example, a project may wish to compare specific proposals for on-board versus ground processing of data. The DTC model is modified to handle these more detailed questions as project definition and preliminary design proceed.

In summary, the DTC model should capture mission and system design knowledge and associated cost information; the model does not create designs, but it does process the underlying technical and cost relationships into new information about the project's top-level metrics of interest. In particular, the model should be capable of producing reliable life-cycle cost projections for alternative mission implementations. Once a baseline mission implementation is established, trade studies using the DTC model should examine the project's design space around the baseline. The DTC model should support the search for new mission implementations—points in the design space—that meet the cost constraint with confidence and are more cost effective. In the DTC process, when such an alternative mission implementation is found, it should become the new baseline, and the search resumed until further improvements are unwarranted.

The project team, not the DTC model, is ultimately responsible for success or failure of DTC. The model is only as good as the equations and data the project team provides to it. That depends on the quality of the project team personnel, their willingness to explore new alternatives, and their openness in the search process.

---

[*] If there are many trades being performed, there are configuration management issues that need to be addressed, for example, maintaining an audit trail of trade studies performed and decisions made, and maintaining the integrity of an archived DTC model and its associated trade study. We handle these issues by assigning configuration management responsibilities to one of the DTC engineers We also use the Jet Propulsion Laboratory's Engineering Data Management System, which provides the necessary configuration management services used by the Jet Propulsion Laboratory projects for drawings and other technical data.

7.4 Two Examples of Design-to-Cost Models for Space Missions

This section presents two examples of DTC models. The first is for Space Station *Freedom*. In building the space station DTC model, we acquired a great deal of valuable DTC experience. The model is too large to present here in its entirety, but the application of our approach to a large program is outlined. The effort required to build the station DTC model was considerable—several dozen workyears—much of which went to develop the custom software so that the model could be run on 386/486 PCs. The model was used to demonstrate potential savings of roughly a billion dollars (in present discounted value), so the return on investment would have been considerable had the recommendations been accepted and the program continued.

The second example deals with the Pluto Fast Flyby[*] project discussed in Sec. 12.2. The effort required to build the Pluto DTC model was only a few workmonths—a considerable reduction from the station DTC model. The disproportional reduction can largely be attributed to the use of commercial spreadsheets to build the model and to the good fortune of having those who built the station DTC model available to work on the Pluto DTC model.

7.4.1 Space Station Freedom

We developed the DTC model for Space Station *Freedom* from 1985 through 1990. It focused on the resources produced and consumed by each subsystem. For example, the power subsystem produced power using large solar arrays, but required substantial logistics resources to provide the fuel for the propulsion subsystem to overcome the resulting drag. These interrelationships were modeled in detail along with the associated life-cycle costs. In the earliest released version, the station DTC model dealt with 19 resources and their interrelationships; later versions modeled 80 resources [JPL, 1990a]. The model's expansion was necessary because program managers asked more detailed DTC questions.

All versions of the model had the capability to automatically resize the space station's subsystems so as to maintain a constant flow of net user services—that is, services available to users after taking into account cross- and self-(parasitic) consumptions. Using the power subsystem example again, larger solar panels increased the net user power, but required more fuel deliveries, which required an offsetting increase in logistics resources to maintain the same net services to users. The model then calculated the costs associated with **all** of these increases. It was then possible to compare the life-cycle cost of two alternative space station designs with the quantities of net user services held constant. This provided an unambiguous discriminator of the two designs. The design with the lower life-cycle cost, properly discounted, was preferred.

The model could also be used to calculate the life-cycle cost of different quantities of net user services for the same configuration, and to calculate the cost when holding some of the 80 resources fixed (in total size). In the latter case, the model would re-calculate the **net** user services of those resources held fixed. The extensive calcula-

---

[*] Since renamed Pluto Express

tions were performed using custom software developed for the Space Station *Freedom* Program Office running on 386/486 PCs. A separately developed model, called *MESSOC* [JPL, 1990b], calculated the operations costs and performance variables. Operations costs made up more than half of the life-cycle cost over the space station's 30-year useful life.

7.4.2 Pluto Fast Flyby

The Pluto Fast Flyby was conceived as a mission to send a pair of identical plutonium-powered spacecraft to Pluto using a direct trajectory so as to arrive before the planet's atmosphere froze. (Pluto's orbit carries it inside Neptune's until 1999, and its atmosphere is believed to collapse as it moves farther from the Sun.) Each spacecraft was to be capable of capturing, storing, and returning one gigabit (Gbit) of science data. The mission design called for the two spacecraft to launch on separate launch vehicles, arrive at Pluto six months apart, and image half of the planet each. Pluto Fast Flyby was intended as a medium-cost project.

Fig. 7-4 shows the basic spreadsheets represented in the Pluto DTC model, but the illustration applies to planetary missions in general. The arrows indicate some of the data and calculations that are passed from one spreadsheet to another. The flight system is represented by a series of spreadsheets, generally one for each subsystem. Each spreadsheet contains equations and data to calculate the mass, power demands by flight mode, DDT&E and flight hardware and software costs, and reliability of the subsystem. Each spreadsheet also contains any governing equations needed to calculate subsystem performance. The telecommunications spreadsheet, for example, contains the link budget and passes tracking time results to the Deep Space Network spreadsheet. Results for mass and power by flight mode are passed to separate accumulation spreadsheets where, for example, the projected wet mass is calculated and passed to the mission design spreadsheet. The mission design calculations are revised and key dates are passed to the operations cost model spreadsheet. Mission operations costs are revised and passed to the life-cycle cost accumulation spreadsheet. The mission design spreadsheet also contains the required ΔV for the mission. To calculate on-board propellant and wet mass, we command the model to iterate through the propulsion, mass accumulation, and mission design spreadsheets to obtain the solution.

In the Pluto DTC model several related risk-based metrics quantify mission effectiveness. These are the probability of obtaining at least one Gbit of returned science data, the expected science data returned, and the *certainty equivalent* science data returned.[*] These are all calculated using the techniques of decision trees, decision analysis, and reliability engineering. Tying in mission risk metrics makes it clear why there are no free lunches in space missions. All other factors held constant, using lower quality parts, or going single-string rather than dual-string (or single spacecraft rather than dual spacecraft), or extending the mission length, all exact a price in terms of these risk-based metrics. The Pluto DTC model calculates the amount to the extent the reliability equations and data are accurate.

---

[*] *Expected* here means the probabilistic mean. *Certainty equivalent* here means the amount of science data (say, x Gbit) that leaves the decision maker (who could be the sponsor, project scientist, or project manager) indifferent between the choice of x with certainty and the uncertain amount they will actually receive from the mission.

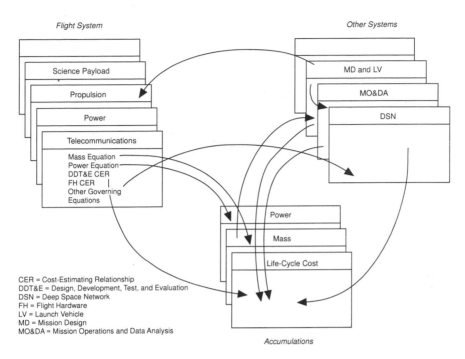

Fig. 7-4. Typical Space Mission Spreadsheets and Representative Passed Data. These spreadsheets are combined into a single "workbook" to form our DTC model. Organizing the relationships among variables, passing calculated values, and verifying the entire DTC model is facilitated by naming variables and using the names in the model's equations. In our Pluto Fast Flyby model, over 500 variables are named and passed as needed from one spreadsheet to another.

In all of our DTC models, a single spreadsheet summarizes the project's top-level metrics. This integrated summary spreadsheet for the Pluto project presents the life-cycle cost (by Level 2 and 3 elements of the work breakdown structure); the flight system's beginning-of-mission power and power margins for critical flight modes; the flight system dry, wet, and injected masses;[*] and the three risk-based mission effectiveness metrics. The integrated summary also shows key programmatic parameters for the mission. The integrated summary is organized so that when trades are performed against the baseline mission implementation, the results for each of the project metrics can be compared item-by-item. A three-column display is used: the first for the baseline mission implementation, the second for the current alternative under consideration in a trade study, and the third for the differences. When a trade study is performed, the baseline numbers do not change. All trade results appear in the second column, and the deltas appear in the third; the baseline numbers change only when the alternative dethrones the baseline.

---

[*] Pluto Fast Flyby doesn't track mass margins because mass is not a constraint for this mission. Instead, additional mass slows the flight system and results in longer cruise time to Pluto.

Navigating through all the spreadsheets (approximately 35 in the Pluto DTC model) is made easier because we have built a series of macro commands that are activated by clicking on the appropriate box in the tree structure shown as Fig. 7-5. This action reveals all spreadsheets associated with that particular system or subsystem. The graphic thus serves as an executive-level spreadsheet for the model as well as its organizing structure

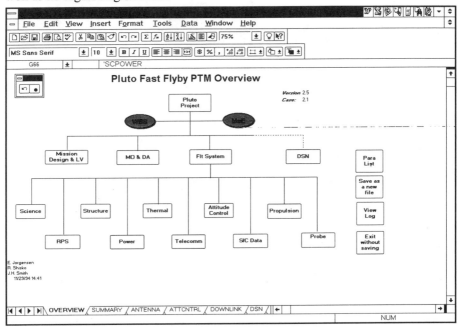

Fig. 7-5. Overview of Pluto Fast Flyby DTC Model Spreadsheet Structure. This structure provides a logical representation of the model's spreadsheets. At the right are macro commands that help in configuration management of the trade studies in which it is used.

Clicking on the top-level box marked *Pluto Project* opens the integrated summary spreadsheet, shown as Fig. 7-6. The baseline mission implementation and the alternative being evaluated in the "current" column show a two-spacecraft mission using two Proton launch vehicles, with the first launch date in January 2001. The Pluto project team considered several trades. What would be the implications of using a different launch vehicle (e.g., a Delta II)? Should the high-gain antenna be larger or smaller? Would the use of a composite structure rather than aluminum payoff? Should the power subsystem output be increased by the addition of another radioisotope thermal generator Pu-238 "brick"? We will illustrate the detail in Pluto DTC model using the last question. In the process, we will show three basic parts of the radioisotope thermal generator subsystem spreadsheet: technical performance equations, equipment list with embedded mass equations, and cost equations. The radioisotope thermal generator subsystem was one of the easier ones to model; the spreadsheets for the attitude and articulation control subsystem, for example, are more difficult to develop fully.

Project name:	Pluto Fast Flyby		
Base year:	$FY93		
Discount rate (%):	0%		
	Baseline	Current	
Launch vehicle:	Proton/DM/3-stage	Proton/DM/3-stage	
Launch date:	1/28/01	1/28/01	
Number of spacecraft:	2	2	
Number of launch vehicles:	2	2	

	Life-Cycle Cost Summary ($FY93)		
	Baseline	Current	Current – Baseline
Project management	$153,672K	$153,672K	$0K
S/C system	$131,024K	$131,114K	$90K
Payload system	$28,209K	$28,209K	$0K
RTG development	$86,419K	$87,419K	$1,000K
MOS development	$14,946K	$14,946K	$0K
Launch adapter	$240K	$241K	$0K
Launch vehicle	$140,000K	$140,000K	$0K
MO&DA	$38,382K	$38,683K	$301K
DSN			$0K
Total ($FY93)	$592,892K	$594,284K	$1,392K

	Power Summary (watts)		
	Baseline	Current	Current – Baseline
Number of modules	6	7	1
Beginning of mission power (watts)	91.59	106.85	15.26
Flight unit 1 power margin (watts)			
Flight unit 1 mode 1 power margin	18.24	30.80	12.57
Flight unit 1 mode 2 power margin	7.74	20.30	12.57
Flight unit 1 mode 6 power margin	8.69	21.25	12.56
Flight unit 1 mode 7 power margin	9.05	21.58	12.53
Flight unit 2 power margin (watts)			
Flight unit 2 mode 1 power margin	18.97	31.66	12.69
Flight unit 2 mode 2 power margin	8.47	21.16	12.69
Flight unit 2 mode 6 power margin	9.42	22.10	12.68
Flight unit 2 mode 7 power margin	9.77	22.42	12.65

	Spacecraft Mass Summary (kg)		
	Baseline	Current	Current – Baseline
Spacecraft dry mass	115.11	118.37	3.26
Spacecraft wet mass	123.67	127.18	3.50
Spacecraft injected mass	171.79	175.67	3.88

	Risk-Based Mission Success Summary		
	Baseline	Current	Current – Baseline
Prob{data volume >= 1.0 Gb}	0.921	0.921	0.000
Expected data volume (Gb)	1.439	1.438	−0.001
Expected utility (data volume)	1.112	1.111	0.000
Certainty equivalent data volume (Gb)	1.213	1.212	−0.001

Fig. 7-6. Pluto Fast Flyby Integrated Summary.

Pu-238 has a half-life of approximately 87 years. The Pu-238 available for the Pluto project was manufactured in July 1983. The thermal output of each brick, $P(t)$, decays exponentially as a function of the time since the date of manufacture, t_m. If t and t_m are in years, then:

$$P(t) = P(t_m) e^{-(\ln 2)(t-t_m)/86.4} = P(t_m) 0.5^{(t-t_m)/86.4} \qquad (7\text{-}2)$$

The radioisotope thermal generator itself degrades from a variety of environmental factors so the available power, $P_{av}(t)$, declines as a function of the time in years since the launch date, t_d, in a manner predicted by the following equation:

$$P_{av} = P_{av}(t_d) e^{-0.0318(t-t_d) + 0.0009(t-t_d)^2} \qquad (7\text{-}3)$$

Together Eqs. (7-2) and (7-3) give the available power for any date during the mission once the launch date is fixed. Mission design strongly influences power margins—the later the launch date, and the longer the cruise time to Pluto, the less power is available in all flight modes. Critical flight modes occurring near the end of the mission will feel this effect the most. Equation (7-4) shows how we computed the power margins for each flight mode, j.

$$\text{Power Margin}_j = \min_{\{T_j\}} \left(P_{av}(t) - \sum_i P_{demand, j, i} \right) \qquad (7\text{-}4)$$

where T_j is the set of times $t > t_d$ such that the spacecraft is in mode j, and the summation of power demands is carried out over all flight subsystems, i, including the science payload. Fig. 7-7 shows the spreadsheet for the computation of power margins for each flight unit and critical flight mode.

Fig. 7-8 shows the mass computation for the radioisotope thermal generator subsystem based on its equipment list. When a Pu-238 brick (the standard increment) is added, the mass is automatically recomputed by equations that adjust the number or size of other components and parts that make up the radioisotope thermal generator subsystem. Fig. 7-9 shows the cost computation for the radioisotope thermal generator subsystem based on the work breakdown structure and the cost categories in Sec. 7.3. When a Pu-238 brick is added, the cost is automatically recomputed by equations that adjust each category. For the radioisotope thermal generator subsystem, this occurs in the flight hardware and integration and test categories. In addition to these direct implications, adding an radioisotope thermal generator brick initiates a series of ripple effects in other subsystems, which ultimately affects the life-cycle cost in ways that cannot always be anticipated, much less superficially calculated.

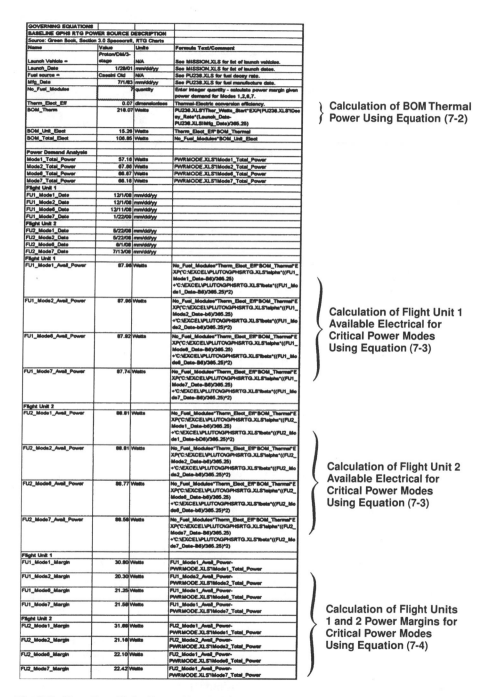

GOVERNING EQUATIONS			
BASELINE GPHS RTG POWER SOURCE DESCRIPTION			
Source: Green Book, Section 3.0 Spacecraft, RTG Charts			
Name	Value	Units	Formula Text/Comment
Launch Vehicle =	Proton/DM/3-stage	N/A	See MISSION.XLS for list of launch vehicles.
Launch_Date	1/28/01	mm/dd/yy	See MISSION.XLS for list of launch dates.
Fuel source =	Cassini Old	N/A	See PU238.XLS for fuel decay rate.
Mfg_Date	7/1/83	mm/dd/yy	See PU238.XLS for fuel manufacture date.
No_Fuel_Modules	7	quantity	Enter integer quantity - calculate power margin given power demand for Modes 1,2,6,7.
Therm_Elect_Eff	0.07	dimensionless	Thermal-Electric conversion efficiency.
BOM_Therm	218.07	Watts	PU238.XLS!Ther_Watts_Start*EXP(PU238.XLS!Decay_Rate*(Launch_Date-PU238.XLS!Mfg_Date)/365.25)
BOM_Unit_Elect	15.26	Watts	Therm_Elect_Eff*BOM_Thermal
BOM_Total_Elect	106.85	Watts	No_Fuel_Modules*BOM_Unit_Elect
Power Demand Analysis			
Mode1_Total_Power	57.16	Watts	PWRMODE.XLS!Mode1_Total_Power
Mode2_Total_Power	67.66	Watts	PWRMODE.XLS!Mode2_Total_Power
Mode6_Total_Power	66.67	Watts	PWRMODE.XLS!Mode6_Total_Power
Mode7_Total_Power	66.16	Watts	PWRMODE.XLS!Mode7_Total_Power
Flight Unit 1			
FU1_Mode1_Date	12/1/08	mm/dd/yyy	
FU1_Mode2_Date	12/1/08	mm/dd/yyy	
FU1_Mode6_Date	12/11/08	mm/dd/yyy	
FU1_Mode7_Date	1/22/09	mm/dd/yyy	
Flight Unit 2			
FU2_Mode1_Date	5/22/08	mm/dd/yyy	
FU2_Mode2_Date	5/22/08	mm/dd/yyy	
FU2_Mode6_Date	6/1/08	mm/dd/yyy	
FU2_Mode7_Date	7/13/08	mm/dd/yyy	
Flight Unit 1			
FU1_Mode1_Avail_Power	87.96	Watts	No_Fuel_Modules*Therm_Elect_Eff*BOM_Thermal*EXP('C:\EXCEL\PLUTO\GPHSRTG.XLS'!alpha*((FU1_Mode1_Date-B6)/365.25)+'C:\EXCEL\PLUTO\GPHSRTG.XLS'!beta*((FU1_Mode1_Date-B6)/365.25)^2)
FU1_Mode2_Avail_Power	87.96	Watts	No_Fuel_Modules*Therm_Elect_Eff*BOM_Thermal*EXP('C:\EXCEL\PLUTO\GPHSRTG.XLS'!alpha*((FU1_Mode2_Date-b6)/365.25)+'C:\EXCEL\PLUTO\GPHSRTG.XLS'!beta*((FU1_Mode2_Date-b6)/365.25)^2)
FU1_Mode6_Avail_Power	87.92	Watts	No_Fuel_Modules*Therm_Elect_Eff*BOM_Thermal*EXP('C:\EXCEL\PLUTO\GPHSRTG.XLS'!alpha*((FU1_Mode6_Date-B6)/365.25)+'C:\EXCEL\PLUTO\GPHSRTG.XLS'!beta*((FU1_Mode6_Date-B6)/365.25)^2)
FU1_Mode7_Avail_Power	87.74	Watts	No_Fuel_Modules*Therm_Elect_Eff*BOM_Thermal*EXP('C:\EXCEL\PLUTO\GPHSRTG.XLS'!alpha*((FU1_Mode7_Date-B6)/365.25)+'C:\EXCEL\PLUTO\GPHSRTG.XLS'!beta*((FU1_Mode7_Date-B6)/365.25)^2)
Flight Unit 2			
FU2_Mode1_Avail_Power	88.81	Watts	No_Fuel_Modules*Therm_Elect_Eff*BOM_Thermal*EXP('C:\EXCEL\PLUTO\GPHSRTG.XLS'!alpha*((FU2_Mode1_Date-b6)/365.25)+'C:\EXCEL\PLUTO\GPHSRTG.XLS'!beta*((FU2_Mode1_Date-bD6)/365.25)^2)
FU2_Mode2_Avail_Power	88.81	Watts	No_Fuel_Modules*Therm_Elect_Eff*BOM_Thermal*EXP('C:\EXCEL\PLUTO\GPHSRTG.XLS'!alpha*((FU2_Mode2_Date-b6)/365.25)+'C:\EXCEL\PLUTO\GPHSRTG.XLS'!beta*((FU2_Mode2_Date-b6)/365.25)^2)
FU2_Mode6_Avail_Power	88.77	Watts	No_Fuel_Modules*Therm_Elect_Eff*BOM_Thermal*EXP('C:\EXCEL\PLUTO\GPHSRTG.XLS'!alpha*((FU2_Mode6_Date-b6)/365.25)+'C:\EXCEL\PLUTO\GPHSRTG.XLS'!beta*((FU2_Mode6_Date-b6)/365.25)^2)
FU2_Mode7_Avail_Power	88.59	Watts	No_Fuel_Modules*Therm_Elect_Eff*BOM_Thermal*EXP('C:\EXCEL\PLUTO\GPHSRTG.XLS'!alpha*((FU2_Mode7_Date-B6)/365.25)+'C:\EXCEL\PLUTO\GPHSRTG.XLS'!beta*((FU2_Mode7_Date-B6)/365.25)^2)
Flight Unit 1			
FU1_Mode1_Margin	30.80	Watts	FU1_Mode1_Avail_Power-PWRMODE.XLS!Mode1_Total_Power
FU1_Mode2_Margin	20.30	Watts	FU1_Mode2_Avail_Power-PWRMODE.XLS!Mode2_Total_Power
FU1_Mode6_Margin	21.25	Watts	FU1_Mode1_Avail_Power-PWRMODE.XLS!Mode6_Total_Power
FU1_Mode7_Margin	21.58	Watts	FU1_Mode1_Avail_Power-PWRMODE.XLS!Mode7_Total_Power
Flight Unit 2			
FU2_Mode1_Margin	31.65	Watts	FU2_Mode1_Avail_Power-PWRMODE.XLS!Mode1_Total_Power
FU2_Mode2_Margin	21.16	Watts	FU2_Mode2_Avail_Power-PWRMODE.XLS!Mode2_Total_Power
FU2_Mode6_Margin	22.10	Watts	FU2_Mode6_Avail_Power-PWRMODE.XLS!Mode6_Total_Power
FU2_Mode7_Margin	22.42	Watts	FU2_Mode1_Avail_Power-PWRMODE.XLS!Mode7_Total_Power

} Calculation of BOM Thermal Power Using Equation (7-2)

} Calculation of Flight Unit 1 Available Electrical for Critical Power Modes Using Equation (7-3)

} Calculation of Flight Unit 2 Available Electrical for Critical Power Modes Using Equation (7-3)

} Calculation of Flight Units 1 and 2 Power Margins for Critical Power Modes Using Equation (7-4)

Fig. 7-7. Pluto Fast Flyby Power Margins Computation.

5 Module RTG Mass (kg)	No.	Unit Mass	Total Mass	Variable Mass Frac. dMass/dFuel Mod.	Fixed Mass Frac.
			15.36	2.294	3.89
Heat Source			7.24	1.448	0.00
Fuel (PuO$_2$)	5	0.596	2.98	0.596	0.00
Capsules (Ir)	5	0.234	1.17	0.234	0.00
Graphitics	5	0.618	3.09	0.618	0.00
Structural Supports			1.07	0.000	1.07
Graphite pressure plates	2	0.115	0.23	0.000	0.23
Load studs & zirconia	1	0.100	0.10	0.000	0.10
Belleville spring (Ti)	1	0.370	0.37	0.000	0.37
Preload hardware	1	0.370	0.37	0.000	0.37
Converter			2.15	0.430	0.00
T E elements	160	0.009	1.51	0.302	0.00
T E fasteners & seals	160	0.002	0.30	0.060	0.00
Alimina insulators	160	0.002	0.24	0.048	0.00
Connectors, terminals	160	0.001	0.10	0.020	0.00
Multi Foil Insulation			1.44	0.140	0.74
Sides	1	1.000	1.00	0.140	0.30
Ends	2	0.220	0.44	0.000	0.44
Support structure	N/A	N/A	N/A	N/A	
RTG Housing			2.90	0.197	1.91
Side wall	1	1.410	1.41	0.197	0.42
End covers, bolts & seals	2	0.300	0.60	0.000	0.60
Pressure release device	1	0.430	0.43	0.000	0.43
Resistance thermometer	1	0.300	0.30	0.000	0.30
Gas mgmt. assembly	1	0.160	0.16	0.000	0.16
Radiator Fins			0.56	0.078	0.17
Fins	1	0.510	0.51	0.071	0.15
Aux. coolant manifolds	N/A	N/A	N/A	N/A	
Emissivity coating	1	0.050	0.05	0.007	0.02
Miscellaneous elements	N/A	N/A	N/A	N/A	

Fig. 7-8. Pluto Fast Flyby Radioisotope Thermal Generator Mass Computations.

Some of the principal effects include an increase in propellant, pressurant, tankage mass, and spacecraft structure mass. Cruise time increases because the spacecraft's wet mass increases, and operations costs climb as a result. The direct gain in power margin during the Pluto encounter is slightly offset by the cruise time lengthening. The subsystem, system, and project-level effects are summarized qualitatively in Table 7-1. The quantitative effects are displayed in the "current" column of Fig. 7-6. On the basis of the analysis, the addition of an radioisotope thermal generator brick was rejected, and the original baseline of six bricks was retained.

While this trade study deals with moving from six bricks to seven, one could study the effects of moving from six bricks to five. (When we did this, power margins went negative, and this alternative was dropped from further consideration.) With the DTC model, it is relatively easy to vary a design attribute over a particular range and trace its effect on life-cycle cost or other system attribute. (The resulting curve applies only to that spacecraft design performing that mission.) In our vision of DTC models for

Option 1 Description
- No. of flight units = 3
- No. of qual units =0
- No. of GPHS/Unit = 6
- GPHS source=old fuel (mfg 7/1/83)

Description	$FY94	$FY95	$FY96	$FY97	$FY98	$FY99	$FY00	Total ($FY93)
JPL Labor	$0K	$96K	$315K	$316K	$321K	$171K	$0K	$1,219K
Procurement (NASA)	$65K	$130K	$6,127K	$8,650K	$9,587K	$1,297K	$144K	$26,000K
Program costs (indep. of prod.)	$30K	$60K	$2,828K	$3,992K	$4,425K	$599K	$67K	$12,000K
RTG converters (3 FU's)	$15K	$30K	$1,414K	$1,996K	$2,212K	$299K	$33K	$6,000K
Qual test & refurbish to flight unit	$8K	$15K	$707K	$998K	$1,106K	$150K	$17K	$3,000K
F5 (Spare) defuel + handling	$12K	$25K	$1,178K	$1,663K	$1,844K	$250K	$28K	$5,000K
RTG Development (DoE)	$100K	$10,800K	$14,000K	$16,000K	$11,000K	$7,000K	$300K	$59,200K
Radioisotope Fuel	$0K	$0K	$0K	$0K	$0K	$0K	$0K	$0K
	$165K	**$11,026K**	**$20,442K**	**$24,966K**	**$20,908K**	**$8,468K**	**$444K**	**$86,419K**

Description	$FY94	$FY95	$FY96	$FY97	$FY98	$FY99	$FY00	Total ($FY93)	Total Adj. ($FY93)
Radioisotope Thermal Generators	$165K	$11,026K	$20,442K	$24,996K	$20,908K	$8,468K	$444K	$86,419K	$87,419K
Design & development	$130K	$10,956K	$17,-43K	$20,308K	$15,746K	$7,770K	$367K	$72,419K	$72,419K
Flight hardware (2 FU's)	$10K	$20K	$943K	$1,331K	$1,475K	$200K	$22K	$4,000K	$4,667K
Integration & test	$25K	$50K	$2,357K	$3,327K	$3,687K	$499K	$55K	$10,000K	$10,334K
Flight software	$0K	$0K	$0K	$0K	$0K	$0K	$0K	$0K	$0K

Standard Incremental Delta (1 Fuel Module) = 2

Total Adjusted Cost Depends on the Standard Incremental Delta

Fig. 7-9. RTG Subsystem Cost Calculation.

TABLE 7-1. Effects of Adding One Pu238 Brick to Pluto Fast Flyby Spacecraft.

System or Subsystem	Projected Cost Effects	Projected Qualitative Technical Performance Effects
Radioisotope Thermal Generator	Higher	Higher subsystem dry mass, more available power
Power and Pyro	None	None
Propulsion	Higher	More propellant mass due to higher total (dry) mass, higher propellant tankage mass, more pressurant gas mass and gas tankage mass (partially offset by fewer control thruster firings)
Structure	Slightly higher	Higher mass and moments of inertia
Thermal Control	None	None
Spacecraft Data	None	None
Attitude & Articulation Control	None	Fewer thruster firings due to higher moments of inertia
Mission Design	N/A	Longer mission cruise time due to higher mass
Mission Operations & Data Analysis	Higher due to longer mission cruise time	N/A
Project (Net Total)	Higher	Higher mass, significantly higher power margins, slightly lower expected science data return due to longer mission time

the not-too-distant future, project teams will be able to visualize a portion of their design space by creating on a computer screen a three-dimensional surface of any quantitative design attribute against mission effectiveness and life-cycle cost.

7.4.3 Lessons Learned

From these and other cases, we have compiled some lessons learned and observations about building the DTC model and about bringing DTC to a culture unaccustomed to it. Some of these lessons learned and observations have implications for project managers as well as for design engineers.

When a project had already completed a substantial amount of conceptual and definition studies before a DTC model was initiated, the process of building the DTC model brought out the project's disconnects. These disconnects took several forms. For example, two different equipments lists were being used for mass and power consumption estimates. Different assumptions were being made about attitude knowledge and control requirements by the attitude and propulsion engineers. Different dates for key operations events were being used. The work breakdown structure contained flaws. We also found that hidden margins were being carried in subsystem designs simply because timely information about another subsystem's technical performance parameters were unavailable. These hidden margins added cost to the spacecraft's design and development.

When projects developed the DTC model from the start of conceptual studies, disconnects were either avoided or resolved much earlier. The DTC model fostered a clearer understanding of the mission across the project team. We found that the project team leader (usually the project system engineer) plays a crucial role in getting the project team to focus on the importance of costs from the beginning and to assign to each team member what "homework" needs to be done before each team meeting.

Developing a DTC model forced project teams to be more rigorous, timely, and forthcoming with data. Under the traditional design and development approach subsystem designer's tended to protect cost data and design margins as long as possible so as to avoid risk to themselves. Not surprisingly, this behavior leads to high cost, but was tolerated in "flagship" missions. In the DTC process, the DTC model is open to all on the project team and it makes the system-level implications of the design visible. This tends to lead to more questioning of assumptions and to earlier revelation of subsystem designs and costs. Project team meetings tend to be more interactive and involve more give-and-take (e.g., "Your subsystem's design causes me to do this, which adds x dollars. Can we find a way to avoid that?").

The other side of this information coin is that program and project managers have to **want to know** the kinds of data the DTC model produces. In order for DTC to work, managers have to demand that project teams report on the top-level metrics, such as mission effectiveness and life-cycle costs. Unfortunately, some managers view calculation of life-cycle cost as dangerous since it can be misused by others to endanger the project. Project managers also have to make it clear that the design engineers on the project team are accountable for the technical performance and cost equations and data they place in the DTC model.

Our DTC process requires some innovations in the way cost estimates are made. Project teams need to make grass-roots estimates earlier in the project cycle. (To assist them, we offer planning tools and a database of available space hardware.) Project teams may be asked to quantify the uncertainty in these grass-roots estimates by providing probability density functions of costs. When new approaches or technology is involved, they also need to make estimates of the cost gradients with respect to subsystem technical performance. We have generally found that when a project's technical definition is weak, so are the cost estimates. The DTC model strengthens the early technical definition and performance requirements so that cost estimates can be made with greater confidence.

Lastly, we have found that education is needed not only in using the DTC model, but also in the concepts behind it. The goal is for project teams themselves to be capable of quickly building a DTC model from an existing model built for an already-completed mission. We anticipate that the library of completed models at the Jet Propulsion Laboratory will grow to a dozen or more over the next few years, and will include: low-cost planetary flyby, orbiter, and lander missions; astrophysics observatory missions; and Earth-sensing missions. The number of studies that we can currently support is limited by the number of people who have sufficient experience in building a DTC model. With previous models to serve as starting points, and with project teams experienced in the process of building and using a DTC model, this constraint will attenuate.

References

Bell, K. D., and L. A. Hsu. 1994. *Balancing Performance and Cost for Cost-Effective Satellite Systems Design Using an Integrated Cost Engineering Model*. Aerospace Report No. TOR-95(5409)-2, AFMC Space and Missile Systems Center Contract No. F04701-93-C-0094. El Segundo, CA: The Aerospace Corporation.

Carraway, John. 1995. "JPL Operations Cost Model for Flight Projects," in *Cost-Effective Space Mission Operations*, Boden, Daryl and Wiley J. Larson (eds.), in preparation.

Defense Systems Management College. 1990. *Systems Engineering Management Guide*. Ft. Belvoir, VA: Department of Defense

Department of Defense. 1983. *Design to Cost*, DoD Directive 4245.3 dated April 6, 1983. Washington, D.C.: Department of Defense.

Department of Defense. 1987. *Report on the Survey of the DoD Application of Design-to-Cost Principles*. Inspector General Report No. 87-109. Washington, D.C.: Department of Defense.

Jet Propulsion Laboratory. 1990a. *System Design Tradeoff Model Version 1.3*. JPL D-5767/Rev. C. Pasadena, CA: Jet Propulsion Laboratory.

———. 1990b. *Model for Estimating Space Station Operations Costs (MESSOC) Version 2.2 User Manual*. JPL D-5749/Rev. B. Pasadena, CA: Jet Propulsion Laboratory.

National Aeronautics and Space Administration. 1993. *Management of Major Systems Programs and Projects*, NASA Management Instruction 7120.4/NASA Handbook 7120.5. Washington, D.C.: National Aeronautics and Space Administration Office of the Administrator.

Shishko, Robert, Robert G. Chamberlain, et al. 1995. *NASA Systems Engineering Handbook*. Washington, D.C.: National Aeronautics and Space Administration Office of Human Resources and Education.

SMAD = Larson, Wiley J. and James R. Wertz, eds. 1992. *Space Mission Analysis and Design,* 2nd edition. Torrance: Microcosm Press and Dordrecht: Kluwer Academic Publishers.

Chapter 8

Cost Modeling

David A. Bearden, *The Aerospace Corporation*
Richard Boudreault, *Centre Technologique en Áerospatiale*
James R. Wertz, *Microcosm, Inc.*

As discussed in preceding chapters, cost drives almost all modern space systems and strongly influences whether programs will proceed. Analyzing and predicting program cost is becoming increasingly important, often critical, to these decisions. At the same time, low-cost programs and contractors are systematically redefining the business of space, driving down cost (often dramatically), and making it more difficult to accurately predict cost or determine which competing system is likely to cost the least.

These trends dictate a changing role for cost estimation. In traditional, requirements-driven programs, cost modeling was primarily used to validate contractor cost estimates or give funding organizations an independent estimate of probable cost. However, traditional cost modeling is inadequate for the more complex, critical tasks of design-to-cost and "go/no-go" decisions. We can no longer just apply a general-purpose cost model and then use the result to decide which programs to fund or how to build spacecraft. Instead, we must develop a deeper understanding of cost-modeling methods and data. To do this, we describe:

- The basis, methods, and use of cost models
- The application and limitations of new cost models designed for small-satellite and reduced-cost missions
- Unmodeled cost drivers and their effect on system cost

In addition to the discussion below, Chap. 7 describes design-to-cost methods and their application to mission design. See also SMAD, Chap. 20 for an overview of traditional cost models and Greenberg and Hertzberg [1994] for a good introduction to space economics.

8.1 Introduction to Space Cost Models

Cost estimation is, to some extent, a self-fulfilling prophecy. Often, a space system will actually cost, as much or more than what the budget allows or the cost model predicts. If you're developing a space system under exactly the same circumstances as before (the same design, organizations, people, technology, requirements, and procedures) you'd expect it to cost the same. But this scenario never exists. First, it's almost impossible to duplicate all of these elements. And second, if a project cost too much before, a program manager certainly doesn't want to repeat the circumstances that led to the excessive cost.

As fiscal pressures continue to drive space budgets lower, cost estimates are being used at virtually all stages of space system development. Early in conceptual design, cost estimates help us assess whether a program will work and identify key design decisions that will influence future costs. Project costs are also monitored throughout the development cycle, and if they move much above budgeted amounts, we often must rescope or even cancel the program. Cost models must be flexible enough to estimate costs at all phases, from preliminary design to much later in the design process, when we're deciding how to re-allocate limited or diminishing resources.

8.1.1 Cost as an Engineering Parameter

Cost is an engineering parameter that varies with physical parameters, technology, and management methods. A system's cost depends on its size, complexity, technological innovation, design life, schedule, and other characteristics. It's also a function of risk tolerance, methods for reducing risk, management style, documentation requirements, and project-management controls, as well as the complexity and size of the project's organizations.

Space systems typically have *specific costs* (cost per unit weight) on the order of hundreds of thousands of dollars per kilogram. Table 8-1 gives approximate specific costs based on historical data. The cost of a human-crewed mission to the Moon or Mars is typically millions of dollars per delivered kg, whereas remote-sensing or communication satellites have been produced for about $100,000 per kg.

TABLE 8-1. Rules-of-Thumb for the Specific Cost of Space Systems. Data from SMAD, Sec. 20.4.

Type of Space System	Typical Range of Specific Cost ($K/kg)
Communications satellites in GEO	70–150
Surveillance satellites	50–150
Meteorological satellites	50–150
Interplanetary spacecraft	>130

It's interesting to compare the specific cost of space systems to other Earth-based systems or devices. Large-scale civil engineering projects cost millions of dollars but are very affordable in specific terms at approximately $1/kg. Cars cost about $10 to

$100 per kg, whereas exotic biomedical products cost thousands of dollars per kg (Fig. 8-1). The more commercial part of the aerospace sector produces aircraft at costs of about $3,000 per kg, or some 30 times cheaper than space systems.

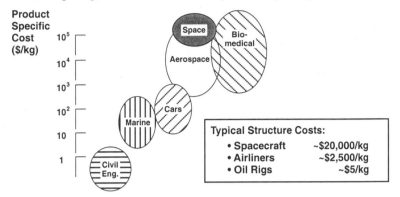

Fig. 8-1. Comparing the Cost of Space Systems with Products of Other Industries. The wide difference between the specific costs of objects relate to economies of scale and production methods and styles. This indicates the cost-improvement range available to space systems.

Specific cost for space products is not sufficient for predicting real cost. Plotting the specific launch cost of a satellite in low-Earth orbit against the mass launched reveals a distinct trend. Figure 8-2 shows the launch business has an economy of scale. [Boudreault et al., 1994] To extract lessons from historical cost events, we need a more refined model than a top-level metric such as specific cost. This leads to cost estimation using parametric relationships.

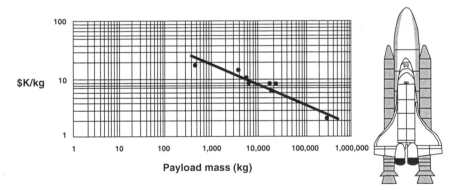

Fig. 8-2. Trends in the Specific Cost of Low-Earth Orbit Launchers. This figure demonstrates that economies of scale can intervene to reduce the cost of space hardware. Space systems haven't been sized based on cost; to make them more economical, we must be able to determine what factors establish the cost.

8.1.2 Parametric Cost Estimation

A *parametric cost model* is a series of mathematical relationships that relate spacecraft cost to physical, technical, and performance parameters. Parametrics is one of three typical ways to estimate costs of future systems, the other two being *engineering buildup* and *analogy*. (See SMAD, Sec. 20.1.) All three methods have advantages and disadvantages, depending on the scope of our estimating effort, amount of design information available, and historical data available for analogy.

Statistically, a *correlation* is a factor that measures the general behavior of a set of data. Figure 8-3 presents three examples of the correlation of data points. In all figures, it's possible to see a trend, but the first two show a more definite trend and less uncertainty about the general direction so that we can estimate more accurately where an unknown data point would lie. Mathematically, the correlation between two variables, *x* and *y*, (such as spacecraft cost and mass) is expressed by the *correlation coefficient, r,* defined by:

$$r \equiv \frac{n\sum xy - (\sum x)(\sum y)}{\sqrt{[n(\sum x^2) - (\sum x)^2][n(\sum y^2) - (\sum y)^2]}} \tag{8-1}$$

where *n* is the number of data points for which both *x* and *y* are known. When *r* approaches 1 or −1, the data is *strongly correlated*—given *x*, one can predict *y* with a high degree of confidence. When *r* is near 0, the data is *uncorrelated*—*x* and *y* are unrelated to each other.

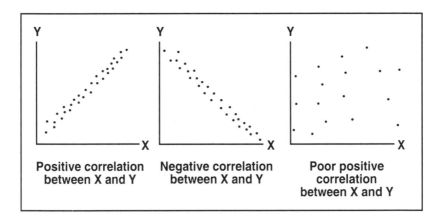

Fig. 8-3. Examples of Statistical Correlations. The correlation, or goodness of fit, in a series of data points relates to the cohesiveness of these points about a definite trend. In this figure, two cases show good correlations (one positive and one negative), and one shows a poor correlation.

Other measures of the goodness of fit include the *coefficient of determination, r^2,* and the *standard error, S_e,* given by:

$$r^2 \equiv 1 - \frac{\sum w_i(\hat{y}_i - y_i)^2}{\sum w_i(y_i - \bar{y}_i)^2} \tag{8-2}$$

$$S_e \equiv \left[\frac{1}{n-m} \Sigma w_i (\hat{y}_i - y_i)^2 \right]^{0.5} \tag{8-3}$$

where y_i are the data value, $\hat{y}$ is the estimated value of y, $\bar{y}$ is the mean value of y, n is the number of data points, and m is the number of coefficients in the equation. r^2 is dimensionless between 0 and 1. S_e is in units of cost. A strong correlation is represented by r^2 near 1 and S_e near 0.

Cost Estimation Relationships (CERs) show how the cost properties of a system or subsystem vary with characteristic parameters. The first ingredient in defining a set of CERs is an appropriate historical database. We must *condition* this database so any correlations will be coordinated. To get to this point, it's important to find out as much as possible about the origin of the data and the reference for measuring the characteristic physical parameters. We must also separate the cost elements into comparable physical subsystems or components, which become the costing elements we use to establish the entire system cost. Other factors, named *wraps*, model non-physical factors not included in the CER. These include system engineering cost, management, and product assurance, as well as the cost related to integrating and testing the space system. Wraps typically account for about 30% of the development cost for space systems.

We can determine trends by plotting the historical cost data against physical parameters such as mass, communication bandwidth, power, volume, or number of lines of computer code. By doing multiple regression with these parameters, we can develop CERs. The precision of this relationship, and therefore the ability to predict cost relates directly to how the data correlates.

CERs are typically expressed in a form such as:

$$Cost = a + b\, M^c P^d \tag{8-4}$$

where M is the spacecraft mass; P is the end-of-life power; and a, b, c, and d are constants derived from the statistical analysis. Other physical parameters, such as data rate or lines of computer code are also used. Often, we set d to 0 in Eq. (8-4) so we $P^d = 1$ and we can drop the power term. Note that this is a purely empirical relationship, not a physical one. The equation isn't dimensionally balanced, so using correct units is critical. (Dollars do not equal kg raised to the 0.357 power.) SMAD, Chap. 20 provides a set of standard CERs for large spacecraft and for subsystems. For example, the CER for the recurring cost of the spacecraft attitude and reaction-control system is:

$$Cost = -364 + 186 M^{0.73} \tag{8-5}$$

where the cost is in FY92$K and the regression parameter is the spacecraft's mass, M, in kg. This CER has an applicable range of 9 to 167 kg, *standard error* (1 σ of a standard distribution model) of $999K and a coefficient of determination of 0.76. Hence, we can interpolate using the CER between these values, but extrapolations are risky. Unfortunately, for cost estimation, technology often drives up performance and reduces mass so the system goes beyond the applicable range.

The basic tenet of parametric cost modeling is to base estimates of what satellites will cost next time on what they cost last time (assuming the system is built the same way). But dramatically reducing the cost requires that, by definition, we won't do it the same way next time. This leads to two key questions:

- How do I estimate the cost if the program intends to dramatically change from previous approaches?
- How do I use parametric cost models to help reduce the cost of future missions, rather than impede cost reduction by helping to institutionalize an expensive way of doing business?

To help reduce cost, rather than impede this process, we must look at new, more accurate ways of modeling cost. We need cost models that capture state-of-the-art technology and procurement practices and evolve as new data and development practices emerge. We need cost modeling that:

- Provides useful information for controlling or reducing cost
- Identifies *cost drivers* in procurement—requirements, subsystems, components, or elements that contribute most to the cost
- Identifies areas in which resources can be better used
- Identifies *soft* costs, such as those associated with management style, level of documentation control, and risk averseness, in addition to the *harder* physical (weight-based, performance-based) costs

Sec. 8.3 addresses these issues.

8.1.3 Learning Curve and Other Factors

We can adjust CERs to correct for special conditions, including cost reduction from learning how to do the job better or the effects of changing technological status. CERs, such as Eq. (8-5), effectively predict costs for developing a prototype or first production unit, often called the *theoretical first unit*, but they poorly estimate the cost for even the most modest production line. The staff who manufactures a second or third similar spacecraft will learn to do the job better and use economies of scale, which quantifiably improve performance. The learning rate for the space and the aerospace industry is such that, on average, the Nth unit will cost between 87% and 96% of the previous unit.

Other adjustment factors correct for uncertainty in the development status of a specific technology. Figure 8-4 presents NASA Marshall Space Flight Center's cost adjustment factors for integrating new technologies. [Hamaker, 1989] This model applies at the CER or system level. The worst-case scenario is to introduce a major new technology using a team unfamiliar with it for which the CER projects a cost of 60% over the nominal value.

The technological risk related to developing a space system depends on how we use the technology and on its degree of "flight qualification." This last factor is a "Catch-22" for reducing cost. If an item has already flown in space, it's more likely to work again, so it represents less risk to the user. A level of uncertainty taints new technologies even if they are less risky. Table 8-2 presents criteria for evaluating cost uncertainty based on a technology's flight readiness.

State of Technology Development	Team Familiarity				
	(A)	(B)	(C)	(D)	(E)
Well within existing state-of-the-art; familiar technology	0.6	0.7	0.8	0.9	1.0
Slightly advancing state-of-the-art; minor amounts of new technology	0.7	0.8	0.9	1.0	1.1
Nominal aerospace project using some new technology	0.8	0.9	1.0	1.1	1.2
Significant amounts of new technology	1.0	1.1	1.2	1.3	1.4
Major new technology; requires breakthroughs in state-of-the-art	1.2	1.3	1.4	1.5	1.6

Fig. 8-4. NASA Marshall Space Flight Center Cost-Adjustment Factors for Technology Development [Hamaker, 1989]. The factors shown in the table for technology state of development and team familiarity should be applied to the cost estimate for a nominal level of technology. Team familiarity levels are:

(A) Team is totally familiar with the project and has completed several identical projects. Team's technical expertise is superior.

(B) Team is very familiar with the type of project and has completed similar projects. Team's technical expertise is very good.

(C) Nominal team has related but not identical project experience. Team's technical expertise is average.

(D) Project introduces many new aspects with which team is unfamiliar. Team's technical expertise is below average.

(E) Team is totally unfamiliar with this type of project. Team's technical expertise is poor.

TABLE 8-2. Technology Readiness Levels (TRL). See text for discussion.

Technology Readiness Level	Definition of Space Readiness Status	Added Cost (%)
1	Basic principle observed	>25%
2	Conceptual design formulated	>25%
3	Conceptual design tested	20–25%
4	Critical function demonstrated	15–20%
5	Breadboard model tested in environment	10–15%
6	Engineering model tested in environment	<10%
7	Engineering model tested in space	<10%
8	Fully operational	<5%

With uncertainty in the cost of 25% or more for integrating new technology, risk-averse project managers may well opt for "space-qualified" technologies that have flown. This trend is opposite to the natural evolution of technology, and it restricts the widespread testing of new technologies that may eventually reduce the cost of space projects. Risk averseness eventually raises the cost of space projects just as quality eventually reduces it.

Other cost adjustments relate to softer issues, such as management style or organization structure, which greatly influence the project cost. D. E. Koelle [1991] developed one such cost factor for spreading the work to many organizations or firms working in parallel. Using the European Space Agency's experience, he was able to show how diffusing the work affects cost. If n is the number of organizations working in parallel, the cost-adjustment factor for the project will be:

$$f = n^{0.2} \tag{8-6}$$

This model testifies to the need for a strong prime contractor organization for large space projects. It suggests that any organization having several co-contractors, with coordination by the customer or an additional external organization, increases costs.

8.1.4 Cost-Risk Analysis

Cost-Risk Analysis is an assessment of the ability of projected funding profile to assure that a program can be completed and meet its stated objectives. Although technical risks are often one of the biggest cost drivers for space systems, many cost-engineering processes and models ignore effects of cost risk in the interest of quick-turnaround estimates. Cost-risk analysis is important because single-point cost estimates, while meeting the top-level needs of budgetary planners, often do not meet the needs of program managers or sponsoring agencies who want to perform more detailed trade-offs between cost, risk, and performance. Program cost is a nebulous quantity, heavily impacted by technological maturity, programmatic considerations, "normal" schedule slips, and other unforeseen events. Thus, point estimates are almost certainly incorrect because every cost element contains uncertainty and "actual" program cost falls within a range surrounding the "best" estimate. The purpose of cost-risk analysis is threefold:

1. Translate qualitative risk assessments into quantitative cost impacts (accomplished in part by having cost analysts work closely with engineers)

2. Assist program directors in managing risk

3. Establish an empirical basis for estimating future programs with confidence

Cost-risk analysis views each cost element as an uncertain quantity that has a probability distribution and attempts to evaluate technical, programmatic, and schedule risks in quantitative terms. Qualitative measures of risks are then translated into cost-estimate adjustments. A key to making quick, consistent, and defensible assessments possible is reducing subjectivity by making assumptions about **sources** and **magnitude** of cost risk. The framework described here relies on conceptual design engineers to assign a Technology Readiness Level (TRL) index, like that shown in Table 8-2, to each spacecraft or payload subsystem, which is then used to estimate a probability distribution for technology risk. The technology risk-based cost distribution is then merged with a probability distribution that describes uncertainty inherent in the parametric cost model to arrive at a composite cost-risk distribution for each cost element. Individual cost-element probability distributions are then merged to form a total-cost distribution, from which percentiles and other descriptive statistics can be obtained. [Burgess, Gobrieal, 1996]

Sources and Magnitude of Cost Risk

At least the following two sources of cost risk should be considered:

- Cost-estimating uncertainty
- Cost growth due to unforeseen technical difficulties

Cost-estimating uncertainty is quantified by computing the standard error of the estimate (SEE). For development of space-system hardware, CERs usually have S_e between 30% and 50%. Cost-estimating uncertainty is therefore quantified by a distribution that has a mean (the estimate), and a variance (square of the S_e).

Cost growth due to unforeseen technical difficulties, is harder to quantify and is the subject of continuing research. Examples of risk drivers include:

- Beyond state-of-the-art technology (cooling, processing, survivability, power, laser communications)
- Unusual production requirements (large quantities, toxic materials)
- Tight schedules (undeveloped technology, software development, supplier viability)
- System integration (multicontractor teams, system testing)
- Unforeseen events

We have made the assumption that potential for cost growth is higher for immature technologies and lower for mature, flight-proven technologies. Other cost-risk drivers include budgetary constraints, design uncertainties, and rigid, beyond state-of-the-art requirements. Spacecraft conceptual design engineers quantify the technology maturity by rating each subsystem on a TRL scale. Shown previously in Table 8-2, TRL is probably better suited for rating individual components (e.g., re-action wheel), but because most parametric cost estimates are generated at the subsystem level (e.g., attitude determination and control), TRL estimates and cost risk must be performed at the subsystem level as well. The TRL rating for each sub-system is converted into a triangular distribution that represents the probability of actual costs being above or below the baseline estimate. Triangular distributions are defined by two points: the best estimate, B, and the TRL-driven, worst-case or best-case estimate, T.

For example, assume the cost model being applied is based on systems with an average TRL of 5, which, as seen in Table 8-2, signifies that breadboard units were tested in relevant environments prior to project start. Subsystems with TRLs higher (more mature) than 5 have $T < B$, and those with TRLs less (more immature) than 5 have $T > B$. The magnitude of T should be determined by engineers and cost estimators familiar with the subsystem in question. For example, we assume that a TRL of 1 (basic principles observed and reported) has a potential cost growth of 100% over the best estimate, or $T = 2B$; and a TRL of 8 (engineering model tested in space) has a potential to cost only one fourth as much as estimated, or $T = B/4$. Other TRLs are determined by linear interpolation between these two points. The resulting expression for T is then

$$T = \frac{9 - TRL}{4} B \qquad (8\text{-}7)$$

Using this approach, cost growth due to unforeseen technical difficulties is quantified by a triangular distribution that has a mean and a variance. The two sources of cost risk are merged into one cost-probability distribution that has a mean equal to the mean of the triangular distribution,

$$MEAN = \frac{1}{3}(2B + T) \tag{8-8}$$

and a variance that is equal to the sum of variances from both sources of uncertainty,

$$VAR = S_e^2 + \frac{1}{18}(B - T)^2 \tag{8-9}$$

These cost-probability distributions, one for each spacecraft subsystem, are then summed into a total spacecraft cost-probability distribution. Research by The Aerospace Corporation has shown that this distribution may be approximated by a log normal distribution. This approximation technique allows confidence percentiles to be computed without Monte Carlo simulation.

The end product of cost-risk assessment in this framework is therefore a total spacecraft cost-probability distribution, from which mean, standard deviation, percentiles, and other descriptive statistics can be read. Sample results are shown in Fig. 8-5.

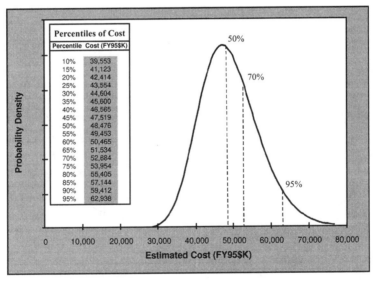

Percentiles of Cost

Percentile	Cost (FY95$K)
10%	39,553
15%	41,123
20%	42,414
25%	43,554
30%	44,604
35%	45,600
40%	46,565
45%	47,519
50%	48,476
55%	49,453
60%	50,465
65%	51,534
70%	52,684
75%	53,954
80%	55,405
85%	57,144
90%	59,412
95%	62,938

Fig. 8-5. Probability Density of Estimated Cost Resulting from Cost Risk Analysis. Percentiles, such as 50%, 70% and 95%, can be used to dictate reserves policy (e.g. "risk" dollars) or guide technology investment.

Concurrent Cost-Risk Analysis and Design Engineering

This framework is not intended to be a definitive answer to cost-risk analysis. It does, however, provide a statistically-valid, flexible, and consistent approach to assessing cost risk. The methodology is a work in progress, and we are currently examining validity and effects of alternative assumptions and correlation among risk elements. Furthermore we have developed a framework for integrating spacecraft conceptual design and cost-risk analyses. This framework is shown in Fig. 8-6.

Our approach to cost estimation and design involves joining the efforts of several engineering specialists and cost analysts into an integrated analysis approach, called the *Concurrent Engineering Methodology*, allowing for concurrent design engineering and cost estimation. This facilitates the mapping of design trades for a complex system through identification of requirements and key parameters that drive cost and performance. This methodology is comprised of integrated detailed engineering and cost-estimating relationships describing the performance, overall characteristics, and design parameters of spacecraft components and subsystems. Automated inter-connectivity between subsystem models and databases using spreadsheet software allows for assessment of system design impacts due to requirement changes. Since the engineering models are directly linked to cost-estimating relationships, the relative cost impact of varying levels of requirements can be assessed through the resulting system design changes. This process facilitates: (1) Quick response to technical design and cost questions; (2) Assessment of the cost and performance impacts of existing and new technologies; and (3) Estimation of cost uncertainties and risk. These capabilities aid mission designers in determining the configuration of space missions that meet requirements in a cost-effective manner. The framework helps system engineers produce consistent, repeatable cost estimates and cost-probability distributions that account for spacecraft physical and performance characteristics as well as subsystem technology maturity. The process is similar in many aspects to the design-to-cost methodology in Chap. 7.

8.1.5 Lessons Learned from Prior Cost Models

We can extract some lessons from our discussion of parametric cost estimation and transfer them to new, reduced-cost, space projects. Cost estimates have shown that increasing risk averseness and complexity have pushed up the cost of space systems. Project managers must feel free to take appropriate risks in order to reduce cost. If an occasional failure is acceptable, we can reduce the complexity of spacecraft and requirements and use established, space-qualified technologies.

The cost adjustment for learning has shown that it may cost less to build many satellites to do the job of a larger, more complex one. One unit can fail without jeopardizing the entire mission. We can try new technologies, even commercial ones, that may allow us to accept off-the-shelf components at much lower cost. Hamaker [1989] claims that the team should know about any new technology being used. This may mean using commercial engineers to work on a space project when adopting commercial technologies.

Limiting the size of organizations and the number of co-contractors can pay off. As we have seen in the case studies, a trend toward smaller satellites constructed by smaller teams could significantly reduce cost. The Skunk Works' rules of Kelly Johnson (Sec. 2.2.7) may be an attractive model for developing new satellites.

Reducing technological risks by early investment in research and development can greatly lower costs. An investment of at least 1.5% of the project's budget in early R&D efforts saves money. This is consistent with Table 8-2; better team awareness of a novel technology reduces the risk of cost overruns.

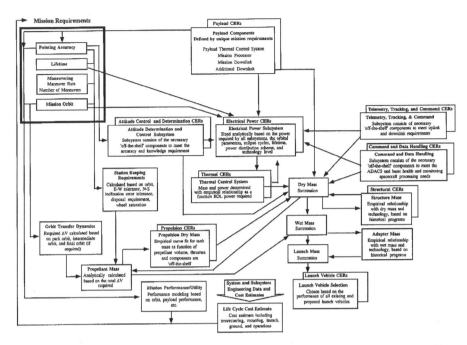

Fig. 8-6. Concurrent Engineering Methodology Framework. Spacecraft design algorithms and components databases are linked to subsystem-level, cost-estimating relationships allowing concurrent design, cost, and cost-risk analysis.

Using a parametric model implies several assumptions. Because parametric models are designed to characterize historical cost trends as mathematical relationships, we assume future costs will reflect these historical trends. In other words, we expect future programs to be similar to past programs. But when technology advances rapidly or fundamental paradigms in system architecture shift, parametric cost models based on old systems may not apply. One example of this problem is using large satellite-based cost models to estimate the costs of today's smallsats. Such a paradigm shift, and perhaps a technology shift, requires a specialized model.

8.2 The Aerospace Corporation Small Satellite Cost Model

Cost-analysis techniques developed over the last 30 years to support NASA and military space programs aren't very useful for analyzing the cost-effectiveness of small-satellite programs. Understanding these limitations and observing an increasing number of small-satellite programs led NASA and DoD to seek better tools for analyzing cost, with emphasis on developing databases and parametric cost-estimating software that address the life-cycle cost of small missions [National Research Council, 1994].

"Small satellite" means something very different to different people and organizations. (See Sec. 1.2 for an empirical definition.) We're interested in the technical and programmatic aspects of small-satellite programs and in collecting information concerning technologies and program-management techniques that decrease costs. In some cases, the characteristics of traditional large-satellite programs (e.g., long schedules, lots of documentation, out-dated procedures, risk abatement) may dramatically increase cost with little added capability.

To better understand these trends, The Aerospace Corporation surveyed small-satellite contractors and designers to collect technical and cost data for programs already launched or late in their development phase. [Bearden, Burgess, and Lao; 1995] We examined several cost models to see if we could apply or adapt them to analyzing small satellites. Case studies and comparisons produced actual and predicted costs for systems in our database.

From the database, we derived parametric relationships based on measures of small-satellite performance and physical characteristics. The result was the *Small Satellite Cost Model* (SSCM), which relates cost, performance, schedule, and risk for this new class of satellite. Of particular interest is the trade-off among cost, risk, and quality and how traditional risk-management philosophies might change if we apply newer small-satellite philosophies.

Figure 8-7 shows the process used to develop the SSCM. We targeted programs either already completed or awaiting launch within a year because we really know how much a spacecraft costs only after it is completed.

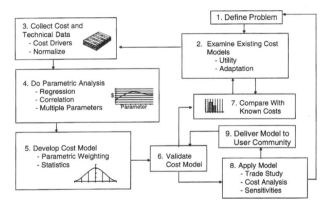

Fig. 8-7. Process for Developing the Small Satellite Cost Model (SSCM). Note that the process is iterative, involving both initial and ongoing collection of cost and technical data, which we use to derive cost-estimating relationships. We compare the model's results to actual known costs and existing cost models, then give it to users for validation and feedback.

The overriding purpose of this ongoing study is to develop a tool to estimate the cost of designing, building, and testing a modern small satellite. We're also interested in cost sensitivity to various measures of small-satellite performance. We can meet these goals by: (1) deriving relationships between cost and various physical and performance parameters for small satellites; and (2) incorporating the relationships into broader tools for analyzing cost effectiveness and improving designs.

The underlying message in Fig. 8-7 is that cost modeling continually evolves. To some extent, the modeler is always playing "catch-up" as the data quickly becomes obsolete in light of new technologies or cost-reduction methods. The key is to close the loop quickly, so the model and its underlying database remain as relevant as possible, and to derive approaches for estimating the cost of systems that use technologies beyond the state-of-the-art.

8.2.1 Comparing Cost Models

We examined several cost models to see whether they were valid, applicable, or adaptable to analyzing small satellites. We determined they weren't very useful because they relied heavily on data from programs for large satellites built with very different rules. We did case studies and comparisons for actual and predicted, recurring and non-recurring costs. Figure 8-8 compares actual costs to SSCM and SMAD model estimates, as a function of spacecraft-bus mass, for several of the case study missions. In both cases, we used a single CER—spacecraft-bus cost versus bus mass. In this case, and for other technical and physical parameters, we observed that cost models based on historical data from traditional, large space programs dramatically overestimate the cost of small satellites by as much as 500%. (See the inside rear cover for a comparison of actual vs. SMAD model costs for all of the case study missions.) Thus, we need a cost model based only on modern small-satellites to credibly analyze cost for the increasing number of small-satellite missions.

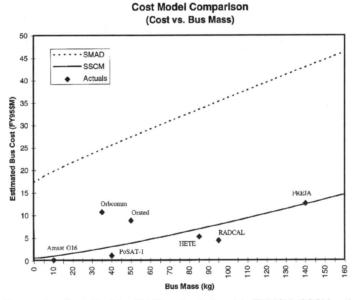

Cost Model Comparison
(Cost vs. Bus Mass)

Fig. 8-8. Comparing Cost Models. SMAD-estimated cost in FY92$M, SSCM-estimated cost, and actual cost are shown as a function of spacecraft-bus mass in kg (excludes payload mass).The SMAD model, as expected, dramatically overestimates the cost of modern (post-1985) small satellites. See the inside front and rear cover for more data.

To develop better cost models for small satellites, we started a four-year program to identify, collect, and analyze cost and technical data on actual, modern small-satellite programs. We agreed to exchange proprietary cost data with most major small-satellite companies. Wherever possible, segregation between sponsorship (amateur and university, commercial, and military) and application (communications, remote sensing, science, space environment) was examined. Because of proprietary agreements, we can't publicly present cost and technical data for individual programs, but general cost trends, CERs, key performance characteristics, and application boundaries are all part of the following discussion. You can find specific cost data in the individual case study descriptions.

We began acquiring data with a concerted effort to separate fact from fiction by researching actually flown or soon-to-be completed small-satellite projects and collecting top-level cost, technical, and performance data. Many of the small satellites launched have been designed and built by universities and amateurs who tend to cut corners, use less documentation, use fewer space-rated components, and devote volunteer (unpaid) effort to the project. Collecting and calibrating cost data from these programs and relating it to commercial and military applications is a large hurdle to overcome. For this reason, we looked at amateur programs and small satellites in the planning or paper-study phase but didn't include them in the analysis.

Another goal was to track programs and observe their cost-reduction techniques, so we could determine what the small-satellite community was really doing and how small satellites could apply to various missions. One of the obstacles in dealing with small-satellite contractors was the minimal cost documentation available. For this reason, many of the programs in the database have a single cost number or only break-out recurring and non-recurring cost. Costs of subsystems or components were available for only a few of these programs. But this did simplify the process of reconciling detailed cost data across different accounting systems. Data was available for about 30 programs, as summarized in Table 8-3.

We were also interested in determining how small-satellite contractors (usually small companies with limited resources) project costs for their systems. Because an all-encompassing cost model doesn't currently exist, companies in most cases base preliminary cost estimates on experience within the company itself. In general, the contractors said they would use SSCM and cooperated with us by providing cost and technical data. In return, The Aerospace Corporation offered to share the model with companies that participated in the study.

8.2.2 Parametric Analysis and CER Derivation

In this section, we present two sets of parametric relationships for estimating cost—SSCM version 7.4 and 8.0. Both were derived from a database of small-satellite performance measures and physical characteristics, but, each set uses a different technique, emphasizes different cost drivers, and may apply to different programs. Version 8.0 CERs use more recently acquired data and a larger sample size.

Each performance and mass-property characteristic is modeled as if it were mathematically related to the total cost of the small-satellite bus, forming a series of CERs. These equations each predict total cost of the spacecraft bus, which includes all costs for planning, designing, building, integrating, and testing it. We didn't include cost of the payload but the cost to integrate the payload to the bus is included. Assessment of additional cost elements require separate analytical tools. [Gross, 1988]. We assumed existing facilities are used and included in the total cost only the cost to operate

TABLE 8-3. Satellites in the SSCM Database as of January 1996. Programs are listed alphabetically by manufacturer. An asterisk (*) indicates a case-study mission elsewhere in this book.

Name	Sponsor	Manufacturer	Launch Date	Launch Vehicle	Application
ALEXIS	DOE	AeroAstro	4/93	Pegasus	X-ray mapping
*HETE	NASA/MIT	AeroAstro	TBD	Pegasus	High-energy exp.
DARPASAT	DARPA	Ball	3/94	Taurus	Classified
LOSAT-X	SDIO	Ball	7/91	Delta 7925	Sensor exp.
CRO	SDIO	DSI (CTA)	4/91	Shuttle	Chem. release
GLOMR II	NASA/DARPA	DSI (CTA)	4/90	Pegasus	Message relay
GLOMR I	DARPA	DSI (CTA)	11/85	Shuttle	Message relay
POGS/SSR	ONR	DSI (CTA)	4/90	Atlas E	Geomag. survey
REX	STP	DSI (CTA)	6/91	Scout	Radiation
TEX	ONR	DSI (CTA)	4/90	Atlas E	Communications
SCE	ONR	DSI (CTA)	4/90	Atlas E	Communications
*RADCAL	STP	DSI (CTA)	6/93	Scout	Radar calibration
MICROSAT	DARPA/Army	DSI (CTA)	11/91	Pegasus	Communications
MACSAT	DARPA/ONR	DSI (CTA)	5/90	Scout	Communications
*SAMPEX	NASA	GSFC	7/92	Scout	Physics exp.
APEX	STP	OSC	8/94	Pegasus	Power exp.
SEASTAR	OSC	OSC	TBD	Pegasus	Ocean color
MICROLAB	NASA	OSC	3/95	Pegasus-XL	Lightning mapper
*ORBCOMM-X	OSC	OSC	7/91	Ariane ASAP	Communications
HEALTHSAT II	Satelife	SSTL	9/93	Ariane	Communications
S80/T	CNES	SSTL	8/92	Ariane ASAP	Communications
MSTI-1	SDIO	JPL/Spectrum	11/92	Scout	Sensor exp.
MSTI-2	BMDO	Spectrum	5/94	Scout	Sensor exp.
*Freja	Swed. Natl. Spc.	Swed. Spc.	10/92	Long March 2C	Particle physics
STEP 0	STP	TRW/DSI	3/94	Taurus	Autonomy exp.
STEP 1	STP	TRW/DSI	7/94	Pegasus XL	Atmosp. physics
STEP 2	STP	TRW/DSI	5/94	Pegasus	Classified
STEP 3	STP	TRW/DSI	6/95	Pegasus XL	Science/comm
BREM-SAT	DARA	U. of Bremen	2/94	STS-GAS	Science exp.

or lease these facilities. Costs to integrate the spacecraft onto the launch vehicle and launch costs aren't included. The only ground-segment and operations costs we included are for on-orbit activation and checkout.

Generating CERs can take several analytical and numerical approaches. We'll describe three here: Ordinary Least Squares, General Error Regression, and General Linear Least Squares and Nonlinear Models. (You should research these approaches before trying to derive CERs.)

Ordinary Least-Squares

One of the most straightforward analytical approaches is an *Ordinary Least-Squares* regression to minimize the sum of squared errors. [Lao, 1995] For example, a linear mathematical formula is

$$y = a + bx + \varepsilon \tag{8-10}$$

where ε is the estimation error. The regression algorithm is:

1. Specify the actual cost, y_i for each data point
2. Estimate the cost for data point i as $a + bx_i$
3. Calculate estimation error for data point i as $\varepsilon_i = y_i - (a + bx_i)$
4. Choose values for a and b that minimize $\sum (y_i - a - bx_i)^2 = \sum \varepsilon_i^2$

The solution is

$$a = \frac{\sum y_i - b \sum x_i}{n} \tag{8-11a}$$

$$b = \frac{n \sum x_i y_i - (\sum x_i)(\sum y_i)}{n \sum x_i^2 - (\sum x_i)^2} \tag{8-11b}$$

We commonly use ordinary least-squares to fit data to a model of the form $y = \varepsilon(ax^b)$ by using a logarithmic transformation to "linearize" the equation before applying the procedure above. This transformation leads to a solution of the form:

$$\log(y) = \log(a) + b \log(x) + \log(\varepsilon) \tag{8-12}$$

Although it appears that the regression can apply to the logarithms of the x and y data points, the problem solved is actually different from the one posed. This approach has a number of weaknesses. Among these, it forces the analyst to assume an additive-error model when historical data suggests a linear relationship between cost driver and cost and a multiplicative-error model when a non-linear relationship is indicated. (These are discussed below.) [Book and Young, 1994]

General-Error Regression

The *General-Error Regression Model*, developed at The Aerospace Corporation, allows us to determine the best solution for any curve shape and to choose the error model independent of the CER's shape. A multiplicative-error model assumes that actual cost equals estimated cost times error (error is the ratio of actual to estimate). Thus, error models for the linear and one possible nonlinear case are

$$y = (a + bx)*\varepsilon \quad \text{(linear)}$$
$$y = ax^b\varepsilon \quad \text{(nonlinear)} \tag{8-13}$$

where ε is the error of estimation. The goal in this case is to minimize

$$\sum (\varepsilon_I - 1)^2 = \sum \left[\frac{y_i - f(x_i)}{f(x_i)} \right]^2 \tag{8-14}$$

where f is the estimating function. Typically, a numerical method is used to minimize this sum.

General Linear Least Squares and Nonlinear Models

We use *General Linear Least-Squares* to fit a set of data points to a model which is not just $ax + b$ (a linear combination of 1 and x) but rather a linear combination of any N specified linear or nonlinear functions of x. These functions could be a polynomial of degree $N-1$, sines and cosines (a harmonic series), or others. [Press, 1988] We define a merit function and pick parameters that minimize it. Several different techniques are available for finding this minimum, including solution by use of normal equations. [Bevington, 1969] Often, a single step matrix solution method is used to solve for the coefficients and a single-step matrix inversion to solve for the standard error. Various off-the-shelf curve-fitting software packages can be used. When the model depends nonlinearly on the set of N unknown parameters, we use the same approach (define a merit function and determine best-fit parameters by minimizing) but must iterate.

8.2.3 Using the Model

Version 7.4's CERs employed an additive-error model, in which true cost is the sum of estimated cost and estimation error. We used general linear least-squares, from COTS curve-fitting software, along with several mathematical models, both linear and nonlinear. We selected CERs based on minimal standard error and the subsystem experts' engineering judgment—whether it made engineering sense that cost was a function of the technical parameter. Table 8-4 summarizes these CERs. The number of data points varies because data wasn't available when we derived the CER or, in limited cases, the data point was considered an outlier due to special program circumstances.

The CERs can be used separately to obtain independent estimates of the spacecraft bus's cost. But it's more effective to use several CERs simultaneously to reduce the effects of statistical outliers and special circumstances that typically occur in any program. Notice that six of the 15 CERs use independent variables that are proportional to the overall system power. They're included for completeness, but you may want to select a subset of these when estimating cost, so you don't bias the results toward being too sensitive to the power budget.

To estimate the cost, C, a weighted-average algorithm is employed, using the standard error of each CER, σ_i, to establish an appropriate weighting factor:

$$C = \frac{\sum C_i / \sigma_i^2}{\sum 1 / \sigma_i^2} \tag{8-15}$$

where C_i is the ith cost estimate, σ_i is the ith standard error and the sum is over the individual CER estimates.

Using this relationship, CERs with smaller standard errors exert more influence on the averaged estimate. The total standard error of the final estimate is also the squared weighted average of all the squared standard errors.

TABLE 8-4. SSCM Version 7.4: Cost-Estimating Relationships for Small Satellites. y is the total cost in FY94$M as a function of the independent variable, x. Applicable range for each parameter and the standard error (a constant dollar value) are also shown.

Independent Variable	# Data Points	CER for Total Bus Cost (FY94$M)	Applicable Range	Standard Error (FY94$M)
Satellite volume (in^3)	12	$y = -34.84 + 4.66 \ln(x)$	2,000–80,000	4.27
Satellite bus dry mass (kg)	20	$y = 0.704 + 0.0235x^{1.261}$	20–400	3.33
ACS dry mass (kg)	14	$y = 6.65 + 0.042x^2$	1–25	5.45
TT&C subsystem mass (kg)	13	$y = 2.55 + 0.29x^{1.35}$	3–30	4.50
Power system mass (kg)	14	$y = -3.58 + 1.53x^{0.72}$	7–70	3.52
Thermal control mass (kg)	9	$y = 11.06 + 0.19x^2$	5–12	5.37
Structures mass (kg)	14	$y = 1.47 + 0.07x \ln(x)$	5–100	5.40
Number of thrusters	5	$y = 46.16 - 41.86x^{-0.5}$	1–8	8.95
Pointing accuracy (deg)	16	$y = 1.67 + 12.98x^{-0.5}$	0.25–12	7.37
Pointing knowledge (deg)	10	$y = 12.94 - 6.681 \ln(x)$	0.1–3	8.79
BOL power (W)	16	$y = -22.62 + 17.9x^{0.15}$	20–480	6.13
Average power (W)	17	$y = -8.23 + 8.14x^{0.22}$	5–410	5.71
EOL power (W)	14	$y = 0.507 + 1.55x^{0.452}$	5–440	6.20
Solar array area (m^2)	13	$y = -814.5 + 825.7x^{0.0066}$	0.3–11	6.37
Battery capacity (A·hr)	12	$y = 1.45 + 1.91x^{0.754}$	5–32	6.01
Data storage cap (MB)	14	$y = -143.5 + 154.85x^{0.0079}$	0.02–100	8.46
Downlink data rate (kbps)	18	$y = 26.0 - 21.86x^{-0.23}$	1–1000	8.91

Recently, SSCM was updated with emphasis on deriving a smaller set of multivariate (two or more parameter) CERs. One of the limitations of SSCM version 7.4 was that, for systems less than $10M, the standard error of the estimate could be in excess of 90% of the estimate. To derive version 8.0's CERs, we used GERM to minimize multiplicative error, so the standard error represents a constant percentage of the cost estimate rather than an absolute dollar amount.

The actual cost is the estimate plus or minus a percentage of the estimate. We regressed many parameters against total cost and selected those with minimum percentage error (i.e., we chose coefficients so the sum of the squared percentage errors was as small as possible). Table 8-5 shows the CERs for version 8.0.

SSCM software was developed by incorporating the individual CERs into a framework that allows estimates of production-development costs and cost sensitivities. We compute the derivative of each CER at the value of the user input so we can measure cost sensitivity around the design point of interest. This information is used to calculate the weighted-average estimate's sensitivity to each input parameter. Therefore, the user gets an estimate of a configuration's total cost and the cost estimate's sensitivity to performance and mass properties around the design point. SSCM software is the basis for a trade-study tool that integrates cost, performance, schedule, and risk into a framework for cost-effectiveness comparisons.

TABLE 8-5. SSCM Version 8.0: Cost-Estimating Relationships for Small Satellites. *c* is the total cost in FY94$M as a function of independent variables *r*, *s*, and *p*. Applicable range for each parameter and the percentage standard error (1 σ) are given.

Independent Variable, *s*	# Data Points	CER For Total Bus Cost (FY94$M)	Applicable Range	Std. Error (%)
r: EOL power (W) *s*: Pointing accuracy (deg)	17	$c = 6.47\, r^{0.1599}\, s^{-0.356}$	*r*: 5–500 *s*: 0.05–5	29.55
r: TT&C mass (kg) *s*: Payload power (W)	18	$c = 0.702\, r^{0.554}\, s^{0.0363}$	*r*: 3–50 *s*: 10–120	35.68
r: Downlink data rate (kbps) *s*: Average power (W) *p*: Prop system dry mass (kg)	21	$c = 1.44\, r^{0.0107}\, s^{0.509}\, 1.0096^{\,P}$	*r*: 1–2000 *s*: 5–410 *p*: –35	35.66
r: Spacecraft dry mass (kg) *s*: Pointing accuracy (deg)	26	$c = 0.6416\, r^{0.661} -1.5117\, s^{0.289}$	*r*: 20–400 *s*: 0.05–5	37.19
r: Solar array area (m²) *s*: ACS type (3-axis or other)	20	$c = 4.291\, r^{0.255}\, 1.989^{\,S}$	*r*: 0.3–11 *s*: 0=other 1=3-axis	38.53
r: Power subsys mass (kg)	25	$c = 0.602\, r^{0.839}$	*r*: 7–70	37.07

The small-satellite CERs were coded into a model hosted in a standalone compiled program that can operate on any IBM-compatible personal computer. It requires DOS 2.0 or higher and is menu-driven. The user is first prompted to enter a series of values that describe the small satellite's configuration for costing. SSCM will run with only a single input, but results are more accurate with multiple entries. Table 8-6 displays a sample input screen for version 7.4.

SSCM can use 17 technical parameters to estimate cost. To estimate the cost of a small satellite, the user must provide at least one of the parameters. SSCM computes the cost estimate based on a weighted average of each of the individual cost estimates. It also shows the individual cost estimate and standard error in FY94$M. Cost estimates are weighted by the inverse square of the standard error as explained above. A separate screen displays the sensitivity analysis, which is the effect on cost (in FY94$M) of changes in the input parameter. Tables 8-7 and 8-8 display two output screens—the single CER results and the weighted-average cost estimate and sensitivities for the case specified in Table 8-8. Graphs of the individual CERs that drive the model can also be displayed or printed.

To apply SSCM and properly interpret its results, we must understand the nature of its underlying data. Design, integration, and testing costs for the spacecraft bus are included, but the cost for the payload, launch vehicle, ground station, and operations and support costs are not. We assume that required resources and facilities (clean rooms, high bays, test stands and chambers, and people) are available. SSCM's database includes actual costs for modern small-satellite programs that embody cost-reduction philosophies. These philosophies call for maximum use of existing hardware, reduced adherence to MIL-STD parts and processes, limited sponsor oversight, and minimal onboard redundancy. Satellite bus weights of 20 to 250 kg and on-orbit lifetimes of 6 to 48 months are typical. We're planning versions of payload cost models tailored to small satellites. We're also developing a tool to simultaneously assess requirements for the ground segment and associated data handling which complement the current SSCM.

TABLE 8-6. SSCM Version 7.4 Input Screen. At least one input is required to run the model.

Small Satellite Cost Model Technical Parameters		
Sat #1: Test	**Value**	**NASA (Civilian) Valid Range**
Satellite volume (in^3)	—	2000–80000
Satellite bus dry mass (kg)	300.00	20–400
ACS dry mass (kg)	15.00	1–25
TT&C subsystem mass (kg)	10.00	3–29
Power subsystem mass (kg)	55.00	7–68
Thermal subsystem mass (kg)	—	0.5–10.5
Structures mass (kg)	—	9–95
Number of thrusters	—	1–8
Pointing accuracy (deg)	1.000	0.25–25
Pointing knowledge (deg)	—	0.1–3
BOL power (W)	—	20–480
Average on-orbit power (W)	300.00	5–410
EOL power (W)	—	5–440
Solar array area (m^2)	4.40	0.3–9.9
Battery capacity (A·hr)	20.00	5–32
Data storage capacity (MB)	100.00	0.02–100
Downlink data rate (kbps)	1000.00	1–1000

TABLE 8-7. SSCM 7.4 Cost Estimation Output Screen. Individual estimates based on single parameters are shown as well as the estimated composite cost and the standard error for the satellite bus. The composite estimated bus cost is calculated as a weighted sum (based on standard error) of individual cost estimates. Estimates for which we haven't provided an input parameter are indicated as N/A (not applicable).

Cost Estimates Based on Single Performance and Design Characteristics		
Sat #1: Test Sensor Platform NASA (Civilian) **Estimated Cost of Satellite Bus = 21.1 $M (FY94)** **Standard Error of Estimate = 5.1 $M (FY94)**		
Cost Driver (Technical Spec)	**Cost Estimate ($M FY94)**	**Standard Error ($M FY94)**
Satellite volume	N/A	N/A
Satellite dry mass	31.9	3.3
ACS dry mass	16.1	5.4
TT&C subsystem mass	9.1	4.5
Power subsystem mass	21.9	3.5
Thermal subsystem mass	N/A	N/A
Structures mass	N/A	N/A
Number of thrusters	N/A	N/A
Pointing accuracy	14.7	7.4
Pointing knowledge	N/A	N/A
BOL power	N/A	N/A
Average on-orbit power	20.2	5.7
EOL power	N/A	N/A
Solar array area	19.3	6.4
Battery capacity	19.7	6.0
Data storage capacity	17.1	8.5
Downlink data rate	21.4	8.9

TABLE 8-8. SSCM 7.4 Cost-Sensitivity Screen. Input values and associated units, provided by the user, are reiterated for reference. The sensitivity of the individual cost estimate to a given parameter is calculated as the slope of a line tangent to the CER at the design point.

Cost Sensitivity to Single Performance and Design Characteristics			
Sat #1: Test Sensor Platform NASA (Civilian) Estimated Satellite Bus Cost = 21.1 $M (FY94) Standard Error of Estimate = 5.1 $M (FY94)			
Cost Driver (Technical Spec)	Input Value	Units	Sensitivity in FY94 $K/unit
Satellite volume	N/A	in^3	N/A
Satellite dry mass	300.0	kg	30.365
ACS dry mass	15.0	kg	111.596
TT&C subsystem mass	10.0	kg	114.754
Power subsystem mass	55.0	kg	68.658
Thermal subsystem mass	N/A	kg	N/A
Structure mass	N/A	kg	N/A
Number of thrusters	N/A	Integer	N/A
Pointing accuracy	1.0	deg	–312.887
Pointing knowledge	N/A	deg	N/A
BOL power	N/A	W	N/A
Average on-orbit power	300.0	W	1.677
EOL power	N/A	W	N/A
Solar array area	4.4	m^2	80.211
Battery capacity	20.0	A-hr	49.950
Data storage capacity	100.0	MB	0.047
Downlink data rate	1000.0	kbits/s	0.034

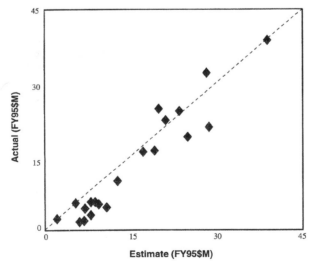

Fig. 8-9. SSCM Version 7.4: Cost Estimates vs. Actuals for Programs in Database. Data represents an in-sample validation, meaning that the programs compared are included in deriving the CERs.

8.2.4 Model Validation

We did a top-level validation of SSCM inputting all available technical parameters and obtaining an estimate for each small satellite that had contributed to SSCM's CERs. Then we inflated the known costs for each small satellite to FY94 dollars and computed the relative differences. Figure 8-9 shows the results for version 7.4. The diagonal line shown in the figure is not a regression, but rather a baseline we can use to compare the actual cost to the estimated costs. Thus, if the actual were equal to the estimated cost, the data point would lie directly on the diagonal. Figure 8.9 demonstrates an interesting result—additional effects appear to drive costs down even further than expected for the very low-cost satellites. This conflicts with our intuition that an irreducible "overhead" cost might exist, below which further cost reduction would be impossible. More likely, though, SSCM's apparent "over-estimation" of extremely low-cost satellites occurs because programs in the database include higher-cost, more sophisticated spacecraft.

8.2.5 Lessons Learned from SSCM

To properly apply a parametric cost model, like SSCM, we must understand the nature of its underlying data. In SSCM's database, two distinct elements appear to drive the cost of small-satellite systems: what is built and how it is procured. In many cases, **how** we buy a system may be as important as **what** we buy because the acquisition environment allows a program to realize its inherent cost and performance advantages. Clearly, technical and non-technical aspects of small-satellite procurement play a part in low-cost, successful programs, leaving a legacy for future efforts. Cost reduction stems from technical and non-technical improvements in the design and operation of space systems. Cost-influencing factors, or cost drivers, include *hard* (physical) factors and *soft* (management) factors. Small satellites have several basic characteristics that lead to dramatic cost reduction. Figure 8-10 shows what we believe to be the keys to cost reduction for the programs in SSCM's database.

• **SCALE OF PROJECT** — Reduced complexity and number of interfaces — Reduced physical size (light and small) — Fewer functions (specialized, dedicated mission)	• **DEVELOPMENTAL AND HARDWARE** — Using modern microprocessors, mass memory — Reduced testing and qualification — Extensive software reuse — Miniaturized command and data handling subsystems — Using commercial electronics whenever possible — Using existing components and facilities
• **PROCEDURES** — Short development schedule — Reduced documentation requirements — Streamlined organization & acquisition — Responsive management style	• **RISK ACCEPTANCE** — Using multiple spacecraft — Using existing technology — Reducing testing & redundancy of subsystems

Fig. 8-10. Cost-Reduction Techniques for Small Satellites. Cost reduction combines technical and program factors.

Small satellites typically use modern microcircuitry and other miniaturized devices much more than large satellites. Spacecraft have become much more capable because of the solid-state revolution. Technology research, sponsored largely by the DoD, NASA, and other agencies, advanced these capabilities. European organizations have also been heavily involved in small satellite projects, drawing on the expertise of the University of Surrey and the Royal Aerospace Establishment in England, for example. The availability of hardware has contributed to the rapid growth of small satellites because we can buy many subsystem components "off-the-shelf." (See Sec. 3.1.3 and 5.2.2.)

Cost estimating and cost reduction require accurate forecasting of technology development. Traditional, large, highly visible programs are restricted by risk-related concerns that often require more established, rigorously tested hardware. Small satellites have become an avenue for getting flight-demonstrated systems much faster. Some small satellites have more ability to test new hardware than the much larger, more expensive versions used in operational space programs today. For example, the small communications satellite Microsat, built by DSI for the DARPA Advanced Space Technology Program, flew the first all-digital subsystem for telemetry and commanding. ALEXIS, built by AeroAstro for Los Alamos National Laboratory, houses a 96-MB mass-memory system. Each of these satellites costs well under $15 million. Introduction of more advanced microtechnology in more subsystem areas could widen the gap in cost effectiveness between traditional large satellites and small satellites. But technology alone may not increase cost-effectiveness. We should couple it with an acquisition environment that allows the inherent cost and performance advantages of the technology to flourish.

The cost of risk abatement isn't trivial. The level of tolerable risk feeds directly into requirements for integration and testing and the number of redundant subsystems and components within the spacecraft. The challenge is to provide the right level of risk while maintaining acceptable costs. The small-satellite arena has tolerated more risk to save money. Small-satellite programs may launch several satellites over time, so the entire mission may not depend on a single large spacecraft. In addition, the unit value of any particular satellite is typically below $100M, including payload and launch. (See inside front cover.) The constellation may be built up incrementally, so success on each launch isn't as critical. Small-satellite projects have had failures, but these are the exception rather than the rule.

Risk and reliability are related but different. The *reliability* of a spacecraft is defined as the probability it will successfully complete the mission's specified objectives. *Risk* depends on both reality and perception. If a particular project has "high" reliability, it will be perceived as "low" risk. Thus, to assess reliability in space programs, we must understand how much risk the sponsor is willing to assume. Assumption of risk strongly drives system cost and encompasses a wide range of activities, including parts qualification, materials and processes, configuration control, and build and verification.

The way we manage risk and acquire and distribute risk-related information drives the cost of space systems. Whether risk is real or not, perception can strongly influence costing, especially at the public and policy levels. Handling the perceptions influences many of the engineering trade-offs involved in a high-profile space system and these trade-offs lead to a growing divergence between the mission and the system's readiness to carry it out. To reduce the cost of access to space, we may need to increase tolerance for risk which may or may not decrease quality. The trade-off between cost and risk is very critical and requires difficult decisions. We must look at each part of the program, identify the risk and quality issues, and set priorities.

Within the small-satellite contractor's development history, many hardware and software systems are carried over from one design to the next, so new development is scarce and reuse is common. The contractor's challenge has been to selectively insert advanced technologies that increase capability and reduce volume and weight, while reducing overall program cost and risk. Figure 8-11 summarizes some of the key small satellite cost-related subsystem developments that will likely continue to drive mass and cost down. Although most small satellites have used "off-the-shelf" technologies—like silicon solar cells, low-cost momentum wheels, low-cost Earth sensors, and simple cold-gas propulsion systems—the manufacturers of high-end buses for small satellites are now trying to get top performance without adding mass or cost. The subsystem technologies listed in Fig. 8-11, among others, will provide a way to increase performance and reduce mass within cost constraints. Using small, off-the-shelf components appears to simplify design, production, launch, and operations at the same time, thus reducing the weight and cost of the entire system. For example, extremely low production costs of microelectromechanical components might result in more redundancy, which in turn could reduce costs for quality assurance, test, integration, system engineering, and analysis.

Attitude Determination, Control & Navigation	Structure
Low-cost, High Performance Gyros Low-cost Star Tracker Low-cost Sun Sensors Miniature Optical IMU's Low-cost Integrated GN&C Subsystems Precision Reaction Wheels Low-cost Attitude Determination & Control Systems	Inflatable Antennas Solar Concentrators Thermal Radiators Propulsion Tanks
Command & Data Handling	**Propulsion & Reaction Control**
Centralized Motherboard Electronics Miniature Microprocessors Large Capacity Data Storage Systems (Solid State) Digital Voice and Video Data Compression	Low-Cost Hydrazine Thrusters Dual-Mode Bipropellant Systems High-Performance Biprop Systems High-ISP Low-Thrust Systems
	Telecommunications
Electric Power	Low-Cost EHF Technologies & Miniaturization Adaptive Uplink Antennas High-Speed, Low-Power Digital Signal Processing Subsystems Lightweight Frequency-Hopping Synthesizers Efficient Solid-state Transmitters
High Performance Solar Cells Small High Energy Density Batteries Low-cost NiCd Batteries Lightweight Solar Arrays	**Ground Stations/TT&C**
	Low-cost Commercial Equipment Miniaturized Man-portable EHF Terminal

Fig. 8-11.   Key Cost-related Subsystem Developments for Small Satellites. Cost savings are often achieved by identifying synergism between components and subsystem architectures.

The flight hardware and ground segment of small satellites are typically based on the most modern hardware and software available, so they require little technology development and flight certification. Recent advances in microelectronics and software allow a system-level downsizing, resulting in a complete satellite that weighs less than 400 kg. Engineering development units are typically not built, and the protoflight unit **is** the actual flight hardware. Wide design margins avoid costly analysis and tests. System-level tests are the rule because the components are readily accessible and design lifetimes are short (typically 1 to 3 years). The systems use little redundancy, and innovative, multifunctional subsystems and software architectures provide for operational work-arounds and modes of degraded performance. Dedicating the payload to one or two specific applications helps simplify the design, which also requires fewer interconnections (such as cables or plug connectors) to avoid interface failures. Fairly standardized structural and thermal designs alleviate the need for costly reanalyzing and validating. Plug-in "trays" and modules allow easy access to hardware components and greatly simplify integration. The relatively small physical scale of small satellites removes the need for large facilities and equipment such as high bays, hoists, large clean-room areas, test facilities, and large, special-handling equipment and containers. Ground stations dedicated to the user are based on workstation-type microcomputers and allow the user direct daily access to the satellite. The systems get the best payback by simplifying ground operations and reducing the time needed for on-orbit activation and check-out after launch.

Good small-satellite programs have aggressive systems engineering policies, plus dedicated lead engineers for subsystems and active principal investigators with full technical and budgetary responsibility. The approach is basically single-string with the goal of graceful degradation wherever practical. For example, independent safe holds may be established, by which software supports the prime mission but hardware handles survival functions. The spacecraft has autonomy for mission operations that are sensitive to operator error. Redundancy applies only to wear-out and critical items (shunts and shunt drivers, gyros, data buses, pyros, deployment actuators for solar arrays, and power-distribution lines). A thorough test program and active quality program apply to the spacecraft, with subsystem components pre-qualified and well screened. Graceful degradation occurs through battery charge control, overlapping of attitude control sensors, solid-state bulk memory, cross-strapping of critical components, and making processors reprogrammable.

Small satellites are typically developed under short, streamlined programs, with much less bureaucratic overhead than traditional programs. These programs tailor performance around what they can do inexpensively on a short and well-defined schedule. They develop a design concept and buy long-lead hardware and software right after preliminary design review. They hold requirement changes to a minimum and avoid costly analysis and reassessment of multiple design options. Engineers and project managers can make decisions at lower levels without heavy oversight from government or management. In many cases, satellite builders have the leeway to do what they think is right and necessary to get the job done well. Greater risk is tolerable because the program expects to launch several satellites over time, and each one isn't too expensive. The entire fully capable constellation, unlike typical "large" satellite systems, can incrementally build up, so it doesn't rely on a single satellite being successful. The sponsoring agency works directly with the service provider, keeping organizational interfaces and paperwork to a minimum. Programs use existing test facilities whenever possible. Adhering to a set schedule is important. The project doesn't delay while panels convene and discuss the effect of every construction decision or test result. Because project time frames are short (typically 18 to 36 months), experiencing real progress and being recognized on a relatively small team tend to inspire high productivity.

8.3 Dealing with a Changing Paradigm

As illustrated in Table 8-9, the cost of a space mission depends on a many variables, associated with what we build and how we build it. All of us who have been in the space business for many years recognize that cost depends strongly on management style and methods, the level of quality assurance, the size and complexity of the organizations involved, and the level of documentation required. Still, as described in the first two sections, cost models deal mainly with physical parameters, such as spacecraft weight or power. This approach to cost modeling is, above all, expedient. Weight and power are unambiguous and easily quantified for cost estimates. This approach is perfectly legitimate. We don't need to take all of the relevant variables into account. For example, we can examine cost as a function of weight and evaluate the historical data for correlations. If this correlation is present, weight is a legitimate measure of cost, even though we recognize that many factors remain unmodeled.

TABLE 8-9. Factors Driving the Cost of Space Systems. Cost will depend on size and other hard parameters but will also be greatly influenced by the project team's management style, organization, risk averseness, and other soft factors.

Hard Cost Elements	Soft Cost Elements
• Physical parameters	• Organization size
• System size	• Levels of management involved
• Performance criteria	• Number of organizations involved
• Reliability	• Risk averseness
• Schedule and timeline	• Management method
• Technology risk impact	• Team's familiarity with the technology used
• Complexity of the system	• Complexity of the project team
• Space qualification of system and components	• Team's experience of the human resources involved
• Initial investment in research and development	• R&D investment
• Quantity produced	• Quality approach
• Production learning curve	• Organization learning curve
• Launch system	• Dependence on external financial source
• Ground operation system	
• Logistic support	
• Sparing policy	
• Quality system in place	

The problem with unmodeled cost drivers, such as management style or performance, is how we use the cost models and draw conclusions from them. It's appropriate to use cost models for broad, general conclusions—for example: a spacecraft that weighs more typically costs more. This is a reasonable and legitimate application. We can also numerically estimate system cost. The problem arises when we attempt to draw specific conclusions about particular programs, such as concluding that increasing the weight of a particular spacecraft by 10 kg will increase the cost, or that cost doesn't depend on soft factors such as management style or levels of documentation. In that case, we're drawing invalid conclusions and not using cost models correctly. In this section, we'll look as some of the main unmodeled cost drivers, as well as how to treat them in an engineering assessment of system cost.

8.3.1 Main Unmodeled Cost Drivers

Although most parametric models omit many cost drivers, the most important of the missing ones are performance, culture, group size, and willingness to accept risk.

Performance

Performance is what the space mission is all about. It's the reason we're there. Thus, what we'd really like is a cost model that treats life cycle cost as a function of system performance. This is the basic objective of analyzing mission utility, as described in Sec. 10.3.2. Yet, creating a model of cost versus performance is typically difficult. Performance is usually hard to quantify. Because most missions have multiple objectives, we also have the problem of characterizing how many of the secondary missions we can achieve and somehow balancing the multiple objectives to create some combined measure. For example, we can compare cost versus weight for a communications satellite and for a meteorological satellite, but comparing these systems in terms of cost versus performance just isn't realistic.

Culture

Perhaps the single most important unmodeled parameter in determining the cost of a space mission is the culture of the organization that produces it. The aircraft industry simply can't use development-cost models from a normal commercial manufacturer to estimate the cost for the Lockheed Skunk Works to create a new high-performance aircraft. The culture and the way of doing business are simply different. The same is true in trying to compare the methods employed by the traditional aerospace prime contractors with those used by AeroAstro, CTA, OSC, Spectrum Astro, the University of Surrey, or any of the other companies involved in the long-term manufacturing of low-cost satellites. The approach to doing business is simply different. We cannot say which is better without understanding the needs of the customer and the community. But we can say that a cost model built from data for traditional large spacecraft won't predict cost for the builders of low-cost satellites.

Group Size

Somehow small organizations (or organizations that "think small") build spacecraft for less money. Other chapters and the case studies discuss the reasons. Yet, small organizations simply can't build some spacecraft. For example, small companies aren't going to build Space Station Freedom or the next copy of the Orbiter. In addition, organizational size is in many respects a fuzzy parameter. What matters is how the organization thinks and behaves rather than how many people it has. In looking at the Naval Center for Space Technology, which produced Clementine, should we consider just the size of the space group, the size of NRL, or the size of the Navy? Each of these organizations and sub-organizations played a part in the process. It's clear that organizational size matters, but it is not at all clear how to take it into account.

Willingness to Accept Risk

Many of the case studies and process chapters have pointed out that a key ingredient to the success of low-cost satellite programs is their willingness to take more risks. If we're willing to be innovative, to try new approaches, and to accept the possibility that they might not all work, we can dramatically reduce a space system's cost. But, once again, there is no way to quantify or model our willingness to accept risk. We have a concept of what it means, but we don't know how to account for it.

8.3.2 Summary: Cost Modeling and the SmallSat Paradigm Revolution

Whether we choose to call them LightSats, SmallSats, or low-cost spacecraft, clearly a major paradigm change is going on in the space industry. The emphasis on cost and need to reduce cost is unparalleled in the history of the space program. Because the business processes are changing, it's difficult to predict spacecraft cost. The fundamental assumption in cost modeling is that, in some respect, spacecraft will be built as they have in the past. But it's precisely this assumption that is breaking down. Spacecraft can't be built as they have been in the past or they simply won't be funded. We may be able to understand the cost consequences, but our chances of successfully modeling it are small.

Certainly the best option for dealing with this circumstance is to develop a new cost model based on spacecraft built under the "new" rules. This is the objective of The Aerospace Corporation's Small Satellite Cost Model, which excludes spacecraft completed and launched before 1990. It's perhaps the most realistic way to adequately model costs and to assess the changing parameters of modern space missions. Making this work depends critically on:

- Keeping the cost model itself up to date, so it can try to capture new processes and methods as they evolve
- Using cost-risk analysis, as described in Sec. 8.1.4, to evaluate the probable variability in the cost when new, unmodeled methods are used to develop spacecraft

Although the small-satellite cost model is effective in the regime it covers, it doesn't cover several critical problems:

- How does the new paradigm affect cost for space payloads or operations?
- How do we judge new low-cost approaches to launch systems?
- How do we evaluate the effect of intangibles, such as group size, organizational culture, or willingness to accept risk?
- How do we judge whether a particular program's or contractor's cost estimate is real?

These problems become critical whenever cost becomes the main driver of a program's survival.

The key to dealing with the intangible, qualitative element, or "new rules," is similar to that of producing low-cost spacecraft. We need to understand the content, applicability, and limits of traditional and newer cost models. We then need to assess, outside of simply running the model, how other factors apply—corporate culture, size of the group, or how they build the system. Unfortunately, we can't simply evaluate formulas and provide the result as an accurate or even crude estimate of cost.

The mechanisms for evaluating cost estimates require judgment from cost estimators on the mission design team and from the organization that must control costs. They must evaluate:

- The organization's experience in creating low-cost missions or reducing cost.
- How much the organization can apply or is applying methods known to reduce cost in other organizations, as well as the engineering basis for the cost estimates.
- The organization's size and overall culture with respect to cost and performance.
- The willingness of the organization building the spacecraft and the funding organization to tolerate risk and continue programs without a guarantee of success.

These factors can be evaluated, but they require engineering judgment and a sense of which ones are important. We can, for example, reduce cost by compressing the schedule. But if the schedule is compressed with no change in requirements, the only result is that we kill the engineers. Schedule compression has to be used intelligently, along with a more rapid decision cycle, reduced documentation, and acceptance of engineering changes, in order to be an effective way to reduce cost.

Accurate cost modeling is at the same time more difficult and more critical than it has been in the past. It requires more engineering judgment, so it needs more reasoning and logic than formulas and regression analysis. Providing appropriate and unbiased methods for doing this will be a major challenge, similar in many respects to the challenge of dramatically reducing space mission cost.

References

Abramson, R. L., D. A. Bearden, and D. L. Glackin. 1995. "Small Satellites: Cost Methodologies and Remote Sensing Issues." SPIE/CNES European Symposium on Satellite Remote Sensing, Paris, France, September 25–28, 1995.

Abramson, Robert L. and Philip H. Young. 1990. "FRISKEM—Formal Risk Evaluation Methodology," ORSA/TIMS Joint National Meeting, Philadeplphia, PA, October 29–31, 1990.

Apgar, H. 1990. "Developing the Space Hardware Cost Model." Paper IAA-CESO-04 (90), AIAA Symposium on Space Systems Cost Methodologies and Applications, May 1990.

Bearden, David A. and Robert L. Abramson. 1994. "Small Satellite Cost Study—Risk and Quality Assessment." 2nd International Symposium on Small Satellite Systems and Services, Biarritz, France, June 27–July 1.

Bearden, D., E. Burgess, and N. Lao. 1995. "Small Satellite Cost Estimating Relationships." 46th International Astronautical Congress, October 2–6, 1995.

Bell, K. D. and L. A. Hsu. 1994. "Balancing Performance and Cost for Cost-Effective Satellite Systems Design Using an Integrated Cost Engineering Model." The Aerospace Corporation, TOR-95(5409)-2, December 18, 1994.

Bevington, Philip R. 1969. *Data Reduction and Error Analysis for the Physical Sciences*. New York: McGraw Hill.

-------January 1994. Access to Space Study—Summary Report, NASA Report Number: NASA-TM 109693, NAS 1.15:109693. Washington, D.C.: NASA.

Book, S. A. 1993. "Recent Developments in Cost Risk." Space Systems Cost Analysis Group Meeting, Palo Alto, CA, June 7–8, 1993.

Book, S. A. and P. M. Young. 1994. "General-Error Regression for USCM-7 CER Development," 28th Annual DoD Cost Analysis Symposium, Leesburg, VA. September 21–23, 1994.

Boudreault, R. et al. 1994. "Space Cost Engineering Seminar Notes." Washington, D.C., AIAA.

Burgess, E. L. and H. S. Gobrieal. 1996. "Integrating Spacecraft Design and Cost-Risk Analysis Using NASA Technology Readiness Levels," The Aerospace Corporation. Presented at the 29th Annual DoD Cost Analysis Symposium, Leesburg, VA. February 21–23, 1996.

Hamaker, J. 1987. "A Review of Parametric Cost Modelling Activities at NASA-MSFC." Paper presented AIAA Economics Technical Committee, Buffalo, N.Y.

Hamaker, J. 1989. "NASA MSFC's Engineering Cost Model (ECM)." Proc. Intl. Soc. Parametric Analysts, 10th Annual Conference, Brighton.

Greenberg, J. S. "Reliability, Uncertainty and Assessment of Space System."

Greenberg, J. S. and Hertzberg, H. 1994. *Space Economics*, AIAA.

Gross, A. September 30, 1988. "Documentation of the Cost and Weight Estimating Relationships for the Space Sensor Cost Model," Version 6.1, The Aerospace Corporation.

Hillebrand, P., et al. 1988 *Space Division Unmanned Spacecraft Cost Model.* 6th edition. USAF Space Division, Directorate of Cost Analysis. El Segundo, CA.

Lao, Norman. 25 January, 1995. "Statistical Methods and CER Development," briefing to NASA's Lewis Research Center, Cleveland, OH.

Law, Glenn W. and Kevin D. Bell. 1996. "Application of Spacecraft System Modeling Techniques for the Assessment of Technology Insertion." IEEE Aerospace Applications Conference, Aspen, CO, February 4–11, 1996.

Press, William H., Brian P. Flannery, Brian., et. al. 1988. *Numerical Recipes in C,* Cambridge University Press, pp. 528–541.

RCA Price Systems. 1987. *An Executive Guide to PRICE.* Moorestoum, NJ.

SMAD = Larson, Wiley J. and James R. Wertz, eds. 1992. *Space Mission Analysis and Design,* 2nd edition. Torrance: Microcosm Press and Dordrecht: Kluwer Academic Publishers.

Stuart, J. R. 1991 "LEO and GSO Communications Lightsats: Developments and Applications," Seminar at Interdisciplinary Telecommunications Program, University of Colorado, Boulder, April 24.

Young, P. H. 1992. "FRISK—Formal Risk Assessment of System Cost Estimates." The Aerospace Corporation. Presented at the AIAA 1992 Aerospace Design Conference, February 3–6, 1992, Irvine, CA.

Chapter 9

Reliability Considerations

Herbert Hecht, *SoHaR, Inc.*

9.1 Interaction Between Reliability and Cost

This section explores the following topics:

- Cost of achieving high reliability
- Cost of not achieving high reliability
- Inherent reliability advantage of a low-cost (low-complexity) satellite

This latter topic is briefly discussed now. Reliability prediction (particularly the practice based on MIL-HDBK-217) has its advocates and its detractors, but an undeniable merit is that it requires accounting for every part that goes into a spacecraft. Hence small, low-cost spacecraft will, by virtue of their low parts count, achieve a higher predicted reliability than more complex spacecraft with a higher parts count. Evidence that simplicity makes for reliability can be seen in Table 9-1 which is a summary of Sec. 4 from a report entitled *Reliability Prediction for Spacecraft* [Hecht, 1985]. There were probably compelling reasons in each case for using the more complex implementation, but the reliability consequences of these decisions must also be recognized.

The first question to be answered is whether a reliability program is required in a low-cost spacecraft development. Alternatives are listed in Table 9-2 in order of increasing direct cost. For very small satellites the low-cost advantage of (A) or (B) will usually outweigh the benefits of the last two alternatives. When subsystems get more numerous and larger, alternatives (C) and (D) may result in overall savings, in spite of their higher tangible cost.

TABLE 9-1. Relationship Between Failure Rate and Complexity. In all systems investigated here, the complex versions have a much higher failure rate than the simple ones.

System	Simple		Complex	
	Type	Failure Rate*	Type	Failure Rate*
Telemetry	Hardwired	0.034	Programmable	0.190
Stabilization	Gravity Spin	0.038 0.216	3-axis active	0.610
Thermal	Passive	0.084	Active	0.320

*per orbit-year

TABLE 9-2. Alternatives for a Reliability Program. Alternatives (A) and (B) are usually preferred for very small satellites; (C) and (D) are more suitable for larger ones.

Alternative	Benefits	Disadvantages
(A) Designers responsible for reliability	• No additional staff • Familiarity with items • Clear responsibility	Difficult to achieve uniformity; also, limitations shown under (B)
(B) Designers responsible with policy guidance from management	• All of the above plus some uniformity	No responsibility for subsystem interactions, opportunity for analytic redundancy may be overlooked, little awareness of reliability tools
(C) Designers responsible with guidance from reliability organization	• Uniform procedures, above disadvantages largely overcome	Requires dedicated reliability function; possible confusion over responsibility
(D) Responsibility in reliability organization	• Responsibility clearly defined, interactions likely to be identified	Lower motivation for designers; expense of reliability organization.

The cost for reliability is incurred a considerable time prior to launch. Decisions to incur that cost frequently must be made in the absence of credible data on failure modes and rates. Investment in reliability database searches and in environmental testing of the candidate components will permit focusing the reliability effort on those subsystems and components that present the greatest risk.

Reliability measures fall into two major groups: those that prevent failures (*fault avoidance*), and those that permit continuation of the mission in spite of failures (*fault tolerance*). Fault avoidance usually involves lower cost than fault tolerance but is effective only when failure modes are known, and a given piece of equipment has only a few significant failure modes. Fault avoidance measures are covered in Sec. 9.3.

Fault tolerance is usually more cost effective when the failure modes are not known or where there is no prior experience with the equipment in a similar application. Fault tolerance techniques are discussed in Sec. 9.4.

The cost of failure depends on the **time** at which it occurs, on the **scope** of the failure, and on available **mitigation measures**. A failure that occurs during launch or orbit injection carries with it a much higher cost than one that occurs in later mission phases because it:

- Affects the entire mission
- Frequently prevents observation of the operation of portions of the spacecraft that have not failed
- Suppresses the demonstration of the mission value (a unique observation opportunity may be lost or alternative means of achieving the mission goals may have become available)

Unfortunately, the probability of failure during launch or orbit injection is much higher than during any comparable period on orbit. A study by the Goddard Space Flight Center showed an average of 1.7 failures per spacecraft during the first 30 days, compared to an average of less than 0.2 failures per month during the following 150 days [Timmins, 1970].

By *scope of failure* is meant the fraction of essential spacecraft systems that are either degraded or made inoperative. A failure of a payload sensor has usually a small scope. A failure of the electric power system will frequently affect all operations and is therefore of very large scope. Much more aggressive reliability measures are warranted for components that can induce failures of large scope than for those that cannot.

Mitigation measures permit the initially large scope of spacecraft failures to be converted to a much smaller scope by "work-arounds," usually initiated on the ground. Putting extensive provisions for ground intervention into the command structure can be one of the most effective ways of enhancing the probability of mission success for satellites that do not employ much redundancy.

Once a failure has occurred, the following cost elements may arise:

Cost of Failure Analysis—Upon occurrence of any significant failure the organization responsible for the development of the satellite will be tasked with defining the scope of the failure, determining possible work-arounds, assessing the remaining mission capability, and identifying both the immediate and the root cause of the failure. All of these activities will be conducted under pressure and without the benefit of advance planning or scheduling, and by senior personnel (the release of inappropriate findings from a failure analysis can greatly compound the cost of failure). These activities will incur the cost of direct labor, applicable overhead, and frequently a "loss-of-opportunity" cost because the senior personnel will not be available to tend to their normal assignments.

Cost of Mitigation—This will include the implementation of work-arounds (e.g., acquisition of additional ground telemetry equipment), re-programming for changed data formats and permanent staffing that may be required when spacecraft functions have to be transferred to the ground.

Loss of Mission Value—The impairment of the mission may be partial or complete (particularly likely for launch and early orbit failures). The amount assigned to this cost of failure element can sometimes exceed the cost of a replacement launch, e.g., when unique observation opportunities were missed.

Loss of Confidence—A major on-orbit or launch failure can create loss of confidence in the spacecraft design, the design organization, the spacecraft mission, and the sponsoring organization. These costs are not easily quantified, but neither can they be neglected. The potential for loss of confidence is well recognized in large spacecraft projects that are open to public scrutiny. The effects on the mission and the sponsoring agency can be expected to be less for low-cost satellites. The effects on the design and the design organization are probably not significantly modified by the size and cost difference unless there is explicit acceptance of a modest reliability goal.

Even when no formal reliability budget is established, the following procedure is useful to control the amount of resources (money and the monetary equivalent for items such as weight or power) that will be devoted to reliability. The theoretical guideline for an optimum budget is deceptively simple: allocate resources to reliability improvement until the incremental cost of reliability exceeds the incremental cost of failure:

$$-dV_R / df = dV_F / df \qquad (9–1)$$

where V_R is the reliability improvement budget, V_F is the expected cost of failure, and f is the total spacecraft failure probability. As shown in Fig. 9-1, dV_F / df is assumed to be constant, and its ordinate value at 1.0 probability of failure is the spacecraft cost. In practice, the derivatives in Eq. 9-1 are replaced by cost increments and failure decrements. A more detailed discussion of the procedure can be found in SMAD. Even in the absence of precise values for these parameters, the relation is useful for sensitivity studies.

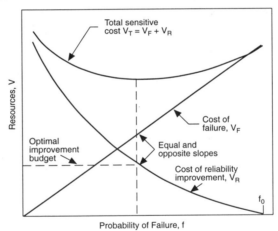

Fig. 9-1. Budgeting for Lowest Total Cost. The total cost curve has a shallow minimum. This indicates that only minor cost increases will be incurred by uncertainty in failure rate and cost data.

9.2 Reliability Program for Low-Cost Spacecraft

The first three sections describe important documents that are usually generated in support of a reliability program: the program plan, the failure modes and effects analysis, and failure reports. The later subsections deal with the assessment of spacecraft reliability and with the identification and evaluation of reliability improvement alternatives.

Reliability Program Plan

A *reliability program plan* adds little to the cost and is recommended even for the smallest spacecraft program. It specifies reliability objectives, assigns responsibility for achieving them, and establishes milestones for evaluating the achievements. It also serves as an agreement with other spacecraft functions regarding their responsibilities in support of reliability. The most significant interfaces usually are with quality assurance, test, configuration management, and thermal control.

Failure Modes, Effects, and Criticality Analysis (FMECA)

FMECA can provide valuable insights into how design decisions affect reliability. Typical benefits are:

- Exposing single point failure modes in a subsystem assumed to be redundant
- Identifying opportunities for functional redundancy (See Sec. 9.4.)
- Permitting components to assume a safe mode in the absence of required signals or power

Sneak circuit analysis is usually considered a part of FMECA. This analysis establishes that explosive or other one-shot devices will not be accidentally actuated, and that they will always be actuated when intended. A good guide to FMECA is MIL-STD-1629. [U.S. DoD, 1983]

Failure Reporting and Corrective Action (FRACAS)

Failure reporting is a key element in any reliability program because it:

- Informs concerned parties that a failure has been observed
- Furnishes a record through which trends and correlations can be evaluated at a future time (an example of a trend is that the probability of failure increases after x hours of use; an example of a correlation is that part y fails during a particular step in the test sequence)
- Permits reassessment of the predicted failure rates and is the basis for consequent modifications of the fault avoidance or fault tolerance provisions

The first two uses listed above require recording of operating time of all units in service; usually by means of an operating log maintained for each part number, with separate sheets for each serial number.

Corrective action is typically recorded on the failure reporting form. This facilitates configuration management in that it establishes at what point a failed component or subsystem has been returned to operational status. Reporting on the same form also facilitates future investigation of the effectiveness of the repair action. Corrective action frequently involves two steps: in the first, a failed part is replaced by a good one of the same design; the second step addresses the root cause, e.g., by tightening limits for the incoming test of this part. The results of retest are included in the corrective action report.

The format used for reporting of failures and corrective actions is not standardized. The following are the most essential data:

- Incident identification (e.g., report serial number)
- Date, time, and locale of the incident
- Part number, name of the failed component, and its serial number
- Higher level part or system identifiers (subsystem or major component)
- Lower level part or system identifiers (usually available only after diagnosis)
- Operation in progress and environmental conditions when failure was detected
- Immediate and higher level effects of failure
- Names of individuals responsible for detection, verification, and analysis
- Diagnosis of immediate, contributory, and root causes of the failure
- Dates and nature of repair and results of retest

9.2.1 Low-Cost Methods of Reliability Assessment

Reliability prediction (usually by using a failure rate handbook) or estimation (based on experience with the component population) are routinely required for major satellite programs. For small spacecraft, they may be required when test or on-orbit incidents indicate insufficient reliability, or when a more expensive payload is to be incorporated (which increases the cost of failure). The small component populations that are typical of space procurements preclude the use of conventional reliability demonstration methods. Even a rather modest requirement, such as reliability of 0.95 for 17,000 hours (approximately 2 years), corresponds to a Mean Time Between Failure (MTBF) of over 330,000 hours and will require over a million component-hours of test to arrive at a statistically meaningful assessment by conventional methods. Yet, experimental verification of the claimed reliability of a component or subsystem is frequently desirable and sometimes required. The following paragraphs explore low-cost methods of accomplishing this.

The major causes of failures are workmanship and design. The first of these can be controlled by quality assurance. Design failures occur primarily because the strength of a component is not adequate for the environment in which it is used, or because the manufacturing process permits too much variability in component characteristics. This is most easily seen in mechanical spacecraft components where reliability depends on the margin between the nominal strength of the component and the maximum service load, and the variability in strength of the delivered product. Since test can characterize the strength of mechanical components fairly easily, strength load margins (design margins) have always played a major role in their reliability assessment. For electrical and electronic components the same relationship holds in principle, but it is usually much harder to define a single failure inducing stress or load. Nevertheless, test data can give valuable insights into potential reliability problems. Important requirements are recording of test results in numerical form (not pass or fail) and statistical evaluation of the probability of failure derived from the numerical test results, e.g., by applying the "6-σ criterion" (the mean value of a parameter is at least 6 standard deviations above the specified minimum or below the specified maximum). The tests need not be specific reliability tests; typically they are the qualification test and acceptance tests.

For low-cost spacecraft, the emphasis is on obtaining high reliability through the normal design and test processes (a bottom-up approach). Targets for reliability improvement can often be identified by using the following questions:

What are the failure-inducing stresses and how are they alleviated?
The nature of the failure-inducing stresses arise from the combination of (a) the launch and orbital environment and (b) the component characteristics. The best source for data are local failure reports (Sec. 9.2.3) or data exchange agreements with other organizations. Historical data are summarized by Hecht [1985].

On previously-used components, what are failure modes and rates?
If the failure rate appears high, investigate how stresses can be reduced. Where failure modes have critical effects on the mission, select a different component or provide protective measures. Stresses can be alleviated by thermal control, regulation of power supplies, EMI protection, and mountings that do not amplify the mechanical forces due to launch and spacecraft separation.

Where are the major reliability risks?
Risks can be associated with new components, one-shot devices, and highly stressed parts. While the normal performance of new components can be observed in the development environment, there is usually little known about failure modes and response to stresses in the launch and space environment. One-shot devices (pyrotechnics, spring-operated deployment mechanisms) pose a risk because of the limited ability to observe their operation. Redundancy, or back-up functions, should be considered for these. Normally, an ample margin between the operating stress and the device rating is the best fault avoidance mechanism, but the weight and power limitations of space missions may dictate that some components be operated close to their rating. When this occurs, very detailed analyses are necessary to determine that ratings are not exceeded under adverse circumstances.

What are consequences of failure?
This is normally addressed when the FMECA is prepared. (See Sec. 9.2.2.) It is particularly important to analyze the consequences of failure for the "reliability risks" defined above. Subsystem and spacecraft design should prevent other components from being affected by failures. We should explore degraded modes, or ground intervention to mitigate the effects of failures.

9.2.2 Design Trade-off Techniques

Trade-offs are normally undertaken only for components identified as targets for improvement by the process discussed above. Thus, almost regardless of the quantitative criteria below, only components with predominant failure modes that have a major mission impact will be candidates. Among these, the following procedure provides "the most bang for the buck." It represents an implementation of Eq. 9-1, with discrete increments replacing the differential terms.

1. For each alternative design, list the increment (from a baseline) in mission reliability, ΔR_{mi}, and the expected cost increment, ΔC_i.

2. Form the ratio, $\Delta R_{mi} / \Delta C_i$.

3. List the alternatives in the order of decreasing $\Delta R_{mi} / \Delta C_i$.

4. If the objective is to achieve a specified reliability increment, list cumulative ΔR in column 5; if the objective is to achieve the highest reliability at a given improvement budget, list cumulative ΔC in column 6. These indicate when the desired reliability increment has been achieved or when the budget has been exhausted.

5. Review the candidates for possible dependencies (selection of one candidate may affect the benefit or budget of another).

Table 9-3 gives an example of the process. We assume that a reliability improvement, ΔR, of 0.0005 is desired, but that at most $20 million is available for achieving this. To keep track of both the reliability increment and the expenditures, two cumulative columns are shown. Note that a reliability increment of 0.00046 can be provided at a cost of $18.5 million. To achieve a reliability increment of 0.00054 would require an additional $20 million, greatly exceeding the available funding. Therefore the improvement process stops after the third entry (addition of a solar panel).

TABLE 9-3. Definitive Improvement Candidates (Costs in $M). Over one-half of the total improvement is due to just the first two candidates.

Alternative	ΔR	ΔC	$\Delta R / \Delta C$	Cum. ΔR	Cum. ΔC
Derated capacitors	0.0005	2.5	0.00020	0.00020	2.5
Cooling for µ circuits	0.0010	6.0	0.00016	0.00036	8.5
Add 1 solar panel	0.0010	10.0	0.00010	0.00046	18.5
Redundant gyro	0.0015	20.0	0.00008	0.00054	38.5
Larger battery	0.0005	10.0	0.00005	0.00059	48.5

9.2.3 Test Trade-off Techniques

The least expensive reliability test is one that is not run at all as a reliability test, but rather as part of a qualification test, lot acceptance test for purchased parts, or as an acceptance test on the spacecraft as a whole or on a major subsystem. To utilize these activities for reliability assessment may require some additional instrumentation and sometimes an extension of the test time, but these are very small resource expenditures compared to those required for even a modest, separately run reliability test. Other alternatives to obtaining reliability data by test are:

*   Use test data from others (including vendors) on the same component

*   Use test or experience data on similar components

*   Stress-strength analysis (particularly for mechanical components)

*   Reliability prediction by MIL-HDBK-217 or similar sources [DoD, 1991]

One of the greatest problems with reliability tests is that the results are usually obtained after many months or even years of test. Once it has been decided that a reliability test is necessary, a suitable scope, and environment of test must be selected.

The scope designates the assembly level; parts and circuit boards usually being designated as small scope, while subsystem and system level tests represent a large scope. Advantages of small scope tests are:

- Low cost and small size of test articles permit testing of multiple items
- Inputs and outputs are easily accessible
- Test environment can be tailored to the requirements of the unit under test
- Test can be conducted early because it does not require integration

Advantages of large scope tests are:

- Interactions between components can be observed
- Test results are easily translated to effects on the mission

The test environment can be quiescent (room ambient) or stressed. Advantages of testing in a quiescent environment are:

- Low cost (no or only simple environmental chambers required)
- Articles under test are easily accessible
- No induction of failures due to unusually high stress

Advantages of a stressed environment are:

- Increased probability of failure (less test time required)
- Can identify environmental vulnerability of the unit under test

Disadvantages of each approach may be inferred from the listed advantages of the opposite approach. These attributes lead to the test recommendations in Table 9-4.

TABLE 9-4. Typical Uses of Reliability Testing. Upper left is the lowest cost alternative.

Environment	Small Scope	Large Scope
Quiescent	Suitability test Critical components—failure-inducing stress unknown	High-risk subsystems
Stressed	Critical components—failure-inducing stress known	

Reliability testing at large scope and in a stressed environment is very expensive and is rarely warranted for low-cost satellites.

The time required for any reliability test can be significantly reduced if **testing by variables** is employed. This means that the value of significant attributes is numerically recorded (as contrasted with the commonly used pass or fail procedure). From these data, the distribution of parameters can be plotted, and the probability of dropping below an acceptance criterion can be assessed. The general technique is similar to that described for screening in Sec. 9.3.3. For reliability assessment, the parameter distributions are of interest, whereas in screening the attributes of individual units are the chief criterion.

9.2.4 Software

Spacecraft operations are becoming increasingly dependent on software both in the spacecraft itself and in the ground segment. There have been spectacular launch and on-orbit failures due to software faults, but many more missions have been saved by software (used for "work-around" solutions) than have been lost due to it. Thus the way to low-cost space missions may not so much lie in minimizing software as in making appropriate use of it.

Reliability assessment of software is not a precise science. The most widely accepted techniques depend on test (very expensive if high levels of reliability are claimed) or on independent verification and validation (even more expensive and less likely to yield a quantitative assessment). Software reliability is enhanced and the cost of evaluation is reduced if it is **simple** and **well-structured**.

Cost effective reliability assessment of software starts with an examination of the ways in which software can impact the mission. A convenient format for this is a fault tree, as shown in Fig. 9-2.

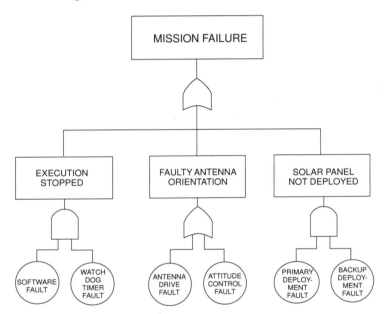

Fig. 9-2. Example of a Fault Tree. "And" gates have rounded tops. "Or" gates are shield-shaped.

In this example, the software failures that are most likely to lead to mission failure are:

- Complete halting of execution, e.g., due to the software being caught in an infinite loop, or being directed to an incorrect memory location
- Faulty antenna orientation due to antenna drive software failure or faulty attitude determination
- Software failure to deploy the solar array

Double failures are required to halt execution of programs and deployment of the solar array, because both of these have back-up mechanisms. Faulty antenna orientation can be caused in two ways, and there is currently no back-up provision for this hazard. Once this deficiency is recognized, alternatives for mitigation will be investigated. In this instance a means of re-orienting the spacecraft by ground command may be the lowest cost solution, provided that ground commands will be received regardless of antenna position.

A study of the space shuttle avionics software showed that causes of high severity failures were overwhelmingly associated with rare events, such as handling of exception conditions, management of hardware failures, and response to unusual or incorrect crew commands [Hecht, 1993]. These data, as well as others cited in the reference, indicate that testing of spacecraft software should emphasize the response to unusual events in the environment.

9.3 Design for Fault Avoidance

9.3.1 S-Level Parts, Design Margins, Derating, and Environmental Protection

Space qualified (S-level) parts came into being because the very high cost of failure on the large satellites of the 1965–1980 era warranted special processes, testing, and documentation of parts that were used for critical components in these satellites. Only a small number of programs mandate the use of these parts today, and their availability is becoming problematic. Among the low-cost satellite programs detailed in this volume only one specifically indicates the use of S-level parts "when available." The reason behind such choices is that the parts cost is only about 10% of the total satellite cost. Therefore, even a large increase in the cost of the fraction of parts for which S-level could be procured would not make much difference to the overall cost. Other programs found that high quality commercial parts, together with other fault avoidance techniques discussed here, were adequate and much more readily available.

Design margins and de-rating accomplish the same goal: prevention of component failure due to higher than expected external stresses. The term *design margin* is mostly used in structure and thermal subsystems and means that a component is designed to carry more than the expected load. *Propellant margins* in propulsion systems are an equivalent concept. The term *derating* is primarily applied to electrical and electronic components and involves the specification of a component that carries a higher rating than is expected to be needed for the application.

The reliability improvement, or reduction in failure rate, by these practices is most significant if a part is used initially near its design strength or electrical rating. As an example (from MIL-HDBK-217F), the predicted base failure rate for a fixed film resistor at 40 °C and used at 90% of rated power is 0.0022×10^{-6}/hr. Selecting a higher rating resistor, for which the dissipated power constitutes only 30% of rated power reduces this to 0.0011×10^{-6}/hr. But further reductions are hard to achieve. A resistor for which the dissipated power is 10% of the rating still has a failure rate 0.0009×10^{-6}/hr. Derated parts not only cost more, but are frequently larger and heavier than the ones that they replace. Derating only reduces the failure probability with respect to the stress that is being derated. In the example of the fixed-film resistor, derating reduces the probability of failure due to power surges but it does not offer any protection against failures due to lead breakage or corroded connections.

Environmental protection can take the form of shock mounting, cooling or heating provisions, and shielding against radiation effects. Whereas derating reduces the failure probability by increasing the strength of the components, environmental protection reduces the failure probability by reducing the stress levels. In many cases environmental protection adds considerable weight and this, rather than cost, limits the amount of protection that can be provided.

9.3.2 Coding Techniques

Coding provides robustness by permitting continued operation in the presence of a defined spectrum of errors; primarily in memory and data transmission. Coding techniques are also available for detection or toleration of errors in arithmetic processors, but are seldom used in this capacity in spacecraft computers.

The prevalent coding techniques are *error detecting code* and *error correcting code*. The former is intended primarily for fault isolation, that is, preventing an incorrect result to be used in subsequent operations. The latter is a fault tolerance mechanism which corrects a class of errors and permits operations to continue normally. All codes require the addition of code, or check, bits to the bit pattern that represents the basic information. If there is agreement between the code and information bits, the data is accepted. If there is no agreement, the data is either rejected (for error detecting code) or corrected (for error correcting code).

The cost effectiveness of error correcting codes is shown in the following hypothetical example for a commercial Earth observation satellite. The payload computer's 4 MB dynamic random access memory has a mass of 400 g. This memory (which does not incorporate error correction or detection) is expected to sustain two "upsets" per orbit-year. Upon detection of an upset by ground monitoring, the memory is reloaded, an operation that typically loses data from two orbits. The expected mission income is $1,000 per orbit. In the absence of error correction, the cost of memory upsets will, therefore, be $4,000 per year. The extra memory and coding and decoding chips will add 100 g to the mass of the memory and will cost $1,000. The cost/mass ratio for this satellite is $5,000 per kg, and thus the extra 100 g will be equivalent to $500. In this example, the cost of the error correction will be paid for in less than one-half orbit-year. The number of upsets to be expected depends on the size and type of memory, the orbit (charged particles are the major cause of upsets), and the amount of shielding provided by the spacecraft structure and the memory enclosure. In most cases coding is found to be very cost effective.

9.3.3 Part Selection and Screening

Screening (selection of parts by test) is a process that eliminates units that have a higher likelihood of failing in service than the other units in the lot [Chan, 1994]. Whereas derating reduces the probability of failure by moving the average strength of the components higher, screening reduces the probability of failure by rejecting the lower tail of the distribution as shown in Fig. 9-3. A typical screening procedure for semiconductors is to measure the leakage current at elevated temperature.

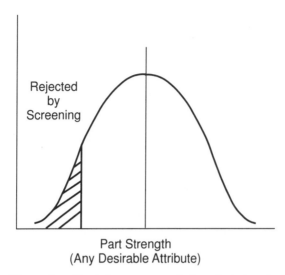

Part Strength
(Any Desirable Attribute)

Fig. 9-3. Effect of Screening. Shaded area represents the rejected product.

The cost of screening is made up of two elements: the cost of the rejected product and the cost of test. The cost effectiveness is high for parts with an initially high failure rate, for modest reliability improvements (generally those in which the cost of screening is not more than 20% of the product cost), and where the cost of test is small compared to the unit cost of the product under test (not over 10%).

Screening does not involve an increase in the size of the components and it is therefore preferred to derating for bulky or heavy parts. Screening is not very effective for reducing the failure probability in a mode for which components have been derated because the failure probability due to external stresses in that mode is already very low. Screening can be applied to assemblies, e.g., by subjecting them to combined temperature and vibration environments, and is thus more versatile than derating. Screening at the assembly level is also likely to result in a lower ratio of test cost to product cost and thus leads to a higher figure of merit.

9.3.4 Comparison of Controlled and Screened Populations

Screening is really a crutch that permits the use of products that do not (as delivered) meet all of the requirements of a given application. A more desirable reliability measure is to tighten the control of the process so that it can be relied on not to produce the outliers that must be screened out. This is not only a philosophical argument, but one with significant practical consequences as shown by the attribute distribution curves in Fig. 9-4.

Both the screened population and the controlled population meet the acceptance criteria in that no parts fall below the lower limit, denoted as L on the figures. However, in the screened population a much larger fraction of the total population is nearer the lower limit than in the controlled population. Environmental effects and aging cause a dispersion of the attributes, and therefore will cause a larger fraction of the screened population to fall below the lower limit than in the controlled population.

This effect can be compensated for by selecting an initial acceptance limit that is higher than the lowest value that can be tolerated in service. However, in the long run costs will be reduced if the process can be improved so that only a small portion of the product will be rejected in screening.

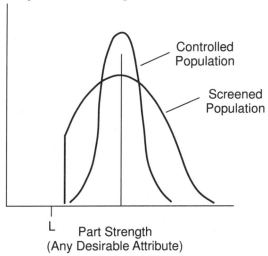

Fig. 9-4. **Comparison of Approaches.** Screening results in a much larger fraction of the population being near the lower limit (L).

9.4 Fault Tolerance

The most attractive features of fault tolerance for space applications are:

- It protects against a wide spectrum of failure-inducing conditions and does not depend on knowledge of failure modes
- The extent of the reliability improvement can be quantified by established mathematical relationships
- There is less risk in the attainment of the reliability objectives than in dependence on product or process improvements; or even on screening (where the reject fraction cannot be predicted)

The cost factors are:

- Increased component cost and weight, sometimes also power and thermal control
- The need for switching or voting mechanisms
- Logic for switching (usually in software or firmware); this is a non-trivial activity because it must work under all failure conditions
- Subsystem and spacecraft-level testing is made more difficult because multiple paths have to be accessed. Operations cost may be impacted by the same factors

Scope of redundancy refers to the assembly level covered by a given fault tolerance provision. Paralleling two relay contacts is a redundancy provision of very small scope. Dual telemetry systems represent redundancy of very large scope. A large scope usually means that some elements are being made redundant that are highly unlikely to fail or the failure of which does not significantly affect spacecraft operation. But redundancy of large scope reduces the number and complexity of switching provisions, and it facilitates system test. Technical and economic studies to select an appropriate scope for redundancy are usually warranted.

In *same design redundancy* two or more identical components are installed together with switching to make one of them active. In a few instances, particularly for power supplies, the outputs can be combined so that switching is not necessary. Voting can also be used for combining outputs of redundant units but this is seldom used for low-cost satellites because at least three identical units have to be installed to make it practicable. Same design redundancy offers very high protection against random failures, and the benefits do not depend on knowledge of failure modes. It is not very effective against failures due to design deficiencies: if one component fails due to insufficient radiation hardness, the redundant one is very likely to fail soon thereafter. Because of the high cost, same design redundancy is used only sparingly in low-cost satellites.

The cost of same design redundancy can sometimes be reduced by employing k-out-of-n replacement, in which a pool of spares can be assigned to replace any one of the pool of active units. An example is bulk memory (for data storage) which usually consists of multiple physically identical modules. Providing one or more spare modules on the same data bus permits replacement of any one failed module. The same scheme can be used for multi-cell batteries, solar panels, and other elements of the electric power supply.

In *diverse design redundancy* two or more components of different design furnish the same service. This has two advantages: it offers high protection against failures due to design deficiencies, and it can offer lower cost if the back-up unit is a "lifeboat," with lower accuracy and functionality, but still adequate for the minimum mission needs. The installation of diverse units usually adds to logistic cost because of additional test specifications, fixtures, and spare parts. This form of redundancy is, therefore, economical primarily where the back-up unit comes from a previous satellite design, or where there is experience with it from another source. Where there is concern about the design integrity of a primary component, diverse design redundancy may have to be employed regardless of cost.

Functional redundancy (sometimes called *analytic redundancy*) involves furnishing a service by diverse means. An example is the determination of attitude rate from a rate gyro assembly (direct), and from observation of celestial bodies (indirect). It is particularly advantageous when the alternate is already installed for another service, e.g., if a star sensor is provided for navigation. In these cases the only cost incurred is for the switching provisions and for data conversion. Both of these can frequently be achieved in an existing onboard computer, thus further minimizing the cost.

Functional redundancy can also take the form of a ground back-up for functions preferably performed autonomously, e.g., navigation, thermal control, or furnishing commands for sensor operation. In the communication subsystem there are frequently omnidirectional and directional antenna systems. For some satellite missions these may also be considered a form of functional redundancy.

The chief benefits of analytic redundancy are: (a) it avoids the cost and weight penalties of physical redundancy, and (b) it is inherently diverse, thus providing protection against design faults. The major limitation is that the back-up provisions usually entail lower performance.

Temporal redundancy involves repetition of an unsuccessful operation. A common example is a re-try after a failure within the computing process. The same technique is applicable to the acquisition of a star, firing of a pyro device, or communication with the ground. This is obviously a low-cost technique. It is most effective when the design incorporates an analysis of the optimum re-try interval and of changes that may improve the success of later operations, e.g., switching of a power supply, reducing loads on the power supply, or reorientation of the satellite. The most important step is to plan ahead for re-try of operations and to incorporate "handles" that permit automatic or ground initiated re-try.

9.5 Summary and Conclusions

The primary objective of this chapter is to achieve high reliability for satellites without exceeding cost constraints. In support of this objective, I have discussed planning activities in Sec. 9.2 and presented a wide spectrum of techniques that can be used to achieve reliability in Sec. 9.3 and 9.4.

Reliability engineers sometimes pride themselves in being "conservative" which means basing predictions on the upper limit of failure rates. Lower costs can be achieved by a "realistic" approach which means use of averages of pertinent failure rates. In either approach good quality data is necessary to arrive at an *overall low-cost design,* by which is meant that the sum of the spacecraft cost and the expected cost of failure is at a minimum. Good data permits designing redundancy of the right scope and the right type (same design, diverse, or functional), and to make the right choices between fault avoidance and fault tolerance.

Many advantages can be cited for early "freeze" of a design, but if this is at the cost of eliminating trade studies in the reliability area, it can easily translate into being penny-wise and pound-foolish. A good trade study evaluates **all** pertinent alternatives, e.g., including replacement launches, and identifies the need for data collection or analysis where the available information does not lead to a rational decision. Early trade-offs usually concern system level design (e.g., dependence on functional redundancy), while later ones address more detailed levels, such as the scope of redundancy within a subsystem, or the degree of derating that should be employed.

For the reliability effort to attain its goal at low cost it must at all times be mission oriented. Any recommended improvements should clearly reduce the expected cost of failure by more than the cost of the improvement. Conversely, where a mission is devoted to observation of a unique event, the reliability measures necessary for achieving a high probability of success need to be recognized in the mission planning. Fault trees and FMECAs are the best tools for focusing the reliability effort into the most beneficial areas. Collection and analysis of failure and operating data from prototyping and testing are the most important tools for assessing whether objectives are being met.

The following check list is intended to steer the reader around the major obstacles to achieving high reliability at low cost:

Organization and Planning

1. Are reliability objectives identified and prioritized?
2. Is responsibility for each objective assigned?
3. Are milestones established?
4. Do participants know budgetary constraints?
5. Is there accessible documentation for all of the above?

Design

1. Are reliability objectives propagated to design partitions?
2. Have trade study objectives and responsibilities been established?
3. Are data needs identified?
4. Have needs for reliability testing of components been identified?

Prototyping and Assembly

1. Is a failure reporting system in place?
2. Are data from reliability and other tests being integrated into design revisions and the assembly process?

Subsystem Test and Spacecraft Test

1. Is a failure analysis process in place?
2. Are test results being analyzed for reliability implications?
3. Are failure modes encountered that are not listed in the FMECA?

Operation

1. Is telemetry data being analyzed to detect impending problems?
2. Are there arrangements to obtain failure data from other satellites and component vendors (again to anticipate problems)?
3. Are "lessons learned" reports being prepared to provide a legacy for future designs?

References

Chan, H. Anthony. 1994. "A Formulation of Product Reliability Through Environmental Stress Testing and Screening," *Journal of the Institute of Environmental Science,* March/April, pp. 50–56.

Hecht, Herbert. 1972. "Economic Factors in Planning Predictive Tests," *Material Research and Standards.* vol. 6, no. 8, pp. 19–24. ASTM, Philadelphia.

Hecht, Herbert and M. Hecht. 1985. *Reliability Prediction for Spacecraft*. RADC Report RADC-TR-85-229.

Hecht, Herbert. 1993. "Rare Conditions—An Important Cause of Failures," *Proc. COMPASS '93,* July. Gaithersburg, MD.

Timmins, A. R. 1970. *A Study of Total Space Life Performance of GSFC Spacecraft*. NASA-TN-D-8017, July.

U.S. Department of Defense. 1991. *Reliability Prediction of Electronic Equipment*. MIL-HDBK-217 (currently 217F, December 1991).

U.S. Department of Defense. 1983. *Procedures for Performing a Failure Mode, Effects and Criticality Analysis*. MIL-STD-1629 (currently 1629A with Notice 1, June 1983)

Chapter 10

Implementation Strategies and Problems

James R. Wertz, *Microcosm, Inc.*

Previous chapters have focused on the philosophy, process, and technology of reducing space mission cost. This chapter is intended to be pragmatic. It provides a recipe for how to implement the process of reducing cost in both new and ongoing programs and identifies major pitfalls to avoid. The chapter is structured around process and data tables which are discussed in the text and which reference other sections or books for details on the techniques people have successfully used to reduce cost.

Much of this chapter deals with implementation problems, partly because cost has always been a concern in space programs. In prior programs, many of which were too expensive to be funded by today's standards, people usually made the best decisions possible within the constraints placed on them. Most often, the external constraints and "rules of the game"—not individual poor decisions—result in high cost. New programs, no matter how well intentioned, will typically encounter the same problems. This implies that, in part, we need to find ways to change the rules. In addition, we need practical recommendations on how to deal successfully with cost problems within the framework of the acquisition environment we live in.

10.1 Techniques Applicable to Most Programs

As summarized in Table 10-1, a number of top-level steps apply to essentially all missions that are interested in creating a program for reducing life-cycle cost. Additional steps for new and ongoing programs are discussed in Sec. 10.2 and 10.3,

TABLE 10-1. **Initial Steps Applicable to All Missions to Initiate a Program for Reducing Life-Cycle Cost.** See Tables 10-6, and 10-11 for follow-on procedures depending on the type of program. Although these are simple top-level procedures, they are also important first steps for getting a full-scale cost reduction program under way.

Technique or Action	Comment	Where Discussed *
1. Determine **real** objectives and constraints	Is the **real** goal to minimize cost, to keep work in-house, to support specific organizations and technologies, or to maximize performance?	Sec. 10.1; SMAD 1.3, 1.4
	1a. Convince the organization that reducing cost is a high priority.	Sec. 1.1, 10.2.1
	1b. Be willing to trade between cost, risk, and performance.	Sec. 1.3, 2.2.8
2. Look for innovative solutions	Major cost reductions rarely come from standard, formal engineering processes	Case Studies; Sec. 2.2 SMAD 2.2, 22.3
3. Make cost data known	The task is hard enough with good cost data and essentially impossible without it	Sec. 1.3; Chap. 8
4. Reward low cost	Provide positive incentives to both people and organizations—if costs are reduced, give them a bonus rather than a smaller budget next year	See text
5. Use the existing knowledge base	Reinventing the wheel is rarely economical. Using new approaches and processes should not mean ignoring 35 years of space experience. There are several concrete approaches to building on existing knowledge:	
	5a. Books and literature	Table 10-2
	5b. Courses	Tables 10-3, 10-4
	5c. Commercial software tools	Sec. 3.2.3, 3.2.4
	5d. Becoming a part of the low-cost community	Table 10-5
	5e. Taking advantage of the knowledge of others	See text

* Also see text for additional discussion of each of the items listed in the table.

respectively. We present them as "steps" to help create a systematic approach to cost reduction. In most cases, they don't need to be sequential. The top-level steps are:

Step 1. Determine Your <u>Real</u> Objectives and Constraints

Most programs are interested in reducing cost. However, achieving a minimum-cost mission may or may not be a reasonable objective for a given program. The funding organization may be more interested in achieving maximum efficiency or performance, supporting an in-house organization, or developing or transferring technology. For example, a country not previously involved in space exploration may want to fly a low-cost mission for which the main purpose is to transfer technology into the country. The lowest-cost approach might be simply to buy the spacecraft intact and on orbit from a manufacturer of low-cost spacecraft, but this would be inconsistent with the fundamental objective of technology transfer and would be the wrong answer for this application. (See Sec. 1.1 for a discussion of alternative cost objectives and SMAD Sec. 1.3 for a discussion of alternative political and technical objectives.) What is critical for each program is that the objectives, including the cost objectives, be clearly established, articulated, and made known to those who must define and implement the mission. It's difficult enough to meet objectives we know about, and nearly impossible to meet those we don't.

One of the most important characteristics of objectives is that they are qualitative rather than quantitative and express what the mission is all about. For example, the objective of the Pluto Express mission might be to obtain as much good observation data over as much of the planet as possible within severe budget constraints. It should not be to obtain imaging data with 500 m resolution covering 80% of the planet for less than $110 million. (Setting explicit numerical objectives would imply, for example, that we should discard a concept that provides 100 m resolution of 75% of Pluto for $50 million.) It is the process of mission engineering that takes broad objectives and defines a mission that can achieve them at minimum cost and risk.[*]

If one objective of the mission is to minimize or substantially reduce cost, then a critical step is to convince the organization that reducing cost is a high priority. This is strongly related to a willingness to trade on requirements and constraints to reduce cost. In general, past programs have been efficiently run so as to minimize the cost while meeting all of the system requirements. If we are to significantly reduce cost, we must be willing to give up something in return, to behave differently, to settle for less than the full "requirements" that we would like, and make saving money more important than in past programs. As discussed further in Sec. 10.3.2, we must be willing to trade between cost, risk, and performance and to make judgments about the level of performance and amount of risk we will accept as a function of overall life-cycle cost.

Step 2. Look for Innovative Solutions

Most space programs are well designed and well engineered. Consequently, they have achieved nearly minimum cost within the requirements and constraints under which they operate. This implies that **major** cost savings will rarely come about from standard, formal engineering processes. In order to make a major cost reduction, something has to change significantly in the program approach, philosophy, organization, or technology. This, in turn, means that to find ways to dramatically reduce cost, we must fund creative people who can devise **innovative** ways to reduce cost and listen carefully to their approaches and suggestions. These may be people from inside or outside the organization, but they must be free to think broadly and propose non-traditional approaches to achieving mission objectives. Most alternative suggestions probably won't work. Nonetheless, without looking at alternatives and questioning objectives, constraints, and methodology, it will be difficult or impossible to dramatically reduce cost. Specific innovative techniques for reducing cost appear throughout the case studies and are discussed explicitly in SMAD Sec. 22.3.

---

[*] We will use these words throughout this section even though minimizing cost and risk simultaneously is probably not possible. Depending on the program and the managers, we may want minimum cost at an acceptable risk, minimum risk at an acceptable cost, or (most likely) to achieve the broad mission objectives making both the cost and risk as low as possible.

Step 3. Make Cost Data Known

Cost data[*] is always sensitive and especially so within the space community. This is due to many causes, including proprietary information and the fact that cost depends more on what is required of a vendor than the actual product being bought. Despite these legitimate reasons for keeping costs private, a key element in reducing mission cost is to make cost data known, both to managers and engineers. Working without data makes a solution nearly impossible to find. As described in Sec. 1.3, to help the whole community, cost data should ultimately be made public, just as it is for pencils, cars, boats, and airplanes.

Step 4. Reward Low Cost

A problem throughout the space community is that there is an incentive to make moderate cost reductions but almost no incentive to make dramatic cost reductions. It's difficult to become a captain of industry, or a leader of the military or civilian space community, by taking a billion-dollar program and transforming it into a hundred-million-dollar program. In the end, such dramatic cost reductions can't support either the organizations or infrastructure put in place to accommodate the larger program. If you try to save money, you'll be blamed for failure if anything goes wrong (such as omitting the very expensive Space Telescope mirror test) and, if you are fully successful, your budget will be cut in all future years. If we genuinely want to reduce cost, we have to incentivize both people and organizations to do so. Too often, the result of being able to work efficiently and economically is unemployment.

Rewards for reducing cost need to be both psychological and physical. We need to share the cost savings with the group or individual that initiated it and give them greater discretion in spending other funds. If an individual or group can dramatically reduce program cost, we should give them the opportunity to work on more and larger programs to try to achieve comparable results. Reducing cost, rather than running large, expensive programs, should become a major avenue for advancement.

There is a potential trap in rewarding low-cost solutions. We need to guard against cost savings that simply push cost into another pot or push it downstream where it will grow larger. This problem tends to make mechanical milestones, such as reducing the current budget by 15%, counterproductive. The easiest way to reduce this year's budget by 15% is to push the problems into next year, when it will cost 25% to fix. Anyone in mission operations, where the problems ultimately arrive, is all too familiar with how this occurs.

Step 5. Use Existing Knowledge

A major failing of some organizations that develop low-cost missions is that, in trying to use new processes and technologies, they too often throw out the experience gained over 35 years of space exploration. Instead, we want to use this vast experience to understand how space missions work, what the key problems are, and how to go about overcoming them in a new, more economical way.

---

[*] In this context, we use "cost" and "price" synonymously. The price charged by a supplier, contractor, or government organization represents the cost to the buyer. Making the profit (i.e., the difference between the cost of creating something and the price for which it is sold) public is **not** critical to reducing cost, is not normally done in commercial systems, and may be counterproductive in that it focuses attention on a relatively small element of the budget that has little potential adjustment.

TABLE 10-2. Principal Publishers of Astronautics Books. See the annotated bibliography at the end of the chapter for a list of the principal books in print related to space systems engineering. Book retailers typically have an extremely limited selection in astronautics but can special order most titles. Microcosm Discount Astronautics Books, Torrance, CA, carries most astronautics books in print.

Publisher	Series or Specialty	Softcover/ Hardcover	Price Range[1]	Where Available
Adam Hilger[2]	Orbits, planetary	Both	Low to moderate	Bookstore special order
AIAA	Largest publisher of space technology books & journals[3]	Mostly hardcover	Moderate to high	AIAA, Wash, D.C.
Artech House	Space telecommunications systems, space-based radar	Hardcover	Moderate	Bookstore special order
Cambridge	Solar system and space science	Both	Moderate	Bookstore special order
Dover	Math and general reference (typically good, older works)	Softcover	Very low	Bookstore special order
Greenwood	Space business	Hardcover	High	Bookstore special order
Jane's	Jane's space directory— updated regularly	Hardcover, CD ROM	Very high	Only from Jane's
Kluwer Academic/ Microcosm[4]	Space Technology Library— general technical series	Both	Low; Some high	Bookstore special order
Krieger	Extensive selection, Orbit series is general technical	Mostly hardcover	Moderate to high	Bookstore special order
McGraw-Hill	General technical	Mostly hardcover	Low to moderate	Bookstore special order
National Academy of Sciences	Space science and policy	Softcover	Moderate	Natl. Academy, Wash, D.C.
Oxford	General technical	Hardcover	Moderate	Bookstore special order
Sky Publishing	Star catalogs, general and observational astronomy	Mostly softcover	Low to moderate	Sky Publishing, Cambridge, MA
University of Arizona Press	Planetary science	Both	Low	Bookstore special order
Univelt[5]	Mostly conference proceedings[6]	Mostly softcover	Moderate; freq. sales	Univelt, San Diego, CA
USGPO	Gov. docs., ephemerides; NASA Special Publications	Both	Very low to low	USGPO bookstores
Wiley	General technical, space telecommunications	Mostly hardcover	Moderate	Bookstore special order

1) Price range is for softcover version where available. Ranges are: Very low: <5¢/pg; Low: 5¢ to 10¢/pg; Moderate: 10¢ to 20¢/pg; High: 20¢ to 40¢/pg; Very high: >40¢/pg. 2) Publisher for the British Institute of Physics. May be hard to find in U.S. 3) Education series—hardcover, general technical; Progress in Astronautics—typically papers selected from conferences. Standards are softcover and are priced **very** high. 20% single-copy discount for AIAA members. 4) Kluwer Academic publishes hardcover version; Microcosm publishes softcover version. 5) Publisher for the American Astronautical Society. 6) Two principal series— Advances in Astronautical Sciences; Science and Technology.

I would like to make Step 5 much more explicit. Throughout the space community, there is a strong tendency to reinvent the wheel or to try to optimize each hardware and software element for the particular task at hand. We tend to modify hardware, reinvent orbit propagators, and re-derive basic laws whenever a new project arrives. Although hard to find, a surprising amount of knowledge exists about what others have learned, about what works and what doesn't work in space, and about reducing cost. Even if this knowledge serves only as a first step, it's valuable in reducing both cost and risk. Specific ways to get at this knowledge are:

5a. Books and Literature. There are only about 500 books in print on space technology, far fewer than for most other technical fields. About half of these are conference proceedings or specialized books of limited interest. Consequently, an investment of approximately $5,000 will buy nearly all of the important works for any subfield, such as spacecraft design. About $15,000 will buy a nearly complete set of books relevant to the overall process of space mission analysis and design. Every project should acquire a reasonable library and make it available where people work. I know of no other comparable investment that can provide as much immediate and direct return in terms of reducing mission cost and risk. At the end of this chapter is an annotated bibliography of the principal books about space systems and mission engineering. Table 10-2 lists the major astronautics publishers, their particular fields of specialization, and the price range of their books.

Although there are currently very few books in space technology, the situation is improving. The U.S. Air Force Academy has undertaken a substantial program to create new space technology volumes, including this book. In addition, many commercial publishers—particularly McGraw-Hill, Krieger, AIAA, and Kluwer/Microcosm—are actively expanding the number of books in space technology. This activity should significantly increase the literature base over the next several years. Unfortunately, this gain is offset by a dramatic loss of experience because many senior engineers are retiring as a result of the downturn in the aerospace industry at the end of the Cold War.

5b. Courses. Formal education and training is a far more efficient way to understand astronautics and space technology than is on-the-job training, which can take many years. Unfortunately, because the knowledge base in space technology resides mostly within industry and government laboratories, there are relatively few university programs in astronautics and space. Many university "aerospace" departments emphasize or focus completely on aircraft design. Table 10-3 lists U.S. colleges and universities and Table 10-4 lists independent organizations which offer courses in space technology. The lack of good texts makes this lack of courses worse, but as books become more available, the number of courses should increase over the next 5 to 10 years.

5c. Commercial Software Tools. In the past, most major aerospace prime contractors developed their own proprietary tools for space mission analysis. This option is not available to smaller organizations and is becoming less cost effective for larger ones. As described in Sec. 3.2.3, a number of commercial software tools are becoming available. Using commercial tools reduces both cost and risk. Also, as more organizations buy them, costs should go down and their number and variety should increase.

TABLE 10-3. Principal Universities Offering Programs in Astronautics or Space Technology. A number of universities have Departments of Aeronautics and Astronautics that specialize mainly in aeronautics. See text for discussion.

Institution	Location	Astronautics Degrees	No. of Astronautics Courses	Enrollment
Auburn U.	Auburn, AL	no	1U 6U/G 7G	(1)
CalTech	Pasadena, CA	no	1U/G 1G	(1)
UCLA	Los Angeles, CA	no	4U/G	Open/Ext.
U. of Cincinnati	Cincinnati, OH	no	5U 1G	(1)
U. of Colorado	Boulder, CO	MS PhD	14G	Open
U. of Colorado	Colorado Spgs, CO	MEng	7U 20G	Open
Embry Riddle U.	Daytona Beach, FL	no	12U	(1)
U. of Florida	Gainesville, FL	no	2U 3G	(1)
Georgia Tech	Atlanta, GA	no	2U/G 2G	(1)
U. of Illinois at Urbana Champaign	Urbana, IL	MS PhD BS in Aero/Astro	4U/G 3G	(1)
Int'l Space U. (ISU)	Strasbourg, France	MS	11G	(1)
Iowa State U.	Ames, IA	BS MS MEng PhD	7U 6U/G	(1)
MIT	Cambridge, MA	SB SM EEA ScD	17 U/G	Open
U. of Michigan	Ann Arbor, MI	no	7G	Open
Nat'l Tech. U.	Fort Collins, CO	MS	7G	(1)
Naval Postgrad. Sch.	Monterey, CA	MS Engineer PhD	24G	(2)
U. of North Dakota	Grand Forks, ND	MS BS minor	15G	Distance Learning & Open
Ohio State U.	Columbus, OH	no	1U 7G	Open
Penn State U.	University Park, PA	no	3U 9U/G	Open
Princeton	Princeton, NJ	no	1U 3G	Open
Purdue U.	West Lafayette, IN	no	2U 4G	Open
San Jose State U.	San Jose, CA	no	1U 6G	Open
U. of Southern Calif.	Los Angeles, CA	MS Engineer PhD	1U 2U/G 8G	Open
Stanford U.	Stanford, CA	MS Engineer PhD	16G	(1)
Texas A&M	College Station, TX	no	3U 3U/G	(1)
U. of Texas	Austin, TX	no	3U 6U/G 3G	Open
U.S. Air Force Acad.	Colorado Spgs, CO	BS	12U	(1)
U.S. Naval Acad.	Annapolis, MD	BS	9U	(1)
Utah State U.	Logan, UT	no	1U 2U/G 2G	(1)
U. of Virginia	Charlotteville, VA	no	2U 2G	Open
U. of Washington	Seattle, WA	no	7U 1G	(1)
U. of Wisconsin	Madison, WI	no	3G	Open

(1) Open only to students enrolled in the institution.
(2) Open only to sponsored government and military personnel.
U = undergraduate, U/G = undergraduate and graduate, G = graduate.

TABLE 10-4. **Organizations Offering Professional Courses in Astronautics and Space Technology.** Includes university "short courses" offered in one week or less. Professional courses typically run one to five days and cost $200 to $350 per day per student with group discounts.

Offerer	Representative Courses	Where Available
American Institute of Aeronautics and Astronautics, Washington, D.C.	Very broad range—"Space Cost Engineering," "Space System Design," "Launch Vehicle Design," plus many technical courses	Nationwide
Athena Educational Group, Colorado Springs, CO	"Space Mission Analysis and Design," "Reducing Space Mission Cost," "Orbit and Constellation Design," plus other technical courses	Nationwide, Canada, Europe
Applied Technology Institute, Clarksville, MD	Broad range—"Spacecraft Systems Design and Engineering," "Launch Vehicle Systems Design and Engineering," plus many technical courses	Mostly Washington, D.C. area
George Washington University, Washington, D.C.	"Low-Earth Orbit Satellite Systems," "Space System Principles and Applications," plus satellite communications and power	Washington, D.C.
Launchspace Falls Church, VA	"LEO Constellations," "GEO Systems" plus many technical courses in launch vehicle design and applications	Nationwide

TABLE 10-5. **Organizations and Conferences Related to Small Satellite Technology.** Many of the professional organizations, such as the American Astronautical Society, the American Institute of Aeronautics and Astronautics, and the British Interplanetary Society, also have conferences and segments devoted to low-cost programs.

Organization/Conference	Comment
AIAA/Utah State U. Small Satellite Conference, Logan, UT	Oldest of the small-satellite conferences; very well attended; held annually in late August or early September
International Small Satellite Organization (ISSO), Washington, D.C.	Publishes monthly newsletter; beginning an E-mail newsletter and database
ISSO Conference, Washington, D.C. area	Small satellite conference with political and business emphasis; held annually in the spring
Space Access Society, Phoenix, AZ	Organization devoted exclusively to reducing space-launch cost; active in grass-roots lobbying
North American Radio Amateur Satellite Corporation (AMSAT-NA), Washington, D.C.	Organization created to build, launch, and use amateur radio satellites; similar AMSAT organizations worldwide have launched more than 40 satellites to date with excellent success (see AMSAT case study)
AMSAT-NA Technical Conference (varied locations)	Technical discussions of practical, low-cost radio satellites; held annually on Columbus day weekend in October

5d. *Becoming a Part of the Low-Cost Community.* One of the most efficient ways to solve specific problems is to become a part of the low-cost satellite community and discuss problems with others who have faced similar issues. People within the low-cost-satellite community have a wealth of knowledge and many want to share it. Table 10-5 lists the principal organizations and conferences related to small-satellite technology. In my view, the two most important steps for becoming an active part of this community are to join the International Small Satellite Organization (ISSO) and to attend the annual AIAA/Utah State University Small Satellite Conference, held in the fall in Logan, Utah.

5e. *Taking Advantage of the Knowledge of Others.* One of the major messages from the appendix and the charts in Sec. 1.2 is that small, low-cost satellites have been developed throughout the history of the space program and reflect a strong and varied experience. If a major goal of your program is to strongly reduce cost, it's probably worthwhile to hire consultants and reviewers who have done it already. You shouldn't expect these consultants to set you up as a competitor to their own business.

Still, most people in the low-cost community recognize that it is in everyone's best interest to dramatically reduce space mission costs. If you want to make major cost reductions, you'll need to bring in review teams that are innovation oriented and are willing to look at new concepts and approaches, but at the same time are sufficiently experienced to foresee problems before they arise.

The above steps apply to essentially all mission phases by both government and commercial organizations. Sec. 10.2 discusses specific steps we can take to reduce cost during concept exploration and on new programs. Sec. 10.3 provides steps ongoing programs can take. Sec. 10.4 addresses problems that occur in implementing techniques for dramatic cost reduction.

10.2 Reducing Cost in New Programs

By far the largest impact on cost occurs during concept exploration, when the program is being defined and when the rules it will live by are made. Even though the cost of many of the individual components are not defined, the tone and scale of the program is established and this, in turn, drives the scale and cost of the rest of the program.

There is something of a paradox in the high cost of space missions. As discussed in Chap. 2, the government is primarily responsible for creating the high-cost environment, but the government cannot fix the problem directly. Even though space-craft are bought by government agencies, corporations, and even individuals, it is largely the government that establishes the environment in which they are built and launched. Contractors will usually respond to what the government expects and requires of them; if they don't, they go out of business. Thus, the framework which created today's high-cost missions largely resulted from the policy, processes, and regulations that have been put in place over the last 30 years.

Unfortunately, the government can't remedy the problem by simply changing the policy or environment. Government actions are limited by rules and regulations intended to protect both the public and contractors. In trying to be fair and even-handed in the acquisition processes, the government has created many of the regulations which now make it extremely difficult to reduce costs. While we want to cut costs, we don't want to give up the goals of fairness, worker safety, open competition, or other aspects that have brought us where we are.

As has been pointed out many times in other chapters, dramatic cost-cutting solutions require innovation and creativity that address practical problems. General rules or policies can't mandate Kelly Johnson's rules for the Lockheed Skunk Works (Sec. 2.2.7). Expressed as government regulations, they probably wouldn't save much money. But the government can (and must, if we are to succeed) create an environment in which reduced cost is both possible and rewarded and then allow the contractors and laboratories which design and build satellites the freedom to solve practical problems in a way that drives down cost. The contractors and laboratories must then find pragmatic solutions that maintain an appropriate level of performance and reliability while reducing cost. The commercial marketplace shows that this can be done. In virtually every commodity there is a wide range of price, performance, and reliability.

TABLE 10-6. Principal Techniques Applicable to Individual Programs for Reducing Life-Cycle Cost During Concept Exploration and Definition. See text for discussion. **See Table 10-1 for Steps 1–5.** See also Secs. 1.3, 2.2, 10.4, and 10.5.

Technique or Action	Comment	Where Discussed
6. Do mission engineering	*Mission Engineering* is the basic process of designing the mission to meet its overall objectives at minimum cost and risk	Sec. 10.2.1, SMAD Chap. 1–4
7. Trade on requirements	Requirements must be based on a balance between what is wanted and what can be achieved within the cost constraints	Sec. 1.3, 2.1 2.2.3, 10.2.1, 10.4.1 SMAD 1.4, 3.2
8. Develop a small-team approach	Perhaps the single most important step management can take	Sec. 2.2.2, 10.2.1
9. Use available hardware and software	Use commercial components wherever possible	Sec. 2.2.6, 3.1.3, 3.2.3, 5.1–5.3, 6.5.2
10. Look for trades among major program elements	Major trades that reduce cost often require moving responsibility and budget between organizations	Sec. 2.1, 10.4.1
11. Design for multiple launch vehicles	Reduces both cost and risk of delays	Sec. 4.2, 10.2.1 SMAD Chap.18
12. Use larger design margins	Reduces cost and risk, increases reliability and flexibility, reduces operations complexity & cost	Sec. 2.2.5, 4.6.4, 6.4.2, 10.2.1
13. Increase onboard processing and **low-cost** autonomy	Replace mechanical functions with software, particularly in multiple spacecraft; avoid AI and other high-cost autonomous systems; develop enough autonomy so the user can operate the ground system	Sec. 3.2.1, 6.4.4 SMAD 11.7
14. Compress the schedule	Do not reduce time or funding spent on up-front mission engineering	Sec. 2.1, 2.2.4, 2.2.7, 5.2.3, 6.3.1, 10.2.1
15. Expedite decision making	Create a process, within both the contracting (or constructing) organization and the funding organization, for rapid, responsive decision making	Sec. 2.1, 10.2.1

10.2.1 What Individual Programs Can Do to Reduce Cost

For the lowest-cost missions, the budget is essentially fixed at a low level, and concept exploration consists of determining what can be achieved for the available funds. For most programs, however, concept exploration represents a series of trades between alternative designs, between risk and performance, and between what we want versus what we can afford. Table 10-6 summarizes the main techniques for reducing life-cycle cost in individual programs during concept exploration. The first entry (Step 6) is the broad, overall space mission engineering process discussed below. The rest are techniques oriented toward **reducing** cost, rather than simply toward designing a low-cost mission. Most are described here. Items 9, 10, and 13 are discussed elsewhere, as shown in column 3.

TABLE 10-7. The Space Mission Analysis and Design Process. Chaps. 1 to 4 of SMAD [1992] describe this process in detail. (Table from SMAD, Sec. 1.1)

Typical Flow	Step	
	Define Objectives	A. Define broad objectives and constraints B. Estimate quantitative mission needs and requirements
	Characterize the Mission	C. Define alternative mission concepts D. Define alternative mission architectures E. Identify system drivers for each F. Characterize mission concepts and architectures
	Evaluate the Mission	G. Identify driving requirements H. Evaluate mission utility I. Define mission concept (baseline)
	Define Requirements	J. Define system requirements K. Allocate requirements to system elements

Step 6. The Space Mission Engineering Process

Space mission engineering is the refinement of requirements and the definition of mission parameters to meet the overall mission objectives at minimum cost and risk. It is the single most important step in creating a low-cost mission. This process, illustrated in Table 10-7, was introduced in SMAD [1990].* It is presented in detail in SMAD Chaps. 1 to 4 and summarized below. In addition, many of the design-to-cost methods in Chap. 7 quantitatively implement this approach. The first step in space mission engineering is to define the broad *mission objectives* and constraints. The objectives reflect the most fundamental characteristics that make the mission what it is. They are drawn largely from a summary *mission statement* and are qualitative rather

---

* *Space Mission Engineering* was introduced in the 1st edition of SMAD [1990]. Although the book was substantially expanded and revised in the 2nd edition [1992], the mission engineering discussion remained essentially unchanged. For consistency, all references in this book are to SMAD, 2nd edition.

then quantitative. For example, the primary mission objective might be to explore Mars, to establish a mobile communications constellation in low-Earth orbit, or to create a surveillance system that can improve federal drug interdiction. Most missions have other objectives as well. *Secondary objectives* come from other activities we can do with the equipment being flown or from additional objectives we must meet to be successful. They frequently come from a *hidden agenda* of secondary, typically non-technical, goals. These may or may not actually be hidden, but are often critically important. If the mission doesn't meet the political and other non-technical objectives, it probably won't be funded. Consequently, everyone responsible for defining and developing the mission must be aware of the full set of mission objectives.

The second step in space mission engineering is to use the mission objectives to estimate the *mission requirements*, which are the quantitative expression of how well the objectives are to be achieved. Trading on requirements is one of the most critical elements of the mission engineering process. This is a central theme of the SMAD book and of many of the process and case-study chapters in this volume. Mission requirements should represent a balance between what we want and what we can afford. Therefore, mission requirements must be a major part of system trades.

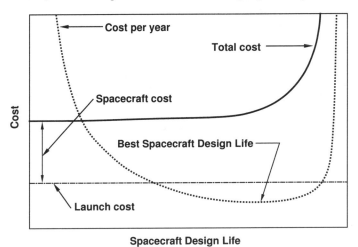

Fig. 10-1. Hypothetical Curve of Cost vs. Spacecraft Design Life. The cost per year is the total cost divided by the design life. See text for discussion. (From SMAD, Sec. 1.4.)

Defining mission requirements is a system trade done throughout concept exploration. As an example, Fig. 10-1 illustrates how we might go about defining a spacecraft's required design life. Assume we have a mission, such as communications or fire detection, that needs to go on indefinitely. We could cover the first 20 years by building one spacecraft with a 20-year lifetime or four spacecraft with 5-year lifetimes that would be launched more frequently. What we want is the lowest cost per year. As shown in the figure, the launch cost will be relatively insensitive to space-craft life. For very short lifetimes, the spacecraft cost will also be insensitive. It doesn't matter too much whether we ask the contractor to build a spacecraft that will last a day, a week, or a month. At some point, however, the cost of the spacecraft will grow dramatically as we demand longer and longer lifetimes. Consequently, the cost per year will be very high for short-lived spacecraft because we're launching too many of them and very high for extremely long-lived spacecraft because we're pushing the bounds of technology. We should select the most economical requirement for spacecraft life based not on philosophy, but on a pragmatic assessment and potentially detailed engineering estimates of the cost of various options. (See Sec. 7.4 for a more detailed example of trading on requirements.)

Analysis Type	Goal	
Feasibility Assessment	To establish whether an objective is achievable and its approximate degree of complexity	Quick, limited detail
Sizing Estimate	To estimate basic parameters such as size, weight, power, or cost	Quick, limited detail
Point Design	To demonstrate feasibility and establish a baseline for comparison of alternatives	Quick, limited detail
Trade Study	To establish the relative advantages of alternative approaches or options	More detailed, complex trades
Performance Assessment	To quantify performance parameters (e.g., resolution, timeliness) for a given system	More detailed, complex trades
Utility Assessment	To quantify how well the system can meet overall mission objectives	More detailed, complex trades

Fig. 10-2. The Mission Analysis Hierarchy. From top to bottom, methods become more detailed, complex, and useful in evaluating and selling concepts. (From SMAD, Sec. 3.2.)

Another key element of the mission engineering process is determining the system drivers and driving requirements. *System drivers* are those mission parameters or engineering characteristics which most strongly affect the system's overall performance, cost, risk, or schedule—such as the altitude for a communications constellation in low-Earth orbit or the aperture diameter for a telescope. The *driving requirements* are the mission requirements which largely dictate these characteristics. Both the system drivers and driving requirements will depend on the mission implementation. For example, a system driver for an observation platform in low-Earth orbit will be the altitude, and a driving requirement will be coverage. If we use a geosynchronous orbit for the same mission, altitude is not a variable, and coverage is essentially continuous for the portion of the Earth the satellite can see. On the other hand, the resolution is much lower from a geosynchronous orbit. Therefore, resolution may become a driving requirement and telescope aperture a system driver.

Most concept exploration studies are limited by resources and schedule. Therefore, we want to concentrate our effort on those issues which have the strongest effect on performance, cost, risk, and schedule. The goal of mission engineering is to adjust the driving requirements (such as coverage and resolution) and the system drivers (such as altitude and aperture) to satisfy the broad mission objectives at minimum cost and risk. This activity is referred to as *mission analysis* or *system trades.* Because we have limited resources during concept exploration, it is equally important to decide what not to evaluate or what to examine only at a top level, as it is to determine what elements to examine in detail. This limitation leads to the mission analysis hierarchy defined in Fig. 10-2. We want to do quick and limited assessments of mission elements that won't have a dominant impact on performance, cost, risk, or schedule and detailed trade studies, performance assessments, and utility analyses on those that will.

The most complex level of mission analysis is evaluating *mission utility*, which is the quantitative expression of how well the system meets the mission's overall objectives. We do this by developing and evaluating numerical *measures of effectiveness*, or *figures of merit*, such as how much science data we can collect, how much revenue a commercial communications system can generate, or how much property and how many lives a fire-detection satellite's early warning can save. Doing mission utility analysis is described in detail in Chap. 3 of SMAD and Chap. 7[*] of this volume. Mission utility analysis helps to quantify the mission design process. It is **not** intended to quantify the technical and political decision making itself, but rather to provide quantitative inputs so that better, more cost-effective decisions can result. (See SMAD, Sec. 3.4.)

Additional Cost-Reduction Techniques for New Missions

Mission engineering is a broad process that provides a framework for minimizing cost. Additional techniques listed in Table 10-6 and described below are used specifically for reducing cost for new missions. (Steps 9, 10, and 13 appear elsewhere in the text as shown in the table)

Step 7. Trade on Requirements

Trading on requirements is probably the single most important action in reducing cost. This will become clear as you review the case studies, the design-to-cost discussion in Chap. 7, or the mission engineering process described in SMAD. Still, trading on requirements often doesn't occur and, when it does, is often less extensive than would be desirable. The reasons for this and specific suggestions for overcoming the problem are given in Sec. 10.4.1.

Ultimately, we need to change the environment in such a way that it's not only reasonable but encouraged to challenge requirements, to identify requirements that drive up cost, and to check mission objectives often to see how we can meet them at lower cost and risk. We all go through this process regularly when we buy a house, a car, a computer, or a television. To reduce space mission cost we must make the government a thoughtful and informed consumer.

Step 8. Develop a Small-Team Approach

One of the most interesting characteristics of this book is the list of organizations that have contributed to the case studies and that have designed and built space missions at dramatically lower cost. The "Who's Who" of low-cost spacecraft is a rather strange list—government organizations such as NRL and APL, companies

---

[*]Although the emphasis is different, the design-to-cost process in Chap. 7 is essentially equivalent to the mission utility analysis described in SMAD, Sec. 3.3.

such as AeroAstro and DSI, academic organizations such as Utah State and the University of Surrey, and perhaps the leader of the community—the Radio Amateur Satellite Corporation, AMSAT. Some organizations that have developed successful low-cost approaches are part of large bureaucratic organizations, such as the Air Force, Navy, and Ballistic Missile Defense Organization. Still, even within these very large organizations, groups that have been successful at dramatically reducing cost have found ways to develop a "small" mentality, with a highly motivated team, localized and rapid decision making, the ability to accept programmatic risk, and a willingness to try alternative approaches.

As discussed in Sec. 10.4.2, I believe there are good reasons why this has occurred. The principal message for reducing space mission cost is to develop a small-team mentality and probably to implement it using a small group, even within a larger organization. The key is to ensure that it **behaves** like a small organization with the capacity to make decisions, spend money (small amounts), change direction, trade on requirements, accept a higher level of risk, and do other things which are necessary to reducing cost.

Step 11. Design for Multiple Launch Vehicles

For individual satellite programs the single most important action to reduce both cost and risk is to design for multiple launch vehicles. This saves money through being able to negotiate. If other launch options are available, any launch-vehicle supplier will be more willing to negotiate and to meet the spacecraft's needs than will one for whom the spacecraft is a captive market. The smallest spacecraft usually fly on a space available basis or "piggyback" with another payload. Consequently, they are typically designed for many launch environments and various orbits. Larger spacecraft tend to optimize their design and select a specific launch vehicle early on.

Designing for multiple launch vehicles also reduces schedule risk. Any launch vehicle failure causes a significant downtime before returning to a normal launch schedule. (For specific values see Table 18-3 in SMAD.) When launches resume, the low-cost program may have been bumped even farther downstream by a higher priority payload. Thus, designing for multiple launch vehicles is also the best way to reduce schedule risk and associated cost overruns.

Designing for multiple launch vehicles is typically not a serious technical problem. As described in Sec. 18.3 of SMAD, we can do so by enveloping the launch environment for the appropriate launch vehicles. Because most launch environments are similar, this usually doesn't significantly increase cost.

The biggest problem in designing for multiple launch vehicles is political. Governments typically want to support launch-vehicle manufacturers within their own country. Because each country has only a few launch vehicles, most government launches typically use a single launch vehicle. This eliminates competition and keeps launch costs high. So long as this policy continues, launch costs will remain high until more or cheaper launchers are developed.

Step 12. Use Larger Design Margins

The problem of spacecraft design margin is strongly related to the launch vehicle problem. Launch costs are high, so we must maximize performance per unit weight. Therefore, we minimize margins, and spacecraft become individually unique designs with very high cost. Essentially, spacecraft are expensive because they were expensive in the past, and we're unwilling to take the programmatic risk necessary to change that.

One mechanism for breaking the cycle of high cost leading to higher cost is to avoid "excessive" weight optimization and provide larger design margins. This has several immediate positive benefits:

- Reduces cost
- Reduces risk
- Increases reliability
- Increases flexibility
- Reduces operations complexity and cost
- Increases potential for standardization, which will further reduce cost and risk and further increase reliability and flexibility

The main disadvantage of large design margins is that they decrease the performance per unit weight and, therefore, are less optimal and less economically efficient. Small, low-cost spacecraft have been willing accept this lower performance and reduced efficiency. (See Fig. 1-1 in Sec. 1.1.) Partly because they have large design margins, small spacecraft tend to be as reliable, or more reliable, than large ones. (See Sec. 2.2.8 and 9.1)

It is largely the continual optimization and intentional minimizing of design margins that has caused standardization to be so dramatically unsuccessful in space applications. Using standard elements implies that missions will fly more capability than needed to meet the mission requirements. But flying more capability than needed is counterproductive in terms of design optimization, so designers typically remove the "excessive" performance of each component. Components are specialized for the mission at hand, and each new mission flies nearly unique hardware.

Launch costs remain high, so, we can't arbitrarily set all of the design margins higher or specify in advance how large the margins should be. Selecting margin requires judgment and a balance between cost and performance. For larger spacecraft, we tend to maximize performance and efficiency; for small, low-cost missions, we tend to minimize cost.

Step 14. Compress the Schedule

Schedule compression is an excellent way to reduce cost, simply because there's not enough time to spend as much money as on a long program. Typically, small, low-cost spacecraft go from concept to launch in one to three years, although launch delays can extend the time.

A key problem with schedule compression is that it is of little use by itself. Mission developers must do it intelligently to truly reduce cost, which means fewer requirements and more rapid and continuous decision making. But we must not use schedule compression as an excuse to reduce up-front systems engineering. Typically, ignorance is not of substantial benefit in reducing mission cost.

Schedule compression works best on a small program, where eliminating many of the communication requirements and excessive documentation makes sense. Within small programs, communication problems are dramatically reduced, and the corporate memory of how and why things were done is much simpler and more direct. In 1 to 3 years, it's much easier to build a small spacecraft than a large, complex one.

Step 15. Speed Up Decision Making

A key to fast, low-cost, innovative programs is a rapid, responsive, yet thoughtful process for making decisions. In a low-cost program, we must be able to make intelligent requirements trades quickly. We need access to someone who knows why specific decisions were made and can rapidly assess when they should be reevaluated. (Documenting **why** these decisions were made would be exceptionally productive, but this may be asking too much of a small, low-cost program.)

It's important to try to develop a consensus among mission designers on how to proceed, but it's equally important to avoid decision making by committee, in which everyone has veto power and no one has the authority to move forward. This process of rapid, strong (yet thoughtful) decision making throughout the mission design is an important element in essentially all of the case studies described in this book.

10.2.2 What the Government Can Do to Create an Environment that Reduces Cost

The government can't reduce cost by itself but it is largely responsible for the environment in which space systems are designed, developed, built, and launched. This environment is primarily responsible for the cost of space missions. Consequently the role of government in reducing mission cost is largely one of creating an environment in which cost reduction is allowed and rewarded. Industry and government laboratories can then determine how to achieve it. Table 10-8 summarizes ten actions the government can take to create the right environment for reducing space mission cost. These fall into three broad categories. Items 1 to 4 are programmatic and management issues, items 4 to 7 deal with space policy, and items 8 to 10 are concerned with R&D spending. As discussed in Chap. 2, all of these approaches have policy aspects and would benefit greatly from policies that foster and reward low cost.

Programmatic Approaches (Table 10-8, Actions 1–4)

Items 1 to 4 of Table 10-8 are management techniques individual programs can carry out, at least to some degree. The first two are concerned with formalizing mission engineering and requirements trading. Both can be done by either the prime contractor or the government laboratory which serves as the program's system engineering organization. However, they're unlikely to be effective without program office direction. The traditional approach is a friendly, yet adversarial, relationship between the government's program office and the principal contracting organization.

In this model the program office establishes the requirements, the contractor determines the cost to meet them, and a change in requirements by the program office will increase program cost. This may prove successful during spacecraft construction but is not appropriate for concept exploration. The purpose of the concept exploration phase is, literally, to explore—to examine a wide variety of alternatives to determine how best to meet the broad mission objectives at minimum cost and risk. This process should be driven not by philosophical principles, such as always minimizing weight, but rather by practical engineering concerns about how we can achieve what we are trying to do at the lowest cost. This means letting engineering assessments flow up into the requirements definition process and examining a wide variety of alternatives and approaches to see what makes sense for the mission as a whole.

TABLE 10-8. Actions the Government Can Take to Create an Environment that Reduces Cost. See text for additional discussion.

Action or Approach	Comment	Where Discussed*
1. Force trading on requirements	Make this a formal process by or with the performing organization or contractor	Sec. 2.1, 10.2.1, 10.4.1 SMAD Sec. 1.4, 3.2
2. Require strong mission engineering	All programs should have strong, formal system trades, ongoing utility analysis, and up-front mission engineering	Sec. 2.1, 6.3.3, 10.2.1 SMAD Chap. 3
3. Provide continuous funding	Force decision making on subsequent phases in parallel with current phase	Sec. 2.1, 2.2.7, 5.2.7
4. Reward low cost	Find mechanisms to reward (and not punish) individuals and organizations which contribute to reducing cost	Sec. 1.3, 10.1
5. Reduce the cost of failure	Recognize the need to allow reasonable risk and failure rates in test and R&D activities	Sec. 2.2.1, 4.4.6, 10.1
6. Make cost data available	Getting the lowest-cost solution is essentially impossible unless the engineers designing the system know what the costs are	Sec. 1.3, 10.1
7. Decentralize space system procurement	Innovation comes from small businesses and "secondary" organizations within the government.	Sec. 10.4.2
8. Sponsor R&D to reduce cost	Make reducing cost an alternative and accept-able objective for R&D, without demanding that it simultaneously "advance technology"	See text
9. Sponsor knowledge preservation and dissemination	Space technology has dramatically fewer books, commercial software, or university programs than any other major discipline. Knowledge is disappearing very rapidly.	Sec. 10.1
10. Revise the SBIR rules	Small companies are a major source of innovative approaches to cost reduction, but current rules discourage them	See text

*See text for discussion of each of these approaches.

The program office should establish a mission's broad objectives and constraints. The example throughout the SMAD book is FireSat, for which the objective is to detect and monitor forest fires for the United States. In today's acquisition environment we might include additional constraints: low cost, rapid development, and real, demonstrated cost savings to the government over the satellite's life cycle. The system engineering organization must then

- Recommend preliminary mission requirements for such a mission
- Identify one or more alternative mission concepts that can meet these requirements
- Identify the system drivers and driving requirements for each of the key concepts and estimate their effect on performance, cost, risk, and schedule
- Quantitatively analyze mission utility to demonstrate whether or not the selected concept or concepts can reduce both the loss of lives and property and the fire-fighting cost to the government

Within the scope of concept exploration, we may not be able to provide accurate quantitative estimates for all of these items, but it's still worthwhile to try to determine the mission's real utility.

Mission engineering, assessing mission utility, and system trades should continue throughout the life of the mission. Whether we still have the right requirements and whether the net benefit remains greater than the remaining cost is something we must examine regularly to ensure the program is on track and remains sold. Today, political concerns may cancel a mission at any time, so we need to do mission engineering and analyze utility at the beginning of each mission phase. This allows the program to reexamine the needs, requirements, and fundamental direction and then to build a stable design that can be defended at the next budget cycle.

Item 3 in Table 10-8 calls for continuous funding. This is critical for small, low-cost programs and is exceptionally difficult to do within the current government environment. The typical programmatic approach is to undertake one program phase and then to evaluate, assess, and debate the results before determining whether or how to proceed to the next phase. This approach dramatically increases the cost, schedule, and risk for any program. People in the program will go elsewhere. In a large company, they may transfer to other programs or leave the company. In a small company, they may start other projects or leave the company. In the worst case, the company itself may not survive. In any case, reassembling the team and reestablishing momentum will be time consuming, expensive, and perhaps impossible. Because the program office doesn't want to assign real cost to the delay, mission capability declines or the risk of not achieving it increases.

The programmatic solution to this problem is straightforward, although difficult to implement. The answer is to have parallel, rather than serial, action on program funding. Decision-making points and processes need to be well defined, so a program can proceed as long as it is making appropriate progress toward its end objectives, staying within defined budget constraints, and continuing to show strong utility. If it becomes clear that the program is proceeding well, but funding is simply unavailable for a period of time, then the best approach is to realistically assess when the next funding increment will become available and to slow the pace of the program, such that it can remain at a uniform level as long as possible. This allows the best compromise between an undesirable stretch-out and maintaining the continuity and knowledge of an ongoing program.

Action 4 in Table 10-8, rewarding low cost, is both a programmatic and a policy issue. As discussed previously in Sec. 10.2.1, the people involved in mission design and development must directly benefit from reducing cost.

Policy Approaches (Table 10-8, Actions 4–7)

Chapter 2 discusses policy approaches to reducing space mission cost. As discussed there, policy can increase cost much more easily than it can decrease cost. The best approach, therefore, is for policy to make it worthwhile to reduce cost and to let creative individuals and organizations determine how to do so in specific situations.

One of the key policy issues is finding ways to reduce the cost of failure. Almost nothing prevents cost reduction more strongly than believing the nation or buyer **must** have the program. The United States lived without space assets for almost two hundred years and can probably survive without any individual program. Programs of perceived critical national importance are indeed cancelled or scaled back if they cost too much. If programs are important to the nation, we need to find a way to achieve them at lower cost.

The fear of failure also has a dramatic impact on cost. The Titan IV is the largest and most expensive vehicle in the United States expendable launch arsenal. Unfortunately, there were several Titan IV failures in 1993. It is widely assumed within the aerospace community that these failures will substantially increase the cost of future Titan IV launches. More tests and safety features will be incorporated, such that the Titan IV will become even more expensive over the next several years. The economic justification is straightforward. The payloads being launched on the Titan IV are very large and dramatically expensive, such that a launch failure may cost several billion dollars. Such a loss is both expensive and embarrassing. Consequently, the launch costs will be allowed to rise in order to try to reduce the failure rate. This in turn will result in even more expensive Titan IV missions for both the spacecraft and launcher in the future.

A second example is the trouble encountered in first deploying the Space Telescope. Many steps were taken during the program to try to hold down cost, including eliminating a test which would have detected the mirror problem. The political analyses of Space Telescope largely forget these cost saving measures while remembering the one key test that wasn't performed. This implies that any potential follow-on programs for Space Telescope will be even more expensive as we try harder and harder to avoid failures.

An alternative approach to the Hubble Space Telescope problem is given in Fig. 2.2-1 (Sec. 2.2.1), which presents the viewpoint of space economist Robert Parkinson of British Aerospace. As Parkinson points out, a key to dramatic cost reduction is willingness to accept some risk of failure, particularly in tests and experimental programs. This is a key ingredient of small satellites and one of the major reasons for their low cost. It is also a reasonable approach for the space program as a whole. In a series of small, low-cost satellites, the failure of a single satellite is not catastrophic. If the entire space program or your entire professional career depends exclusively on a single satellite, you're much more dependent on that satellite's working correctly. In part, being able to accept failure implies being able to innovate, to try new techniques, and to use newer technology, which makes small satellites more affordable and productive.

It is easy to blame this problem on government program managers or policy makers, but I believe we are all to blame. The press concentrates on failures, particularly spectacular ones, far more than successes because that is what sells. It may be critically important but is hardly exciting that a program manager saved $50 million by cutting a series of tests designed to increase confidence in the spacecraft. In turn the White House and high level decision makers respond to the press and public pressure by pursuing a policy that is risk averse—even to the detriment of the space program as a whole by driving cost to beyond the level that we can reasonably afford. In this sense, Clementine may have made the most important breakthrough in the modern space program. As pointed out in the case study, Clementine was not particularly low cost, but was widely perceived as an example of "faster, better, cheaper." It had a dramatic, mission-ending failure, but was still regarded as a success by most of the community. We need that willingness to accept some failure to extend to far more space missions.

It is important to point out that accepting programmatic risk does not necessarily make the resulting spacecraft less reliable. In 41 space launches to date AMSAT has had only one mission failure. This is a remarkable record for the builders of what are probably the lowest-cost satellites in the world. Smaller spacecraft usually have far fewer components, many terrestrial components, larger margins, and more robust structures—all of which make them more reliable. The principal risks in small spacecraft are programmatic and psychological. If 20 engineers are building a spacecraft, a traffic accident involving two of them can strongly affect the program. Historically, loss of key personnel does not appear to have been a significant problem in small satellite programs. This may be due in part to the high loyalty and personal involvement when only a small number of people are building a satellite. Nonetheless, while small spacecraft are highly reliable it is clear that there is a larger level of risk associated with the process used to build them.

As discussed in Secs. 1.3 and 10.1 cost data for specific programs and components should be public if we're going to reduce cost. I don't know whether, in the long run, it's good or bad to publicize cost data, but I do believe it's impossible to design the lowest-cost system without it. We've tried to follow that example in this book by providing, wherever possible, cost data for everything from spacecraft to software, courses, and books.

Lastly, decentralizing the space procurement process is a key method of fostering both competition and innovation. The reasons for this are discussed in some detail in Sec. 10.2.3. Note that this is contrary to what many individuals within the government are trying to do. Nonetheless, I believe it is an important element in obtaining the lowest-cost space systems.

R&D Spending (Table 10-8, Actions 8–10)

The government controls much of the R&D spending in the United States, particularly for space technology. This has the legitimate role of advancing technology, but there should be an equally legitimate path associated with reducing cost. The government currently shows a strong bias toward those elements of R&D that are the most challenging, rather than those which are capable of being implemented and that can drive down near-term cost. Consequently, major strides in reducing space mission cost could be achieved if the government would choose to sponsor R&D specifically oriented toward reducing cost, without requiring that it advance technology at the same time.

A second major spending objective should be to preserve and expand our knowledge. First, we should take advantage of existing knowledge and build on the experience we've gained from 35 years of space exploration. Unfortunately, much of this knowledge base has been lost. The Saturn V provided the lowest cost per pound of any launch vehicle in the U.S. arsenal. However, the Saturn V is not only not being built, it probably cannot be built at the present time. We've lost much of the knowledge gained at great expense during the Apollo program. Knowledge is disappearing rapidly as the defense and space industries downsize and many of the most senior and knowledgeable engineers retire or are laid off, so nearly every program re-invents the wheel. Of course, many of these programs also re-invent past mistakes.

Part of the problem with the space technology knowledge base is that both DoD and NASA are program-oriented organizations. Their main objective is to carry out specific programs; not to foster development of the knowledge that is crucial to making new programs better and cheaper than old ones. In other disciplines, universities and academic institutions undertake most of this activity. However, in space technology, industry and government laboratories have most of the knowledge. Unfortunately, companies are unlikely to assign their most senior engineer to take two years off to write a book about how to do what they do best and thereby inform the world and their industrial competitors. As a practical matter this simply does not occur. Without an active government program to sponsor preserving and distributing knowledge, it simply disappears.

The lack of a strong knowledge base is reflected in the lack of college and university programs in space technology. Many schools have aerospace departments, but only a few of these teach space technology. Table 10-3, at the front of this chapter, is a nearly complete list of colleges and universities teaching astronautics and space technology—far fewer than those teaching almost any other major field of science or engineering.

The implication of the lack of a strong knowledge base is that we will continue to re-invent the wheel and re-invent our mistakes. To prevent this, the government must spend at least some portion of its resources on preserving and distributing knowledge. This includes developing books, journals, commercial software, university programs, and other means of recording, retaining, and passing on the lessons the space program has taught us. The U.S. Air Force Academy's program to develop books in space technology is one example of how this can be done. More books will lead to more courses, and more universities will be willing to teach them. The U.S. Air Force Academy has made an excellent start, but it needs long-term, stable funding to progress. As budgets for space programs shrink, it's relatively easy to eliminate small activities that preserve and expand knowledge. But, these programs are now more important than ever if we are to succeed at dramatically reducing space mission cost over the long term.

Finally, another straightforward and economical step to reduce space mission cost would be to revise the rules associated with the Small Business Innovative Research (SBIR) program to allow innovative research oriented specifically toward reducing cost and not necessarily advancing technology. Small companies are a major source of innovation and new ideas, particularly with respect to reducing cost. Large corporations do excellent research, but small companies and individuals tend to introduce dramatically new approaches.

As an example of this problem, it is not clear whether the development of the personal computer would be allowed under current SBIR rules. It did not advance the state-of-the-art in computing—in the sense of allowing more memory, speed, or throughput—and indeed the first personal computers were extremely limited in terms of their processing capacity. What it did do was dramatically reduce the cost and increase the availability and flexibility of the computing environment. This revolutionary invention probably wouldn't have qualified for funding under the government's SBIR program. The current SBIR rules foster details rather than broad ideas. They strongly emphasize advancing the state of the art, rather than trying to find lower-cost ways to do things (see Table 10-9). But the SBIR program has some bright spots. The Ballistic Missile Defense Organization has provided very wide-ranging topics, which allow innovative solutions to be proposed and evaluated. However, neither NASA nor DoD has as yet asked small businesses for ways to dramatically reduce the cost of access to space.

TABLE 10-9. Evaluation of the Principal Objective of 221 NASA 1993 and 1994 SBIR Space Topics. Reducing cost is rarely a primary concern even though we could expect strong contributions in this area from small businesses.

Principal Objective	No. of Topics	Percentage
Develop new software	33	15%
Improve existing software or technology	67	30%
Develop new technology	113	51%
Reduce cost	8	4%
Total	221	100%

The objective of revised rules for the SBIR program would be to let small businesses do what they do best—create and invent. As always, we'll need to tell people what we want (lower-cost ways to explore Mars) and not how to do it (higher I_{sp} or storable bipropellants). A key problem in revising the SBIR rules will be to actually allow innovative proposals—that is, ones that may have a low probability of success but which should still be examined. Thus, proposals should be judged in part by people searching for innovative solutions and new approaches to problems, rather than fostering research in areas of particular interest to the sponsoring organization. By simply changing the interpretation and not the rules, we could bring significant resources to bear on the problem of reducing cost. In addition, we could fund groups which have been the most active and most successful in dramatically reducing cost.

10.2.3 Places Where Cost Cutting is Frequently Counterproductive

Our general goal should be to cut cost in all phases of the program and in all components and elements. However, a mechanical cut of 30% in all budgets is certainly not the right approach. This section discusses specific components for which cutting cost may increase, rather than decrease, the life-cycle cost of the entire program. Table 10-10 lists five specific aspects of a program which typically should receive more funding, rather than less funding, to reduce the program's life-cycle cost. (Sec. 10.4.2 discusses counterproductive approaches to system acquisition as a whole.)

Our cost objective should be to reduce the program's life-cycle cost or, perhaps better, to achieve the government's or bill-payer's long-term objectives at minimum cost and risk. This may imply spending more of the money on planning and exploring options, rather than on physically building hardware. However, this approach requires balance. Unlimited planning and re-planning as budgets change simply drive costs up, not down. The history of the Space Station is an example of large amounts of money largely wasted because of continual vacillation in planning and budgeting. The solution is to define the broad objectives clearly, do a strong mission engineering analysis (as described in Sec. 10.2.1 and SMAD Sec. 2.3), and then proceed smoothly and deliberately to full-scale program development. However, none of these tasks is easy to achieve in bureaucracies or large companies.

There is a good analogy here with software development. Software is extremely expensive and many attempts have been made to reduce cost. However, lowering software-development costs by reducing the up-front planning and design has nearly always proven counterproductive. Beginning to write code in the first day or month

TABLE 10-10. Areas Where Cutting Cost is Often Counterproductive. In several program elements cost reduction is not effective or has a high probability of increasing costs at a later stage. Cutting costs here is similar to trying to begin software coding without having done an acceptable software design, and the results may be comparable. See also Sec. 10.4.2 on "Counterproductive Approaches to System Acquisition."

Area	Comment	Where Discussed
1. Up-front mission engineering	Strong mission engineering is essential. Ignorance is rarely of value in reducing life-cycle cost.	Sec. 1.3, 10.2.1, 6.3.3
2. Operations planning	Frequently not sufficiently taken into account during mission design—this oversight can significantly increase life-cycle cost.	Sec. 6.3.3, 6.4.3
3. Exploring options	Can be done in parallel with ongoing engineering. Explore options for reducing cost throughout the mission life.	Sec.10.3.1, 10.3.2
4. Selling the program	Sell the program to the funding groups and continue to sell it throughout the program life. Canceled programs aren't cost effective.	See text
5. Program end-of-life	Virtually all of the money has been spent. A very small investment in capturing and retaining knowledge can be extremely valuable in reducing cost for future programs.	See text

of a software project may be psychologically satisfying, but in the end it will substantially increase cost for development and maintenance. See Sec. 10.2.1 for a further discussion of the value of up-front mission engineering.

In most programs, additional resources should also be allocated to operations planning. This should be done in two distinct stages. First, operations planning should be a major part of the mission engineering process. How are we going to realistically operate this mission to achieve our end objectives? Not taking operational issues into account can significantly increase the life-cycle cost. (See Chap. 6 for more a extensive discussion.) Second, detailed planning and ground operation system development needs to begin well in advance of the launch of the first satellite. Representative timelines are in Chap. 6 and in SMAD Chap. 14. In many programs, there is a strong tendency to put off problems. Costs are constrained at essentially all stages and it is frequently seen as more productive to simply proceed and resolve problems at a later stage. Ultimately, these become operations problems which tend to dramatically drive up the cost of mission operations.

Another area which typically receives inadequate support is exploring options. Too often, we curtail looking at options as "serious" engineering begins. However, options for reducing cost should be explored throughout mission life. The pragmatic approach is clear. Given the current state-of-the-program, is it cheaper to proceed with the current plan or to choose a lower-cost option, taking into account the cost of making the change to a new baseline? As we go further downstream in a program, re-engineering becomes more and more expensive. The cost of looking at options, however, does not change significantly and if costs can be reduced, changes may be worthwhile, even late in the program development cycle. Even if the cost of implementing changes is judged too expensive, it may be appropriate to evaluate options such that follow-on programs can take advantage of the knowledge gained.

What we have called by the fancy title of mission utility analysis is really a matter of selling the program. This is an extremely important process, which does not necessarily reduce cost directly but does so indirectly in a number of ways. First, it keeps the effort focused on the mission objectives. Is the program achieving what the customer intended? Second, keeping the program firmly sold helps reduce the start-stop syndrome of many programs, which is a major source of cost increases and schedule and program overruns. Third, keeping the program sold keeps the program alive so it can achieve the original objectives, assuming they are still worth doing. Programs can be cut or eliminated at any stage, and cancelled programs are rarely cost effective.

In the government arena, selling a program can take on a negative connotation of convincing the government or the public to do something that isn't in their best interest. However, analyzing mission utility is really about establishing, as quantitatively as possible, the benefits of the program and how well it can meet the objectives of the group providing the funding. This is a positive and critical step for creating any new mission.

Finally, we should provide adequate funding at the very end of mission life. At this time basically all of the money has been spent and there is no further potential for cost savings. Still, spending some money on assessing and recording "lessons learned," while they're still fresh, can cut cost and improve performance of subsequent programs.

10.3 Reducing Cost in Ongoing Programs

For most space programs the early conceptual design phase was lost in antiquity.[*] Most engineering and economic resources are being devoted to programs well beyond conceptual design. Consequently, to reduce space mission cost, it is critical to find ways to reduce cost in these ongoing programs.

There are strong economic and political pressures on a space program throughout its design and operational life. Programs are continuously in danger of being cut or having their budgets reduced. It is not sufficient to sell the initial program to either the government or private investors and then methodically proceed with implementation. In the current environment, we need to conduct programs differently than in the past:

- Conduct the program to minimize cost and risk at all stages of the mission.

- At all stages, sell the program and keep it sold, if it is still worth doing.

Most of the cost is locked in very early in mission definition. (See, for example, NASA's *Systems Engineering Handbook* [Shishko, 1995] or SMAD, Fig. 19-2.) However, we can still reduce cost later in the program. We would like to lower costs

---

[*] The longest running continuous spacecraft design and development that I am aware of is the AXAF program. The concept design for AXAF, as a follow-on to the HEAO-B Einstein Observatory, was done in the mid-1970s. (HEAO-B was launched in 1976.) AXAF was a high-priority program throughout the astronomy and space science community, but was continually delayed because of cost overruns in Space Telescope and overall reductions in space budgets. The spacecraft is now being built and is scheduled for launch in 1998.

throughout the mission's life cycle, even if the main result is just to contain cost growth. In this section we will first discuss how to start a cost-reduction program and then describe ways to continue the mission engineering and system trade process throughout mission life.

10.3.1 Starting a Cost-Reduction Program

In very-low-cost programs, the program duration is short enough (typically less than 18 months) that they don't need regular reviews, and the least expensive approach is typically to define a low-cost solution and proceed to build the spacecraft. Although it costs somewhat more, we can significantly reduce risk in such programs by incorporating timely, senior-level review to ensure that we haven't overlooked major potential problems and that we've considered lessons learned from prior programs.

TABLE 10-11. Process for Initiating a Cost-Reduction Program for Ongoing Missions.
Many of the techniques discussed in Sec 10.2 for the "Concept Exploration" phase may also apply, depending on the stage of the program and the aspects being examined. **See Table 10-1 for Steps 1-5.**

Step	Comment	Where Discussed
6. Initiate an ongoing mission engineering program	The Mission Engineering activity should conduct system trades, reassess mission utility, and develop explicit cost-reduction strategies	See text, also Sec. 6.3.3 SMAD Sec. 3.2, 3.3
	6a. Identify driving requirements	SMAD Sec. 3.1
	6b. Can driving requirements be reduced?	SMAD Sec. 3.2
	6c. Identify system drivers	SMAD Sec. 2.3
	6d. Look for major alternatives	SMAD Sec. 2.1, 2.2
7. Review list of techniques in Table 10-6	Ask which of the concept design cost reduction techniques might be applicable to the mission being evaluated	Sec. 10.2
8. Create cost-reduction incentives	Need to reward cost reduction **and** encourage reasonable risk	Sec. 1.3, 10.1, 10.2.2
9. Look for alternative sources of "income"	Reduce the effective cost by increasing the utility, expanding the number of users, sharing the cost, or charging for data or services	See text
10. Obtain independent review and feedback	A key element of innovation is to obtain the ideas and opinions of others and to thoughtfully review the approaches proposed **and** the reasons for them	See text
11. Look for ways to reduce operations costs	Because operations is the last major program element, we may reduce cost even when the program is well underway **if** we plan properly	Chap. 6
12. Document reasons for key decisions	Others need to understand key decisions as the program evolves so we can reassess as needed	See text

In programs of normal duration, we can reduce cost even after the program definition phase by starting a cost-reduction program as shown in Table 10-11. Table 10-1 at the front of the chapter discussed steps 1–5. Step 6, continuing the mission engineering process, is the most critical and is described in more detail in Sec. 10.2.1. Steps 7, 8, and 11 are discussed elsewhere, as indicated in the table. The remaining steps are described below.

The key to success is starting a pro-active program that looks for cost-reduction methods. Individual recommendations and approaches resulting from this search may or may not be worth the cost of implementation, but substantial cost reductions are unlikely without a deliberate and serious review of the major alternatives. People who don't have key assignments in the program itself should do most of this work. Otherwise, cost-reduction effort may tie up key people and increase, rather than decrease, costs. However, it requires the active endorsement and participation of program management and system engineering personnel, since these individuals will ultimately be responsible for deciding what is implemented.

Step 9. Look for Alternative Sources of "Income"

One way to reduce the net cost of a program is to find alternative mechanisms to offset the cost—i.e., alternative sources of "income" in the broad sense of the word. In other words, space programs should behave in a more business-like fashion. They should continually be looking for "customers" and for ways to satisfy their customers' needs at low cost, while also meeting the needs of the system's buyer. Customers do not necessarily mean outside groups to whom your "product" is sold. For example, other branches within the government (or other divisions of a company) might use the services and share the cost if they see a net savings. For example, LandSat images that show extensive cloud cover might be valuable to NOAA. Space-based radar data, created for surveillance purposes, could benefit civil aviation. GPS data could significantly impact the cost of map making by the USGS.

There is also the possibility that some satellite products and services can be sold. LandSat images are sold commercially. Data on manufacturing in space could be of interest to pharmaceutical or chemical firms. GPS access could be sold to some users —particularly foreign users who didn't help pay for the system. Typical responses from government-run programs are that charging for services is not in keeping with the government tradition or that the cost of such an effort would exceed the income. Both may be true. However, like any corporate activity, to assess the potential requires a serious, thoughtful investigation and, perhaps a business plan. In addition, the government may not be particularly good at implementing the business. The government might choose instead to franchise the commercial aspects or hire a company to manage the sale of information or products.

The government may, as a matter of public policy, choose not to charge for services. Libraries are usually free. However, other public assets, such as bridges, toll roads, and national parks frequently carry a usage charge. I believe that most users of national parks regard the usage fee not as a burden but as a means of supporting environmental protection. The key point is that alternatives should be addressed. In times of strong fiscal constraint, it might be more appropriate to charge for services than to see the services disappear, particularly if the people who use these services didn't help pay for them.

Step 10. Obtain Independent Review and Feedback

No matter how clever, a few program people won't invent all of the innovative approaches nor will they have available the full experience of past programs. This is the reason for obtaining program reviews from experienced people.

This is a cost-reduction area in which traditional, large programs are much better than small ones. Formal, independent reviews are a part of life on major programs. We may not like someone looking over our shoulder and challenging our best ideas, but we recognize it as inevitable and try to make the best of it. In the small-satellite community, serious and thorough reviews are, at times, regarded as something of a threat to the program's independence. But, this review process may be even more critical in small programs, where the team may be innovative and talented, but lack experience on what has happened on prior programs.

The key to strong reviews for cost-reduction is having experienced, but innovation-oriented, reviewers. The goal is not to fill binders with responses to action items, but to listen to the key issues identified by the reviewers and to understand how to apply their experience in the context of the current program.

The dramatic downsizing of the space industry in the early 1990s may end up helping the review process. Many senior engineers with extensive and valuable experience are recently retired and ready to use their experience to benefit future programs. It's also important for the space program as a whole to use and internalize this experience, before it is permanently lost.

Step 12. Document Reasons for Key Decisions

I am not advocating more formal paperwork, but I am suggesting the need to increase the information content in the documentation that does exist. Specifically, it is important to document the **reasons** for decisions, even reasons which are non-technical. Unless the engineering process stops entirely, people in the future will need to understand the basis of the trades and decisions in order to intelligently determine whether to re-evaluate them. Thus, if we understand **why** a program chose a 650 km altitude or set the resolution requirement at 100 m, we can determine whether these conditions are still valid and whether an appropriately thorough trade was done to establish them. This allows us to make intelligent engineering choices in the future without redoing work done in the past or overlooking the potential to reduce cost by revising either system drivers or driving requirements.

10.3.2 Continuing the Mission Engineering and System Trade Process (Step 6)

The objective of this step is to start a mission-engineering program within an ongoing project. Our goal is to ensure that at each stage we meet the broad mission objectives at minimum cost and risk. To do this, we must regularly revisit the objectives, requirements, key trades, and mission utility. The best time to do this, as described in Sec. 10.4.2, is at the beginning of each mission phase. Specifically, the first formal review at the beginning of a new phase should start with

- Broad mission objectives, including non-technical ones

- Principal system requirements and constraints

- A formal review of the source or derivation of the principal requirements and which requirements or constraints mainly drive performance, cost, risk, and schedule

- An analysis of mission utility as a function of the principal system drivers

- An assessment of the state-of-the-art for any new technologies that could potentially significantly reduce mission cost

As discussed in Sec. 10.2.1, mission utility analysis is simply a way of quantifying how well we're meeting the broad mission objectives. Looking outward, it's a way to sell the program to those who must decide how best to use limited resources. Looking inward, it's a way to decide which of a program's features are worth the cost.

Throughout the program, we need to maintain a strong systems engineering function responsible for continuing the system trades. We need to continually look for new technologies and new approaches that will reduce cost. And we should try to continue to **loosen** requirements as needed to drive down cost. Although implementing change is more difficult in the later stages of the program, it is never too late to do something easier and cheaper.

The overall process of mission engineering is outlined in Sec. 10.2.1, and is described in detail in SMAD Chaps. 2 and 3. After initial concept exploration, much of the mission has been defined. Although solutions may become more difficult to carry out, the mission engineering process itself is easier:

- Identify the driving requirements

- Look for ones that can be relaxed to drive down cost

- Identify the system drivers

- Look for major alternatives that can meet the relaxed requirements at reduced cost

The system drivers and driving requirements may change as the program itself evolves. Typically what these are will become apparent as the program proceeds. They will be the requirements and system parameters that are causing never-ending frustration among the system engineers and that everyone is trying to re-interpret to make more achievable. If they are not immediately apparent, we identify them by looking at the top-level requirements and system parameters to see which ones pose fundamental limitations on the system design, performance, cost, risk, and schedule. It is also important to look for hidden requirements, such as the use of a particular technology, as well as hidden drivers, such as the altitude, which may affect many system parameters. Table 10-12 shows the most common driving requirements for space missions; Table 10-13 shows the most common system drivers.

Most are the parameters that we would expect to have a major impact. A less obvious one is the minimum elevation angle at which systems can operate. Payloads often set this parameter rather arbitrarily. However, the coverage of the satellite system and, therefore, the revisit interval, number of satellites, and amount of service are all extremely sensitive to the minimum elevation angle. Consequently, this is a key parameter, which often isn't evaluated as carefully as desirable.

A key issue for mission engineering is to define requirements in terms of **what** must be done, rather than **how** to do it. This approach leaves the way open for creative

TABLE 10-12. Common Driving Requirements for Space Missions. Which requirements are the most stressing will depend on both the requirements themselves and the system design. (from SMAD, Table 2-8)

Driver	What Limits Driver	What Driver Limits
Size	Shroud or bay size, available weight, aerodynamic drag	Payload size (frequently antenna diameter or aperture)
On-orbit weight	Altitude, inclination, launch vehicle	Payload weight, survivability; largely determines design and manufacturing cost
Power	Size, weight (control is secondary problem)	Payload & bus design, system sensitivity, on-orbit life
Data rate	Storage, processing, antenna sizes, limits of existing systems	Information sent to user; can push demand for on-board processing
Communications	Coverage availability of ground stations or relay satellites	Coverage, timelines, ability to command
Pointing	Cost, weight	Resolution, geolocation, overall system accuracy; pushes spacecraft cost
Number of spacecraft	Cost	Coverage frequency, and overlap
Altitude	Launch vehicle, performance demands, weight	Performance, survivability, coverage (instantaneous and rate), communications
Coverage (geometry and timing)	Orbit, scheduling, payload field of view and observation time	Data frequency and continuity, maneuver requirements
Scheduling	Timeline & operations, decision making, communications	Coverage, responsiveness, mission utility
Operations	Cost, crew size, communications	Often the main cost driver, main error source, pushes demand for autonomy (can also save "lost" missions)

TABLE 10-13. Common Space System Drivers. Hidden drivers include those, such as altitude, which have multiple impacts on the system design and performance. (from SMAD, Table 3-1)

Requirement	What it Affects
Coverage or response time	Number of satellites, altitude, inclination, communications architecture, payload field of view, scheduling, staffing requirements
Resolution	Instrument size, altitude, attitude control
Sensitivity	Payload size, complexity; processing, and thermal control; altitude
Mapping accuracy	Attitude control, orbit and attitude knowledge, mechanical alignments, payload precision, processing
Transmit power	Payload size and power, altitude
On-orbit lifetime	Redundancy; weight, power, and propulsion budgets; component selection
Survivability	Altitude, weight, power, component selection, design of space and ground system, number of satellites, number of ground stations, communications architecture

and innovative solutions and allows technology to flow up, so we can meet the overall mission objectives at minimum cost and risk. For example, many space systems have requirements on attitude and orbit or position accuracy, but, almost no end users are interested in the spacecraft's attitude or position. The end user wants accurate mapping and pointing. Thus, we should express requirements in terms of mapping and pointing (the "what"), rather than in terms of attitude and position knowledge (the "how"). This would allow creative new ways to achieve, for example, the desired mapping accuracy, at less cost than pointing with ultra-high accuracy.

A second element that applies throughout mission life is to try to reduce the demand for high-tech solutions. We should always be looking for innovative non-technical, or less technical, approaches. An example is the development of the "space pen" early in the American space program. At considerable cost, the United States developed a pen that can write in a weightless environment. The Russians use a pencil.[*] These are the processes we are looking for throughout the new space program.

TABLE 10-14. Principal Problem Areas in Implementing Dramatic Cost Reduction Methods. The table summarizes the cause and possible actions to alleviate the problem. See text for a more extended discussion of each.

Problem	Most Common Cause	Positive Steps
1. Failing to trade on requirements	Don't want to be non-responsive or identify a requirement as difficult	Make formal requirements trades a part of the system exploration process
2. Constraining trades to too low a level	Politically sensitive—may involve shifting cost and responsibility among major groups	Conduct trades between elements explicitly and early in the program
3. Postponing or avoiding assessment of alternatives	Budget constraints plus problem of giving a program the appearance of instability	Maintain a strong systems engineering organization with the responsibility for assessing alternatives
4. Failing to take advantage of prior knowledge base	Lack of adequate books and training; "Not invented here" syndrome	Bring in experienced review team; make adequate resources readily available—books & COTS software; establish professional training program
5. Insufficient trades during concept exploration	Low budget; need to make the program look advanced and well-defined	Undertake systematic mission engineering
6. Poor requirements definition	Traditionally requirements are too detailed and specify **how** rather than **what** is needed	Review the requirements with these problems in mind
7. Poor data processing trades	Perceived need to lock in computer hardware selection prior to software design or requirements definition	Almost unavoidable in today's environment; provide large (400%) margin in computer sizing (See SMAD Sec. 16.2)

[*] With thanks to Wiley Larson, who has made this example well known in his lectures on space system design.

10.4 Problem Areas in Implementing Dramatic Cost Reduction

In my experience, most problems in implementing mission engineering and cost-reduction programs result from the environment or the acquisition process, rather than poor technology or erroneous decisions. Sec. 10.4.1 summarizes the main problems in implementation and Sec. 10.4.2 looks at approaches to system acquisition which are intended to reduce cost but which may have the opposite effect.

10.4.1 Principal Implementation Problems

Principal problem areas in implementing mission engineering and in dramatically reducing mission cost are listed in Table 10-14. Because these are recurring problems on multiple programs, they typically are a result of the environment in which we work or the acquisition process. Consequently, I will try to explain why each problem occurs and offer positive suggestions on how to overcome them.

Problem 1. Failure to Trade on Requirements

Mission requirements should represent a balance between what we need and what we can afford. Consequently, the definition of requirements should be a major element of system trades and should be a continuing trade throughout the design and development cycle. However, this frequently does not occur for two principal reasons. First, the government wants to deal fairly with contractors by giving them a chance to bid against firm and definitive requirements. In the traditional approach, the government establishes the requirements defining the needs of the program, and contractors compete on their capacity to meet them at the lowest possible cost. The requirements are defined by the government with, at best, rough guesses on the impact of these requirements on system cost, risk, and schedule. This process prevents the technology from flowing back up and prevents us from having the information needed to determine what the requirements should be.

The second difficulty is that once requirements are defined, contractors and government laboratories will typically go to great lengths to meet them. The group which either hopes to be or is in charge of implementing the system does not wish to be non-responsive and does not wish to identify a given requirement as difficult. An organization which identifies all of the proposed requirements as reasonable and achievable is more likely to be assigned the task of implementing the system than is the organization which identifies some of them as being difficult or driving up cost. In reality, the task will probably be just as difficult for both organizations, but identifying problems is not seen as the most productive way to win business. Nonetheless, this is what is required if we are to succeed in driving down cost. We have to understand which requirements drive most of the cost, so we can determine whether they must be at the specified level.

We can take several positive steps to overcome the natural barrier to trading on requirements. First, we can begin concept exploration with formal and defined trades on system requirements. This is a part of the mission engineering process summarized in Sec. 10.2.1 and defined in detail in SMAD. We can also re-visit the top-level trades on system requirements at each major milestone, as summarized in Sec. 10.3.2. The contractor's or government's system engineering organization can usually identify the requirements that drive most of the system's cost. But the program must direct this investigation and use the information to help refine system requirements.

A second positive step is to specify what we want to achieve, not how to achieve it. By doing so, we can let the system-trade process flow down to where the detailed technical knowledge resides. This can be done productively at nearly all levels of system design. The program office can specify functional requirements for the system engineering organization or prime contractor. The prime contractor can specify functional requirements for subcontractors. For example, a subcontract for the attitude control system might specify levels of attitude knowledge, rather than a particular equipment list. Once we have specified the use of star sensors for attitude determination, we have closed the option of using precision interferometry with GPS signals, even if this could produce the needed results at much lower cost.

Finally, it is important to maintain open communications. At each level in the process, people and groups at a higher level are working with the data we supply and, at a lower level, are trying to meet the needs we define. By maintaining open communications we can begin to understand the problems and difficulties at both levels. The prime contractor should try to understand the government program office's problems in order to provide appropriate information to assist in resolving them. Looking downward, the prime contractor or system engineering organization needs to understand the subsystem developers' main difficulties, so that they can determine whether or not requirements trades or changes in the rules are warranted. In this process, there is no substitute for strong informal communications, so that individuals recognize the needs and problems of others working on the same project.

Problem 2. Constraining Trades to Too Low a Level

The most effective cost-reduction techniques frequently involve shifting cost between operations, spacecraft, and launch. For example, using a low-thrust system onboard the spacecraft to "fly the spacecraft," has the potential to eliminate an expensive upper stage and to greatly reduce launch costs. However, doing so will add cost for spacecraft hardware so funding and responsibility will transfer from the launch organization to the organization responsible for the spacecraft. Transferring funds and responsibility between departments, organizations, or even government centers, can be very sensitive. As a second example, more autonomy can potentially reduce operations cost. But autonomy transfers funding and responsibility from the operations group to the spacecraft group. The political and organizational sensitivities tend to make these trades very difficult to implement.

The single most positive step we can take is to conduct high-level trades (i.e., trades that might involve transferring funding and responsibility between organizations) as early as possible in mission design. These early trades would ensure that major options are considered and not ruled out by the program structure. A second step is to look **explicitly** for trades between elements of the system. Thus, the program office might ask what trades can be done between the top-level organizations, looking at positive and negative effects for each group. If these trades are explicitly identified, the resistance to making them decreases somewhat.

Problem 3. Postponing or Avoiding Assessment of Alternatives

The cost of making changes escalates dramatically as a program proceeds. During early concept exploration, changes cost almost nothing. During the definition phase, changes involve more people and reassessing and reevaluating a more detailed design. During the design phase, changes involve formal paperwork, change controls, and some level of reinserting and reanalysis. Changes during hardware build greatly magnify cost and typically are nearly impossible.

Nonetheless, it is often expedient to put off assessing alternatives. There are budget constraints at every phase. Concept exploration has less funds available than mission definition and so on. Thus, we often need to put off assessing some alternatives until later, when more funds are available. In addition, looking at alternatives gives the program an appearance of instability and may make it more difficult to obtain the funding needed to proceed. It's easier to sell a program that is well defined and includes substantial engineering detail, than one in which the program office is still trying to evaluate whether there are two spacecraft or seven and whether they are spinning or 3-axis stabilized. Consequently, evaluating alternatives may reduce the momentum to proceed. All of this tends to postpone looking at alternatives until they become too expensive to implement.

On the other hand, it is the evaluation of alternatives and the examination of new and innovative approaches that can lead to lower-cost methods of achieving mission objectives. Therefore, it is critical to continually assess alternative approaches. One way to do this is to regularly review major trade decisions. This should be done at the **beginning** of each mission phase. This allows the program to stabilize before the next round of decision making and avoids giving the appearance of instability, while still allowing us to assess potentially lower-cost options.

As described in Sec. 10.3.1, two more positive steps are to document the reasons for design decisions (even non-technical ones) and to bring in independent, innovation-oriented reviewers during concept exploration. We must recognize that change is likely and value flexibility in the design. Usually we can do so easier in small spacecraft than in large ones, which tend to be much more optimized. Small programs often must fly on a space-available basis and must be able to accommodate a wide variety of orbits and mission conditions and, therefore, are better able to accommodate alternative solutions.

Lastly, we must avoid decision making by committee. This means maintaining a strong systems organization with well defined responsibility for undertaking, reviewing, and documenting system trades. With a strong systems-engineering activity, it should be possible to determine quickly whether particular alternatives have been addressed in the past and what the top-level implications are likely to be. This is valuable both for assessing alternatives to reduce cost and for resolving problems that may arise during mission definition and development.

We need dictators for the sake of efficiency, the ability to take calculated risk, and to trade away performance. However, the dictator must listen carefully to the condemned before calling in the lions.

Problem 4. Failing to Take Advantage of the Prior Knowledge Base

In trying to create new and better solutions, we often forget the painfully acquired knowledge of others, who have been through the process before. This doesn't imply we should use only experienced engineers. After all, new and innovative solutions frequently come from people who don't know how hard the problem really is. However, it can be difficult for the small, innovative contractor to recognize the value of an experienced senior specialist who has worked space problems for many years. At the same time, large, professional organizations find it difficult to overcome the "not invented here" syndrome and recognize they can learn new ways of doing business from younger and smaller organizations.

The most positive step we can take is to appreciate that we can all learn from the experience of others. Small organizations need the extensive experience large organizations have acquired. That store of knowledge can be extremely valuable in avoiding pitfalls that have caused major problems. Similarly, large organizations that haven't faced the problem of trying to work on dramatically constrained budgets can learn much from those who've had to live that way for a number of programs. Both types of organizations need to learn that the experience of others is valuable and that knowledge comes both from extensive experience on a large number of major programs and from having successfully built spacecraft for a few million dollars. Working collectively, we can dramatically reduce space mission cost. Working independently, we will almost certainly reproduce the mistakes that have been made before.

Problems 5–7. Other Potential Problem Areas

One of the most common problems is insufficient trades during concept exploration. This comes about for many of the reasons previously discussed, including short schedules, limited budget, and the need to establish a design in order to be able to sell the program. This results in a strong tendency to accept a point design as an optimal solution and to give insufficient initial consideration to elements other than space hardware—e.g., political, organizational, economic, operational, and regulatory issues.

One solution to this problem is to provide more resources during concept exploration. Funds spent here are perhaps the most important in terms of ultimately driving down system cost. But, it's also where funds are the most severely limited, such that the most realistic approach may be to examine as many alternatives as possible and to identify those we need to address during the next mission phase.

Another major problem that occurs throughout the mission design process is poor requirements definition. Specifications are often too detailed and specify methods rather than desired results or functions. We should try to keep the requirements at a high level and as functional as possible, thus allowing the next lower level of engineering organization as much latitude as possible to meet their objectives at minimum cost and risk.

Lastly, a particularly difficult technical area for mission engineering is poor data-processing trades. Specifically, there is a strong tendency to lock into computer hardware before designing the software. Space missions are necessarily hardware oriented. The objective is to design, buy, integrate, and launch a spacecraft capable of achieving particular objectives. Therefore, the initial requirements tend to be in terms of those things that we can build and assign weight and power budgets to. In almost all programs, it is simply unacceptable to argue that you are defining the processing requirements now and will get to the hardware requirements sometime during the next mission phase. Nonetheless, specifying the computer hardware first and then later deciding what that hardware must do is a prescription for high risk, cost overruns, and schedule delays. The main way to resolve this problem is to require a very large margin in computer sizing. Specifically, Hansen and Pollack recommend a margin of 400% in both memory and throughput at the time of System Requirements Review. [SMAD, Sec. 16.2] Large processing margins are difficult to "sell," but are also the single most important protection against major software cost overruns.

10.4.2 Counterproductive Approaches to System Acquisition

As discussed in Sec. 10.2.1, one of the most striking characteristics of this book is the nature of the organizations that have contributed to it. Historically, much of the most innovative and important work in reducing space mission costs has come from small companies or small enclaves within large organizations, just as Lockheed's Skunk Works has produced much of the innovative work in aircraft design.

In the automotive industry the large manufacturers are responsible for low-cost, high volume automobiles and small businesses produce small numbers of highly capable, but dramatically expensive vehicles. The situation in spacecraft is nearly opposite. I believe there are good reasons for this circumstance. The main advantages large organizations bring to the engineering and manufacturing process are often irrelevant to major cost reductions in spacecraft. For example, well-financed, multi-tiered organizations can bring massive resources to bear on major engineering programs: extensive R & D budgets, many senior experts, large facilities, and massive capital resources. While these are important assets, they are not dramatic advantages in producing low-cost spacecraft and may be counterproductive in some circumstances, such as the difficulty of making rapid decisions and the requirement to consult with many divisions and suborganizations. In contrast, the principal assets for small organizations or independent enclaves are that they tend to be innovative, have minimal infrastructure, have individuals that are highly motivated, allow rapid decision making, and work each problem as it comes along in a pragmatic fashion. These assets can frequently counterbalance the advantages of larger organizations and enable smaller groups to achieve what bigger ones cannot.

There is another "advantage" that most small organizations have. They have to do things to reduce cost or they cannot play. On the whole, driving down cost dramatically is difficult and, in many respects, not as psychologically rewarding as pushing the bounds of technology and working on huge programs, such as Space Telescope, Apollo, or the Space Station. The small business understands very well how to succeed with limited resources.

An industrial example of this process is the development of the personal computer. IBM was the dominant leader in the community, with virtually all of the capital and engineering resources. But a few motivated engineers at Apple Computer developed the personal computer, which revolutionized the cost and availability of computing.

I believe this example provides both an important message and a critical warning for acquisition policymaking intended to reduce space system cost: we must encourage, not discourage, the small-business or small-organization mentality and environment. The litmus test for any new approach, regulation, or policy intended to reduce cost should be: "Would it encourage or discourage participation by people from NRL, AeroAstro, the University of Surrey, and AMSAT?" If it would discourage these organizations from participating, then there is a high probability that rather than reducing cost, it may serve to support expanding the existing infrastructure and methodology and thereby discourage genuine cost savings. Viewed from this perspective, many of the policies and regulations that have been proposed in recent years have the potential of being counterproductive. While intended to reduce cost, they may, in fact, increase cost or solidify the high costs that currently exist.

For example, creative, low-cost organizations rarely have any internal funds to carry any additional burden. This implies that they cannot share in program cost and have little potential for undertaking large, time-consuming, and expensive proposals. Organizations that are willing to cost share generally intend to recover their cost plus any prior cost sharing that was not eventually funded. In addition, because cost sharing represents risk, the source of the capital expects a much higher return than would otherwise be the case.

I don't want to imply that small businesses are the only ones who know how to cut costs. Clearly, many large corporations are extremely competent at reducing cost, but their size and money may actually be a hindrance in the small, low-cost satellite market. An exception is the Lockheed Skunk Works, which in the airplane world, has strong financial backing and yet has achieved low-cost results. Most knowledge on low-cost satellites is in small companies and "secondary" government organizations, at least in terms of preeminence in space research, so we should look there for experience in innovation and low cost.

Based on this assessment, Table 10-15 lists system acquisition approaches to cost reduction which I believe may be counterproductive. That is, they may lock in the current cost structure or increase costs, rather than reduce them. Because these approaches were introduced in part to contain costs, they can be effective in some situations. But, they may force out those who have good ideas but whose financial resources or political power is too limited to bring them about.

TABLE 10-15. Potentially Counterproductive System Acquisition Approaches to Reducing Space Mission Cost. See text for a more extended discussion of each approach.

Approach	Potential Benefit	Disadvantages
1. Consolidated, centralized acquisition and engineering	Provides greater accountability; gives appearance of reducing waste	Likely to force out small organizations and innovative approaches; likely to lock in high-cost approaches
2. Contractor cost sharing of development costs	Reduces cost to the government **if** major customers are in the private sector (e.g., computer development)	Forces out the small player; investment economics drive acquisition—will require very large ROI, because the government is a high-risk customer
3. Contractor cost sharing in up-front studies	Reduces study cost	Forces out the small, innovative contractor; contractor costs will be recovered in higher indirect rates
4. Cost guarantee on R&D programs	Limits government cost commitment; good for achieving accountability but **not** for reducing cost	Forces out the small contractor and doesn't permit taking risks that could dramatically reduce cost
5. Doing work in-house	Eliminates subcontracting cost; can be effective **if** the group has experience with low-cost, efficient production	May drive up costs due to lack of efficiency and knowledge; may be largely an excuse to maintain a large infrastructure
6. Reducing the level of up-front systems engineering	Shortens program schedule and avoids over-engineering a strawman design	Ignorance is rarely of value in reducing cost or improving performance

1. Consolidated, Centralized Acquisition and Engineering

There is frequently discussion of the need to set up a single person or organization, to run government procurements in a particular area, such as a launch czar or single government agency in charge of all missions of a given type. The intent is to produce efficiency, eliminate duplication, and establish a more streamlined procurement approach. Unfortunately, this is likely to minimize innovation in favor of a "more of the same" approach, because the main players will be those major companies, laboratories, and government organizations which are established in the space business. What will disappear in this approach are the innovative, new, small businesses, government enclaves, and creative individuals who have new approaches and ways to reduce cost, but who don't have the political connections necessary to obtain a hearing in a new, large, and important bureaucracy. In the name of efficiency, we may cut out precisely those people who have the most knowledge, experience, and ability to drive down system cost.

2. Contractor Cost Sharing of Development Costs

If the government does not have sufficient resources and must reduce cost, one of the most compelling approaches is to require cost sharing by the contractor. This can be effective **if** the major customer base is in the private sector. For example, this could be effective in reducing the cost of R&D for communication satellites. This is a well-established commercial commodity with principal players who have substantial financial assets, who devote a significant fraction of those assets to R&D, and who can recover much of that R&D by selling products on the commercial market.

This approach will be counterproductive and is likely to significantly increase cost if the government is the primary customer. In this case, an entirely different set of economics takes over. The investment community (either the large corporations themselves or those who put up the money for the R&D) will provide the investment capital only if there is potential for a very high return on investment to cover substantial risks. This includes the risk that the government will change its mind about what it wants. The nature of political decision making is such that the government is an untrustworthy, high-risk customer and may or may not choose to purchase products developed at private expense. Providing adequate return on investment could easily double or triple the cost of the end product. The government isn't particularly motivated to buy such products because it hasn't invested either funding or intellectual effort in their development. The investor wants to obtain a large enough return to cover the loss for products the government decides not to buy. If the government is the primary customer, the end price will be much higher than if the product had been developed at government expense.

3. Contractor Cost Sharing in Up-Front Studies

Much the same problem occurs here as in development cost sharing. The large, well-financed organizations will be willing to invest their resources in up-front studies assuming they'll recover that cost through indirect rates and later development activity. This expectation nearly guarantees that the individual who has a novel approach won't be allowed to take part.

4. Cost Guarantee on R&D Programs

Establishing cost guarantees and contractor responsibility is excellent for achieving cost accountability, but doesn't reduce cost. It can work successfully in production programs for which costs are well established, but almost by definition, it doesn't permit taking risks that can dramatically reduce cost. Innovative, new approaches intended to drive down costs may or may not be fully successful. One of the major ways of reducing cost is to accept a higher level of programmatic risk. Trying to drive this risk to zero will tend to significantly increase costs.

In addition, much like the processes above, establishing cost accountability forces out the small organizations that don't have the financial backing to provide cost guarantees. Yet, these same organizations are most likely to complete the tasks at low cost. I do not mean that small organizations can't design to cost or meet cost ceilings. However, in these circumstances, they don't necessarily guarantee full performance.

5. Doing Work In-House

Doing the work in-house and not using contractors and subcontractors can sometimes reduce cost. DSI, the University of Surrey, and the Lockheed Skunk Works all have done a lot of work in-house, but they also have substantial experience with low-cost, efficient production. The organizations were established in large part with that in mind and have a management philosophy, corporate culture, and history of doing so. The main danger here is that organizations will do work in-house to support an existing large infrastructure, rather than to reduce cost. Maintaining infrastructure is a reasonable and legitimate goal. If a funding organization chooses to do the work in-house explicitly for this purpose, then it is appropriate to do so. However, this should not be confused with cost-cutting. If the organization does not have the culture and experience of low-cost spacecraft production, then it is more likely to raise costs rather than lower them.

For prime contractors who choose to do work in-house, the principal issue is what this does to the network of low-cost suppliers that you would like to have cooperate on the next major program that becomes available. For the government, the more important question is whether doing the work within government laboratories is forcing significant numbers of small players out of business and expanding the bureaucracy to conserve government jobs. While there certainly are many exceptions, government laboratories are not traditionally known for the highly motivated, long hours, highly efficient work force that is the trademark of the small, dynamic company. Consequently, any savings in eliminating profit are likely to disappear through the lack of efficiency of the large bureaucracy.

6. Reducing the Level of Up-Front Systems Engineering

"Shoot the engineers and get on with the program" has been used as a motto on some programs and is one way to hold down cost and compress schedule. This can work in some circumstances. It is a mistake to over-engineer a program. We don't want to optimize a strawman design that will change dramatically later. It is certainly counterproductive to do a detailed spacecraft design while we're still trying to determine whether there will be one spacecraft in geosynchronous orbit or seven in low-Earth orbit.

Still, we shouldn't use the danger of over-engineering as an excuse to minimize the up-front systems engineering or to avoid trades on objectives, requirements, major system parameters, and mission utility. Expediency can seem to be a good excuse, but ignorance is rarely of value in reducing cost or improving performance.

10.5 Summary—Maintaining Balance and Perspective

In reviewing the methods that organizations have successfully used to significantly reduce space mission cost, I have been struck by the lack of any common rules or characteristics. Many of the suggestions in both the process chapters and case studies are essentially contradictory:

- Do the work in-house vs. subcontract to efficient organizations

- Use the latest software tools and structured processes vs. move promptly with a small team and just proceed with the design

- Compress the schedule as much as possible vs. build and test, build and test

Some programs, such as MSTI, try to do both simultaneously. The key to reduced cost doesn't appear to lie in any particular technology, process, or government policy. Rather, it appears to be more a matter of maintaining a balanced approach and looking for solutions that work within the environment and constraints of each individual program. If the case studies have any feature in common, it is that they are all strongly committed to dramatically reducing cost and are willing to give up something and change the way they do business in order to achieve it. The technology matters, but often not as much as how it is applied.

A second element of reducing cost is to always look for new and innovative solutions, rather than trying to defend previous approaches. I don't mean things have been done badly on prior programs. It's simply that, if we're going to dramatically reduce cost, we must, by necessity, do something differently. We must look for new technology and for new ways to apply both new and old technology. In some cases, this will mean developing new components or materials. More often, however, it will be applying existing technology in a new way. Looking for these alternatives is a key ingredient of the mission engineering process. Work with others in terms of **what** you want to achieve, and not **how**, in order to remain open to alternative solutions.

Finally, several of the more thoughtful participants in this book project have identified perhaps the most important key to reducing cost: hire the right people and the job will get done; hire the wrong people and it won't. This includes having the right people to do the work, the right leadership, **and** the right customer. Everyone involved in the program and their capacity to work together are critical to its success. Unfortunately, I have no idea how to create a process to ensure that the right people get the key roles, have authority to do their jobs, and work together successfully. But we can control one element. If a successful team is assembled, it is valuable and should be preserved. This is perhaps the most important argument for maintaining program continuity. People aren't interchangeable pegs. Disassembling a successful team while waiting for the next round of funding can harm cost, risk, schedule, and performance more than nearly any other single action. To reduce cost and risk, find the right people and give them the capacity and authority to get the job done.

It is interesting to speculate how future missions will change in response to both new technology and to an apparently unrelenting demand to reduce cost. First, it appears nearly certain that onboard processing will increase. This will result in a major move toward more autonomy and toward more complex spacecraft, at least in terms of what they undertake. Depending on how autonomy is implemented, it could drive costs up or down. As discussed in Sec. 3.2.1, to drive costs down we need "autonomy in moderation," in which we automate simple, repetitive functions and leave complex tasks, such as problem solving and repair, to ground operations.

More on-board processing has another major advantage. A large problem for most spacecraft is that once they've been launched, they're no longer available to us. We can't get at them to fix, change, adjust, revise, or upgrade. This becomes a major advantage for spacecraft that are driven predominantly by software. Software is the one element we can economically change, revise, and upgrade while the spacecraft is on orbit. There have been some software upgrades on orbit, but they remain relatively rare. In the future, it's likely to be much more common. Just as we continuously upgrade software for office and engineering use, we'll upgrade the software in active spacecraft to reflect new capabilities, needs, and missions.

Clearly we'll also continue to emphasize reducing the cost of launch and operations, and building low-cost, light-weight spacecraft. It is certainly the intention of the authors and editors of this volume to accelerate the move toward lower cost. Low-cost and small spacecraft aren't the same, but they are strongly related. LightSats are becoming a major player for future missions and will be more so as communications constellations appear in low-Earth orbit and as smaller planetary, science, and military spacecraft become more popular. In a broad sense, LightSats represent a new way of doing business in space that, if used effectively, can dramatically reduce space systems cost. This approach is important to the space program as a whole, because we have far more things to do in space than we have resources to do them.

Lastly, the limits on space exploration are shifting away from technology to policies, politics, and economics. Virtually any solar system mission is technically feasible. It's clear that, if we choose to do so, we can have people exploring Mars, establish a lunar base, industrialize space, and send probes to any of the planets. Our limits depend on what we choose to do, or what we can afford. This, in turn, puts more emphasis on why we choose to use or explore space and strongly challenges us to bring costs down so they're consistent with other areas of exploration and development. This provides the substantial challenge for future space mission engineering.

10.6 Annotated Bibliography on Space Systems and Mission Engineering

All books listed are in print. List prices are quoted. Actual prices may vary slightly. The cost of the entire set is approximately $1,400 plus $110 per year for updates to *Space 2000*.

Agrawal, B. N. 1986. *Design of Geosynchronous Spacecraft.* Englewood Cliffs, NJ: Prentice-Hall. 459 p. $110.00.
Emphasis on the analysis, theory, and design of geostationary communications spacecraft, bus subsystems, and communications payloads; includes discussion of geosynchronous orbits and stationkeeping.

Brown, C. D. 1992. *Spacecraft Mission Design.* Washington, D.C.: AIAA. 187 p. + software diskette. $69.95.
The title of this volume is misleading—the book is about orbit design with an emphasis on interplanetary missions and some discussion of both general and specialized Earth orbits. Includes a copy of ORB, a PC program for computing spacecraft orbits for Earth and planetary missions.

Burger, J. J. [updated annually]. *Space 2000.* Colorado Springs, CO: Space Analysis and Research, Inc. $295.00 + $110.00 for 4 quarterly updates.
A complete electronic data base of all space launches, including most launch failures, and many planned launches. Includes both Space Command and international designations; launch date, time, vehicle, and origin; spacecraft name, owner, weight, frequencies, and mission; orbital elements; and current spacecraft status. Can be sorted on nearly all fields. (PCs)

Chetty, P. R. K. 1991. *Satellite Technology and Its Applications, 2nd edition.* Blue Ridge Summit, PA: TAB Books. 554 p. $59.95.
Practical discussion of the design and characteristics of spacecraft subsystems; includes substantial hardware discussion and data; discusses satellite applications with an emphasis on geostationary communications satellites.

Davidoff, M. 1990. *The Satellite Experimenter's Handbook, 2nd edition.* Newington, CT: ARRL. 339 p. $20.00.
This AMSAT book provides a detailed recipe for designing and operating very low-cost communications satellites; practical, relevant data and methods that apply to many missions; lots of references.

Fleeter, R. 2000. *The Logic of Microspace.* El Segundo, CA: Microcosm Press and Dordrecht, The Netherlands: Kluwer Academic Publishers, 447 p. $34.95.
A remarkably readable discussion of the design and application of small spacecraft. The author has his own unique style that makes the book fascinating while, at the same time, covering very well a topic that is important to those interested in low-cost and reduced-cost missions.

Fortescue, P., and J. Stark. 1995. *Spacecraft Systems Engineering, 2nd edition.* New York: John Wiley & Sons. 581 p. $56.95. (softcover)
Broad overall discussion of space technology and spacecraft systems engineering. Discusses specific techniques used for most of the major subsystems. Good textbook for an introductory technical course on space systems.

Greenberg, J. S., and H. R. Hertzfeld. 1994. *Space Economics.* Washington, D.C.: AIAA. 446 p. $79.95.
The only current treatment of economic analysis and methods for space systems; includes cost analysis, cost-effectiveness models, standard cost elements, and implications of economics for space programs and major programmatic issues.

Griffin, M., and J. R. French. 1991. *Space Vehicle Design*. Washington, D.C.: AIAA. 465 p. $61.95.
An excellent summary of the spacecraft design process. Well-written, practical orientation. Also has discussions of mission design, launch, orbits, the space environment, and re-entry.

London, J. R. 1994. *LEO on the Cheap—Methods for Achieving Drastic Reductions in Space Launch Costs*, Maxwell AFB, AL: Air U. Press. 213 p. free.
The most definitive study available of launch system cost and methods of reducing it. Summarizes existing and proposed launch systems with particular emphasis on the reasons for high cost and the methods required to drive down cost.

Maral, G. and M. Bousquet. 1993. *Satellite Communications Systems—Systems, Techniques and Technology, 2nd edition*. New York: John Wiley & Sons. 688 p. $89.95.
General text on essentially all aspects of communications satellite systems, including both spacecraft and ground stations. Extensive discussion of link analysis, transmission techniques, satellite networks, and communications payloads. Some discussion of support subsystems, the space environment, orbits, and reliability.

Morgan, W. L., and G. D. Gordon. 1989. *Communications Satellite Handbook*. New York: John Wiley & Sons. 900 p. $114.95.
Principally discusses communications satellite systems and multiple-access techniques. Has one of the best available discussions of satellite technology and spacecraft subsystems. (Covers all subsystems well, not just communications.) Extensive, practical discussion of orbits and viewing geometry appropriate for satellite communications.

Pisacane, V. L., and R. C. Moore. 1994. *Fundamentals of Space Systems*. New York: Oxford U. Press. 772 p. $84.95.
Defines the spacecraft design process as done at APL, a long-standing member of the low-cost spacecraft community. The book emphasizes the analytical and physical basis for designing spacecraft systems and subsystems, with some treatment of related issues such as operations, systems engineering, and the space environment.

Pocha, J. J. 1987. *An Introduction to Mission Design for Geostationary Satellites*. Dordrecht, The Netherlands: Kluwer Academic. 222 p. $95.50.
Extensive and complete discussion of orbits and orbit operations for geostationary missions; includes launch window, orbit transfer, stationkeeping, orbit propagation, tracking, and orbit determination.

Rechtin, E. 1991. *Systems Architecting*. New York: Englewood Cliffs, NJ: Prentice-Hall. 333 p. $74.95.
A broad discussion of the process of designing and building complex structures. Provides a general conceptual framework for systems engineering as applied to space systems, launch vehicles, aircraft, communications, and defense systems.

Shishko, 1995. *Systems Engineering Handbook.* Washington, D.C.: NASA. 154 p. free.
   Summarizes NASA's formal approach to systems engineering, systems management, trade studies, and defining and controlling the process for developing and producing space systems. Defines the terms and methods NASA uses to acquire most space systems.

Wertz, J. R. and W. J. Larson. 1999. *Space Mission Analysis and Design, 3rd edition.* El Segundo, CA: Microcosm Press and Dordrecht, The Netherlands: Kluwer Academic Publishers. 969 p. $49.75. (softcover)
   Referred to as SMAD throughout this book. Has become a standard reference for space mission engineering and the mission analysis and design process. Process-oriented with many data tables. Broad coverage of all mission aspects—including topics not covered in other volumes, such as mission engineering, mission geometry, orbit selection and design, cost modeling, and low-cost spacecraft. Also available: *Space Mission Analysis and Design Workbook.* El Segundo, CA: Microcosm. 172 p. $15.00 —Problems and solutions for all of the SMAD III chapters.

Williamson, M. 1990. *The Communications Satellite.* New York: Adam Hilger. 420 p. $88.95.
   A largely non-mathematical, yet technical introduction to satellites with an emphasis on communications satellites. Aimed primarily at undergraduates and young engineers. Discusses all of the spacecraft subsystems plus Earth stations and launch vehicles. Discussion of space insurance not found in most books. Includes an extensive case study on designing a geosynchronous direct-broadcast satellite.

Part II
Case Studies

Explanation of Facing Page Cost Data

A summary of actual and expected costs is given on the page facing the first page of each case study. All costs are in FY95$M. The source of this data is as follows.

Expected Cost

All expected costs are based on the SMAD cost model [Wong, SMAD, chap. 20]. The inflation factor from FY92 (for the SMAD model) to FY95 (RSMC tables) is assumed to be 1.106. These costs represent the best estimate of what the mission would have cost had it been developed under the traditional "rules of the game."

Spacecraft Bus: This is the sum of the non-recurring cost estimate (SMAD, Table 20-4, entry 2.0) and the theoretical first unit cost estimate (SMAD, Table 20-5, entry 2.0). Costs are based on the spacecraft bus dry mass. When the payload actual cost is included as part of the spacecraft cost (see below), then the same is done for the expected cost and the entire spacecraft dry mass is used.

Payload: Because the instruments are typically smaller than for traditional spacecraft, we have applied the spacecraft bus cost model defined above to the payload. This will be done as part of the bus or separately, depending on how actual costs are broken down.

Launch: Launch costs are based on the smallest appropriate launcher listed in SMAD Chap. 20. Launch vehicles used are listed in Table 1. **Note that the assumed launcher is from SMAD and may or may not correspond to the actual launcher used.** For launch cost estimation, any orbit above 60 deg inclination is modeled as "polar."

Ground Segment: All costs are ratios of the software cost as defined in SMAD Tables 20-8 and 20-9. Parameter assumptions are listed in Table 1.

Operations and Maintenance (Annual): Cost estimates from SMAD Table 20-10, using the parameters from Table 1. Maintenance cost is 10% of equipment + facilities + software = 0.10 * 1.99 * (software cost). Operations cost is based on contractor labor rates from SMAD Table 20-10, operations team size from SMAD Table 14-3 and the parameters in Table 1, and an assumption of 4.5 teams per spacecraft. (See SMAD Table 14-4.)

Actual Cost

All costs in the table are in FY95$M. Inflation factors are approximate based on middle of the development period. Currency exchange rates are approximate values at the time of development. Costs include non-recurring development cost (or proportional share of non-recurring cost for multiple spacecraft programs), plus recurring cost for building the first flight unit, plus launch cost, plus ground segment cost (if a new ground segment is built), plus operations cost for 1 year. Explanations of the cost for each specific mission are contained in the respective case study discussions.

Spacecraft Bus: Cost of the spacecraft bus, including the cost for systems and mission engineering, and spacecraft integration and test where these are separately identified. Includes payload cost for simple store-and-forward communications satellites and small test satellites.

Payload: Cost of the instruments where separately identified. Includes estimates of the value of instruments provided by outside organizations. (Where noted in the table, payload costs are included in the spacecraft bus cost.)

Launch: Cost of the launch and launch operations.

Ground Segment: Cost of the ground station facilities and equipment if acquired separately for the program.

Operations and Maintenance: Operations cost for first year of operations or for entire program if less than a year.

Small Spacecraft Model Cost

The intent of this column is to provide a cost estimate more in keeping with the small satellite philosophy, approach, and experience. To date, the only small satellite component for which a complete cost model has been developed is the spacecraft bus. (See Sec. 8.2.)

Spacecraft Bus: These cost estimates are from The Aerospace Corp. Small Satellite Cost Model described in Sec. 8.2. Many of the Case Studies are themselves part of the SmallSat cost model database.

Payload: Consistent with the "Expected Cost," the SmallSat spacecraft bus algorithm was also used for the payload and was computed separately or as part of the spacecraft bus, depending on how actual cost data was collected.

Launch: We assume that Small Satellites will follow a ride-sharing approach and, therefore, apply the cost/kg for the appropriate size launcher from Table 20-12 of SMAD. We have assumed a FY95 cost of Pegasus of $15.0 million.

Ground Segment: Most small satellites use predominately commercial software and equipment in existing facilities. There is not yet sufficient experience to establish a cost model for this component, so it has been left blank.

Operations and Maintenance (Annual): Small satellites have frequently streamlined and automated operations activities so as to reduce operations cost commensurate with reductions in spacecraft cost. Nonetheless, there has been insufficient data collection and analysis to date to establish a cost model, so this field has also been left blank.

TABLE 1. Data and Assumptions Used for Development of Case Study "Expected Cost" and "Small Spacecraft Cost Model." See text for explanation. Note that the launcher is the vehicle assumed based on the SMAD model and is not necessarily the launcher used or planned. Any orbit above 60 deg inclination is modeled as "Polar" for launch cost estimation.

Mission	Bus mass (kg)	Payload mass (kg)	Launch mass (kg)	Lines of Code (KLOC)	Language	Ops Complexity	Assumed Launcher	Assumed Orbit
Ørsted	47	13	60	25	C	Simple	Pegasus XL	Polar
Freja	141	73	256	150	Pascal	Standard	Delta 7920	Polar
SAMPEX	109	52	160	50	C	Standard	Pegasus XL	Polar
HETE	84	41	125	116	C	Standard	Pegasus XL	LEO
Clementine	226	8	1693	75	mix	Complex	Delta 7920	IP
Pluto Express	67	7*	103	n/a	n/a	Complex	Titan 4/Cent.	IP
RADCAL	92	in spc	92	60	mix	Standard	Pegasus XL	Polar
ORBCOMM	33	in spc	47	280	C	Standard	Pegasus XL	Polar
AO-13	84	in spc	140	12	Basic	Simple	Delta 6920	GTO
AO-16	9	in spc	9	4.8	C	Simple	Pegasus XL	Polar
PoSAT-1	41	9	49	28	C/Pascal	Simple	Pegasus XL	Polar

*Excludes 15 kg Russian Zond probe that will be released at Pluto.

Chapter 11

Science Missions

Reducing the cost of science missions has proven to be feasible in a number of scientific disciplines. In addition to the case study missions in the areas of magnetospheric exploration and high energy particles, recent reduced-cost, small, scientific satellites include:

- WIRE, Wide field Infrared Explorer, SMEX mission to be launched in 1998

- FAST, Fast Auroral Snapshot Explorer, near-term SMEX mission

- SWAS, Sub-millimeter Wave Astronomy Satellite, near-term SMEX mission

- TRACE, Transition Region and Coronal Explorer, near-term SMEX mission

- GFZ-1, German geodetic laser reflector, launched in 1995

- Astrid, Swedish magnetospheric satellite, launched in 1995

- BREM-SAT, German microgravity and particle experiment, launched 1994

- ALEXIS, American X-ray explorer, launched in 1993

- SROSS-C2, Indian gamma ray astronomy satellite, launched in 1992

- SARA, French radio astronomy satellite, launched in 1991

- AMPTE-UKS, UK magnetospheric subsatellite launched in 1984

Although a common mechanism for reducing mission cost is to reduce the number of payload instruments, most of the small science missions carry multiple instruments. Cost reductions frequently come from scaling back the size of the instruments and complexity of the associated support functions. Nonetheless, most of the reduced cost science missions have made new discoveries and contributed significantly to the advance of their respective fields. This may come about, in part, due to the potential for creating fast-paced, responsive programs that use relatively modern sensing and processing. For example, the Advanced X-ray Astrophysics Facility, AXAF, began conceptual design in the late 1970s and is scheduled to be launched in the late 1990s. In contrast, ALEXIS went from concept to launch in about 5 years.

Ørsted Summary

Ørsted was built by a consortium of Danish space companies, universities, and research institutions for the Danish government. It carries five magnetospheric and auroral research payloads.

Spacecraft dry mass:	60 kg	No propulsion system	
Average power:	31 W	TT&C:	256 kbps

Launch: "Piggyback" launch with the U.S. Air Force's Argos satellite on a Delta II from Vandenberg Air Force Base. Launched February 23, 1999.

Orbit: 450 km × 850 km at 96.1 deg inclination

Operations: All elements of the ground segment are in Denmark

Cost Model (FY95$M): See page 348 for explanation.

	Expected Cost	Small Spacecraft Model	Actual Cost*
Spacecraft Bus	$27.7M	$3.8M	$8.8M
Payload	$21.0M	$1.3M	$5.7M[†]
Launch	$16.6M	$4.0M	$2.0M
Ground Segment	$26.3M	—	$1.1M
Ops. + Main. (annual)	$2.4M	—	$0.8M
Total *(through launch + 1 yr)*	**$94.0M**	—	**$18.4M**

*An inflation factor of 1.000 has been used to inflate to FY1995$ [SMAD, 20-1].

[†] Includes $3.2M of science support that would not be a part of traditional mission budgets.

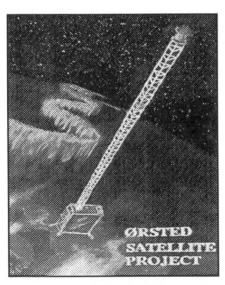
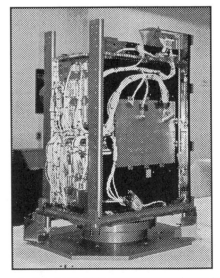

Fig. 11.1-1. Ørsted. This first Danish satellite uses its 8 m deployable magnetometer boom to enhance the magnetic cleanliness and provide gravity-gradient stabilization. The electronic boxes are mounted for easy access during integration and test.

11.1 Ørsted

Kim Leschly[*],
Computer Resources International

11.1.1   Focused Mission Objectives
11.1.2   Cost and Funding
11.1.3   Significant Precursors
11.1.4   Implementation Approach
11.1.5   Design Overview
11.1.6   Organization and Management
11.1.7   Conclusion

Ørsted is a small 60 kg Danish satellite, with the science objective of providing very accurate, state-of-the-art measurements of the geomagnetic field and global monitoring of high-energy particles in the Earth's environment. The project started as a cooperative effort among a group of Danish research institutions, universities, and space industries. The principal intent is to develop a national, well-focused, science satellite, in an area where the Danish science community has a strong interest and at a cost compatible with the limited resources of the Danish space program.

The total direct cost of the Ørsted satellite project is about $15 million (100 million Danish Kroner[†]), jointly funded by the Danish Government and the participating Danish industries. These costs include post-launch operation and science data calibration and analysis. Excluded are the indirect costs of some of the participating scientists provided by their respective research institutions, the launch provided by NASA, and two of the instruments provided by NASA and CNES. Including all the direct and indirect costs, the total price tag for Ørsted is about $20 million.

July 1993 marked the project start, following funding approval by the Danish Government and 2 years of pre-project studies. After 21 months of development, integration and test of the Ørsted satellite began in April 1995, almost 2 years[‡] before launch from Vandenberg Air Force Base, California, in March 1997. Ørsted is designed to be launched as a secondary payload on a Delta II launch vehicle, with the U.S. Air Force's 2,700 kg Argos satellite, into a low-Earth, elliptical, polar orbit.

Ørsted has generated a strong interest in the international solar-terrestrial and geomagnetic science communities. The ambitious and focused science objective and relatively short schedule are key motivating forces behind this first Danish satellite, which is named after the Danish Physicist Hans Christian Ørsted, 1777–1851, who discovered electromagnetism in 1820.

The selected design and management approach described in this case study may be unique to the Ørsted project. However, I believe it represents the can-do culture

---

[*] Currently with Jet Propulsion Laboratory.
[†] All costs have been converted from Danish Kroner to U.S. FY95$.
[‡] Including 6 months of storage.

that is so essential for making the successful shift towards lower-cost space missions, and as such it will hopefully provide encouragement for similar initiatives to follow. The worldwide interest in smaller, more focused satellite missions arises not only from a need to reduce cost and avoid putting all our eggs in one basket, but also because these missions enable a reduction in the time from initial concept to mission data return, hence revitalizing science enthusiasm, technical creativity, and management ingenuity.

11.1.1 Focused Mission Objectives

The primary science objective of the Ørsted mission is to provide very accurate measurements of the geomagnetic field and global monitoring of high-energy charged particles in the Earth's environment, for one year. The Ørsted science data will be used for studies in geomagnetic and solar-terrestrial physics, within five main research areas:

- Modeling of the geomagnetic field and its secular variation
- Properties of the fluid core and the electrical conductivity of the mantle
- Lithospheric structure and history
- Solar wind and magnetosphere interaction
- Magnetospheric field aligned currents and their relationship to ionospheric conditions

The science data from the magnetometers and the charged-particle detectors will be correlated with ground-based observations, including data from Denmark's extensive magnetic-research facilities in Greenland and similar facilities in Canada and other auroral regions in the northern hemisphere.

As a secondary science objective, Ørsted will collect GPS occultation measurements to obtain ionospheric electron density data, stratospheric temperature profiles, and tropospheric humidity profiles for weather models. The GPS data may also be used for studies of the Earth's gravity field, depending on how well the gravitational forces can be distinguished from the satellite drag.

The geomagnetic field is continuously changing in a way which is only poorly understood. It is believed to be generated by self-sustaining dynamo effects, driven by complex physical processes and fluid motion in the Earth's inner fluid core. Preferably, comprehensive and accurate mapping of the geomagnetic field should be done on a continuous basis. The recent report on the National Geomagnetic Initiative [National Research Council, 1993] suggests mapping at least every 5 to 10 years. Ørsted is long overdue: the first mission to provide geomagnetic field data with the required accuracy since NASA's Magsat in 1979–80.

Ørsted has generated considerable interest among internationally recognized solar-terrestrial and geomagnetic physicists. A total of 77 science groups from 17 countries have expressed their interest in participating in the planning and utilization of the Ørsted science data, in response to the Ørsted research announcement [Danish Meteorological Institute, 1993].

In addition to the above science objectives, the following underlying mission objectives have been defined for Ørsted:

- To contribute to the technological advancement of small satellites and demonstrate their feasibility as a means of performing well-focused, world-class space science exploration within a relatively short schedule and a cost conscious environment.
- To establish a forum for cooperation and mutual inspiration among Danish universities, research institutions, and industry in space-related basic research, education, and applied technology.

11.1.2 Cost and Funding

Ørsted's total cost of $20 million includes post-launch costs for flight operation and science data processing, as well as foreign, in-kind contributions covering launch and two of the flight instruments. Table 11.1-1 summarizes Denmark's direct costs for Ørsted, totaling about $15 million. The Danish government funds about 70% of this cost; participating Danish industries pay for the rest.

TABLE 11.1-1. Overview of Project Cost (in U.S. FY95$M). The cost includes post-launch operation and science data analysis, as well as indirect cost of researchers and industry contributions. Excluded are launch, supplied by NASA, and two science instruments, supplied by NASA and CNES at a cost of $1.0 M.

WBS	Project Element	Labor ($M)	Procurement ($M)	Total ($M)	Total (%)
1	Project management & planning, system engineering	$1.0M	$0.2M	$1.2M	8
2	Science management & planning, science data center & analysis	$2.0M	$1.2M	$3.2M	21
3	Science instrument development	$1.2M	$0.3M	$1.5M	10
4	Satellite development	$4.6M	$2.0M	$6.6M	43
	Structure & mechanisms	($1.1M)	($0.1M)	($1.2M)	
	Electrical power & cabling	($0.8M)	($0.9M)	($1.7M)	
	Command & data handling	($1.8M)	($0.2M)	($2.0M)	
	Attitude control	($0.6M)	($0.1M)	($0.7M)	
	Communication	($0.2M)	($0.6M)	($0.8M)	
	Thermal control	($0.1M)	($0.1M)	($0.2M)	
5	Ground segment development	$0.9M	$0.2M	$1.1M	7
6	Satellite integration & test	$0.9M	$0.1M	$1.0M	6
7	Flight operation	$0.7M	$0.1M	$0.8M	5
	TOTAL ($M)	**$11.3M**	**$4.1M**	**$15.4M**	—
	TOTAL (%)	**73%**	**27%**		**100%**

Funding provided by the Danish government is budgeted under three ministerial departments: the Ministry of Industry (60%), the Ministry of Research and Technology (30%), and the Ministry of Transportation (10%). The funding is provided with a fixed-budget ceiling, with no specific provisions for covering cost overruns, and no direct government customer involvement in designing, fabricating, and operating Ørsted. The customer role is vested with the Ørsted steering committee, which represents the six major participating organizations and is ultimately responsible for the mission. This unique arrangement sharply reduces administrative oversight and allows the project team to make critical technical and management decisions faster and more effectively than normal.

As for most conventional, one-of-a-kind space projects, the principal cost driver is labor, which amounts to 73% of the total cost. One of the main attributes which make Ørsted a low-cost mission is the relatively short schedule and the limited and empowered engineering team. The focused satellite and mission design goals limit the multitude of often complex and time-consuming design trades inherent to larger satellite missions. Numerous design trades and implementation choices have been made along the way, to keep the cost low and within the original budget.

Decisions about making or buying the various satellite elements were made at the onset of the project, based on simple cost trades and the availability of the required expertise within the group of participating organizations. Table 11.1-2 lists the selected supplier for all key satellite and ground system equipment. All external procurements were based on fixed price and schedule, with minimal special hardware performance testing and documentation requirements. Competitive bids were obtained internationally for all the major procurement items: solar panels, battery cells, RF transmitter-receiver, and GPS receiver. The prices for these items varied greatly, indicative of the dramatic difference between business-as-usual (high cost) and cost-conscious (low cost) proposals. Surveying the entire international market and not being constrained by national boundaries increased the available choices significantly, despite the added complication of long-distance communication and export licenses.

Currently, Ørsted is in integration and test. The satellite's overall electrical design has been verified through extensive interface testing with a complete set of subsystem engineering models, and the structural design has been qualified through modal testing, finite element modeling, and qual model vibration testing. The main cost uncertainty at this time is related to the recent 10-month launch slip (to March 1997), and the resulting stretch-out of the Ørsted work force and added temporary storage of the satellite following integration and test. If Ørsted is able to avoid further schedule delays, it will most likely be able to keep the total project cost to within 10% of its original $20 million budget.

11.1.3 Significant Precursors

Prior to the project start in July 1993, a series of events took place which were significant to the overall viability of the project. These significant precursors consisted of a mix of events initiated by the project participants and events reflecting the general trend in the international space community toward smaller and cheaper space missions. The main external precursors were:

- The availability of low-cost launch opportunities for secondary payloads on the Ariane Structure for Auxiliary Payloads, and on other launch vehicles—e.g., Delta II, Long March, and Proton.

- The technological advancement in microelectronics towards smaller and cheaper components with increased power efficiency and reliability.

- The successful launch and operation of several smaller satellites in the 10 to 50 kg class, initiated by various national and university-based organizations—e.g., AMSAT, University of Surrey, University of Berlin, and others.

TABLE 11.1-2. Equipment List and Suppliers. The make or buy decisions were made at the start of the project, based on cost and the available Danish expertise. M (making) indicates elements provided by one of the participating organizations listed in Table 11.1-5. B (buying) indicates external procurements. C (contribution) indicates in-kind contributions.

Acquisition			
Subsystem	**M**	**B**	**C**	**Supplier or Subcontractor**
Overhauser magnetometer			X	CNES/Leti, Grenoble, Fr.
CSC fluxgate magnetometer Electronics fabrication	X		X	Electro Physics Dept., DTU, Lyngby, Den. Royal Tech. University, Stockholm, Swed.
Charged-particle detectors	X			Danish Meteorological Inst., Copenhagen, Den.
Star imager Optical elements Optical bench (SiC)	X X	X		Electro Physics Dept., DTU, Lyngby, Den. Copenhagen Optical Co., Copenhagen, Den. Inst. of Engineering Design, DTU, Lyngby, Den.
GPS TurboRogue			X	NASA/JPL, Pasadena, CA, U.S.
Structures, solar panel substrate	X			Per Udsen Co, Grenå, Den.
Mechanisms and boom Separation mechanism	X		X	Per Udsen Co, Grenå, Den. NASA/McDonnell Douglas, Huntington, CA, U.S.
Electrical power Solar panel (GaAs) Battery assemblies (2x6 cells) Battery cells (NiCd)	X X	X X		Terma Electronic, Lystrup, Den. FIAR, Milan, Italy Innovision, Odense, Den. Gates, Gainesville, FL, U.S.
Command and data handling Flight software (ADA)	X X			Terma Electronic, Lystrup, Den. Computer Resources Intl., Birkerød, Den.
Attitude control (software) Electronics and coils	X X			Aalborg University, Aalborg, Den. Innovision, Odense, Den.
S-band receiver/transmitter Antennas GPS TANS receiver Design support	 X	X X X		Satellite International Ltd., Guildford, U.K. Electromagnetic Inst., DTU, Lyngby, Den. Trimble, Sunnyvale, CA, U.S. Rescom, Vedbæk, Den.
Cabling (satellite & boom) Cable wiring Cable connectors (mini D)	X	 X X		Terma Electronic, Lystrup, Den. Habia, Stockholm, Swed. Cannon
Thermal control (MLI) Thermal analysis	X X			Per Udsen Co., Grenå, Den. Dept. of Electro Physics, DTU, Lyngby, Den.
Magnetic calibration (instrument) Satellite calibration		 X	X	Technical University, Braunschweig, Ger. IABG, Munich, Ger.
Ørsted control center	X			Computer Resources Intl., Birkerød, Den.
Science data center	X			Danish Meteorological Inst.,Copenhagen, Den.
Ground station (Copenhagen) Aalborg ground station Ballerup ground station RF equipment (common)	X X X	 X		Danish Meteorological Inst., Copenhagen, Den. Aalborg University, Aalborg, Den. Copenhagen Engineering College, Ballerup, Den. Information Proc. Syst., Belmont, CA, U.S.

In early 1991, Danish universities and industry formed a study group to determine the feasibility of building and operating a small Danish satellite. Although other mission objectives were considered, it was decided early on to focus on a geomagnetic mapping mission, due to the strong interest and experience in this field within the Danish science community, including prior instrument experience from sounding rocket flights. Together with the industrial experience base from previous and ongoing space projects, mainly within ESA, the study group organizations encompass nearly all aspects of developing, integrating, and operating Ørsted. Based on the resulting *Ørsted Satellite Project Feasibility Study* [CRI, 1991], the Danish Space Council provided partial funding of approximately $200K for a one-year system definition study.

The Ørsted system definition study was focused on defining the overall system design for the satellite and ground segment. Two external reviews, a science review in March 1992 and a system design review in October 1992, involved an international review board to provide an independent, expert evaluation of the scientific merits and technical feasibility of the project. The reports from both reviews were very supportive, based on the project's scientific significance and timelines, as well as its overall technical feasibility and demonstrated teamwork. The main technical outcomes of the system definition study were high-level specifications for Ørsted and the ground segment, detailed reports from each of the technical study areas (including key design trade-off studies), and a draft project implementation plan. A *Summary Report for the Ørsted System Definition Study* was also produced [CRI, 1993]. These documents formed the basis for the detailed budget and funding plan which led to funding approval by the Danish Parliament in June 1993.

In October 1992, NASA decided to provide a launch opportunity for Ørsted as a secondary payload on the Delta II launch of the U.S. Air Force's Argos mission. The decision was motivated by the strong alignment of the Ørsted science objectives with the geopotential fields mission program within the NASA Solid Earth Science Branch. The official letter of agreement between NASA and the Ørsted project was finalized in the summer of 1993. Agreements were also made with CNES to provide an Overhauser scalar magnetometer, and with the Jet Propulsion Laboratory to provide a TurboRogue GPS receiver.

The one-year system definition study also helped to establish and solidify the teamwork essential to the overall project viability. Nearly all the key technical people who contributed to the system definition study stayed on after the project start. However, one of the initial industrial companies decided to withdraw from the project a few months after funding approval, creating a minor panic and ultimately a slower than preferred project start-up due to replanning of the affected work areas.

11.1.4 Implementation Approach

The diversified institutional cultures inherent in a project with so many different organizations made the need for a project implementation plan paramount. This plan, developed partly before the project start, defines the organization and management approach, the project master schedule, the general design guidelines, and the project work breakdown structure. The plan includes a one-page description of each work element, identifying the key deliverables and receivables and their due dates. The plan provided an essential focus on "who is doing what and when" and enabled the detailed planning within each organization to be consistent with the rest of the project.

The Ørsted project master schedule defines the major project phases and milestones consistent with the planned launch in March 1997 (Fig. 11.1-2). It is kept simple, to imprint it in everybody's mind and ensure that everyone works to the same overall schedule. This is particularly important for a project with a relatively short schedule, where slipping of significant interim milestones may jeopardize the entire project. The schedule has only a little slack, which in some cases has directly driven the design decisions. A more detailed 15-page schedule containing all interim milestones identified within each work element tracks the month-to-month progress.

	ØRSTED PROJECT MASTER SCHEDULE	1993			1994				1995				1996				1997				1998		
	Project Phases and Milestones	Q2	Q3	Q4	Q1	Q2	Q3	Q4	Q1	Q2	Q3	Q4	Q1	Q2	Q3	Q4	Q1	Q2	Q3	Q4	Q1	Q2	Q3
1	**Subsystem Development and Test**																						
2	Research Announcement				▲																		
3	Satellite System CDR				▲																		
4	Science Mission CDR					▲																	
5	Ground System & MO CDR							▲															
6	Satellite Subsystem Interface Tests																						
7	Satellite Integration & Test Review								▲														
8	1st Int'l Science Team Meeting									▲													
9	Subsystem Qual-tests & Burnin																						
10	Subsystem Hardware Delivery																						
11	**Satellite Integration and Test**																						
12	Mechanical Integration																						
13	Electrical Integration																						
14	Functional Testing												□										
15	Dynamic Testing and Burnin													□									
16	Thermal-Vacuum and Boom Testing													□									
17	Magnetic Calibration													□									
18	Ground System Test													□									
19	Launch and Mission Readiness													△									
20	**Satellite Storage (6 months)**																						
21	**Ship and Launch**																						
22	Preship Review															△							
23	**Flight Operation (primary mission)**																						
24	Launch																△						
25	Science Observation																						

File: C:\WINPROJ\ORST1.MPP	Activity	Progress	Milestone △	Completed ▲
Date: 8/22/95				

Fig. 11.1-2. Project Master Schedule. The original tight schedule (planned for an October '95 launch) has been stretched due to the 17-month launch delay of the Argos mission to March 1997.

Documentation is kept to a minimum. The high-level specifications developed during the system definition study are contained in a single ring binder, the project document book. Everybody working on the project has a copy and the project office keeps these key documents current. Detailed design documents and drawings are generated by the organizations responsible for the associated hardware and software, and copies are placed in a project design file together with other pertinent technical notes, accessible to everyone on the project.

The following list of implementation guidelines summarize the selected approach and are intended to keep the Ørsted design focused and within cost and schedule:

* *Keep the design simple.* Use off-the-shelf equipment and conservative and known design solutions, rather than having to embark on expensive and time-consuming analysis or developmental testing to prove a new design.

- *Don't try to fool Mother Nature.* It can't be done. Make sure that the underlying physics of the design is well understood.

- *Use common sense*, rather than relying on formal schemes for evaluating the overall reliability and risk rating (e.g., FMECA). The designer's time is better used on making the design work than on providing reliability numbers.

- *Communicate frequently with the other team members.* Ask questions and discuss issues openly and often. Cross-fertilization creates surprising new ideas. If somebody else has a bright solution to one of your design problems, use it. Don't let the not-invented-here syndrome get in the way. Bimonthly project design team meetings provide an excellent arena for this type of interaction, including lunch breaks and hallway conversations.

- *Eliminate multiple design margins.* For example, the mass and power estimates provided by each subsystem reflect best estimates, allowing all the margin to be maintained on the project level.

- *Define and freeze the detailed subsystem interface design at the start of the project.* It took the full-time dedication and hard work of two brave system engineers to lead this task to completion, and resulted in surfacing and resolving many design issues which would otherwise have been overlooked or ignored until much later in the design process, when fixes are so much more expensive.

- *Use the best components available within cost and schedule.* Use space-qualified hi-rel parts when readily available. Otherwise, use MIL-SPEC parts, or high-grade, high-volume industrial parts with a dependable production history and 1,000 hours burn-in. Derate parts according to ESA standard PSS-01-301 [ESA, 1992].

- *Test the hardware design and interfaces whenever possible*, provided the test is quick, easy, and dirt-cheap (affectionately labeled QED testing). One example is the early vibration testing of the star imager hardware, using a commercial sander as a vibration table. Testing must be executed with care to avoid running the wrong test. Explaining why bad test results have to be discarded can sometimes be very time-consuming and costly.

- *Check the magnetic cleanliness along the way.* Inadvertent use of highly magnetic components could be very costly to replace late in the assembly process. Every fabrication area is equipped with a sensitive magnetic moment monitor for easy inspection of flight hardware.

- *Use common software for satellite ground testing and control center operation*, as much as feasible, by using the same software platform and database and co-locating the involved software and operation engineers.

- *Adhere to the local company quality assurance standards and procedures as much as possible*, to ensure familiarity and prevent retraining of QA personnel.

- *Follow deliverables through integration and test.* Flight hardware and software elements are not considered delivered by their respective organizations until fully integrated and tested on the satellite.

11.1.5 Design Overview

The Ørsted mission design life is 14 months, allowing for up to two months for initial on-orbit satellite commissioning, including boom deployment, check-out, and instrument calibration. This relatively short mission duration imposes a less stringent reliability criteria on the flight hardware consistent with the allowed use of commercial-grade parts and basically single-string design. As a goal, the mission may be extended to 3 years.

Ørsted will be launched into a 450 km × 850 km elliptical polar orbit with a nearly Sun-synchronous inclination of 96.1 deg. The orbit has a slow nodal drift rate of 0.77 deg/day and a corresponding shift in local time from about 2:30 pm to 9:00 am over the 14-month mission. Following separation from the Delta II launch vehicle, the satellite attitude will be stabilized using magnetic torquers to ensure adequate solar illumination and battery charging, before powering on the transmitter-receiver unit and establishing initial radio communication with the ground stations in Denmark. When the satellite is stabilized and all subsystems appear to operate as expected, the lightweight, 8-m magnetometer boom will be deployed in response to ground command. The deployed boom provides adequate separation between the sensitive, boom-mounted magnetometers and the satellite body, as well as ensuring gravity-gradient stabilization of the satellite attitude, with the boom pointing away from the Earth, as shown in Fig. 11.1-3.

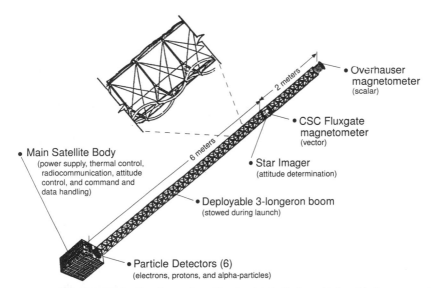

Fig. 11.1-3. Satellite Flight Configuration. The Danish-built, 8 m deployable boom provides adequate separation of the magnetometers and the satellite body, to ensure magnetically clean measurements.

Ørsted weighs 60 kg and provides, on average, 31 W to power its onboard instruments and supporting subsystems. It contains five science instruments and a set of conventional satellite platform subsystems: structure, mechanisms, electrical power, command and data handling, attitude control, radio communication, and thermal control. The satellite's main body is box-shaped, with rigidly mounted solar panels on five sides, as depicted in Fig. 11.1-4. The bottom side contains the satellite separation mechanism and the two turnstile antennas used for radio communication with the ground.

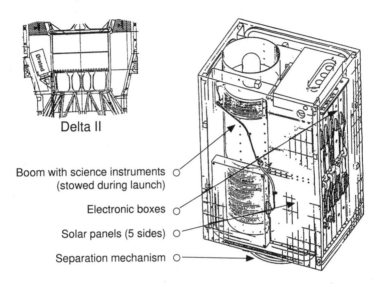

Fig. 11.1-4. Satellite Cut-away View (Launch Configuration). The H-beam, primary structure divides the satellite in two main compartments. One contains the stowed boom, and the other holds the satellite electronics. Solar panels are body-mounted on four sides and the top. The attitude is controlled by 3-axis magnetorquer air-coils.

The satellite key design features and the subsystem mass, power, and data rates, are listed in Tables 11.1-3 and 11.1-4. The satellite functional block diagram is shown in Fig. 11.1-5. Several papers describing various detailed aspects of the Ørsted design, have recently been published [Nielsen, et al., 1993; Liebe, 1993; Baron, 1994; and Jørgensen, 1994].

Science Payload

The complement of science instruments consist of two magnetometers and a star imager to map the geomagnetic field, six high-energy charged-particle detectors to measure the charged-particle environment, and a dual-band GPS receiver to obtain atmospheric temperature profile data. In total, they weigh about 13 kg (22% of the satellite mass) and they use about 10 W (31% of the average available power).

The two magnetometers and the star imager are mounted on the deployable boom, away from the satellite body, to limit the magnetic disturbance from the satellite as much as possible. (See Fig. 11.1-3.) In addition, the satellite is built with minimal use of magnetic material. The charged-particle detectors and the TurboRogue GPS receiver are located on the main body of the satellite. A second multichannel Trimble GPS TANS receiver with a resolution of 50 m or better determines position for the magnetic field measurements. The TurboRogue GPS has a separate low-power mode which can provide position data with reduced accuracy (similar to the GPS TANS accuracy) as a backup to the GPS TANS.

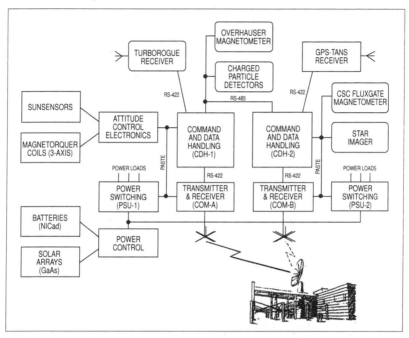

Fig. 11.1-5. Satellite Functional Block Diagram. Ørsted is mostly single-string, except for the two fully redundant S-band transmitters and receivers, and the partially redundant computers (CPU and memory) and power switching.

The key to providing the required high-accuracy vector measurements is the co-location of the fluxgate magnetometer and the magnetically clean star imager on a stable optical bench. This unique configuration avoids the complexity and inherent inaccuracies due to boom oscillations resulting from a conventional attitude-transfer system with a boom-mounted magnetometer and a satellite-mounted star imager. The design and calibration of the Magsat attitude-transfer system was both technically challenging and costly.

The five instruments all use state-of-the-art technology, and none of them have flown before. They are:

- *Overhauser magnetometer*, a proton-precession scalar magnetometer with a measurement accuracy of one nT and a data rate of one sample/sec. It contains coils for proton-resonance excitation and detection, and a resonator for electron-spin resonance pumping of a nitro-oxide solution. The Overhauser magnetometer is mounted on the tip of the boom to minimize the disturbance from the rest of the satellite.

TABLE 11.1-3. Key Design Features. Ørsted is a low-Earth orbit, gravity-gradient stabilized satellite with a design life of 14 months. Co-location of the vector magnetometer and the star imager on the 8 m long deployable boom provides the required high-accuracy angular resolution for the magnetic field measurements, critical to the mission success.

Element	Design Feature
Body size	• Height: 70 cm; base: 45 cm × 34 cm
Mass	• Total mass: 60 kg, including 13 kg science payload
Science payload	• Fluxgate vector magnetometer, with 1 to 2 nT accuracy (10/100 samples/sec) • Overhauser scalar magnetometer, with 1 nT accuracy (1 sample/sec) • Star imager, with 20 arcsec accuracy (1 sample/sec) • Solid-state charged-particle detectors, measuring electrons (30 keV to 1 MeV), and protons and alpha-particles (200 keV to 100 MeV) • TurboRogue GPS P-code receiver, with 5 to 10 cm accuracy (50 samples/sec)
Structure	• H-beam primary structure and mounting platforms, resonance frequency >70 Hz
Mechanisms	• 8-m deployable boom, which is coiled up and latched during launch • Mechanisms for satellite separation, boom release, and boom deployment
Electrical power	• User load: 31 W; provided by body-mounted GaAs solar panels (64 W output) • NiCd batteries: 2 packs of 6 cells each, with 6 A·hr capacity total. 20% DOD • Power regulation at ±5 V, ±8 V, +15 V, and +16 V unregulated.
Command and data handling	• Main computer with two 80C186/16 MHz CPUs • Mass memory (16 MB) for storage of up to 13 hours of science data • ESA standard packet format for telemetry and commands • Autonomous fault protection for selected failures • Software (ADA code) development based on formal methods
Attitude control	• Gravity-gradient attitude stabilization using the 8-m deployable boom • Autonomous attitude correction using 3-axis magnetic torquer coils • Attitude determination using vector magnetometer, Sun sensors
Position determination	• GPS receiver (Trimble TANS), with resolution of 50 m or better (3σ) • TurboRogue GPS has a low-power mode with same resolution (as backup)
Communication	• S-band transmitter (2.2 GHz) and receiver (2.0 GHz) • Left-hand circular polarized and BPSK modulated • Downlink at 4 or 256 kbits/sec, using Reed-Solomon and convolutional encoding • Uplink at 1 or 4 kbits/sec, using BCH encoding • Transmission of data 3 to 4 times per day
Cabling	• Non-magnetic miniature Cannon-D connectors
Thermal control	• Passive thermal control, using MLI blankets and thermal control paints
Orbit	• Elliptical polar orbit: 450 km × 850 km and 96.1 deg inclination • Orbit period: 97 minutes, with maximum eclipses of 35 minutes per orbit • Slow nodal drift rate: 0.77 deg per day, providing a shift in local time from 2:30 pm to 9:00 am during the 14-month mission
Ground software	• Significant (90%) commonalty between test and operations software (EasyMap)

TABLE 11.1-4. Satellite Allocations. The low mass, power, and data-rate is consistent with current technology and the launch constraints of secondary payloads.

Subsystem	Mass kg	Power W (avg)	Science Data Rate bits/sec (avg)	bits/sec (peak)
Overhauser magnetometer	2.5	2.8	20	20
Fluxgate magnetometer	1.8	1.2	1,600	5,400
Charged-particle detectors	1.9	1.5	480	1,700
Star imager	4.6	5.6	90	90
GPS TurboRogue[1]	2.1	(8.5)	(480)	(480)
Structures and solar panels	12.0	—	—	—
Mechanisms and boom	4.2	—	—	—
Electrical power	9.5	2.0	—	—
Command and data handling	4.2	6.7	—	—
Attitude control	4.1	1.4	—	—
Communication and GPS (TANS)	5.8	9.0	90	90
Cabling	3.1	0.3	—	—
Thermal control	0.5	0.0	—	—
Margin	3.7	0.5	—	—
TOTAL	**60.0**	**31.0**	**2,280**	**6,820**

[1]The GPS TurboRogue will be powered on only during periods with adequate power margin.

- *Vector magnetometer*, a compact spherical coil triaxial fluxgate vector magnetometer with a measurement accuracy of 1 to 2 nT and a data rate of either 10 or 100 samples/sec. It consists of three orthogonal sensors, each containing a small ferromagnetic ring core with an excitation winding. The fluxgate creates a true null field at the three ring-cores, compensating the homogeneous external field in three orthogonal directions. The sensor coils are placed on a spherical surface and are concentric along the three orthogonal axes.

- *Star Imager*, a wide-angle CCD camera with a 30 deg field of view, providing attitude determination for the vector magnetometer measurements with an accuracy of less than 20 arcsec and a data rate of 1 measurement/sec. The star imager measures the attitude by matching star constellations of magnitude 4.5 or brighter in the camera's field of view with an onboard star catalog. The 752 × 582 pixel CCD is passively cooled and operates with an integration time of about 0.8 sec. The pixel size is 8.6 × 8.3 micron. The star imager uses a 80C486 type CPU for image processing and attitude determination.

- *Charged-particle detectors*, consisting of six solid-state, high-energy particle detectors, to detect electrons with energies from 30 keV to 1 MeV, and protons and alpha-particles with energies from 200 keV to 100 MeV. The sampling rate ranges from 6 to 300 samples/min. The detector package contains two collimated electron detectors, two collimated proton/alpha detectors, and two wide-angle proton and alpha detectors.

- *TurboRogue GPS receiver*, a dual-band, eight-channel GPS receiver, with a position measurement accuracy of 5 to 10 cm and a data rate of 50 samples/sec. It uses the P code on L1 and L2 frequency carrier for occultation measurements of radio signals from GPS satellites. The TurboRogue also provides velocity data within ±0.5 m/s and UTC time within ±1.0 μs. It can operate in a low-power mode, using only the C/A code on the L1 frequency carrier with a reduced position accuracy, as a backup to the GPS TANS receiver. The zenith- and aft-facing ceramic patch antennas are mounted on the satellite body.

Structures

The primary load-carrying structure is designed to accommodate the Ørsted science instruments and supporting subsystems within the constraints of the launch-vehicle envelope and dynamic environment. It consists of a sheet metal aluminum H-beam structure, with upper and lower platforms. The solar cells are mounted on aluminum honeycomb substrates which attach to the four sides and the top of the satellite structure.

The central web of the H-beam structure divides the satellite into two major vertical compartments. One side is occupied by the boom and boom canister. The other side contains two rows of standardized electronic boxes separated by a cable tray, providing easy access to each subsystem during satellite integration and test. Each electronic box holds two to three large circuit boards with a connector panel facing out. In addition to the structural stiffness requirement (resonance frequency > 70 Hz), the principal design drivers for this configuration were to

- Optimize the utilization of the allowed envelope, dictated by launch vehicle
- Maximize the boom diameter and thus the boom-deployment torque
- Accommodate boom instruments within the boom diameter and envelope
- Provide easy access to subsystems during integration and test

Mechanisms

The lightweight, 8-m deployable boom consists of three coilable longerons, radially separated by spacers and stiffened by cross-wires that attach to the longerons approximately 10 cm apart. During launch, the boom is coiled up inside the boom canister, occupying 3 to 4% of its extended length. This hingeless spacer-longeron design acts as a spring which provides adequate deployment forces without the need for additional mechanisms, except for a retaining wire in the center of the boom, which is slowly released by a small motor to control the deployment rate. The boom deployment takes 5 to 10 minutes. This relatively stiff, lightweight, and thermally stable boom design was chosen after carefully studying a multitude of boom designs, ranging from stiff, mechanically intricate, scissors booms to flexible, telescoping booms. The boom design was one of the key design uncertainties early in the project.

Since one of the participating companies had the required advanced materials experience needed for making this boom it was decided to embark on making the boom in Denmark, and save precious procurement funds. The risk of this new development was considered moderate because a similar boom design had flown successfully on the Japanese Akebono satellite, launched in 1989.

Cables to the boom-mounted instruments attach to each of the three longerons, eliminating concern about entangling the cables during deployment. Non-explosive initiators release the boom hold-down latches, in response to ground commands.

The separation mechanism, which will separate the satellite from the launch vehicle when in orbit, is a clamp-band design. The mechanism is provided by the launch vehicle manufacturer, McDonnell Douglas, and activation is controlled by the launch vehicle.

Electrical Power

The electrical power subsystem design is based on conventional space technology, using GaAs solar cells, NiCd batteries, and two partially redundant power control units. The solar panels are assembled from standard 2 cm × 4 cm GaAs solar cells, with an average conversion efficiency of 18%. Careful material selection and conventional back-wiring techniques reduce the magnetic fields generated by the panels in order to avoid disturbing the magnetic-field measurements. The two battery packs contain six cells each and have a total capacity of 6 A·hr, providing enough power during eclipses using 20% depth of discharge. The power control electronics provides battery charge regulation, DC/DC power conversion, and power distribution for all the satellite users. Regulated power is provided for ±5 V, ±8 V, and +15 V. The unregulated bus operates nominally between 13.2 V to 17.4 V.

Command and Data Handling

The satellite command and data handling uses two identical onboard computers, each containing an Intel 80C186 16 MHz CPU with prior flight experience from other small-satellite projects. Each computer handles the commands and telemetry from various instruments and subsystems using three different data buses, as shown on the functional block diagram in Fig. 11.1-5. The approach of using multiple data buses appears less optimal even though it eases the demand for intelligence within each of the satellite's other instruments and subsystems. The satellite data is collected and stored in the 16 MB of solid-state RAM, sized to hold at least 13 hours of science data. The data is downlinked three or four times per day to one of the ground stations in Denmark. Based on in-house experience, the Ørsted Project chose to develop the flight software in ADA code using formal methods and rapid prototyping techniques to allow early testing and verification of the software code and to reduce the risk and cost associated with changing the code late in the development cycle. The memory is protected against single event upset using hardware Error Detection And Correction (EDAC) circuits.

Onboard autonomy is implemented for selected failure modes critical to the satellite's survival, such as low power, loss of command link, and temperature extremes. All of the satellite's application software can be reconfigured and uploaded from the ground. Both telemetry and command uses the ESA's standard packet formats.

Attitude Control

Ørsted is gravity-gradient stabilized, using the zenith pointing 8-m boom. In addition, three sets of orthogonal magnetic torquer coils are used in a high power mode to initially detumble and stabilize the satellite following launch vehicle separation. Later, once the boom is deployed, the torquer coils are used in a low power mode to maintain a fixed angular position of the satellite relative to the boom axis (within ±10 deg in yaw), and ensure optimal illumination of the solar panels for power and preclude Sun exposure of the star imager. Once the satellite is stabilized, the yaw angle correction is kept to a minimum, even though the disturbance to the magnetic measurements due to activation of the torquer coils in this low-power mode is almost negligible (less than 1 nT) and well within the error budget established for the vector measurements (4.5 nT RSS).

To provide adequate visibility for the bottom-mounted antennas, the satellite bottom must stay nadir pointed to within ±5 deg in pitch and roll, which is within the normal pendulum-like movements of gravity gradient stabilized satellites. The magnetic-torquer coils are sized to be able to invert the satellite with the boom deployed if the satellite is captured upside down as can happen with small gravity-gradient stabilized satellites when the boom is first deployed.

The satellite's attitude is determined by eight hemispheric Sun sensors mounted at four of the satellite's corners. These sensors provide attitude information to within 1 to 2 deg during the sunlit portion of the orbit, which is accurate enough to control the satellite body to within the above attitude deadband. The star imager or the vector magnetometer measurements, in conjunction with the onboard map of the geomagnetic field, may also be used to determine the satellite attitude. Position determination is provided by either of the two GPS receivers.

The attitude control software resides within one of the two onboard computers. The attitude control laws are tested in a satellite simulator laboratory environment.

Communication

Telemetry and commands are communicated by either of two redundant S-band transmitter-receiver units, when the satellite is in view of one of the ground stations. The telemetry downlink uses Reed Solomon and convolutional encoding with a maximum rate of 256 kbits/sec and RF output power of 1 W. The uplink uses Bose-Chaudhuri-Hocquenghem encoding with a maximum rate of 4 kbits/sec. Power is preserved by switching to a low-power mode when not in view of one of the Ørsted ground stations, and continuing to transmit a satellite-heartbeat signal containing spacecraft ID, position, and key status information.

Two turnstile circular polarized antennas mount on the satellite's bottom, nadir facing, for communication with the ground stations. The GPS TANS antenna mounts on the satellite's top, zenith facing.

Thermal Control

The satellite's thermal design is based on passive thermal control using multilayer insulation blankets and radiative surfaces and paints. The boom-mounted instruments have small heaters for infrequent use, including decontamination of the star imager's optics. All heat-dissipating equipment is thermally coupled to the satellite's primary structure, except the boom-mounted instrument sensors. Most of the satellite equipment is designed to operate within –20 °C to 40 °C, except for the battery, which prefers a narrower temperature range of 0 °C to 25 °C. The solar panels are thermally coupled to the satellite body. The satellite body's internal temperature is expected to vary from about –5 °C to 15 °C, whereas the solar panel will vary from about –40 °C to 70 °C. Thermal modeling of the satellite and the boom instruments uses ESARAD and ESATAN software.

Partial Redundancy

The Ørsted design is driven to a large extent by the severe power and mass constraints inherent in small satellites. One significant consequence of these constraints is the satellite's single-string design. However, partial (smart) redundancy is implemented to improve the reliability in selected mission-critical areas, both on the component level (e.g., the A/D converters in the fluxgate magnetometer electronics) and on the box level (e.g., transmitter-receiver, GPS receiver, and torquer coils). Another example is the star imager, which may also operate using the main satellite computer for image processing to provide attitude information at a degraded rate, in case the star imager's CPU fails. The phrase "real Danes do not depend on redundancy" was coined during some of the more heated design trade-off discussions concerning redundancy.

Launch Vehicle Interfaces

As a secondary payload, there is a need to keep the launch vehicle interfaces and interaction to the essential minimum. Ideally, safety should be all that matters. On Ørsted, the overriding concern about the launch vehicle is to ensure structural integrity of the satellite's primary structure and the boom-latch restraint during launch. Since the Ørsted satellite separation system is provided by the launch vehicle, the project has only limited involvement with its design. However, the work needed to provide the required documentation, analysis, and test results specified in the *Delta Launch Vehicle Secondary Payload Planners Guide for NASA Missions* [GSFC, 1993], should not be underestimated. Launch vehicle interface meetings are scheduled every 6 months, with follow-up telecons about once a month.

Satellite Testing

Ørsted is a protoflight satellite, meaning that only one complete satellite is fully assembled, tested, and launched. To reduce the risk of failures during satellite integration and test, box-level qualification testing is implemented for most of the subsystem elements. Also, flight spares are available for almost all flight hardware (unless they're too expensive) to reduce the schedule risk in case of unit failures during the relatively short integration and test phase.

A complete satellite subsystem interface test, using engineering models, is implemented in advance of the protoflight satellite integration. This test is intended to verify the electrical interface design (including EMI), and hence reduce the risk of finding potential interface problems later, during satellite integration, when resolving them would take much more time and threaten the project schedule. After the planned 9-month satellite integration, the assembled protoflight satellite will undergo environmental testing, including dynamics, solar thermal-vacuum, and magnetic calibration at suitable test facilities in Denmark and Germany.

Ground Segment and Operations

The Ørsted ground segment consists of a control center, which will communicate with the satellite through either of three ground-station antennas (one prime and two back-up) and transfer the data products to a science data center. All elements of the ground segment are located in Denmark. Within two months after launch, the Ørsted science instruments will begin to collect science data. The satellite science and house-keeping data will be downlinked three or four times per day.

The Ørsted control center will have two or three full-time operators to uplink routine command sequences (about once per week), archive and distribute data, monitor the satellite's health, and analyze and predict flight dynamics. The operational software is developed on the same platform (EasyMap) as the satellite test software, ensuring a high degree (90%) of commonality. Key people from the satellite development team will be on call, in case of anomalies. The Ørsted science data center will provide quick-look science data, as well as calibrated science data products for distribution to the various science investigators, over the Internet and on CD-ROM.

11.1.6 Organization and Management

The 12 organizations participating in the Ørsted project, and their respective principal areas of responsibility are listed in Table 11.1-5. They are an almost even mix of space-related industry, science institutions, and technical universities. The industry group is responsible for the satellite platform development, integration, and test, as well as mission operation and the overall project management. The science group is responsible for the instrument development and the overall science planning, calibration, and analysis. Finally, the technology group provides selected design support and two of the ground stations.

The strong science participation from the very start has been essential to the over-all satellite and mission design process, involving the trade-off between often conflicting requirements and constraints. Similarly, the diverse industry participation has been important to ensure that the satellite design, fabrication, and testing is based on proven industrial practices and cost-effective approaches.

The day-to-day project management and planning is provided by the lead industrial company Computer Resources International, while all science related activities are managed by the project scientist, located at the Danish Meteorological Institute. The major management challenge is to be a good team builder and to ensure close coordination between the geographically distributed project participants. Colocation of key technical people is not considered paramount, except during satellite integration and testing. Actually, it may be counterproductive, because it removes lead people from the more critical day-to-day work at their home base. The personal hardship associated with temporary extended relocation can be detrimental to morale. Telephones, telefaxes, and E-mail are a magnificent and important medium for this type of distributed project organization. Denmark is fortunately small enough to be contained in one time zone, which allows one-day trips to meetings anywhere in the country.

TABLE 11.1-5. Key Project Organizations and Principal Responsibilities. Ørsted is a cooperative effort between Danish industry, science, and research institutions, encompassing nearly all aspects of developing, integrating and operating the satellite. International support is provided by NASA, CNES, ESA, and others.

Industry	• Computer Resources International A/S*:
	   − Project management and planning, external interfaces, system engineering
	   − Flight and ground support equipment software, S-band radio & GPS receivers
	   − Satellite integration and test, Ørsted control center and flight operation.
	
	• Per Udsen Co. Aircraft Industry A/S*:
	   − Primary structure and analysis, solar panel substrates, boom & boom canister
	   − Boom release/deployment mechanism, mechanical ground support equipment
	
	• Terma Electronic A/S*:
	   − Command and data handling computers, and electrical power subsystem
	
	• Innovision A/S:
	   − Attitude control electronics, magnetorquer coils, and battery modules

Science	• Danish Meteorological Institute*, Solar-Terrestrial Physics Department:
	   − Science management and planning, International science working groups
	   − Science data center, charged-particle detectors, Copenhagen ground station
	
	• University of Copenhagen, Geophysics Department at Niels Bohr Institute:
	   − Main field analysis, science calibration
	
	• Danish Technical University, Department of Electro Physics*:
	   − CSC fluxgate magnetometer, star imager, Overhauser magnetometer (support)
	   − Instrument calibration tests, thermal modeling
	
	• Danish Space Research Institute:
	   − Project engineering support

Tech-nology	• Aalborg University, Institute of Electronic Systems*:
	   − Attitude control software, Sun sensors, Aalborg Ground Station
	
	• Copenhagen Engineering College, Department of Electronics:
	   − Ballerup Ground Station
	
	• Danish Technical University, Institute of Engineering Design:
	   − Optical bench design, boom & mechanism design
	
	• Danish Engineering Academy, Mechanical Engineering Department:
	   − Thermal analysis, structural & dynamic analysis

* Together, these six organizations are responsible for more than 80% of the work.

The detailed technical coordination is executed through three design teams, which meet at regular intervals: the project planning and design team, the science team, and the instrument team. Twice a month the project team holds full-day meetings with empowered participants from all the key organizations, enabling effective discussions and decision making across the project. Decisions concerning the high-level allocation of the available funding resources, and resolution of potential conflicts between the participating organizations is made by a steering committee. This committee has the overall management and fiscal responsibility for the Ørsted project. It has six members, representing the six major participating organizations, and meets quarterly.

Product responsibility for each hardware or software element is placed with the respective organizations, as defined in work package agreements developed for each of the 54 work breakdown structure elements. The formal reporting to the project office is reduced to a monthly report from each work element, containing a one-page technical report and a one-page resource and cost report. Cost summaries are prepared by the project office based on the monthly resource and cost reports. Each organization has designated a lead technical person, who participates in the bimonthly project team meetings, and is the single-point contact to the rest of the project.

Teamwork across the organizational boundaries and within each organization is essential to Ørsted's success. Most of the engineers and scientists are working full-time on the project, with a peak of about 50 people total and 5 to 10 people at each of the six main locations. Each person is directly responsible for one or more key technical products, normally visible to the entire team. The teamwork is based on mutual professional respect and open communication between the various technical disciplines, characteristic of Danish engineering culture and educational background, which emphasizes a broad generalist view on technical problem solving. Most team members are capable of spanning multiple disciplines and are therefore able to tackle the shift in work tasks as the project moves along, ensuring valuable continuity in the workforce. Furthermore, the key organizations are large enough to keep "critical" people available to solve ad hoc problems, even after they have transferred to other projects within the organization. Commitment by the participating organizations to ensure continuity in the workforce was secured by developing a workforce plan for the entire project duration, listing every team member by name. Deviation from this plan is not treated lightly, and fortunately only a few team members have left the project.

11.1.7 Conclusion

Ørsted is headed toward launch in March 1997. It has been going through challenging times, with critically small mass and power margins, delays in delivery of key engineering model hardware for the interface test, and intense workload. Budget pressures have arisen partly because of technical problems in selected areas and partly because of increased pressures to ensure that the mission is successful. The international importance and visibility of Ørsted does not come free. As a consequence, the cost is expected to grow above the $20 million price tag by less than 10%. The launch slip to March 1997 has provided some needed schedule relief, but further delays may start to be costly. The can-do attitude and ingenuity of the technical project team, the management commitment of the participating organizations, and the strong support and cooperation from the international space science community are pivotal ingredients that make this first Danish satellite look promising for the future of small satellites.

People who argue that small, low-cost spacecraft will never be able to generate world-class science will have to rethink their position. One of the main reasons behind the growing worldwide enthusiasm for smaller satellite missions is the relatively short project life-cycle, from the initial concept to mission data return. With smaller and scientifically focused missions, the scientists are able to see the results of their contribution in a few years, rather than "after retirement." There is a whole class of space missions which should be carried out by small spacecraft, based on their science merit. The key characteristics are a short development schedule, use of the latest technology, and better science data quality due to lack of competing requirements imposed by other unrelated science experiments. Ørsted is just one example.

I believe that curiosity and exploration of the unknown must continue to have a key place in space missions and that scientifically exciting and spectacular results aren't necessarily expensive. Missions involving small, low-cost satellites provide a stimulating opportunity for bringing the fun back into space science. I had a lot of fun working on the Ørsted mission.[*]

---

[*] I greatly admire the dedication and professional teamwork of the people who are working on the Ørsted Project. Without them, this chapter wouldn't have been possible. I will always treasure my involvement with Ørsted during its initial project phases (1992–94). Without this experience, I wouldn't have been inspired to write this chapter. A special thank you to the visionary minds who originated the Ørsted concept: Jens Langeland-Knudsen, Ray Baron, Eigil Friis-Christensen, Torben Risbo, and Fritz Primdahl.

Freja Summary

Freja, built by the Swedish Space Corporation for the Swedish National Space Board, is a magnetospheric and auroral research satellite. It carries eight payloads from four countries.

Spacecraft dry mass:	214 kg	Propulsion:	6 solid prop. thrusters
Average power:	95 W BOL	TT&C:	524 kbps

Launch: "Piggyback" launch with Chinese FSW-1 satellite on a Long March 2C from Jiuquan Satellite Launch Center, China, on Oct. 6, 1992

Orbit: 601 km × 1756 km at 63 deg inclination

Operations: Ground stations for operations and science in Sweden and Canada

Status: Operational as of January 1996

Cost Model (FY95$M): See page 348 for explanation.

	Expected Cost	Small Spacecraft Model	Actual Cost*
Spacecraft Bus	$44.4M	$12.8M	$12.5M
Payload	$32.4M	$6.0M	$6.0M
Launch	$66.4M	$6.6M	$4.8M
Ground Segment	$118.2M	—	$0.8M
Ops. + Main. (annual)	$9.8M	—	$0.4M
Total *(through launch + 1 yr)*	**$271.2M**	—	**$24.5M**

*An inflation factor of 1.000 has been used to inflate to FY1995$ [SMAD, 20-1].

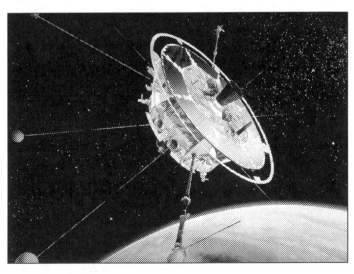

Fig. 11.2-1. Freja. The satellite is 2.2 m in diameter. It is spin-stabilized with its spin axis pointing at the Sun. Boom systems keep sensors away from the satellite body.

11.2 Freja

Sven Grahn, *Swedish Space Corporation*

The Freja[*] magnetospheric research satellite was launched on October 6, 1992 as a piggyback payload on a Long March 2C (CZ-2C) rocket from the Jiuquan Satellite Launch Center in China. The satellite is in an orbit between 601 and 1,756 km at 63 deg inclination. Freja is a Sun-pointing spinner with a 2.2 m diameter and 214 kg mass. The Swedish Space Corporation is the prime contractor to the Swedish National Space Board. Freja images the aurora and measures particles and fields in the upper ionosphere and lower magnetosphere. Swedish, German, Canadian, and U.S. instruments fly on the satellite (Table 11.2-1). This chapter overviews Freja and the design, procurement, engineering, and management methods used to achieve a low program cost ($19 million FY95 including satellite, launch, ground system, and 30 months of operation, but excluding experiments).

11.2.1 Is Freja Really "Low Cost"?

Freja began in 1987 when the CZ-2C launch reservation was "left over" from the canceled Mailstar project, a store-and-forward, low-orbit communications satellite. The launch reservation resulted from a competition for launching Mailstar; it wasn't particularly adapted to an auroral mission. However, the 63 deg inclination provided by the CZ-2C was high enough to stimulate the interest of Swedish scientists into using it for an auroral research mission. The launch price was very attractive, so building a satellite was feasible despite the acute lack of funds in the Swedish space budget. The Freja project's budget was less than half the cost of the low-cost Swedish VIKING project (not to be confused with the NASA Viking mission to Mars). The VIKING satellite [excluding experiments] and launch cost $40 million. Grahn [1987]

---

[*] Freja is the goddess of fertility in Nordic mythology.

TABLE 11.2-1.   **Mass, Power, Data Rate, and Cost of the Freja Program.** Costs are for the spacecraft, ground system, launch, and operations for 30 months. The payloads were provided at no direct cost to the program. We estimate their cost at $6 million.

Item	Mass (kg)	Power (W)	Data rate (kbps)	Cost (FY95$K)
Payload total	**73.08**	**66.1**	**250.70**	
F1 electric fields (Sweden)	5.37	9.0	30.70	
F2 magnetic fields (USA)	3.97	4.0	14.33	
F3 cold plasma (Canada)	5.79	6.2	16.38	
F3H hot plasma (Sweden)	9.52	15.6	42.98	
F4 waves (Sweden)	8.40	11.6	75.73	
F5 auroral imager (Canada)	9.94	6.7	44.01	
F67 electron beam & particle correlator (Germany)	8.49	13.0	26.61	
Wire & stiff booms	21.60			$791K
Spacecraft bus total	**182.77**	**29.4**	**11.28**	
Attitude control (equipment mostly powered down)	12.92	0		$399K
Structure	48.68	0		$332K
Computer	13.07	7.0		$765K
TT&C/Communications (peak power indicated)	5.60	22.4		$495K
Power system	40.60	0		$811K
Propulsion	56.80	0		$575K
Thermal	5.10	0		$124K
Electrical ground support equipment				$368K
Mechanical ground support equipment				$186K
Management & systems engineering				$3,697K
Integration & test (including test site fees)				$2,500K
Transportation & travel				$720K
Other services				$720K
Satellite total (excluding instruments)	**255.85**	**95.50**	**261.98**	**$12,483K**
Ground system total				**$800K**
Software				$500K
Hardware				$100K
Management				$200K
Operations (30 months)				$1,000K
Launch				$4,756K
GRAND TOTAL (estimated total with payloads)				**$25,000 K**
Excluding payloads				$19,039K

describes the VIKING satellite, its development, and its management philosophy in detail. The Freja and VIKING missions are quite comparable in scope and level of ambition. Even with almost identical launch costs, Freja is much less expensive. To achieve the cost goal, the Swedish National Space Board and the project scientists agreed to let the Swedish Space Corporation use unconventional methods to cut costs.

Table 11.2-1 summarizes the design and development costs of the Freja satellite (excluding experiments) plus the launch cost. To determine whether the launch cost is low or high, we must remember that Freja represented about 12% of the CZ-2C's capability. The Chinese main satellite weighed about 1,800 kg at launch, and Freja weighed 256 kg. So Freja was a companion payload requiring extensive modifications of the CZ-2C. (See Sec. 11.2.3.)

The financing for development came from the Swedish taxpayers through the Swedish National Space Board; donations from Wallenberg Foundations; price reductions by the Swedish Space Corporation and FFV Aerotech (the integration contractor); and a generous contribution in cash, hardware, and services from the German Ministry for Science and Technology, amounting to approximately 25% of the above costs. Canada contributed the tracking services of the Prince Albert tracking station. Table 11.2-1 doesn't include costs for developing experiments, which were paid by the space agencies of the principal investigators' home countries. It's difficult to estimate these costs, but they could amount to almost 30% of satellite-development costs, or $6 million. Thus, the whole project cost about $25 million.

11.2.2 Project Schedule

The Freja project started in August 1987 with feasibility studies on using the Long March 2C launch opportunity. During the first half of 1988, we completed the system design, wrote equipment specifications, and collected equipment bids. The protoflight satellite structure was complete in the fall of 1989. The vibration test to qualify this structure occurred in China in March 1990. The qualification model of the System Unit (see Sec. 11.2.5) was delivered for integration on the satellite in February 1991. Satellite integration was complete in August 1991. During the rest of 1991, we tested for system electromagnetic compatibility, acoustic noise, satellite balancing, boom deployment, and acceptance vibration. After doing the solar-simulation test, magnetic survey, and ground-station compatibility tests in the spring of 1992, we shipped the satellite to China for launch in August 1992.

This is a long development schedule for a small satellite, but several things contributed to it. Developing the science instruments, which included cutting-edge technology in packaging, image sensors, and software, took an extra year. The lack of test facilities in Sweden required time-consuming travel to other countries. Also, the project team strongly emphasized a long system-test phase because Freja was a new design from the bottom up. This long test phase with the integrated satellite was very helpful in finding various hardware and software problems. Running the satellite 8 hours a day for a year certainly builds confidence into the system. Another factor was the long lead time for parts. Because this was the Swedish Space Corporation's first satellite built in-house, no space-grade parts were in stock, so critical parts typically took 40 weeks for delivery. By using previous designs and keeping key parts in stock, we could build a satellite platform of this size and type much faster—perhaps in 24 to 30 months. This faster development could greatly reduce manpower requirements for assembly, integration, and test. Of course, the design staff may have to increase under a compressed schedule, but still, we can expect 10 to 15% lower labor costs if we can cut a schedule could from about 4 years to 30 months. The ASTRID microsatellite, based extensively on the Freja design, has been designed, built, tested, and launched in 14 months.

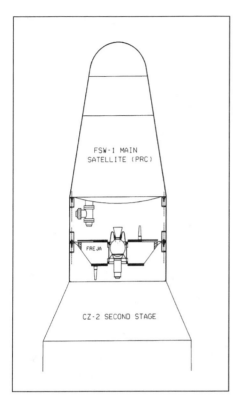

Fig. 11.2-2. Piggyback Launch Arrangement. Freja is placed in a "piggyback cabin" between the second stage of the CZ-2C and the FSW-1 Satellite.

11.2.3 Launching Arrangement

The Long March 2C (CZ-2C), is a 2-stage, liquid-fuel rocket with a lift-off mass of 191 tons and a lift-off thrust of 2,747 kN. The rocket is 34 m long and has a diameter of 3.35 m. Both stages are propelled by N_2O_4 and UDMH.

For launching Freja, the CZ-2C was modified by adding a cylindrical transition bay (or piggyback cabin) between the Chinese FSW-1 satellite and the top of the CZ-2C's second stage. The transition bay was split into two cylindrical sections. Freja has an interface ring which was clamped between the outer rims of these two cylinders. Figure 11.2-2 shows this arrangement. The main satel-lite was released first while the CZ-2C stage was under 3-axis control. After a maneuver to put the separation vector parallel to the orbital tangent at the orbital apexes, the top half of the piggy-back cabin was separated by firing four explosive bolts. These bolts acted only on this part, and Freja gently moved away from the CZ-2C stage, which was braked by solid rockets. A few seconds after separation, solid-propellant spin rockets fired to give Freja a 50 rpm spin and prepare it for ignition of orbit-adjust motors.

We put Freja into a 213.7 by 317.4 km parking orbit at 63 deg inclination. This orbit was raised to reach scientifically interesting regions and to avoid drag decay. A Thiokol STAR 13A fired at the southern apex of the parking orbit, 37 minutes after launch, to give Freja an apogee of 1,756 km. Fifty-three minutes later a STAR 6B motor fired to raise perigee to 601 km. The satellite will decay in 250 years.

11.2.4 Summary of Satellite Design

Grahn [1993] describes Freja's design in detail, so I'll just summarize it here. Figure 11.2-3 shows the satellite configuration; Fig. 11.2-4 gives a block diagram of it; Table 11.2-2 lists the experiments. The structure consists of a central tube machined from cast magnesium. The solid rocket motors mount on a Kevlar adapter inside this tube to limit heat soak into the satellite from the spent motors. Four radial walls connect the central tube with the CZ-2C interface ring, and the top and bottom platform decks mount on the top and bottom of the walls. Platform and science equipment mount on the walls and the platforms.

TABLE 11.2-2. Freja Technical Characteristics.

Launch	• Piggyback on Long March 2C, Jiuquan Satellite Launch Center, Oct. 6, 1992
Orbit	• 600 × 1756 km, 63 deg inclination
Mass	• 255.9 kg at launch, 214 kg on orbit
Payloads	• Electric fields instrument • Magnetometer • Cold plasma analyzer • Hot plasma detectors • Auroral UV imager • Electron beam instrument • Wave/particle correlator
Structure	• Magnesium central tube 0.5 m high, 0.44 m diameter • Top and bottom aluminum honeycomb platforms, 0.9 m diameter • Four radial aluminum honeycomb walls • Launch vehicle interface ring (2.2 m diameter) supported by the radial wall
Electric Power	• Approximately 75 W orbit average power requirement • Solar panels on top, Sun-facing, side of satellite, 130 W end-of-life, 20 × 40 mm Si cells • +28 V ±2 V bus • Two 6 A·hr NiCd batteries • Dedicated solar array charge strings • Charge termination by software amp-hour meter • Solar array voltage limited by Zener-stabilized shunts • Pyros powered directly from the battery and fired by onboard computer
Attitude Control	• Spin-stabilized at 10 rpm, spin axis approximately Sun-pointing • Magnetic torque rods for spin axis & spin rate control (142.5 A·m^2, 80.5 A·m^2) • Viscous nutation damper • Two 3-axis magnetometers • Two Sun sensors • IR horizon-crossing indicator • Torquer commands for spin axis control uplinked from the ground • Torquer commands for spin rate control generated onboard
Thermal Control	• Passive control with paints and multilayer insulation • Ground-controlled battery heaters
C&DH	• Telemetry formatting performed by hardware • Command interpretation and queueing by onboard computer • Battery charge control & spin torquer commutation by onboard computer
Communications	• S-band phase-modulated telemetry link at 524 kbps & 2 W transmitter power • UHF FM command link at 1200 bps
Propulsion	• Two solid propellant rockets for spin-up to 50 rpm • Two solid propellant rockets for subsequent spin-down to 8 rpm • Rocket motor with 32 kg solid propellant for raising apogee • Rocket motor with 7 kg solid propellant for raising perigee • All rocket motors fired by onboard computer

Notice the unusual mechanical interface to the launch vehicle and the need to locate wire boom assemblies and scientific instruments on the largest possible radial distance from the spin axis. These features make it impossible to locate solar panels on the satellite's outside perimeter. Instead, the panels attach to the top surface, leaving the backs of the panels free so heat can be radiated. A Sun-pointing attitude gives maximum electrical power (130 W end-of-life) and acceptable, though not optimum, viewing angles for the scientific sensors. Despite these limitations, the system can meet instrument viewing requirements, but may have to wait days or weeks for the correct attitude. To allow a completely arbitrary orientation, the science instruments would probably have had to require roughly 50% less power.

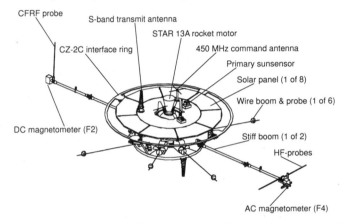

Fig. 11.2-3. The Freja Satellite. The diameter is 2.2 m. (Wire booms are up to 10 m long and not drawn to scale.)

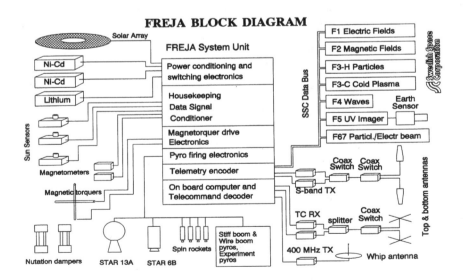

Fig.11.2-4. **Block Diagram of the Freja Satellite**.

Figure 11.2-5 below shows the motion of the spin vector relative to the Sun (at the origin of the plot) during the first 670 days of the mission. As you can see from this plot, the spin vector stays within 30 deg of the direction to the Sun. Spin-stabilization is needed to keep the so-called wire booms for measuring electric fields extended. It's also suitable for a satellite making a 2-impulse Hohmann transfer. We can place the two rocket motors along the spin axis with nozzles pointing in opposite directions and control the spin vector's attitude using well-proven, quarter-orbit, magnetic torquing. For Freja, this torquing makes the spin axis approximately track the Sun.

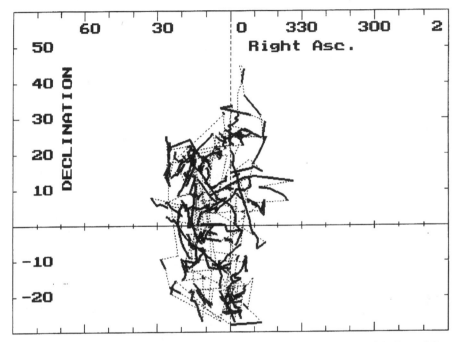

Fig. 11.2-5. Spin Vector's Orientation Relative to the Sun (at plot origin). Dotted lines represent attitude maneuvers.

Scientists planned attitude maneuvers to provide appropriate pointing of particle sensors and imagers. The spin axis needs to be controlled to better than 5 deg—preferably, to 2 deg. On-orbit operations have usually met these approximate requirements. A 3-axis magnetometer, redundant solar-aspect sensors, and an infrared Earth sensor collect data from the best location along the orbit and store it in an onboard memory. The ground station uses this data to determine an attitude vector at least once a day. The attitude determination accuracy requirement is 0.5 deg (1σ). However, because of the long wire booms, the satellite undergoes strong gravity-gradient torques, so the spin vector drifts a few degrees per day. Therefore, we must interpolate between consecutive attitude vectors.

Freja uses redundancy whenever possible. It has redundant transmitters, receivers, batteries, solar-array shunts, telemetry systems, telecommand decoders, computers, pyro circuits, and attitude sensors and actuators. Reliability analyses didn't drive all of this redundancy. The Freja System Unit (see Sec. 11.2.5) needed some internal redundancy to achieve a reasonable overall reliability for the satellite. But other redundant elements were introduced to be on the safe side and boost the confidence of the project's sponsors and development team. We introduced redundant elements very early in the program, so they weren't particularly costly. They caused a rather small and entirely tolerable mass penalty. Table 11.2-1 lists mass, power, and costs for the satellite's subsystems. However, the integrated housekeeping unit (called the Freja System Unit) combines the functions of many subsystems in one electronics box. The mass, power, and cost for the System Unit are under the computer item in Table 11.2-1. The power item includes only the solar array, NiCd batteries, and current shunts. Power-conditioning equipment and bus-voltage regulators are part of the System Unit and, therefore, of the computer item. Thus, it's difficult to break down Freja into traditional subsystems. Of course, we've made *mechanisms* part of the payload because most mechanisms are.

11.2.5 Designing Subsystems for Low Cost

Simplicity was a general guideline for designing Freja subsystems in order to reduce costs. Using small subcontractors, simple contracting, reduced documentation, and other factors probably contributed more to this reduction than simple design. Still cost-reducing configurations were important for two subsystems: the radio system and the System Unit.

Before the advent of spaceborne GPS receivers, the most accurate orbit determination for satellites in low-Earth orbit came from an S-band transponder compatible with STDN/SGLS, which provided range and range-rate data. But such a transponder is costly for a small satellite program—on the order of a million dollars per copy. In the Freja program we therefore carefully analyzed whether or not we really needed a full-blown transponder. We simulated the orbit-determination process using the ORAN software (delivered by EG&G). These simulations showed we could determine the orbit to within 10 km, as required, with weekly tracking campaigns using only angular data on azimuth and elevation from the 9 m tracking dish at the main ground station. Using angular tracking made a range and range-rate transponder unnecessary. Instead, we could use separate transmitters and receivers. Freja has S-band transmitters of the type normally used on sounding rockets but with protection against corona added in the output-amplifier stage. Because only one ground station is used for uplinking commands, the command receiver doesn't need to be compatible with tracking networks. Instead, a standard, low-cost receiver is used in missile tests to destroy the missile if it gets out of control. Thus, we spent about $100K for a separate transmitter and receiver (without range and range-rate functions) instead of $1 million for a transponder.

Platform functions are normally provided by several units connected by the harness and interface specifications, but Freja uses a scheme from AMPTE/IRM and other satellites—a common unit does most platform functions. This System Unit, or integrated housekeeping unit, performs the functions listed in Table 11.2-3. Outside this unit are only batteries, solar panels, radios, attitude sensors, attitude actuators, and propulsion units. It is a 13 liter box weighing approximately 15 kg and containing 13 printed circuit boards (220 × 260 mm).

TABLE 11.2-3. What Freja's System Unit Does.

• Encodes telemetry
• Conditions signals for all housekeeping channels
• Decodes telecommands
• Processes onboard the stored commands for attitude control and other purposes and provides "intelligent" control of the power system
• Regulates bus voltage
• Distributes power
• Controls battery charge
• Controls pyro firing circuits
• Handles drive circuits for magnetic torquers

Combining all these functions in one unit saves a lot of money for interface engineering, perhaps even a full engineering position in the team, or about 10% of the labor costs. One person handles the interface work as an integral part of the design process. Of course this advantage is offset by slightly more complex testing of these functions. The System Unit communicates with experiments by a single connector, carrying only a 2-way serial link through optocouplers and 28 VDC for powering the experiment and a heater power circuit. All telemetry data going to the ground are also available to all experiments, and commands go to the experiments only as part of this serial bitstream. Freja System Unit provides no high-level commands except pyro firings. We simulated this very clean interface to each experiment by an extension board to a personal computer—a satellite emulator against which we could design and test each experiment. When the experiments arrived for integration on the satellite, the data-handling interface rarely presented any problems. Using this emulator possibly shortened the project's system-test phase by several months.

11.2.6 Procurement Methods in Satellite Development

In addition to the size and organization of the project team, the way we procure equipment and services is a key to limiting costs in a small program. The main procurement concept for the Freja project was to eliminate subsystem contractors. The Swedish Space Corporation, as the prime contractor for Freja, assumed subsystem responsibility and procured only on the equipment level. Because many units are off the shelf and the System Unit combines several subsystems, such an approach is possible for a very small project team (nine people). It saves money by avoiding the costly layers of management that subsystem contractors would introduce, as well as a layer of documentation and, very often, of specification margins. Typical qualification levels for vibration on Freja equipment were 15 g's from 10 to 100 Hz. As Freja subsystems were not contracted out, they didn't carry the typical extra-specification margins. Such extra margins at each contractual level result in higher specifications for equipment vibration, which eventually could lead to overtesting and costly failures of the specimen.

Another cost-saving procurement method used for Freja was to limit subcontractor responsibility. That is, subcontractors didn't have to accept any significant technical or financial risks. Buying "risk-taking" is costly, so the prime contractor should absorb technical risk if they have the necessary technical expertise. Another potential advantage of limiting the suppliers' technical responsibility and financial risk is that we can use smaller companies. These smaller suppliers may be very skillful, be unbureaucratic, have small overhead costs, and offer quick turnaround times, but they may not be able to absorb much risk. Of course, we used competition whenever possible, except when a country participating in the project contributed in kind with hardware or services. All equipment contracts and most service contracts were at fixed-prices.

Suppliers of off-the-shelf equipment were encouraged to propose minimum deliverable documentation while adhering strictly to their normal quality and test procedures. Preparing documents is very costly. By asking the supplier to use the methods of QA and test employed in delivering equipment to ESA, NASA, or DoD, the project essentially piggybacks on the QA requirements of these organizations. Often, suppliers insisted on delivering certain documents; otherwise, their system got upset. The Swedish Space Corporation then replied, "OK, but we do not want to pay for it—we trust you guys to deliver a good product." We usually got the documents anyway—at no cost. Table 11.2-4 lists major suppliers to the project.

TABLE 11.2-4. Major Suppliers for the Freja Project.

Item	Supplier	Country
Orbit-adjust rocket motors	Thiokol	U.S.
Spin rocket motors	Chinese Academy of Space Technology	P.R.C.
Solar array	Telefunken System Technik	Germany
NiCd batteries	Xin-Yu Power Supply Factory	P.R.C.
Battery pack	AeroAstro	U.S.
Magnetometer	Schonstedt	U.S.
Primary Sun sensor	SME GmbH	Germany
Acquisition-phase Sun sensor	ACR	Sweden
Transmitters and receivers	Aydin Vector	U.S.
S-band antennas	Saab Ericsson Space	Sweden
UHF antennas	FFV Aerotech	Sweden
Primary structure	ACR	Sweden
Magnetometer & antenna booms	Max-Planck-Inst. für extraterrestrische Physik	Germany
Wire booms	Weitzmann Consulting	U.S.
Thermal-blanket design	Stapf Scientific	U.S.
Assembly integration and test operations	FFV Aerotech	Sweden
Solar simulation, acoustic-noise test, magnetic survey, mass-properties test	Industrie-Anlagen Betriebs-Gesellschaft (IABG)	Germany
Satellite-vibration tests	Beijing Inst. of Structure & Environment	P.R.C.
Freja System Unit	Swedish Space Corporation	Sweden

11.2.7 Using Engineering Analysis to Save Costs

To reduce costs by assuming subsystem responsibility and absorbing technical risk, prime contractors must be able to do most engineering analyses in-house. Therefore, computer-aided engineering tools (structural, thermal, mass properties, radiation, orbit, and others) were directly available to the Freja project team, who ran analyses within hours of finding problems. No work orders or contract negotiations were required. We just did the work when it was needed. Our main numerical tools were a VAX-based ABAQUS (switched later to a PC-based ANSYS) for finite-element modeling of the structure; the NASA-developed thermal analyzer, SSPTA; and ESABASE, a computer-aided engineering tool developed for the European Space Agency and its member states, which we used for such tasks as placing equipment and analyzing gravity gradients or radiation. We used our own PC software, Freja Operations Support Software, to do mission analysis.

Finite-element modeling of structures is now so sophisticated that we don't need a separate test model of the satellite structure and the protoflight-model concept is possible on the system level. For Freja, validated analytical tools replaced costly tests because CPU time is cheaper than test-crew labor. Validated methods of analysis also replaced tests in developing Freja equipment. For example, we never tested the damping action of the nutation damper, thus saving up to $30,000. The damper was built, vibration tested, and flown. It worked just as expected.

11.2.8 How Cost and Schedule Affected Selection of Electronic, Electrical, and Electromechanical Parts

Parts for the System Unit represent a small percentage of Freja's total cost. Therefore, component quality and associated cost aren't issues in programs with Freja's scope. Availability determines the choice of electronic, electrical, and electromechanical parts. Flight-grade parts were needed quite early in the program, so delivery times between 40 to 50 weeks were really the maximum tolerable. Within such a schedule constraint, we should buy the best parts possible, but what constitutes a good part? Are plastic microcircuits produced in gigantic batches better than ceramic circuits produced in small batches? Perhaps. In the Freja program, we used ceramically encapsulated microcircuits whenever available. Some plastic parts were used but only in an Ericsson DC-to-DC converter delivered as an integrated hybrid circuit.

Early in the project we decided to buy the best parts available without going to full-blown, space-qualified parts for the System Unit because of their intolerably long delivery times. We bought critical components (CPU and RAM) to MIL-STD-883C level B made with epitaxial CMOS processes, which often have better radiation-tolerance (total dose, single-event effects) characteristics than bulk CMOS. Level B has a failure-rate factor for parts quality (π_Q defined in MIL-HDBK-217E) $\pi_Q=1.0$. We bought other logic circuits to MIL-STD-883C level B-2, with $\pi_Q=5.0$. The reasons for going to this quality level were the high costs and long lead times for higher quality levels. For discrete components used in large numbers, such as transistors and diodes, reliability analyses showed we needed JANTXV parts ($\pi_Q=0.12$). When they weren't available, we used JANTX parts ($\pi_Q=0.24$) instead. The differences in price and delivery time between these two levels are slight. Reliability analysis also showed that the failure rate of capacitors was most critical to the System Unit's overall reliability. Therefore, we bought ceramic capacitors to level S and tantalum capacitors to level D. These qualities have a failure rate of 0.001% per 1,000 hours. The next lower level (R for ceramic, C for tantalum) of quality has a failure rate ten times higher. The difference in price is less than 10%, and they have the same delivery time. The situation is the same for resistors. We used S parts but were sometimes forced down to R for values over 100 kohms.

Projects like Freja must have their own test program for incoming parts because we can't trust parts suppliers to deliver uniform quality. We used receiving inspection and test (parameter testing) on all electronic parts and successfully detected a bad FET batch. Such a test program costs only a minute fraction of the total program cost but protects the program from costly, schedule-wrecking failures. Thus, even a low-cost program can't live without testing incoming parts. Table 11.2-5 lists electronic, electrical, and electromechanical parts in terms of manufacturer, types, and specification. In general, a major concern was to keep all project workers aware of the problems of radiation tolerance. Often, a circuit designer focuses too much on functionality and forgets radiation tolerance. In the Freja program, many team members have a long background in sounding-rocket programs for which radiation and thermal characteristics are often unimportant. Even though sounding-rocket projects are valid models for low-cost satellite programs, we must remember the technical differences between sounding rockets and satellites. The key part in the Freja system is the microprocessor in the main spacecraft. The philosophy we used in selectioning it was that it should have very good radiation tolerance. High performance was not so important because its tasks didn't require high speed or wide address space. Thus, we used a rather simple processor—a version of the 80C31 microprocessor manufactured by Matra-Harris (France). This unit has a documented total dose tolerance of 16 krad. An epitaxial version of the chip is available and has a single-event upset cross section=1.35×10^{-4} cm^2 (rate 8.3×10^{-3}/min) and a single-event latchup cross section=4.5×10^{-7} cm^2 (rate 2.81×10^{-5}/min). The Swedish Space Corporation procured this processor to MIL-STD-883C level B (SCC9000 level C). The processor uses a Matra-Harris HM65641 8K $\times$ 8 SRAM as external memory. The manufacturer indicates 30 krad total dose tolerance, single-event upset cross section 5.13×10^{-2} cm^2 (7.8×10^{-7} cm^2/bit), single-event latchup cross section 1.06×10^{-6} cm^2 (6.6×10^{-5}/min). All single-event data were generated with a Californium 252 source at linear-energy transfer = 43 MeV/(mg/cm^2). During the first two years, the CPU-board watchdog hasn't "barked" once. So the processor itself hasn't suffered any single-event upset. The RAM could have experienced such upsets because it's used for many temporary variable values, but we haven't seen one. Total dose effects have not been detected so far. Thus, even though the 80C31 is not a particularly capable processor, it seems to be highly dependable. From our experience in Freja, we can recommend it for critical uses in space.

11.2.9 The Right Philosophy Model and System Testing Can Help Save Costs in Satellite Development

A small program shouldn't compromise the scope of tests on the hardware, but it can simplify the number of development models at the system level.

On equipment level, off-the-shelf equipment normally requires only acceptance testing (or delta-qual if your project's qualification levels exceed previous qualifica-

TABLE 11.2-5. Selection of Electronic, Electrical, and Electromechanical parts for Freja.

Component	Category	Manufacturer	Type	Quality Specification
Capacitors	Ceramic Met. lacquer Tantalum	AVX Siemens Sprague	CKR MKU CSR13	MIL-C-39014 level S ESA/SCC 3006/009 level B3 MIL-C-39003/01 level D
Connectors	External Internal	ITT/Cannon Hypertac	D*M HPF	ESA/SCC 3401 level B3 MIL-C-55302/161
Diodes	Signal Zeners Schottky	Microsemi Microsemi Solid State Devices		MIL-S-19500 JANTXV MIL-S-19500 JANTXV. MIL-S-19500 JANTXV
Microcircuits	Digital Linear CPU,RAM EE-PROM	Texas Instr. RCA/Harris Analog Dev. Harris PMI NS Matra MHS SEEQ	HCMOS CMOS ACMOS Epi-CMOS	MIL-STD-883C level B-2 MIL-STD-883C level B-2 MIL-STD-883C level B-2 MIL-STD-883C level B-2 MIL-STD-883C level B-2 MIL-STD-883C level B-2 MIL-STD-883C level B-2 ESA/SCC 9000 level B MIL-STD-883C level B-2
Optocoupler		HP	6N134	TXVB
Relays	Signal Power	Teledyne LRE	V432 M210	MIL-R-39015 Hi-Rel MIL-R-6106J ER, Hi-Rel
Resistors	Metal film Network Power	Dale Dale Dale	RNC55 RLR07 MSM08 MSM09 RWR81 RWR84 RWR89 RER60	MIL-R-55182 level S MIL-R-55182 level S MIL-R-83401 MIL-R-83401 MIL-R-39007 level R MIL-R-39007 level R MIL-R-39007 level R MIL-R-39009 level R
Transistors	Signal FET	Raytheon Intern. Rectif.		MIL-S-19500 JANTXV MIL-S-19500 JANTXV
Miscellaneous	X-tals Fuses Inductors Oscillators	Quartzkeramic Littlefuse Belfuse Sprague Spectrum Technologies	 FM08A 125 FM04A 125 0446 66Z 4116-283	ESA/SCC 3501 level C MIL-F-23419 MIL-F-23419 MIL-D-23859 MIL-T-21038 MIL-O-55310/01

tion levels), whereas new equipment requires an engineering model, a combined qual-model flight spare, and a flight unit. For a small project build only one complete satellite model *on the system level*—the protoflight model. Use it for qualification tests and for actual flight. The flight structure with dummies is useful for thermal-blanket manufacturing and fitting. This protoflight scheme is nowadays not so risky. because computer tools for structural analysis are quite accurate in computing loads.

Despite this, you'll still need to build up a bench-test model of the satellite in the lab early in the program. It contains:

- Breadboard models of new equipment like satellite computers, attitude sensors for the data-handling system, a copy of the flight harness

- Commercial grade batteries

- Simulators for solar panels and wire booms

- Engineering-model or breadboard experiments

- Engineering models of new equipment (attitude sensors)

- Flight spares or flight units for off-the-shelf equipment (e.g., receivers and transmitters, antennas)

- Pyro simulators

Use the bench-test model to verify hardware and software design, electromagnetic interference, and test onboard and ground software. The bench-test model saves money because you can do time-consuming troubleshooting tests unhurriedly on it—in parallel with satellite testing. Even a microsatellite needs a high-fidelity wooden mockup with equipment mass dummies for harness manufacturing. A full-scale, sheet-metal, RF-test model of the satellite with prototype TM/TC antennas is essential even for the smallest project.

11.2.10 Technical Problems Encountered in Satellite Development

I won't list all technical problems that appeared during the development of Freja; rather I'll describe some typical problems and how the project team handled them.

A piggyback payload like Freja needs to go through a joint vibration test with the main payload and associated launch-vehicle adapters. Freja, the piggyback cabin (see Sec. 11.2.3), and the main Chinese FSW-1 satellite were vibration tested together mounted in flight configuration on the vibrator. We did this test twice, first to qualify the Freja structure and then for flight acceptance of the assembled Freja satellite. For the first test, the FSW-1 test article was a supposedly representative mockup. For the acceptance test, we used the actual flight model of the FSW-1. During initial low-level sweeps to search for resonance before the qualification test, we found resonances in the FSW-1 that overlapped perfectly with the frequency of an important mode of vibration of the beams that support Freja's solar panels. Between the qualification and acceptance tests, we made the support beams less stiff to avoid resonance overlap. However, the Freja team decided not to entirely trust the results of the qualification test. Therefore, we made the resonance frequency of the beams tunable by installing detachable mass elements at their tips. During the low-level sine sweep preceding the acceptance test, we discovered the flight model of the FSW-1 had resonance frequencies different from the mockup. Again, these resonances overlapped with those of the support beams, but we solved this problem in five minutes by removing the masses from the tips of the beams. The rest of the test went smoothly. Making critical structural elements tunable in terms of resonance frequency could be useful in other projects.

I've already mentioned the problem of keeping all project workers aware of radiation tolerances, made worse by the lack of radiation-tolerance data for many parts. We got advice from ESA's technical center (ESTEC) and from the Space Department of the Applied Physics Laboratory of Johns Hopkins University. However, for some parts, we had to do our own tests. About a year before launch, total dose tests were run on a DC-to-DC converter hybrid from Ericsson and on a audio-frequency-shift modem chip. The DC-to-DC converter survived very high radiation levels and obtained a clean bill of health, but the modem chip failed dismally at 1 krad total dose. The modem chip was intended for the command decoder; our test showed we needed to redesign this decoder. We quickly decided to change the uplink-modulation scheme, and the Swedish Space Corporation developed a bi-phase bit synchronizer (using discrete parts with known radiation tolerance) to replace the modem chip within two months. The bit synchronizer became much more bulky than the modem chip and had to be piggybacked on the System Unit box because it was too late in the project to redesign the printed circuit boards of the command system. This problem was a serious threat to the project schedule, but we could solve it rather quickly because the Swedish Space Corporation, as prime contractor, completely controlled the System Unit.

Using solid rocket motors on small satellites poses some problems. These motors generate a lot of heat, which conducts from the motor casing to the satellite structure. Motor-casing temperatures are known only approximately, so our design of the rocket-motor mount had to take this into account. The extent of the exhaust plume is also poorly known, which forced us to take extra precautions to protect sensors and subsystems from the plume's heat flux. The heat from the casing and associated thermoelastic effects could be the dimensioning case for the structure's static loads unless the motor is thermally isolated from the structure. In Freja, we used a Kevlar mounting cone. But Kevlar softens at high temperatures, and our predicted motor temperatures suggested the Kevlar would fail at the upper temperature limit but possibly retain stiffness at the lower limit. Clearly, we needed a thermal washer between the motor and the Kevlar motor mount. The Swedish Space Corporation and our structure subcontractor, ACR, tried various ceramic materials, but they were all too brittle. Finally, ESTEC helped us by proposing solid, glass-fiber-reinforced Kapton as the washer material. We found the plume impinged mostly on the main antennas for S-band telemetry. The solution was to make the antennas deployable. The Max-Planck-Institut, one of the principal investigators, quickly built the deployment mechanisms based on their well-proven, standard, space hinge.

Combatting electromagnetic interference is especially important in projects that use VHF or UHF command frequencies. The noise from fast data buses typically lies in the VHF and low-UHF frequency bands. The data bus reaches out to all instruments through the cable harness and, despite careful shielding and grounding, the RF noise from the data-bus noise leaks out and couples into the command-reception antenna. The data-bus noise could be especially high with ACMOS drivers.

This logic family has very short switching times and therefore generates much VHF and UHF noise. During EMC testing of the System Unit, intolerably high, radiated, electrical-noise levels appeared at 450 MHz, the command frequency. The noise threatened to make the command link inoperable. Our solution was to put ferrite sheets under the DIL ACMOS drive circuits for the data bus. These sheets reduced the noise at the command frequency by 15 dB and essentially solved the problem. You can find such ferrite sheets in commercially available EMI-suppression kits containing ferrites of all possible shapes. Never build a satellite without one!

Looking back at the 5-year development of Freja, we tend to remember only the technical horror stories and all the anxiety they created. The examples here are typical of the technical problems we encountered. During the development, more or less severe technical problems surfaced every week and had to be handled promptly by the project team—suspected NiCd battery leaks, a critical batch of transistors rejected during incoming inspections, the super-expensive conductive paint not sticking to the satellite structure—a humbling experience.

11.2.11 Ground Segment and Mission Operations

To save costs, as well as technical problems, the satellite-control center and the electrical ground-support equipment must be identical in architecture and software. We verified the control-center software during system testing of the spacecraft and needed only minor extra testing of specific functions within the control center. The same group must design the control center's electronics system for ground support and the satellite's onboard data-handling system. Modern personal computers (PC) can easily support sophisticated satellite-control centers. By using separate computers for telemetry readout, command transmission, and other tasks we can simplify software development compared to using a single minicomputer for all tasks. The control center at Esrange, Freja's operations center, is essentially a PC-LAN with PC workstations (Fig. 11.2-6). Lundin [1994] describes this center. Software for the operations center was developed mainly by one person, who designed the command-handling and telemetry-readout software. The project manager developed subsidiary software packages for attitude control (the control loop was closed on the ground) and mission planning, whereas the engineer in charge of satellite hardware for controlling and determining attitude developed the attitude-determination software. Another person developed the onboard software at the same time. We tested the onboard and ground-station software against each other extensively by using a breadboard version of the System Unit. Ground software (150,000 lines of code) was developed using Borland Pascal (MS-DOS version) and assembler language, whereas onboard software was written in PL/M and assembler language (5,000 lines of code). The operations center interfaces directly with the existing antennas at Esrange: 9 m, S-band receiving antennas and an 8 m, 450 MHz uplink dish.

Normally, only one operator is in the Freja operations center. The operator uses the telemetry and telecommand workstation during passes and the tape-copying work station between passes. The operations team has four people. Because the orbit node drifts 24 hours in local time over 102 days, the operators' working hours change around the clock. Data are received at Esrange, Kiruna, Sweden, and at the Prince Albert Satellite Station in Canada's Saskatchewan province. Prince Albert is well located to receive real-time data with its 24 m tracking antenna when the satellite traverses the auroral oval. Commands are only transmitted from Esrange, and most commands are stored onboard for later execution, even such mundane tasks as switching the transmitter on and off. The use of stored commands eases the operator's burden during the 20-minute passes and eliminates most of the risk of erroneous commands being transmitted when the operator is rushed. The satellite can store up to 500 commands. Sometimes the command queue needs to be refreshed several times per day. While over the North American continent, the satellite operates entirely under stored commands.

The most critical housekeeping task—onboard battery management—is handled entirely by onboard software; the operator normally doesn't need to worry at all about

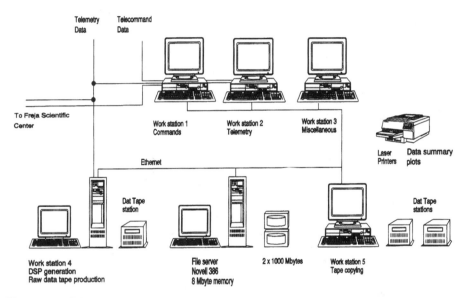

Fig. 11.2-6. PC-LAN Configuration of the Computer System at Freja Operations Center at Esrange, Sweden.

this task. The onboard software has several charge-control methods, but the one we selected uses an amp-hour meter the software implements using a k-factor of 1.05. The operator just checks weekly that the end-of-charge voltages don't go up or down If they do, a telecommand can change the k-factor. During the first two years of operation, we've had no reason to change the k-factor.

The two ground stations (Esrange and Prince Albert) receive and distribute science data on DAT (Digital Audio Tape) cassettes. During the first two years, about 300 cassettes with 700 MB on each one (about 400 GB) have been distributed in eight copies. So-called Freja summary plots are generated at Esrange and distributed over the Internet in PostScript format. At regular intervals a Japanese station in the Antarctic, Syowa (69°S, 39.35°E) also receives data. The scientific work with Freja has several modes. A Freja Interactive Science Center is available at Esrange where scientists can operate their instruments in real time when the satellite passes near Sweden and work with the satellite operators face-to-face. Another mode is to receive data and send commands over the Internet directly from computer terminals at their institutes. A third mode is the *workshop,* when principal investigators and co-investigators gather at the Freja Interactive Science Center to analyze data together and also run the instruments in real time.

Freja is tracked weekly using the 9 m tracking dishes at Esrange. The monopulse, autotrack antennas generate azimuth and elevation data from three passes spread over two days. Freja doesn't have a transponder, so the project only does angle tracking. The tracking antennas at Esrange have an azimuth/elevation (az el) bias of 0.05 deg (1σ) and az el noise of 33 arcsec (1σ). Simulations show that after one week the maximum uncertainty in satellite position is 10 km (due to random variations in the Solar UV Flux). This accuracy in satellite position is needed both for scientific reasons and to enable the Prince Albert Satellite Station to find the satellite using its 24 m dish with a very narrow beamwidth. Esrange uses software from the General Orbit Determination System developed by the German Space Operations Center to compute the osculating element sets and the state vector. For normal planning tasks in the control center, this element set is converted to the equivalent two-line elements format employed by the U.S. Space Command and propagated using the SGP algorithm (SpaceTrack Simplified General Perturbations). This analytical orbit generator runs fast on PCs. Element sets for Freja generated by U.S. Space Command (NORAD) are also downloaded from NASA Goddard's bulletin board and other sources. Space Command tracks Freja at roughly five-day intervals. A numerical test shows that, if the SGP algorithm is used with Space Command's two-line element sets, the satellite's position as computed seven days after the element-set epoch deviates about 9 km from the position determined by Esrange tracking. Thus, if Space Command keeps tracking Freja every five days, the two-line element sets and SGP provide the orbital-position accuracy needed for all our purposes. So small programs could rely entirely on two-line element sets and save costs.

11.2.12 Cost-saving Management Methods

The Freja program has shown that management methods are most important to cost reduction. Ballooning labor costs are the gravest threat to total program costs. If the project team's core is small, and part-time labor is used for various tasks, there is a substantial risk that the "part-time" workers exceed their budgeted labor hours. The management effort needed to monitor all part-time tasks will become huge. Instead, the Freja program used a relatively small but full-time project team of nine people (satellite assembly was subcontracted to a company that employed a team of five). This eliminated the risk of catastrophic labor-cost overruns, but each team member needed multiple skills in order for the project team to handle all technical, management, and clerical matters. Thus, each team member is important and not immediately replaceable, which presents a risk to the project in case a team member leaves the company, becomes ill, or is otherwise incapacitated. In Freja, we were lucky in that nobody left the company, became ill, or retired during the project. Other people in the company could take over some functions, but some people weren't immediately replaceable. The Freja team contained mostly people with a long background within the company—in some cases up to 25 years. Another problem with a small team is that the knowledge it gains in building a satellite may be lost if there is no continuity between programs. Luckily, the Freja team was able to transfer directly into two new satellite programs, eliminating the risk of having to reinvent the team's knowledge later.

Despite a few risks associated with a small project team, the many advantages greatly outweigh them. A small team is by necessity a flat organization. With few people, it's difficult to get much depth in the organization chart. The flat organization is well in tune with the egalitarian values of most young people in Sweden. A flat organization is also highly efficient because it permits quick face-to-face communications instead of slow communication by paperwork. Of course, we have to watch for hidden hierarchies because an organization always looks flat from the top.

The small team is cruel but efficient in separating the "doers" from the "talkers." It enforces personal accountability and is a very demanding work environment in which some people may not be comfortable. These people must move out of the team to avoid damaging morale. The project manager must also do technical work instead of remaining in an "ivory tower." A small project needs leadership at the front, not from the back. As mentioned above, every team member must do more than one task. For example, the same person often handles two interfacing subsystems as a single unit, eliminating the need for many interface specifications. Nearly all team members also manage subcontractors in some way, so they'll identify with the project, actively participate in daily management, and be acutely aware of schedule constraints.

A small team is also the basis for quality. Every team member can participate in project meetings and get an overview of all activities and how their own work is related to the project's overall progress. In these meetings, the team does not just review progress and nag about action items; it should also discuss and decide on design, test, and other technical problems. With everyone participating in the discussion, every system aspect is represented, and high-quality technical decisions can be made quickly with very little paperwork. The team has an overview of the system and maintains the integrity of the system concept.

Note that the team's overview included the satellite's ground-control center and electrical ground-support equipment, which were designed by the same people who designed the satellite's onboard system for handling data. Satellite operators were brought into the project when system-level testing started. They actually helped operate the satellite during critical tests, such as the week-long simulation of a solar vacuum. This cooperative effort transferred system know-how from the builders to the operators. Of course, operating the satellite from a site 1,000 km away from the design team's office still can pose problems. But these concerns haven't much affected the cost of the overall program.

A simple sponsor interface saves costs. Every project has a sponsor who pays the bill and who is worried about the project's success. The sponsor must be kept informed, involved, excited, and happy—but not through a big, formal, costly reporting system. In the case of Freja, the Director General of the Swedish National Space Board and the project scientists required mostly informal, oral reports each day. Formal (but still brief) written reports went in every two months or so. For Freja, the Swedish Space Corporation's project manager interacted directly with the customer's highest-ranking person. The government had no intermediate program manager. The project scientists were the technical interface with the Swedish National Space Board. The Swedish Space Corporation's project manager negotiated directly with the project scientists about system requirements trade-offs. (Because the program was a joint Swedish-German effort, there were two project scientists: Professors Rickard Lundin and Gerhard Haerendel.) Such matters were also freely addressed by scientists and engineers from the principal investigators and coinvestigators at the science team's meetings every 6 months. Swedish National Space Board administrators, scientists, and engineers from the experiment teams have attended design review. The Swedish Space Corporation as project manager also issued a Project Newsletter every 3 months to discuss progress and problems.

Thus, the Swedish Space Corporation managed all aspects of the Freja program, even complex administrative tasks such as export-license applications and drafting agreements between the Swedish National Space Board and other participating space agencies. In this way, the project team never had to sit idle and wait for someone else to solve a problem. They could solve all problems themselves, which is vital to keeping the program cost low. This means avoiding external constraints on the project team, such as "you must buy from this supplier" or other detailed technical, administrative, or reporting instructions. The project team must have a very sharp focus for its work—design, build, test, and launch the best spacecraft possible within the available cost and schedule constraints.

11.2.13 Conclusions

The Freja satellite has far exceeded its design lifetime of one year and is working well after 30 months in orbit. The satellite platform performs nominally and only one redundant element has been activated. We brought a backup Sun sensor into operation after 23 months. The experiments have delivered very valuable scientific data already published in Geophysical Research Letters [1994] and other publications. After more than two years in orbit, some experiments aren't operational, but those remaining make it worthwhile to continue mission operations. Therefore, the project must be regarded as a resounding success.

The technical and scientific success, coupled with just a 10% overrun on a budget estimated 5 years before launch, shows that the small-team approach and the management and procurement schemes worked well for a mission of this type and scope. The project team's total control over the project was a special key to keeping costs low. We're using the same methods to develop future small satellites in Sweden. A microsatellite based on Freja was launched in 1995. ODIN is being developed   for launch in 1997. It's a 250 kg radio-astronomy satellite with submillimeter wavelength, 3-axis attitude stabilization, and a 15-arcsec pointing accuracy.

Early perceptions that a low-cost program can accept higher risk are not completely valid. It's true that having a very small project team, for which losing one person could be critical, is more risky. Using somewhat unusual suppliers and the protoflight-model philosophy are also tolerable risks. But, as the project progresses, we become much less willing to accept the risks involved in reducing test work. Actually, the tendency is to increase testing and throw in new, confidence-building, tests as the launch date approaches. In building low-cost, reliable space systems, thorough testing is essential, especially on the system level. Compromises in this area cost money rather than saving it.

By using previous designs for key units, keeping key parts in stock, and choosing parts at the right levels of quality, we can build a satellite platform of Freja's size and type in 24 to 30 months instead of more than 4 years. Reducing development time could considerably reduce in manpower requirements for assembly, integration, and test. Of course, the design staff may need to increase if we compress the schedule, but that would still leave us with much lower labor costs. As mentioned earlier, a microsatellite based extensively on the Freja design has been designed, built, tested, and prepared for flight in 14 months.

In summary, streamlined management, total control by the project team, using previously qualified design solutions, keeping long-lead items in stock, and extensive testing are the key ways to obtain rapid development, low cost, and high reliability.

SAMPEX Summary

SAMPEX (Solar Anomalous and Magnetospheric Particle Explorer) is designed to detect solar and interplanetary charged particles, galactic cosmic rays of energies from 0.4 MeV to hundreds of MeV, and magnetospheric electrons.

Spacecraft dry mass:	160.7 kg	No propulsion system	
Average power:	102 W EOL	TT&C:	900 kbps

Launch: Scout G1 launch from Vandenberg Air Force Base, July 3, 1992

Orbit: 512 kg × 687 km at 81.7 deg inclination

Operations: Goddard Space Flight Center via downlink at Wallops Island, VA

Status: Operational as of January 1996

Cost Model (FY95$M): See page 348 for explanation.

	Expected Cost	Small Spacecraft Model	Actual Cost*
Spacecraft Bus	$38.8M	$9.4M	$31.9M
Payload	$28.6M	$4.2M	$11.7M†
Launch	$16.6M	$10.6M	$15.6M
Ground Segment	$52.6M	—	$7.5M
Ops. + Main. (annual)	$5.5M	—	$6.0M
***Total** (through launch + 1 yr)*	**$142.1M**	—	**$72.6M**

*An inflation factor of 1.198 has been used to inflate to FY1995$ [SMAD, Table 20-1].
†Includes estimated cost of the instrument provided by Germany.

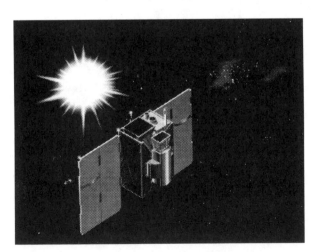

Fig. 11.3-1. SAMPEX. The illustration depicts SAMPEX with its solar arrays pointed toward the Sun and the four scientific instruments (LEICA, HILT, MAST, and PET) pointed in the zenith direction as the spacecraft crosses the Earth's magnetic poles.

11.3 SAMPEX

Orlando Figueroa, Gilberto Colón
Goddard Space Flight Center

The *Solar Anomalous and Magnetospheric Particle Explorer* (SAMPEX) is the first of the recurring Small Explorers (SMEX) Missions. This highly autonomous scientific satellite rode into a nearly polar Earth orbit on July 3, 1992 on a Scout launch vehicle from Vandenberg Air Force Base. The 160 kg satellite carried four instruments, which weighed a total of 52 kg and were designed to study the energy composition and charge states of particles from supernova explosions, solar flares, and nearby interstellar space. In addition, the SAMPEX instruments monitor magnetospheric particles, which occasionally plunge into the Earth's middle atmosphere and can alter it by ionizing neutral gases [Baker, et al., 1994].

NASA started the SMEX program in 1988 to provide frequent flight opportunities for low-cost scientific missions that could launch on small, expendable launch vehicles, such as Pegasus and Scout. The Announcement of Opportunity leading to SAMPEX required the average cost of the missions selected to be less than $30 million (FY88). This amount excluded the costs of ground and launch segments, as well as the mission operations and data analysis beyond the first 30 days in orbit. NASA selected three missions: SAMPEX, the Fast Auroral Snapshot (FAST), and the Submillimeter Wave Astronomy Satellite (SWAS). Each mission was to take less than 3 years from the start of detailed design to launch. Following a 3-month definition phase, SAMPEX took 35 months for detailed design, development, and launch. The mission's main scientific objectives were completed in 1995 (3 years after its launch), and the mission is now entering an extended science campaign expected to last another 4 years if the budget allows it. The total cost of SAMPEX, from the start of definition to the end of its primary mission in orbit is about $67 million. This amount includes its amortized share of the SMEX program's startup and technology-development costs. NASA announces opportunities for new SMEX missions approximately every 2 years and selects two missions so the program can meet its goal of launching one mission per year.

The Announcement of Opportunity, along with the selection of SAMPEX, FAST, and SWAS as the first three missions, complements to the larger ($150 to $250 million) astrophysics and space physics missions. It was also a transition to a faster, better, and therefore cheaper means to conduct scientific investigations. With SAMPEX, FAST, and SWAS, the SMEX program committed to launching the three missions in 5 years for no more than $121 million. This total excluded the costs of the ground and launch segments, as well as mission operations and data analysis beyond the first 30 days in orbit. At the beginning, the program focused on taking 3 years or less to launch, using far reaching technology that enabled diverse missions in a recurring program, and training scientists and engineers. Having fixed-cost caps and designing to the lowest cost possible while completing at least the "minimum science mission" is prevalent in today's science missions using small spacecraft. But that didn't become the highest priority in the SMEX program until the design phases for FAST and SWAS were complete. A *minimum science mission* is the "rock bottom" option beyond which the science mission can't be justified for the agreed upon cost cap. Table 11.3-1 summarizes the SAMPEX mission's costs.

TABLE 11.3-1. **Summary of the SAMPEX Mission Cost in Real Year $K.** The non-recurring cost consists of the SAMPEX share (approximately 25%) of the SMEX program startup and technology development cost.

	Non-recurring Cost	Recurring Costs	Total Costs
Mission management/system I&T and L&EO support[1]	$480K	$8,272K	$8,752K
Spacecraft	$3,700K	$14,215K	$17,915K
Instruments[2]	—	$6,284K	$6,284K
Ground segment	$840K	$5,400K	$6,240K
Launch segment	—	$13,000K	$13,000K
MO&DA (3-year mission)	—	$15,000K	$15,000K
TOTAL	$5,020K	$62,171K	$67,191K

1) Includes project office support, observatory I&T, Goddard Space Flight Center tax, flight assurance, transportation, and other mission integration costs. 2) Does not include $3.5M U.S. equivalent of Heavy Ion Large Telescope instrument provided by Germany.

As the first mission, SAMPEX provided the foundation for recurring SMEX missions of diverse scientific and technical requirements. If SAMPEX had been a standalone one-of-a-kind mission development for the lowest cost possible, while still ensuring mission reliability, it may have saved $10 to $15 million in life-cycle costs. As a part of a mission set that included FAST and SWAS which had far greater requirements for state of the art technologies in spacecraft and ground system capabilities, we believe that savings of the order of $25 million were realized to meet the total budget for the three missions. In developing SAMPEX, we took a series of steps to reduce cost while establishing the infrastructure for the future SMEX missions. The key factors that allowed us to reduce cost are:

- Use of dedicated project team
- Shared institutional project support
- Distributed architecture with standard interfaces
- Engineering test bed for verifying interfaces and developing software
- Reduced paper and greater emphasis on testing
- Leverage on technology developments
- Dedicated parts engineering and procurement

11.3.1 The SAMPEX Mission

SAMPEX's instruments measure energetic electrons as well as the ion composition of particle populations from 0.4 MeV/nucleon to hundreds of MeV/nucleon. They're designed to operate from a zenith-oriented satellite in near-polar orbit. They consist of a composition analyzer for low-energy ions, a heavy-ion large telescope, a mass-spectrometer telescope, and the proton/electron telescope. The science investigation team for SAMPEX consists of eleven co-investigators from five U.S. institutions and one German institution (Table 11.3-2).

TABLE 11.3-2. The SAMPEX Science Investigation Team.

Investigators	Institution
G. M. Mason (Principal Investigator), D. Hamilton	University of Maryland
D. Baker, T. Von Rosenvinge	NASA Goddard Space Flight Center
J. Blake	The Aerospace Corporation
L. Callis	NASA Langley Research Center
D. Hovestadt, B. Klecker, M. Scholer	Max-Plank Institute, Germany
R. Mewaldt, E. Stone	California Institute of Technology

A key part of the SAMPEX observations is to use the Earth's magnetic field as a giant magnetic spectrometer to separate different energies and charge states of particles as SAMPEX executes its near-polar orbit. SAMPEX's orbit has an altitude of 520 km by 670 km and an 82 deg inclination. Some of the most significant discoveries and "firsts" from the SAMPEX mission are as follows:

- Determination that anomalous cosmic ray nitrogen, oxygen, and neon are singly charged

- Discovery of the precise location of trapped anomalous cosmic rays in the magnetosphere

- Excesses (factor ~4) of neutron rich isotopes of Ne and Mg in ^{3}He-rich solar particle events

- Discovery that magnetospheric electrons are globally accelerated in association with the impact of high-speed solar wind streams

- Discovery that the inner radiation belt at $L = 1.2$ consists of roughly equal amounts of ^{3}He and ^{4}He at 10 MeV per nuclei [Baker, et al., 1996]

- Evidence that relativistic precipitating electrons provide odd nitrogen to the middle atmosphere and can affect middle-atmospheric ozone

SAMPEX is a momentum-biased, Sun-pointed spacecraft that maintains the experiment-view axis in a zenith direction across the Earth's polar regions and keeps its solar arrays pointed at the Sun. It does so by maintaining the momentum vector toward the Sun and rotating the spacecraft at one revolution per orbit about the Sun and spacecraft axis. Schematics of the SAMPEX spacecraft and the orbit concept are in Fig. 11.3-2 and Fig. 11.3-3, respectively.

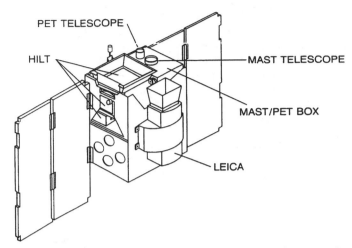

Fig. 11.3-2. The SAMPEX Spacecraft. The top of the spacecraft in the figure remains pointed toward the zenith.

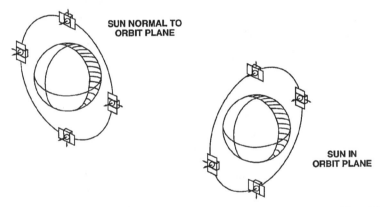

Fig. 11.3-3. SAMPEX Orbit Attitude Control Maneuvers About the Sun Line. The pay-load axis remains pointed toward the zenith. The angular momentum vector is Sun pointed. The spacecraft rotates once per orbit about the Sun-spacecraft line.

11.3.2 Description of the Spacecraft Design and Architecture

Table 11.3-3 lists the technical characteristics of the SAMPEX spacecraft. Figure 11.3-4 shows the SAMPEX spacecraft architecture. The heart of the SAMPEX architecture is the Small Explorers Data System (SEDS), shown in Fig. 11.3-4 [Smith and Hengemihle, 1990; Fairchild, 1989]. The SEDS receives commands from the ground and delivers them to onboard subsystems. It collects engineering and science data for telemetry transmission to the ground, records data for playback when out of ground contact, and provides autonomous spacecraft operation. The SEDS uses the DOD-STD-1773 standard interface to connect to other onboard systems. Space to ground communications use the standard international communications protocol of the "Consultative Committee on Space Data Systems" (CCSDS). Internally, the system uses standard interfaces for software communications and backplane interfaces, as well as standard packaging concepts. With this base of standard interfaces, the system uses a distributed architecture to provide modularity and increase its performance and reliability. Within the distributed environment, the individual components monitor each other's performance and can support autonomous fault recovery. Even in SAMPEX's single-string configuration, the distributed system allows degraded modes of operation if components fail. SEDS incorporated advanced microprocessors, gate arrays, fiber-optic interfaces, surface-mount technology, and other advances in spacecraft technology and concepts.

TABLE 11.3-3. SAMPEX's Technical Characteristics.

Launch	• Dedicated Scout from Vandenberg Air Force Base, July 3, 1992
Orbit	• 520 x 670 km, 82 deg inclination
Payloads	• Low Energy Ion Composition Analyzer (LEICA) • Heavy Ion Large Telescope (HILT) • Mass Spectrometer Telescope (MAST) • Proton/Electron Telescope (PET)
Structure	• 4-sided • Designed to 15 g's axial, 9 g's lateral acceleration
Electric power	• 102 W average power at end-of-life • 200 W peak ±28 V power • 2 deployable (2 panels each) GaAs solar arrays • 28 V, ±15 V, ±5 V DC-to-DC converters • 9 A·hr super NiCd battery • Voltage-controlled charge regulators • Pyrotechnic release mechanisms controlled by onboard computer
Attitude control	• Solar-pointed, momentum-biased system • Attitude-control accuracy < 0.5 deg except when the direction to the Sun and the Earth's magnetic field vector are co-aligned within 10 deg • 3 electromagnetic torquers • 3-axis magnetometer sensor • 5 coarse Sun sensors • 1 digital fine Sun sensor
Thermal control	• Passive control with multilayer insulation blankets and coatings • Thermostatically controlled heaters
C & DH	• Attitude determination and control • Spacecraft and payload scheduling • Telemetry data formatting • Command interpretation • Housekeeping • Communications system control • 30 MB (after EDAC) solid-state recorder • Payload and satellite control from the ground
Communication	• 2 quadrifilar helix antennas • 2 90-deg hybrid junctions • Power divider • Near Earth S-band 5 W transponder • Programmable rates of 4, 16, 900 kbps for data downlink telemetry • 2 kbps command rate • 3 modes for store-and-forward and "bent-pipe" communication • Coherent Doppler reflection and range tone-broadcast

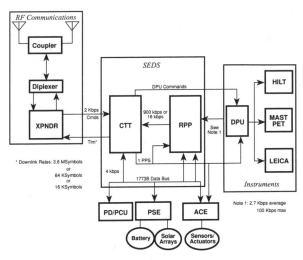

Fig. 11.3-4. SAMPEX Spacecraft Architecture. The Small Explorer Data System (SEDS) consists of the Command and Telemetry Terminal (CTT) and the Recorder, Processor, and Packetizer (RPP). See also Fig. 11.3-4.

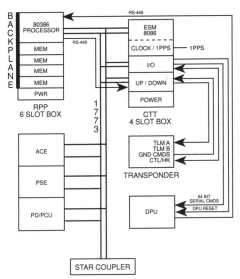

Fig. 11.3-5. Block Diagram of the SAMPEX Small Explorers Data System. See text for discussion. See also Fig. 11.3-3.

The SEDS hardware consists of two components: the Recorder Processor and Packetizer and the Command and Telemetry Terminal. The Terminal provides a central processing platform, stores data between ground contacts, accepts and packetizes serial data from the instruments, controls the spacecraft's attitude, and generates the telemetry data stream for transmission to the ground. The packetizer uses an Intel 80386 as the spacecraft's main processor which, combined with an 80387 numeric processor extension, provides three million instructions per second of processing power for spacecraft command and data handling. Although the packetizer can handle larger memory requirements, the one flown in SAMPEX had 30 MB of solid-state memory (after error correction and detection) available for science data. The Terminal provided the spacecraft a second processor, a timing and synchronizing reference, a transponder interface, and an onboard engineering I/O. Each function is accessible through the 1773 bus.

The attitude-control system combines three orthogonal magnetic torquers to react against the Earth's magnetic field, and one momentum reaction wheel to provide the biased momentum. A 2-axis, digital Sun sensor, a 3-axis magnetometer, and a set of five coarse Sun sensors are used for attitude determination at better than 2 deg accuracy except when the Sun's and Earth's magnetic vectors become co-aligned. When this occurs, the spacecraft simply coasts until the co-alignment condition passes. This control approach provides the "anti-nadir" pointed attitude required by the science investigation.

TABLE 11.3-4. SAMPEX Spacecraft Weight, Power, and Cost. SAMPEX was developed in-house at Goddard. The SAMPEX spacecraft design was intended to satisfy the requirements of a variety of future SMEX missions.

Element	Weight (kg)	Power (W)		Cost ($K)
		Average	Peak	
Instruments	52	21	29	$6,284K
Structure and mechanisms	37	N/A	N/A	$1,387K
Attitude-control system	10	11	32	$1,815K
Command & data-handling system	12	17	30	$6,898K
Communications system	5	8	33	$1,031K
Power system	32	13	14	$4,910K
Thermal system	4	3	16	$84K
Harness	8			$109K
TOTAL	160	73	154	$22,628K

SAMPEX's power system transfers energy directly from two deployable, fixed, solar arrays. Those arrays contain 1.7 m² of GaAs solar cells that provide an average power of 100 W at the end-of-life orbit to the observatory, a 9 A·hr battery, and the power control and distribution electronics.

Two quadrifilar helix antennas provide hemispherical coverage for ground communication. SAMPEX uses a near-Earth transponder with 5 W of RF power. It operates in S-band and can support data rates up to 3 Mb/sec. The average science data rate for the mission is 3 kbps. The spacecraft is configured to operate with two ground contacts per day, each typically 10 minutes long. The stored data transfers to the ground stations at the downlink rate of 900 kbps. Commands uplink at 2 kbps transmission rate. For orbit determination, the transponder also does coherent Doppler reflection and range tone-broadcast.

11.3.3 Information Acquisition and Operational Data Flow

The SAMPEX spacecraft operates through a multimission operation center at Goddard Space Flight Center (GSFC). Science data transmits twice per day as a series of CCSDS-compliant transfer frames to a 9 m dish antenna at the GSFC Wallops Flight Facility, Wallops Island, VA. Commanding is CCSDS compliant and is done once per day. SAMPEX is the first NASA mission to fully exercise the CCSDS's recommendations for packetized command and telemetry.

The onboard unit that processes instrument data oversees control of the science payload. It directly controls the data acquisition from each instrument and forms packets of science data for recording by the Small Explorers Data System. Technicians calibrate instruments once a month. Requests for special operations originate from instrument representatives, and go to the Flight Operations Team from the Science Operations Center at the University of Maryland by Internet or FAX. Data transfers through electronic links between the ground station and the Mission Operations Center, between the ground station and a level-zero processing facility at GSFC, and between this facility and the Science Operations Center. The data resides on analog tapes at the ground station and on optical disk (short term) and magnetic tape (long term) at the level-zero processing facility. The Science Operations Center distributes the science data to all the co-investigators and to the National Space Science Data Center. The Science Operations Center delivers data to the instrument team using a SAMPEX team standard called TENNIS. TENNIS is compatible with the Standard Formatted Data Unit of NASA and the CCSDS.

11.3.4 Mission Development and Cost

The SAMPEX mission was developed "in-house"—GSFC developed the spacecraft and integrated the mission. The SMEX/SAMPEX Project Team:

- Managed the mission
- Supported project scientists
- Provided mission systems engineering
- Managed mission resources (scheduling, financial, etc.)
- Defined, developed, and qualified the spacecraft
- Integrated and tested the observatory
- Ensured spacecraft performance
- Ensured observatory performance
- Handled the mission's safety program
- Supported the launch-vehicle interface
- Transported the satellite to the launch site
- Operated the launch site
- Prepared for mission operations and managed the ground-segment interface

The team of co-investigators developed and qualified the instruments at the Science Operations Center. The average civil-service staffing for the SAMPEX mission was 45 to 50 full-time equivalents per year through the 35-month definition phase. Amortized startup and technology development added 10 to 15 full-time equivalents per year to this civil-service staffing.

The SAMPEX mission included the following technology developments. These technology developments—in particular the distributed-architecture concept—

allowed for a robust, capable, and adaptable spacecraft design that saved a lot of money in later missions.

- Distributed architecture with standard interfaces
- Single-board, 80386/80387 processor with miniaturized ASIC support logic
- Fiber-optic data bus (DOD-STD-1773)
- Data system fully complies with CCSDS
- Low-power, low-weight S-band transponder
- GaAs solar arrays
- Super NiCd batteries
- Autonomous, self-protecting spacecraft

The SAMPEX mission launched 35 months after a 3-month definition phase. At the time of selection, the SAMPEX instruments had matured for space flight through Get-away Special experiments and other flight projects, or by using hardware qualified for other flight programs but never flown. The pre-SAMPEX cost is estimated at $3.0 million for the LEICA, MAST/PET, and DPU instruments, and $1.7 million (U.S. equivalent) for the HILT instrument funded by Germany.

11.3.5 Approach to Low Cost

More than anything else, schedule drove SAMPEX's development. We took a series of programmatic and technical steps that allowed us to be cost effective, while establishing the infrastructure needed for the missions that followed SAMPEX. SAMPEX validated the projected future reductions. The paragraphs below describe some of these initiatives.

Dedicated Project Organization and Shared Support for Multiple Projects

The Small Explorers Project Organization oversees multiple missions, each with unique requirements, developed roughly one year apart. To accomplish this, all the developing missions shared many of the project functions, such as overall management, engineering of software and hardware systems, flight assurance, missions operations management, general project and institutional support, configuration management, and planning and scheduling. This organization allowed us to keep the program and technology uniform across the SMEX missions, as well as to share the costs of common activities. Every mission in a set of three missions, for example, would share a third of the cost.

Within this environment SAMPEX was assigned a team with mission-unique management and technical responsibility. The mission and instrument managers were permanently assigned to the project staff. Core lead engineers remained with the mission from the start through launch and validation in orbit. The core team obtained needed support from their matrix home organization for the time required to complete given tasks. Figure 11.3-6 shows the organization of the SAMPEX mission. Continual interaction and close communication between the team members was crucial, especially because the large team of investigators lived in widely separated locations. We relied on frequent face-to-face meetings at the investigators' institutions and at Goddard Space Flight Center.

The approach described above reduced costs, improved communications, and created a sense of ownership in the team. The pooled resources and expertise of the matrix organizations were assets, but to be efficient we had to negotiate their guaranteed availability because of their support to multiple projects. Likewise, during times when the matrix organization does not have enough work, other mechanisms must be provided to off-load personnel from the missions when the tasks finished. In general we were quite effective in managing personnel. However, with our technology developments driving the schedule, in many cases we could not off-load personnel and risk "losing" them to other projects.

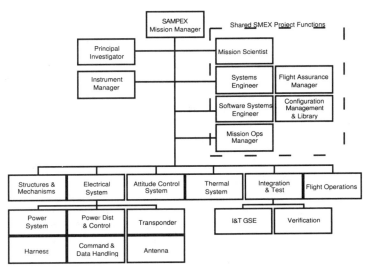

Fig. 11.3-6. SAMPEX Mission Organization. The program used a mix of dedicated SAMPEX personnel and share SMEX personnel. This approach was successful but required some attention to ensure that it would work. See text for discussion.

Distributed Architecture With Standard Interfaces

The Small Explorers Data System was intended to allow us to develop and integrate future spacecraft quickly and inexpensively. It uses standard interfaces in many areas including those in major hardware, software, and ground systems. SAMPEX demonstrated the expected benefits of standard interfaces by decreasing documentation, integration time and effort, test effort, system maintenance, and system weight and power, as well as increasing system reliability.

The standard databus for the spacecraft hardware was the SEDS' important design feature. We used readily available MIL-STD-1553 specifications and tutorials, plus low-cost, commercial, bench-top equipment, to completely check out how software and hardware interacted in each subsystem before integrating them. By using the standard interface protocol, we didn't have to document as much as is sometimes necessary for unique interfaces, and we simplified the spacecraft harness.

Using the standard interface reduced the weight of the harness by about 1 kg, and the fiber-optic MIL-STD-1553 interface (or 1773) reduced power by several watts. Fewer harness connections allowed for easier inspection and reduced the possibility of mistakes. The MIL-STD-1553 interface is inherently redundant and therefore added reliability even in SAMPEX's mainly single-string architecture.

The Consultative Committee on Space Data Systems standards for commanding and telemetry gave organizations and developers of hardware and software a common framework to follow, thus reducing unique documentation. Its recommendations reduced integration efforts because the subsystems' peculiarities didn't affect SEDS.

The onboard flight software also benefitted from using the command and telemetry standards of the Consultative Committee on Space Data Systems. In SAMPEX we applied this standard to the communication between software applications. The different tasks send and receive messages following the standard, and thus, individual applications don't need to specify how they communicate except to say they sent or received a message through the "software bus." Using this concept, we could independently load applications without affecting existing ones, which added flexibility and reduced the integration efforts.

The one-of-a-kind nature of space science missions make it difficult to estimate the projected cost savings from SEDS. However, based on observations of the SMEX/SWAS mission and other Explorer missions, such as the X-ray Timing Explorer, savings of 10% to 20% (or more) are possible. Reducing cost depends mainly on less interface documentation, modular, robust hardware and software, flexibility in early testing, electromagnetic cleanliness of the 1773 interface, and a standard format for packetizing of data.

Engineering Test Bed for Verifying Interfaces and Validating Software

An engineering test bed verified interfaces between the spacecraft components, between the instruments and the spacecraft, and between the spacecraft and equipment for integrating, testing, operating, and controlling it. We used the test bed extensively for acceptance and validation testing of off-line system and flight software. We built engineering test units for power, SEDS, and attitude-control electronics that replicated the flight units, except for using commercial-grade parts. After the launch, the test bed was dedicated to maintaining flight software for SAMPEX and for other SMEX missions.

Having a stable test bed provided much needed flexibility to verify and validate hardware and software before delivering them to the spacecraft. We also were able to troubleshoot off-line any of problems encountered during the spacecraft integration and test. The engineering test hardware replaced the spacecraft's flight hardware whenever problems arose and allowed us to integrate and test them with little interruption.

Leverage on Technology Development

SAMPEX benefited from the technology developments funded by others. An example of this leverage was the 1773 optical-interface modules, in which the communications industry and the military had a substantial investment. Using the 1773 allowed us to capitalize on their expertise, facilities, and procedures. Because SAMPEX was the first spacecraft to fly the 1773 interface in space, our accelerated schedule allowed this technology to mature quickly. Without that outside expertise, we would have developed the 1773 more slowly for SAMPEX. The 1553 would have been a backup, but we'd have lost the reduced power and the fiber-optic bus's inherent immunity to noise, which we considered very important for applications beyond SAMPEX.

Advances in surface-mount technology developments in industry also provided a vehicle for rapid design and part verification, automated part population and testing of the finished boards. For recurring use, surface-mount technology can reduce development time for electronic boards by an order of magnitude. Industrial advances in solid-state memory also helped us reduce development costs.

Dedicated Parts Engineering and Procurement

As a new program introducing new designs and technologies, SAMPEX depended on efficiently procuring EEE parts to keep construction of the spacecraft and flight instruments on schedule. We had a parts engineer who was expert in the parts industry and procedures. This dedicated expert kept us on schedule by reviewing, expediting, and quickly finding suitable replacements for parts.

Streamlined Flight Assurance and Reduced Documentation

The quality-assurance program was tailored to the mission and the abilities of the principal investigator's institution. We emphasized reducing documentation throughout the mission. Quality parts and materials, a limited program for ensuring reliability and quality, and relying heavily on the test program were key to balancing reliability against cost and complexity. We eliminated the traditional "plan-type" documents and, instead, focused on procedures for integration and tests. Lead engineers or scientists kept log books, instead of the more elaborate test reports and travelers.

We analyzed failure modes and effects across interfaces between only the instruments and spacecraft, thus eliminating extensive documentation at the part level for multiple boxes. The SMEX project provided a reliability engineer to help efficiently and uniformly analyze failures for all subsystems.

We could have reduced or simplified the program, data-management, and implementation plans. We know some data in these documents is valuable, but estimate they can be reduced by 80% to 90%. Programs should avoid complex, extensive documentation because it requires elaborate infrastructures for configuration management.

11.3.6 Summary

The development of "one-of-a kind" scientific missions such as SAMPEX, especially with the introduction of the new technologies was a challenge. Key lessons learned were:

- The detailed planning, end to end systems evaluation, cost, and risk assessment could have benefited from a longer than 3 months definition phase.

- Funding constraints at the beginning of the program took away flexibility to make early decisions to preserve schedule and therefore cost. A wise additional investment upfront to preserve schedule would have been worth ten times its cost later in delayed integration. Timely delivery of fully qualified components to integration and closure of associated non-essential tasks are key to a low cost.

- Concurrent engineering and the parallel development of components worked well, especially when a stable engineering test bed of core spacecraft components and interfaces was made available. It is strongly recommended that this test bed of engineering test units be in place before the end of the definition phase so that its fabrication does not compete with the development of the flight system. This test bed was invaluable for the timely development and validation of the flight software.

- A small dedicated team somewhat insulated from the matrix organization was also key. It is important that this relationship be defined as one where the organization acts as an enabler. It is also important that a skilled experienced core team, especially early in the planning phases of the mission be committed to the mission from start of definition to deployment in orbit.

- Emphasis on early planning and the reduction of duplication of efforts and reduction of paper are essential to low cost approaches.

As the first of the SMEX mission, SAMPEX was the test bed and foundation for the development of an international multi-organization space science mission in less than three years, and at the same time develop technology designed to increase capability, and provide flexibility for future missions in a cost effective manner. The thrust of the developments was to re-design a spacecraft architecture that allowed to shorten the schedule by simplifying the development and integration process through noise immune robust standard interfaces, increased technical capability and flexibility, and reduced paper. This technology is viewed as one of the most significant cost reduction initiatives for the spacecraft of the future.

HETE Summary

HETE is to carry out the first multi-wavelength study of gamma-ray bursts with UV, X-ray, and gamma-ray instruments.

Spacecraft dry mass:	125 kg	No propulsion system	
Average power:	60 W	TT&C:	250 kbs

Launch: To be launched by Pegasus from Wallops Island, VA; now scheduled for 1996.

Orbit: 550 km × 550 km at 38 deg inclination

Operations: Operated from MIT's control center utilizing primary ground stations in MA, Japan, and France

Status: *HETE* was lost due to launch failure November 4, 1996. Replacement *HETE-2* launched October 9, 2000.

Cost Model (FY95$M): See page 348 for explanation.

	Expected Cost	Small Spacecraft Model	Actual Cost*
Spacecraft Bus	$34.4M	$7.0M	$5.6M
Payload	$26.6M	$3.3M	$15.8M†
Launch	$16.6M	$5.5M	$7.0M
Ground Segment	$121.8M	—	$1.7M‡
Ops. + Main. (annual)	$10.1M	—	In spacecraft payload
***Total** (through launch + 1 yr)*	**$209.4M**	—	**$30.1M**

*An inflation factor of 1.066 has been used to inflate to FY1995$ [SMAD, Table 20-1].
†Includes estimated foreign contributions, initial operations and maintenance cost, and substantial test and software integration cost that are usually included in the spacecraft bus cost.
‡Includes estimated foreign contributions.

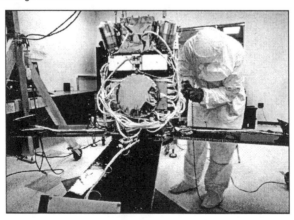

Fig. 11.4-1. HETE. In integration and test with solar arrays deployed. Momentum wheel is in the large circular object in the center. Tilted canisters at the top are the UV cameras, at the corners of the X-ray monitor support structure. The gamma-ray detectors are located between the UV cameras.

11.4 HETE

François Martel, *AeroAstro, Inc.*

The High Energy Transient Experiment (HETE) is an astrophysical observatory intended to localize and help identify the sources of cosmic gamma ray bursts [Ricker, et al., 1988]. It was built by the MIT Center for Space Research (MIT-CSR) and AeroAstro, Inc., with a fixed total budget from NASA of $14.7 million and contributions from foreign partners worth over $5 million. HETE is to be launched at an additional cost of approximately $7 million on a Pegasus XL along with the Argentinian satellite SAC-B. Due to technical and programmatic difficulties, including two launch vehicle failures, launch has slipped from 1994 into 1996.

A similar low-cost astrophysics spacecraft, ALEXIS, was previously built by AeroAstro and Los Alamos National Laboratory with DoE funding of approximately $18 million [Priedhorsky, et al., 1993a]. ALEXIS was launched April 25, 1993, by the Air Force on a standard Pegasus at a cost of $7 million. Although damaged during launch, it is still operational and provides continuing science data through a small ground station in Los Alamos operated by the project science team [Priedhorsky, et al., 1993b]. ALEXIS and HETE are both examples of lower cost systems that provide extensive capabilities due to the use of small teams, shorter schedules, modern commercial technologies, less paperwork, and a far more streamlined bureaucracy.

11.4.1 HETE System Overview

HETE's mission is to detect and localize cosmic gamma ray bursts using onboard gamma, X-ray, and ultra-violet instruments, so we can identify accurately the direction of the observed events [Ricker, et al., 1988]. Satellites have observed bursts of cosmic gamma rays for the last 30 years, but their sources and physical mechanisms are still mysterious. The direction of these high-energy bursts is difficult to pinpoint precisely. Deep searches of their sources from ground telescopes have been fruitless because the search areas are too large.

HETE's goal is to help localize the bursts and identify their sources. The system functions are to:

- Detect bursts in the gamma-ray range

- Use an X-ray imaging monitor to image the bursts in the X-ray range

- Locate, whenever possible, the direction of their origins within a few arc minutes

TABLE 11.4-1. HETE's General Specifications.

Launch	• Dual launch on Pegasus XL with Argentine spacecraft SAC-B, from NASA Wallops
Nominal orbit	• 550 km circular • 38-deg inclination
Payload	• Four NaI gamma-ray detectors with an energy range of 5 to 350 keV • Coded mask, X-ray imaging monitor with a field of view of 1.8 sr and 2.0 to 2.5 keV energy range • Four UV (250 nm) cameras, with 1024 × 1024 pixel CCDs • Radiation belt monitor
Digital processor	• Imbedded system shared by platform and payload instruments • 12 microprocessors (100 MIPS) • 96 MB of mass memory
Power system	• Four panels, stowed at launch, all Sun-pointing in flight • Aluminum honeycomb structures • Silicon solar cells • 175 W BOL • 60 W average regulated power to payload and system • Digitally controlled power point tracker • NiCd batteries
Attitude control	• Sun-pointing, momentum-bias stabilization • Instruments include: momentum wheel (1), torque coils (3), magnetometers (2), coarse Sun-sensor array (12), medium and fine Sun sensors • Payload's UV cameras provide night drift rate from star tracking • 0.5 deg Sun-pointing
Structure	• Instrument housings provide mechanical structure • 125 kg total mass • Dimensions 0.9 × 0.5 × 0.5 m with panels stowed • Passive thermal control, with localized heating capabilities
Communication	• Two S-band, 250 kbit/sec, 2 W transceivers • Five S-band antennas • VHF 0.5 W, 300 bit/sec burst-alert/status transmitter • 8-channel, cold-boot GPS receiver, providing microsecond UT time and orbital parameters • VHF launch beacon
S-band stations	• 1.8 m dish in Miyazaki (Japan) and Haystack, MA (U.S.) • 3.8 m dish in Toulouse (France) • Commands and data managed from MIT-CSR through internet
VHF stations	• 15 low-cost, receive-only stations operating at 137 MHz • Spread around world for good coverage of HETE's footprint and linked to the Internet for alerts to observation teams on the ground

- Try to localize them within a few arc seconds from UV transients that may be associated with the events. HETE has four UV cameras that cover the same overall field as the X-ray monitor.

HETE processes data onboard and broadcasts the detections and computed locations within seconds through a VHF channel. A global network of receive-only

VHF stations alerts ground observatories through Internet within minutes after HETE detects an event. Successful HETE observations will lead to much smaller search areas for ground observatories that are looking for the sources of these bursts. Thus, HETE contributes significantly to the international efforts to understand the phenomenon. HETE will also provide flux counts at small time scales and spectral information of additional value for studying gamma-ray bursts, as well as other valuable "secondary science" results such as the detection of UV transients of stellar origins.

HETE is an innovative system, breaking new ground in using current technologies for space. Innovations include a particularly powerful, imbedded computer system on board the spacecraft (100 MIPS, 100 MB); using the Internet for sending commands and acquiring data through three remote ground stations in Japan, France, and the U.S.; Internet-based worldwide alert using more than 15 small VHF receive stations. Table 11.4-1 lists the system's general specifications.

Figure 11.4-2 shows the spacecraft block diagram. Although redundancy isn't required, some critical devices have back-ups, and the digital system is cross linked to provide graceful degradation in case a processor fails. The diagram shows only one of two RF chains.

Figure 11.4-3 shows the ground-station network. MIT-CSR operates HETE's ground segment through the Internet. Three S-band primary stations in the U.S., Japan, and France send ground commands and receive high-resolution data from instruments.

11.4.2 Budget

NASA provided MIT-CSR with a budget of $13.5M for HETE. This money covers costs for the spacecraft platform, UV instruments, the U.S. ground station, project management, system integration and tests, support of the science teams for data analyses, and a first year of operation. NASA also funded Los Alamos through a NASA/DOE transfer of funds of $1.2 M to develop the coded mask and software for the X-ray monitor.

AeroAstro built the platform and one primary ground station, as well as eight secondary ground stations, under a Cost Plus Fixed Fee contract with MIT, amounting to close to $5.9M. RIKEN, in Japan, funded the Miyazaki ground station which cost close to $500K. The Toulouse station at the French Aerospace school ENSAE (also known as "Sup'Aero") is a facility originally built for the ARSENE mission, now modified for HETE.

RIKEN built the X-ray monitor, with funding from the Japanese agency, STA. The gamma detectors built by the Centre d'Etude Spatiale des Rayonnements (CESR) in France and a GPS receiver developed by the French space agency, CNES, were funded by CNES. These foreign contributions, worth several million dollars, helped reduce NASA's cost for HETE.

Figure 11.4-4 breaks out development costs ($5.9M) for HETE's platform. Digital and software labor take a remarkable 35% of the overall costs. Including costs for materials drives this number up to 50%. Thus, HETE is as much about developing a space-qualified computer system as it is about small-spacecraft development. That's not surprising considering the power, complexity, and sophistication of HETE's digital system and software.

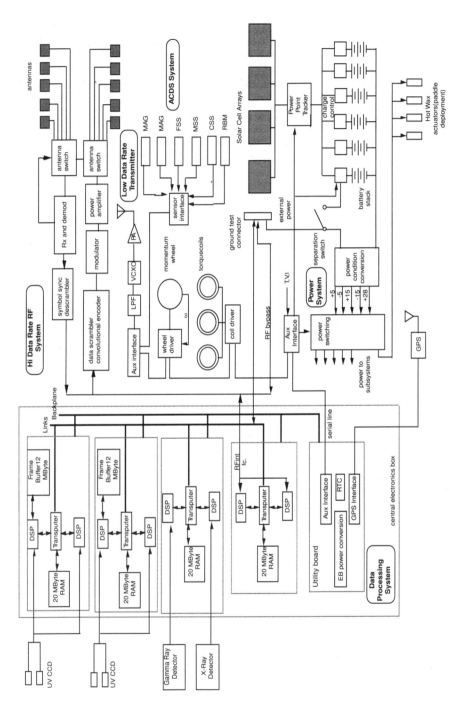

Fig. 11.4-2. HETE Block Diagram. Platform and scientific instruments share a 12-processor system with 96 MB of memory.

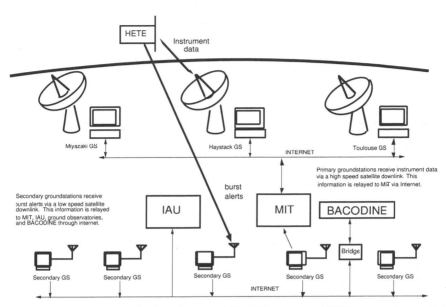

Fig. 11.4-3. HETE Ground Station System. MIT-CSR operates HETE's ground segment through the Internet. About 15 secondary ground stations provide immediate global coverage for the burst alerts.

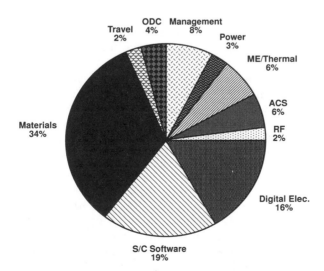

Fig. 11.4-4. HETE Platform Costs. Distribution of development costs for the HETE platform and onboard data processor excluding costs for the scientific payload's instruments. Design and development of the onboard computer system, related development platforms, and software were a major part of the effort.

11.4.3 Key Reasons HETE Can Succeed with a Low Budget

Clear Ground Rules

Keeping HETE's ambitious project within a limited budget required certain operating procedures:

- The Principal Investigator, who forms and leads the project team, is responsible for developing the full system.
- The project team manages its limited budget under an agreed-to funding profile.
- Except for safety, usual standard specifications aren't mandatory.
- The project team handles technical decisions and trade-offs.
- The project team reports to NASA's project manager at NASA headquarters.
- Significant foreign contributions of hardware and services help reduce NASA's costs.

Memorandum of Understanding between NASA and MIT-CSR

MIT and NASA drafted a Memorandum of Understanding (MOU) before starting the HETE contract to clarify the ground rules listed above. The other HETE partners agreed to similar MOUs with MIT. The MOU establishes these important relationships:

- HETE's scientific mission is defined as a "goal", undertaken as a "best effort" by MIT-CSR, who can "descope" it within limits, if necessary. MIT-CSR does its best to meet the goal within the budget and schedule.
- MIT-CSR handles the space and ground segment of the project. NASA provides the launch.
- HETE's budget is fixed for the mission, with the understanding that MIT handles internal schedule or budget problems by descoping or other means, within limits.
- NASA oversees the project from headquarters but MIT need not follow NASA's standard technical specifications except those concerning safety.
- NASA must scrupulously follow the funding profile.

The MOU is important in clarifying the HETE project's rules, especially because they depart from classical programs. They define HETE as a class-D program that operates at high risk and low cost. External requirements are limited to safety regulations and documentation, plus the needed oversight for NASA managers. The project team has wide authority to define designs, resolve problems, select parts, and make rapid technical decisions, saving both time and money.

But an MOU only clarifies an understanding between those who start the program. It doesn't necessarily define the administrative flow and work is needed to keep its spirit alive. Also, the actual contract, which carries funding, associated obligations, and legal rules, obviously supersedes the MOU and some conflicts are unavoidable.

For example, procurement regulations impose administrative approvals, often slowing down the acquisition of necessary hardware and affecting project costs and schedule. Buying equipment that costs more than a certain amount often depends on approvals including a mandatory search of government surplus for possible equivalent hardware. In most cases this search achieves nothing more than a few weeks of delay and more administrative costs.

Some administrative delays can be devastating. Contracting difficulties caused delays for HETE's initial funding which affected the entire project. When money isn't available on time, project people need to find work on other programs. This is true in particular for small organizations with low overhead, few reserves, and a critical need for consistent cash flow. Most people who have left don't come back immediately when funding is reinstated; and some don't come back at all. Continuity and morale break up, experience may fade, and the added delays affect the work of other teams. Funding perturbations reverberate throughout the project. After a difficult start, NASA provided HETE with well-timed funding for the rest of the program.

Organization of Partners from Several Nations

Figure 11.4-5 outlines the overall organization, which includes significant participation from Japanese and French partners. For HETE, these foreign contributions far outweighed the costs and difficulties of managing a development project across international boundaries.

MIT-CSR and the other HETE partners signed Memoranda of Understanding specifying the responsibilities of each team; they are similar in spirit to the memorandum between NASA and MIT. NASA and agencies in Japan and France signed similar agreements at the government level.

Science teams in the U.S., Japan, and France are concentrating on the same goal with complementary instruments. HETE is focused on a single mission which, in general, helps harmonize system requirements and interfaces.

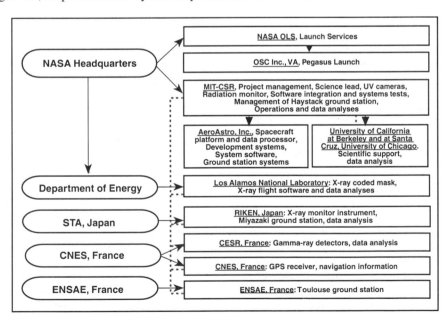

Fig. 11.4-5. HETE's General Organization. Contractual and funding relationships are shown with plain arrows. MIT-CSR coordinates the activities of the different teams (dotted lines).

11.4.4 Developing a Low-Cost System: The HETE Experience

NASA sometimes categorizes programs in four classes from A, the most reliable human flights, to D, associated with low-cost, high-risk experimental projects. For class D, the only formal requirements concern safety. The project team determines management, documentation requirements and controls, formal meetings, and technical specifications.

HETE has the flexibility of a class D project. MIT handles most project decisions. That means HETE's Principal Investigator, George Ricker, and his science team, who are the primary "customers" for the mission and are most interested in its scientific success, are also directly involved in system trade-offs and in balancing budget and schedule. This participation helps us quickly reach reasonable requirements and intelligent compromises while keeping in sight the overall scientific goal.

System designers decide on reliability issues based on their engineering judgment and the Principal Investigator's guidance. In particular, each scientific instrument team manages and makes key decisions about their hardware, within specified interfaces and envelopes, but with minimal external control. This approach worked especially well on HETE because the scientific teams (see Fig. 11.4-3) are experts on their instruments and bring to bear decades of experience in space hardware.

The possibility of single-string failures is acceptable, so redundancy isn't required. But the HETE designers design conservatively and include layers of protection on key components whenever possible. For example: current-limiting circuitry protects from catastrophic latch-ups; either one of two digital-signal processors can operate the platform instruments; and any of four transputers can operate the spacecraft controls, allowing "graceful degradation" of the system.

In selecting parts, we don't have to follow the government's preferred parts lists or component classes. Commercial components are acceptable, and often preferred. Experience seems to show that, all else being equal, "non-screened" parts may be more reliable than screened ones, because they've been handled less during production and testing.

Of course, environmental constraints do apply. HETE's designers selected the unusual combination of the system microprocessors after searching literature and reviewing industrial data on radiation hardness for the candidate parts.

Digital Hardware and Software Development

Digital hardware and software development were a major source of costs for HETE. The digital system has an original embedded architecture (see Fig. 11.4-2) that combines eight Motorola 56001 DSP processors and four INMOS Transputers. We developed this architecture to read four CCD cameras with a resolution of 1024 × 1024 pixels and to process data in real time. Other instruments, and the platform itself, run in parallel on the same system. Developing HETE meant developing a new, powerful, space-qualified computer at the same time—in a sense, doubling our challenge.

Developing the software was a special challenge because the digital system was itself in development, tools were inadequate or unavailable, different teams were using different platforms (Macintoshes, NeXTs, SPARCs), and, initially, we had too little money to harmonize all the development hardware. With requirements at first poorly defined, teams were developing embedded real-time software at more than nine locations and on three continents.

To solve these problems, we defined a message-passing protocol and relevant functions very early to specify interfaces between the processors and processes. The entire system reliably uses that protocol—from real-time processes onboard to data transfer, ground commands, and data archiving.

We built copies of the spacecraft's processor as soon as possible (but later than desired) and shipped them to the software developers as realistic development platforms. We modified these copies to resolve discovered problems. EEPROM technology allowed remote upgrading of software for development platform.

The HETE project used programmable gate arrays so we could rapidly develop and change hardware that could do complex tasks. It was a key to successfully developing the digital hardware and the CCD camera electronics.

Communications within the HETE Team

The HETE team communicates extensively through electronic mail. We do tests on the live spacecraft remotely through the Internet, and software developers work in "real time" from remote locations, taking into account their work schedules in different time zones. But we still had "face-to-face" meetings, which are most productive while integrating and testing the spacecraft.

The HETE contract requires some formal meetings, such as NASA's annual review, as well as monthly teleconferences. We also organized other formal meetings, such as PDR and CDR, software reviews, and science team meetings. They contributed to the communication between teams and individuals, especially to clarify requirements and interfaces. They often turned into lively debates resulting in significant decisions.

Launcher Interface

HETE is to launch with an Argentinian satellite, SAC-B, on a Pegasus XL rocket. HETE's Marmon ring is bolted close to the launcher's avionics section. A closed canister, with a cylindrical body and a truncated conical head, covers HETE. SAC-B attaches to the top of the canister. This means releasing HETE in its orbit requires four successful operations besides the successful launch of the rocket: opening of the rocket fairing, releasing SAC-B, ejecting the canister top, and finally releasing HETE through the cylindrical part of the canister.

The canister wasn't fully designed until late in the project and the new PEGASUS XL's characteristics weren't fully known. Important launcher interface remained opened for years: What vibration levels should we design and test to? Might the carbon-fiber structure or other fairing materials contaminate the UV cameras? Meetings, reviews and conference calls helped deal with interface and schedule issues among OLS, OSC, SAC-B and HETE.

Import-Export Authorizations

The International Traffic in Arms Regulations (ITAR) required us to get export licenses for the development and spacecraft hardware and software we sent to and from Japan and France. These licenses needed joint authorizations from the Department of State, Department of Defense, Commerce Department, and NASA. As of October 23, 1993, NASA's civilian scientific satellites don't have to follow ITAR, but the regulations still apply to other kinds of projects. The International Affairs office at NASA Headquarters helped us file the required licenses.

We had to transport test equipment, engineering units, and flight units for the X-ray and gamma instruments to and from France and Japan. Custom duties could be waived because we built the hardware under funding within the different countries, with no exchange of money. Government to government Memoranda Of Understanding (MOUs) between the agencies involved (NASA/RIKEN and NASA/CNES) defined the responsibilities. However, there were delays in drafting and signing the MOUs due to minor legal disagreements between countries on some liabilities. Ad hoc measures, such as carnets (occasionally paying duties on some equipment to save time), resolved these problems.

NASA's international office at Headquarters provided needed support and occasional emergency help. One such emergency occurred when the CNES GPS flight receiver was blocked from passing through customs for 48 hours because the international GATT agreements had annulled the previous authorizations and the renewed authorizations weren't entered in the customs system yet.

Frequency Allocations

NASA's support was essential to get RF-band allocations. Obtaining frequencies and bandwidth allocations through FCC (commercial projects) or NTIA (government-sponsored projects) is difficult and lengthy. It requires detailed analyses of broadcasting patterns, transmission-system specifications, orbit, ground station locations and specifications, and potential interferences with other systems worldwide. The process includes reviews and negotiations between U.S. agencies, as well as reviews and approvals through international conferences and with individual countries, some of whom register objections and get responses.

Several possible S bands and VHF bands were floated in the first year. Of course, not knowing the final frequencies impeded development of the spacecraft's and ground station's RF hardware, especially because components that depend on frequency require long lead times.

Eventually, the NASA office interfacing with NTIA felt confident enough about a chosen band to provide the HETE project with its final set of frequencies. Final official approval was to require many more months and included filing the characteristics measured in testing the hardware after development. Clearly, NASA and its contractors gave us significant support; without it HETE would have had real difficulty producing the needed paperwork and sustained technical and bureaucratic expertise. HETE's foreign ground stations also required licensing in those countries according to their administrative rules.

Orbital Information

The initial orbital elements for HETE will come to the project team from the launch range and NASA stations. Later, we'll track the orbital elements using independently the onboard GPS receiver and range measurements from HETE's ground contacts with the three primary ground stations in Japan, France, and the U.S. The onboard GPS receiver itself computes a good set of orbital elements which can be down-loaded to the S band and VHF ground stations. In addition, raw range data will go to CNES through the Internet for processing, and CNES will promptly return to MIT updated orbital elements in NORAD-type, "two-line" format. CNES will also use the raw range data to analyze the GPS receiver's performance and evaluate related navigation algorithms.

Growth in Schedule and Cost

Some preliminary design studies for HETE began in 1991. The development contract (NASW-4690 to MIT's Center for Space Research) turned on in April 1992. The initial launch date was July 1994. The delay in negotiating and initiating the original contract slowed HETE by six months, so the launch was postponed to December 1994. Other difficulties, in particular launcher failures, pushed the launch date to 1996. NASA increased the HETE total budget to about $15.7 million to cover these launch delays.

In parallel, HETE faced several problems, especially in developing the digital system, a powerful but complex architecture.

Differences in funding cycles and schedules in the three countries involving HETE also caused some problems. For example, interface questions are better resolved when the two sides are at comparable levels in system analysis and design. In other words, a late starter may slow down other team members. But the HETE teams have many years of experience in international space projects, so these difficulties never became major crises.

HETE is a Research and Development project, which means scheduling and cost are inherently uncertain. HETE is not only a scientific experiment but also a technology experiment, with many instruments and subsystems designed and built for the program. The project has had its share of technical problems; HETE's aggressive, "success-oriented" schedule, reflecting the limited budgets, was mismatched with the complex system development. Arguably, it put excessive stress on the designers.

All these difficulties cost money. We had to redirect this money from other items in the budget, such as operations and data analysis. Thus, we handled our problems within our budget, but at the cost of affecting the Science Team's goals for extended operations and data analyses.

"Real life" always introduces the unexpected in a research project such as HETE. Money and schedule margins are necessary to handle the inevitable unforeseen difficulties, and we wish the budget would have included some "reserve" to reduce the need for descoping.

11.4.5 Conclusion

The key to HETE's relatively low cost is in the set of ground rules defining it as a class D program managed by the science team. That approach allows for rapid and efficient decision making with a deep understanding of the fundamental science goals. It cuts through excessive bureaucratic rules and permits using state-of-the-art technologies. It does require a small and knowledgeable team who are self-motivated and disciplined.

The HETE experience reminded us that keeping things simple, not compounding development efforts, can save a lot of grief. Using current technology for electronic communications, the Internet, programmable gate arrays, and carefully selected commercial components was crucial in the low-budget environment. Foreign contributions helped keep costs down without burdening the project.

HETE's spacecraft and ground system were completed within their limited budget, despite the usual difficulties for an R&D program. NASA's cost for HETE is remarkably small, especially in view of its capabilities and the scientific interest in this international mission.

References

Baker, D. N., G. M. Mason, O. Figueroa, G. Colon, J. G. Watzin, R. M. Aleman. July, 1994. "An Overview of the Solar Anomalous, and Magnetospheric Particle Explorer (Sampex) Mission," *IEEE Transactions on Geoscience and Remote Sensing.*

Baker, D. N., J. B. Blake, L.B. Callis, J. R. Cummins, D. Hovestadt, S. Kanekal, B. Klecker, R. A. Mewaldt, R. D. Zwickl. March, 1994. "Relativistic Acceleration and Decay Time Scales in the Inner and Outer Radiation Belts," SAMPEX, *Geophysical Research Letters*, vol. 21, no. 6.

Baker, D. N., J. B. Blake, S. Kanekal, B. Klecker, G. Rostoker. August, 1994. "Satellite Anomalies Linked to Electron Increase in the Magnetosphere," EOS Transactions, *American Geophysical Union*, vol. 75, no. 35.

Baron, W. R., M. Houghton-Larsen, P. L. Thomsen, Computer Resources International (CRI). 1994. "Development of the Ørsted Satellite Project." Paper presented at the 8th Annual AIAA/USU, Logan, UT, August/September 1994.

Computer Resources International (CRI). September 1991. *Ørsted Satellite Feasibility Study.* CRI, Bregnerødvej 144, 3460 Birkerød, Denmark.

Computer Resources International (CRI). March 1993. *Ørsted System Definition Study Summary Report.* CRI, Bregnerødvej 144, 3460 Birkerød, Denmark.

Danish Meteorological Institute (DMI), November 1993. *Ørsted Research Announcement.* DMI, Lyngbyvej 100, 2100 København Ø, Denmark.

European Space Agency. 1992. *Derating Requirements and Application Rules for Electronic Components*, ESA PSS-01-301 Issue 1. ESA-ESTEC, Noordwijk, Holland.

Fairchild Space Company. November, 1989. *Small Explorers Data System Description Document.*

Grahn, S. 1987. "VIKING and MAILSTAR, Two Swedish Small Satellite Projects." In *Proceedings of the 1st Annual AIAA/USU Conference on Small Satellites."* Logan, UT, Oct 7–9, 1987. Center for Space Engineering, Utah State University.

Grahn, S. 1993. "The Freja Magnetospheric Research Satellite, Design and Flight Operations." In *Proceedings of the 7th Annual AIAA/USU Conference on Small Satellites.* Logan, UT, September 13–16, 1993. Center for Space Engineering, Utah State University.

Geophysical Research Letters. 1994. vol. 21, no. 17. Special Section, "The Freja Project." Twenty-six papers.

Jørgensen, C., Computer Resources International (CRI), 1994. "On-board Software Development Approach for the Ørsted Micro Satellite." Paper presented at the Eurospace On-board Software Management Symposium, Italy, January 1994.

Liebe, C. C., 1993. "Star Trackers for attitude determination," Internal Report; Danish Technical University, Department of Electro Physics, 2800 Lyngby, Denmark.

Lundin, S., S. Grahn, B. Holmqvist. 1994. "A Reliable High Performance Control Centre For Scientific Satellites Using Personal Computers" in *Proceedings of the International Symposium on Spacecraft Ground Control and Flight Dynamics, SCD1*. Sao José dos Campos, Brazil, 7–11 February, 1994. Published in the *Journal of the Brazilian Society of Mechanical Sciences*, vol. XVI. Special, Issue. 1994. ISSN 0100-7386.

National Research Council, 1993. *The National Geomagnetic Initiative*, National Academy of Science. Washington D.C.: National Academy Press.

Nielsen, O. V., J. R. Petersen, F. Primdahl, B. Hernando, A. Fernandez, J. M. G. Merayo, and P. Ripka. 1993. "Development and Construction of the Ørsted Fluxgate Magnetometer," Internal Report; Danish Technical University, Department of Electro Physics, 2800 Lyngby, Denmark.

Orbital Launch Services, GSFC. 1993. *Delta-II Launch Vehicle Secondary Payload Planners Guide for NASA missions*. Greenbelt, MD: GSFC-OLS.

Priedhorsky, W. C., J. J. Bloch, S. P. Wallin, W. T. Armstrong, O. H. W. Siegmund, J. Griffee, and R. Fleeter. 1993a. "The ALEXIS Small Satellite Project: Better, Cheaper, Faster Faces Reality," *IEEE Trans. Nucl. Sci.* v. 40, p. 863.

Priedhorsky, W. C., J. J. Bloch, D. H. Holden, D. C. A. Roussel-Dupré, B. W. Smith, R. Dingler, R. Warner, G. Huffman, R. Miller, B. Dill, and R. Fleeter. 1993b. "The ALEXIS Small Satellite Project: Initial Flight Results," *Proc. SPIE 2006*, p. 114.

Ricker, G. R., J. Doty, S. Rappaport, K. Hurley, E. Fenimore, D. Roussel-Dupre, M. Niel, G. Vedrenne, D. Lamb, S. Woosley. 1988. "The High Energy Transient Experiment—HETE—A Multi-Wavelength Survey Mission for the 1990's." *Nuclear Spectroscopy of Astrophysical Sources*. New York: American Institute of Physics.

Smith, B. S., J. Hengemihle. 1990. "The Small Explorer Data Subsystem. A Data System Based on Standard Interfaces," *American Institute of Aeronautics and Astronautics*.

Chapter 12

Interplanetary Probes

12.1 Clementine

12.2 Pluto Express

Interplanetary missions are traditionally among the most expensive.They are long duration, high profile missions with extremely high launch costs because of the large ΔV required. Because of the high launch cost there is even more than the usual demand for optimization and minimizing weight which directly opposes the usual process of using larger margins to reduce cost. Nonetheless, there has been an unrelenting demand to reduce the cost of interplanetary missions that appears to have made major progress. Clementine, while not low cost by many standards, has begun a new process for creating rapid turn-around missions with limited objectives. Pluto Express (née Pluto Fast Flyby) has specifically drawn on the experience of AMSAT (Sec. 13.3), and other low cost spacecraft, to help to understand how to reduce the cost of missions that are inherently very expensive.

Among the missions intended to change the way interplanetary exploration is carried out are:

- StarDust; interplanetary material sample return to be launched in 1999

- Planet-B; Japanese Mars orbiter to be launched in 1998

- Lunar Prospector; Lunar orbiter for mineral mapping to be launched in 1997

- Mars Pathfinder; Mars lander and rover to be launched in 1996–97

- Mars Global Surveyor; replacement for Mars Observer to be launched in 1996

- NEAR; an APL asteroid mission to be launched in 1996

- Mars Observer which failed just before Mars orbit insertion in 1993

- Muses-A; Japanese lunar mission launched in 1990

A key issue for these missions is to reduce the size and complexity of the spacecraft and to limit the scope of the activity and the support required on any one spacecraft. This is, of course, particularly important for interplanetary spacecraft because of the high launch cost. In the future, a high level of miniaturization such as proposed for New Millennium can reduce the launch cost, although it may be at the expense of high cost in the spacecraft itself.

Clementine Summary

Clementine was built by the Naval Research Laboratory for the Ballistic Missile Defense Office to test lightweight imaging sensors provided by the Lawrence Livermore National Laboratory.

Spacecraft dry mass:	232 kg	Propulsion:	Mono & biprop. ($\Delta V = 4,865$ m/s)
Average power:	360 W	TT&C:	0.125 to 128 kbps in 8 steps

Launch: Titan IIG Launch from Vandenberg AFB, January 25, 1994

Intended Orbit: 2 months in Lunar orbit, followed by 4 month transfer to asteroid 1620 Geographos

Status: Software error ended the mission 3 days after leaving lunar orbit after discovering water ice at the lunar poles.

Cost Model (FY95$M): See page 348 for explanation.

	Expected Cost	Small Spacecraft Model[*]	Actual Cost[†]
Spacecraft Bus	$58.8M	$22.7M	$52.0M
Payload	$20.0M	$1.0M	$4.8M
Launch	$66.4M	$66.4M	$21.3M
Ground Segment	$79.0M	—	$1.5M
Ops. + Main. (annual)	$9.9M	—	$5.3M
Total *(through launch + 1 yr)*	**$233.9M**	—	**$85.0M**

[*]The Small Spacecraft Model does not include interplanetary spacecraft and is not intended for this application.
[†]An inflation factor of 1.066 has been used to inflate to FY1995$ [SMAD, Table 20-1].

Fig. 12.1-1.   Clementine. The satellite has an octagonal shape constructed of aluminum honeycomb. It has a cross section of 1.1 m diameter and a height of 1.2 m, excluding the 489 N thruster which adds an additional 0.5 m. The solar panels, seen deployed here, are wrapped about the spacecraft and deployed only after leaving low-Earth orbit.

12.1 Clementine

Donald M. Horan, *Naval Research Laboratory*
Bruce D. Berkowitz, *Carnegie Mellon University*

Clementine was launched on January 25, 1994 by a Titan IIG rocket from Vandenberg Air Force Base. It demonstrated the space-worthiness of several new technology components, algorithms, and software [Regeon, et al., 1995; Horan and Regeon, 1995]. Clementine spent 10 weeks in lunar orbit collecting multispectral images of the entire lunar surface and generating a coarse surface profile for nonpolar regions of the Moon [Nozette, et al., 1994]. On May 4, 1994, it left lunar orbit in preparation for a flyby observation of the asteroid 1620 Geographos. Shortly afterward, an onboard malfunction effectively ended the mission. Approximately 1.8 million digital images at visible and infrared wavelengths and the laser altimetry data have been delivered to NASA's Planetary Data System to support scientific analyses and for general distribution. The Clementine mission is considered a resounding success by the sponsors, the lunar science community, and the media, but of course not by the asteroid science community. It has received several awards as an example of a "cheaper, faster, better" space mission.

Clementine was a joint DoD and NASA mission whose primary objective was to test hardware developed for the Strategic Defense Initiative Organization (SDIO) [Nozette, 1995]. Its secondary objective was to collect scientific data for NASA. The Naval Research Laboratory's (NRL) Naval Center for Space Technology (NCST) designed, developed, integrated, tested, and operated the spacecraft. It took just over 22 months from formal acceptance of the mission by NRL on March 17, 1992 to launch.

12.1.1 Cost of Clementine Mission

As part of the mission-feasibility evaluation in early 1992, SDIO requested that NCST provide an estimate, not to exceed $50 million, of Clementine's cost, excluding launch vehicle costs. NCST estimated $55 to $56 million. SDIO accepted this estimate but urged NCST to look for additional economies to approach the $50 million goal. Considering the haste with which NCST generated this estimate, it proved to be quite accurate.

Table 12.1-1 shows the Clementine program's spending in major categories. Allocation of costs to the categories is somewhat arbitrary. Lawrence Livermore National Laboratory developed the cameras and laser transmitter for SDIO. The $4.8 million shown as the cost of the cameras and laser transmitter doesn't include development costs for those instruments. Other major flight hardware includes both readily available components, such as the solid-state data recorder, and uniquely developed components, such as the data-handling unit. We estimate that 75% of the components were readily available, needing little or no modification, but we don't have any hard facts to back this up. Integration and testing include flight- and ground-software development, as well as heavy costs for staffing. At NRL, the salary, benefits, and overhead costs of civil-service employees are charged against the project, not against an institutional-support account. Ground hardware includes the spacecraft simulator, the mission operations center, and the NRL ground station that provided about 50% of the coverage for the Clementine flight. Operations and data preparation exclude costs of support from NASA's Deep Space Network, which provided the rest of the coverage for the flight, and the use of NASA's communications links. They also exclude costs associated with the NASA-appointed science team, which NASA covered. NASA will also cover costs for analyzing data. However, the Clementine program paid for preparing the data for deposit in NASA's Planetary Data System. Clementine also paid for orbit determination and trajectory analysis by NASA.

TABLE 12.1-1. Summary of Clementine Costs (real year $).

	Cost ($M)
Cameras and laser transmitter	$4.8M
Other major flight hardware	$12.6M
Systems engineering and program management	$4.5M
Integration and testing	$31.7M
Ground hardware	$1.4M
Operations and data preparation	$5.0M
Launch vehicle and launch support	$20.0M
TOTAL	$80.0M

Clementine has been widely acclaimed as an example of a "cheaper, faster, better" mission. However, the popular mantra during early 1992 was not "cheaper, faster, better," but "Cheap, fast, good—pick two!" Although these slogans aren't necessarily contradictory, NCST was more attuned to the latter during Clementine's formative weeks. NCST's inability to provide a cost estimate within SDIO's $50-million goal is evidence of refusal to pick "cheap" over "fast" or "good." Thus, cost-saving heroics didn't drive Clementine. In preparing its cost estimate, NCST focused on a spacecraft and mission that would be fast and good, while not wasting any money. Therefore, although cost was important, delivering a quality product quickly was more important, and the $55 to $56 million estimate reflected that.

Was Clementine really a "cheaper, faster, better" mission? Let's look at each of these elements.

Cheaper

Clementine was clearly identified as a bargain by the media and the general public, perhaps for two key reasons. First, Clementine's cost appeared acceptable compared to missions more commonly featured by the media, with costs ranging well into the hundreds of millions, or even billions, of dollars. Possibly this is a reaction to "sticker shock" associated with the higher-priced missions. Second, Clementine produced a flood of data, especially images, which were quickly made available to, and excited the interest of, the media and the public. On the other hand, some have pointed out that, at $60 million (excluding launch costs) for a 463 kg spacecraft, the resulting $130 thousand/kg is close to normal cost density. Thus, Clementine was cheap because it was small. Of course it was! Both SDIO and NCST recognized they had to keep Clementine small in order to keep the cost low. In fact, SDIO put great effort into developing small, lightweight components precisely to try to make a space-based missile defense system affordable. Small, lightweight spacecraft can be launched by small, inexpensive launch vehicles. As spacecraft grow in size and mass, we need larger and more expensive launch vehicles. This makes the mission cost increase disproportionately because, after we transition to a bigger launch vehicle, we then need a larger, more capable, more expensive spacecraft to use the additional launch capability. A metric of $/kg may be useful, but dollars are a useful metric in themselves. Cheaper is easy to measure, and Clementine clearly cost less than many other space missions.

Faster

Although Clementine wasn't the most complex spacecraft ever built and didn't involve the safety issues of manned missions, it was complex enough so taking less than 23 months from approval to launch is a remarkable achievement. The lead time required to obtain hardware and develop software would make it a noteworthy accomplishment for all except the simplest of missions. Most of the work done before mid-March 1992 had to be redone, or at least carefully reexamined, after the program was approved to ensure that it was done with enough rigor. Therefore, 23 months is the proper time span to use in claiming that Clementine was a "faster" program.

Better

Clementine's primary objective was to test hardware developed for SDIO. By extensively and systematically using this hardware to collect scientific data, coupled with doing specific tests, Clementine fully met its primary objective. Ironically, the spacecraft's failure in early May caused the hardware under test to encounter a harsher environment than planned, and its survival provides unexpected, useful information. Clementine's secondary objective was to collect scientific data for NASA. On the grand scale of things, the quantity, quality, and completeness of Clementine's lunar data exceeded reasonable expectations and compensates for the failure to obtain the asteroid data. Ask the lunar scientists! (But not the asteroid scientists!)

12.1.2 Background

In late 1990, senior people at DoD (especially SDIO) and NASA were interested in exploring the pros and cons of joint missions. As a result, they ordered a study of a few candidate missions in early 1991 as the Joint SDIO and NASA Study of SDIO Technology Applications to NASA Solar System Exploration. The results of the study were briefed in April 1991 and included

- Potential science objectives for a flyby or rendezvous involving a near-Earth asteroid

- Potential science objectives for a lunar-orbiting or flyby mission

- SDIO technology developments of interest to NASA

- Outlines of flyby flight profiles for six near-Earth asteroids, including an extended version of the Clementine mission

During the summer and autumn of 1991, proponents of the joint mission in SDIO worked to stimulate programmatic and financial support within SDIO and DoD. SDIO wanted to test the new space-technology components it had sponsored as part of its charter to develop a space-based missile defense. In particular, Lawrence Livermore National Laboratory had developed extremely small and lightweight sensors to detect and track the relatively cold bodies of coasting missiles. Man-made targets to test these spaceborne sensors would be expensive and would violate a narrow interpretation of the Anti-Ballistic Missile treaty. Using suitable natural targets, such as the Moon or asteroids, would be cheaper and wouldn't violate treaties. If they could obtain useful scientific data, that would be a real bonus. In addition to the sensors, SDIO had other lightweight components to test.

Space scientists interested in data from the Moon or asteroids worked to define achievable, scientific goals that would truly advance the state of knowledge and be based on the sensors and other hardware that SDIO wanted to fly. They analyzed six candidate flight profiles from the joint study to identify one as the best combination of achievability and scientific utility. The selected flight profile had a launch in January 1994 with several months in lunar orbit, departure from lunar orbit in July 1994, a flyby of Geographos in August 1994, and a flyby of the comet Giacobini-Zinner in November 1998. They believed they could launch using a Pegasus, or even a Scout, because of SDIO's progress in reducing the mass of spacecraft components.

The name Clementine arose during this period, but the true source of the name is unclear. Some say that Clementine was an obvious name for the mission because, after the flyby of Giacobini-Zinner, the spacecraft would be "lost and gone forever," as described by the well-known folk song. Others say that Clementine was an informal, unofficial, randomly-selected code word used to shield those promoting the project from scorn and derision from those of lesser faith.

By November 1991, those promoting the program had found enough higher-level support and worked through enough mission details and objectives so they were ready to officially begin the program. SDIO wanted NCST to be the agent to carry out the

TABLE 12.1-2. Technical Overview of the Clementine Spacecraft.

Mass	• Dry spacecraft 232 kg • Fueled spacecraft 463 kg • STAR 37FM solid-fueled rocket 1,155 kg • Interstage adapter 47 kg • Launch vehicle adapter 28 kg • Total mass launched 1,693 kg
Structure	• 8 sided, 1.1 m diameter by 1.2 m high • Primarily aluminum • Some titanium, beryllium, composites
Electrical power	• Average power requirement 280 W at 30 VDC • GaAs/Ge solar array, 2.3 m^2, 360 W • Autonomous tracking of Sun by solar arrays • 15 A·hr common pressure vessel NiH$_2$ battery
Attitude control	• 3-axis stabilized • 0.05 deg control, 0.03 deg knowledge (3 σ) • 2 star-tracker cameras • Stored star catalog • 2 inertial measurement units • 4 reaction wheels of 2 N·m·s each • 12 thrusters, 10 at 4.4 N and 2 at 22.2 N
Reaction control	• STAR 37FM solid-fueled rocket, 1,155 kg, 47,260 N • Monopropellant (hydrazine) for 12 attitude control thrusters • Bipropellant (N$_2$O$_4$ and hydrazine) for delta-V • 1 delta-V thruster, 489 N
Thermal control	• Heaters with thermostatic control • Multilayer insulation (Kapton™/Dacron™) • Radiators • Heat pipes: diode, constant and variable conductance • Beryllium thermal capacitor
Command, telemetry and data handling	• Main processor: RH-1750, radiation hardened, 16 bit, 1.7 MIPS • Sensor image processor: R3081, 32 bit, RISC, 18 MIPS • Data-handling unit • Solid-state data recorder, 1.6 Gb, 20 Mbps I/O • Data-compression chip
Communications	• 2 S-band transponders, Deep Space Network compatible • Downlink 0.125 to 128 kbps in 8 steps • Uplink 1 kbps • 1.1 m diameter directional high-gain antenna • 2 omnidirectional antennas
Payload	• New technology hardware and software • Ultraviolet/visible camera • Near-infrared camera • High-resolution camera • Long-wavelength infrared camera • Laser transmitter • Charged particle telescope • Radiation and reliability assurance experiment • Dosimeters

mission, so NCST representatives were first briefed on the Clementine mission on November 12, 1991.

NCST had designed, built, and operated SDIO's Low-power Atmospheric Compensation Experiment (LACE) satellite, a 1,445 kg spacecraft. Its payload mainly consisted of a triple array of calibrated sensors on an extended target board to test atmospheric-compensation techniques applied to low-powered, ground-based lasers, as well as an instrument that tracked and imaged plumes of launch vehicles by their ultraviolet emission. LACE was a $130-million program begun in 1985, with launch in 1990 after a development cycle marked by several changes of launch vehicle and payload. LACE was the 79th satellite built and launched by NCST since June 1960. Clementine and its interstage adapter would be the 81st and 82nd.

NCST contacted developers of the lightweight, new-technology components to verify their state of development and availability to support the early launch. They did calculations to verify the trajectory and its required velocity changes, to identify the characteristics of the spacecraft's subsystems, and to estimate the launch mass. Although this quick look didn't support launch by a Scout or Pegasus, it did show that launch using the next larger class of vehicles (Titan II, Taurus, Conestoga) might be possible. Therefore, in mid-December 1991, NCST began a 3-month effort to verify the feasibility of the Clementine mission by working through a preliminary mission plan and spacecraft design. The mission concept which evolved included a launch in late January 1994, 10 weeks in lunar orbit, cruise to 1620 Geographos starting in May, and flyby of the asteroid on August 31, 1994. It didn't include the flyby of Giacobini-Zinner for several reasons, such as fuel constraints and the cost of extended operations. After 2 months, NCST and SDIO believed the concept was possible, but risky. However, NRL didn't approve NCST's accepting the program until March 17, 1992.

12.1.3 Technical Description of the Spacecraft

Table 12.1-2 provides a technical overview of the Clementine spacecraft. See Regeon, et al. [1995] for a more complete description. The total mass launched was 1,693 kg. The total mass of the Clementine spacecraft was 463 kg. The interstage adapter and the casing of the spent solid-fueled rocket became an independent spacecraft in a highly eccentric Earth orbit to measure space debris and radiation.

The spacecraft's primary structure (Fig. 12.1-2) was an aluminum-alloy frame with enclosing panels of aluminum honeycomb covered by aluminum-alloy sheets. The spacecraft, excluding solar arrays and interstage adapter, had an 8-sided cross section of 1.1 m diameter and 1.2 m height. The externally-mounted nozzle of the 489 N thruster added 0.5 m to the height. While in low-Earth orbit, the solar arrays wrapped around the spacecraft. After firing of the solid-fueled rocket to leave low-Earth orbit, the solar arrays deployed as two wings which could independently rotate about a single axis. Each solar array wing had a composite isogrid substrate and was mounted to the spacecraft using titanium hardware. The interstage adapter contained the solid-fueled rocket and carried structural loads during launch and firing of the solid-fueled rocket. The interstage adapter's longerons and skin were made of carbon fiber/epoxy composite. Another major structural component was the sensor bench, which held the primary science sensors in alignment and within acceptable temperature limits. The sensor bench was an aluminum honeycomb panel with beryllium standoffs.

The attitude-control subsystem stabilized the spacecraft on three axes to 0.05 deg absolute pointing accuracy (3σ). It mainly used reaction wheels and changed to thrusters for a more rapid response. The combined Clementine spacecraft and interstage adapter were spin stabilized using thruster control for the firing of the solid-fueled rocket. Clementine carried four reaction wheels, with three mounted orthogonally and one skewed to provide redundancy. Twelve monopropellant hydrazine thrusters provided more force than the wheels for faster response or for control during firing of the bipropellant thruster. They also unloaded the momentum built up in the wheels. Star-tracker images with a 29 deg by 43 deg field of view were used to determine the spacecraft's orientation. Software provided by Lawrence Livermore National Laboratory compared the angular relationships of stars in a star-tracker image to those provided by a stored catalog of only 600 stars distributed over the celestial sphere. Typically, Clementine could calculate its attitude to better than 150 μr in pitch and yaw and 450 μr in roll within 100 ms from a single star-tracker image. Clementine took star-tracker images about 10 sec apart to calibrate the attitude information provided continuously by two inertial measurement units. Kalman filtering smoothed the attitude information obtained from the star-tracker images.

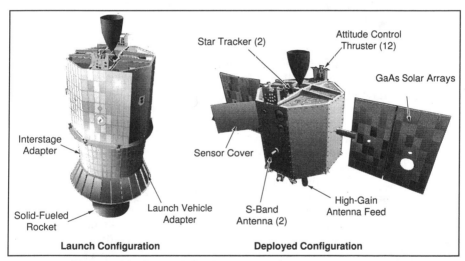

Fig. 12.1-2. Clementine's External Configurations. Conversion from launch to deployed configuration began after firing of the solid-fueled rocket to leave low-Earth orbit. The solar panels unfolded immediately after the firing ended. Approximately 24 hours later, the interstage adapter and spent casing were released as a separate satellite, and the high-gain feed deployed to complete the conversion.

The reaction-control subsystem provided the 12 monopropellant thrusters for attitude control, the solid-fueled rocket for the 3,115 m/s to leave low-Earth orbit, and a bipropellant thruster for multiple, smaller changes in the spacecraft's trajectory. The 489 N bipropellant thruster provided nearly 1,750 m/s velocity increment for entering and leaving lunar orbit and for trajectory adjustments.

Thermal control had to cope with the diverse challenges of heat dissipation during continuous sunlight lasting for months, and heat retention during periods in shadow of more than 2 hours. Multilayer thermal blankets covered most of the spacecraft's external surface to isolate it from its environment. About 1.6 m² of surface were covered with silver tape to radiate heat away from the spacecraft. Internal surfaces were also taped to enhance movement of thermal energy to the radiating panels. For shadow transits, 60 thermostatically controlled heaters warmed boxes, tanks, and propellant lines. Operation of the primary science sensors provided a special thermal challenge. The sensors were designed for SDIO with the expectation that they would be active in space for only about 15 minutes. During the imaging sequence of each lunar orbit, the sensors were on continuously for 2 hours. Even though they were of low-power design, their compact size decreased their thermal capacity, and it was difficult to keep their focal planes within acceptable temperature limits for proper calibration. Also, the continuous cycle of 2-hour imaging runs at 5-hour intervals allowed only 3 hours to restore the sensors and their thermal-control elements to their initial state before another imaging run started. A 4-kg beryllium block was a thermal capacitor to store thermal energy from the sensors that would otherwise overwhelm the radiating panels. Constant- and variable-conductance heat pipes carried heat from the sensors to the radiators and the beryllium block. Diode heat pipes carried heat from the block to the radiators. The system needed about 11 kg of heat pipes, radiators, and thermal capacitor to keep the 7 kg of primary science sensors within acceptable temperature limits.

The command, telemetry, and data-handling subsystem used an RH-1750 processor for primary control of the spacecraft. It is a radiation-hardened device with a 16-bit architecture and capable of 1.7 MIPS, which handled telemetry and commands. An R3081 RISC processor, with a 32-bit architecture and capable of 18 MIPS, processed images from the spacecraft's cameras for various spacecraft-control functions. The R3081 could easily have done the RH-1750's command and telemetry tasks, but it was secondary for these tasks because it wasn't radiation-hardened. The RH-1750 could also do image processing, but much more slowly than the R3081, so the spacecraft's autonomous performance would have been much poorer. A data-handling unit controlled the lunar-latitude-dependent timing and sequencing of images from the primary science sensors to ensure that the entire lunar surface was imaged in 11 pass bands without excessive overlap between images. This unit controlled the taking of between 5,000 and 6,000 images every 5-hour lunar orbit, as well as the flow of data into and out of the solid-state data recorder. The solid-state data recorder provided 1.6 Gbits of storage capacity with 20 Mbits/s data input and output rates and very low error rates. It had a radiation tolerance of 30 krad(Si) plus active error detection and correction. A data-compression chip with highly controllable data-compression levels, under control of the data-handling unit, compressed image data before storage in the solid-state data recorder.

The communication subsystem used a pair of redundant, Deep Space Network-compatible, S-band transponders for communication and tracking. Two omni-directional antennas supported low-gain transmission from the spacecraft and commanding. A fixed, parabolic reflector, with a diameter of 1.1 m, supported high-gain transmission and commanding. The spacecraft could transmit data at 0.125, 0.250, 0.500, 1, 2, 8, 64, or 128 kbps using the omnidirectional antennas or the high-gain antenna, as the link margin allowed.

Because Clementine's primary objective was to test new-technology hardware and software in space, its payload includes many components which we wouldn't ordinarily consider part of a spacecraft's payload. For example, the NiH$_2$ battery flown aboard Clementine was the first use in space of an NiH$_2$ battery with a common pressure vessel. It offers the NiH$_2$ battery's greater depth of discharge while cycling, plus a significant mass advantage over NiCd batteries or NiH$_2$ batteries with individual pressure vessels. Relative to an NiCd battery, its volume penalty is much smaller than that of an NiH$_2$ battery with individual pressure vessels. Also, the charge state of a NiH$_2$ battery with a common pressure vessel is linearly related to the battery's pressure, which we can measure directly. This simplifies operations because we must determine the charge state of NiCd batteries and NiH$_2$ batteries with individual pressure vessels by continuously monitoring the current into and out of the battery. The solar cells Clementine carries are the thinnest GaAs/Ge cells ever flown, and thus offer a mass advantage. Clementine's solid-state data recorder provided four times more data-storage capacity than any similar flight-qualified recorder. Thus, these components and Clementine's inertial-measurement units, reaction wheels, composite materials, lightweight release devices, R3081 processor, data-compression chip, and the lightweight sensors are all part of Clementine's payload. The new Spacecraft Command Language flight software and the first use in space of a commercially available operating system also contribute to the payload. See Horan and Regeon [1995] and Regeon, et al. [1995] for more information on these non-standard payload items.

TABLE 12.1-3. Camera Characteristics.

Camera	Technology	Pass Bands	Array Size (pixels)	Field of View (deg)	Power (W)	Mass (g)
UV/Visible	Si CCD	6	288 × 384	4.2 × 5.6	6.7	410
Near IR	InSb	6	256 × 256	5.6 × 5.6	5.7	1,920
Long IR	HgCdTe	1	128 × 128	1 × 1	5.1	2,100
High-Resolution	Intensified Si CCD	5	288 × 384	0.3 × 0.4	9.2	1,120
Star Tracker	Si CCD	1	384 × 576	29 x 43	4.7	290

More conventional payload items include the five cameras and the laser transmitter developed for SDIO through Lawrence Livermore National Laboratory. The five cameras are the UV/visible, near IR, long-wavelength IR, high-resolution, and star-tracker. Table 12.1-3 gives their characteristics. Clementine's UV/visible camera decreased mass by a factor of 2 to 4 and power consumption by a factor of 3 over comparable sensors. The masses given for the two IR cameras include their mechanical coolers, but the power values exclude the coolers. The high-resolution camera uses a microchannel plate intensifier. The laser transmitter is a diode-pumped Nd:YAG laser. It emits at 1,064 nm with 180 mJ per pulse and a pulse duration of 10 ns. Clementine's laser transmitter and its power supply weighed 1.25 kg and were very compact. They improved mass and volume by a factor of 10 compared to earlier units. The high-resolution camera also contained the avalanche photodiode receiver for the laser ranger.

The primary science sensors for this mission were the UV/visible, the two IR, and the high-resolution cameras and the laser ranger. Another science sensor was the charged-particle telescope which measured fluxes of electrons with energy between 25 and 500 KeV in six channels and protons with energy between 3 and 80 MeV in three channels. A radiation and reliability-assurance experiment measured total radiation, recorded single-event upsets in devices, and distinguished single-event upsets caused by protons, alpha particles, and heavy-cosmic-ray particles. Finally, dosimeters measured total radiation.

12.1.4 Design, Development, Integration, and Testing

Figure 12.1-3 identifies the major phases of NRL's Clementine program and the timing of these phases and major events. This schedule is noteworthy for two aspects: the schedule compression, which results in activities occurring out of normal sequence, and an innovative approach to spacecraft integration, which greatly shortened the time required for integration and test.

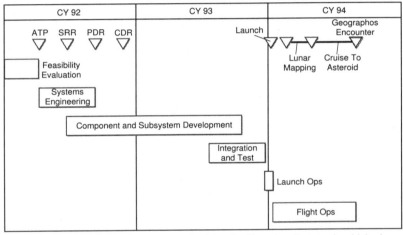

Fig. 12.1-3. Clementine Schedule. The schedule significantly shortened, which decreased cost, by starting procurement of long-lead components before the preliminary design review. These components then became fixed points in the design at a very early stage.

After NCST gained approval from NRL in March 1992, to proceed with the Clementine mission, they started a systems-engineering phase to design the spacecraft's subsystems. Initial subsystem-design work during this phase was based on the preliminary spacecraft design and mission plan that NCST generated while evaluating Clementine's feasibility from December 1991 through March 1992. Thus, the systems-engineering phase had to start with a major effort to more precisely define the system and subsystem requirements. A formal review of system requirements took place in June 1992, several months into the systems-engineering phase. The systems-engineering phase ended with a formal preliminary design review in late July. Component and subsystem development normally would begin after the critical design review, but that was incompatible with timely delivery of long-lead items. Therefore, component and subsystem development involving long-lead components began shortly before the review of system requirements, as soon as requirements and design were adequately specified. Starting component procurement before the preliminary design review was risky because these components then became fixed points in the design, but it was the only way NCST could reasonably expect timely delivery. A formal critical design review took place in November 1992. A major task during component and subsystem development was continuous interaction with suppliers to prevent delays in delivery. On a few occasions, a team of engineers and sponsor's representatives went to a supplier's site to help overcome obstacles to timely delivery.

Integration and test formally began in the summer of 1993, when NCST started integrating flight hardware into the flight spacecraft. However, preparation for integration began several months earlier, when they integrated prototype and even breadboard versions. By integrating these non-flight components, NCST identified early and eliminated many of the large-scale, time-consuming electrical and software problems, such as grounding and commanding problems, which commonly plague the start of integration. As flight components became available, they replaced their non-flight versions. We believe this approach to integration saved between one and two months. A significant number of flight components were late by days or even a few weeks in some cases. Therefore, NCST changed the integration and test schedule almost daily to minimize time loss caused by late deliveries.

Spacecraft integration was done at NRL. All testing took place at NRL, including such major tests as thermal, thermal-vacuum, vibration, acoustic, and electro-magnetic. On December 30, 1993, the spacecraft was flown to Vandenberg Air Force Base to begin launch integration.

12.1.5 Operations

The general operations concept took shape during the system-engineering phase and had several key points. All operations were to be conducted from a mission operations center in Virginia, near Washington, D.C., known as the BATCAVE.* NASA's communications links would connect the mission operations center to ground stations of NASA's Deep Space Network in Spain, Australia, and California; to an NRL ground station in Maryland; to the Jet Propulsion Laboratory in California; and to NASA's Goddard Space Flight Center in Maryland. A control center within the mission operations center would handle spacecraft monitoring and commanding. Commands would flow from the control center over NASA's communications links to the ground stations, which would strip them of formatting bits associated with transmission on NASA's links and transmit them directly to Clementine without any checking or additional processing on the ground. The ground stations wouldn't use

---

* The name BATCAVE for the mission operations center arose during the LACE program. Bendix Field Engineering Corporation, a contractor supporting LACE, was directed to find accommodations that could be used as a control center for the LACE satellite. They found a building that had been vacant for several months in Alexandria, Virginia. While inspecting the building, a group entered an open bay area whose high ceiling had exposed steel trusses. The room was dimly lit. Large cobwebs hung from the steel trusses and the overhead lighting fixtures. The floor and other flat surfaces were covered with a carpet of dust. One of the group (several people claim this honor) exclaimed: "This looks like a real bat cave," and the name stuck. After extensive cleaning and remodeling, LACE used the building. The name BATCAVE was displayed on cover sheets for the facsimile machine, but higher authorities said they couldn't use BATCAVE in this manner because it wasn't a meaningful acronym. In rebuttal, the words Bendix Alexandria Technical Center for Aerospace Vehicle Experiments (BATCAVE) were quickly assembled to justify the acronym and the BATCAVE became legitimate.

data from the spacecraft, but would put it directly onto the NASA links for transmission to the mission-operations center. NASA's Science Team would be extensively involved in operations. Key engineers, after being deeply involved in designing, developing, integrating, and ground testing the spacecraft, would assume active roles in operations. The mission operations center would immediately process all data from the spacecraft, whether it is telemetry providing the spacecraft's status or image data from the cameras. The processed data would be available within 15 minutes to people on duty in the control center, and to sensor experts and representatives of NASA's Science Team in the mission-operations center. A hardware-based spacecraft simulator would be available to verify commands, test flight software, and resolve anomalies.

The operations team helped establish the preliminary definition of spacecraft and mission that led to NRL's acceptance of the Clementine mission. After March 1992, several members of the operations team routinely participated in meetings of NASA's Science Team to fully understand what the Science Team wanted, discuss candidate operational approaches for achieving these results, brief the Science Team on operational issues, and even point out new opportunities to acquire data for the Science Team's consideration. Working groups were established to address major operational issues, and participants included representatives of the Science Team as well as members of the operations team. All major spacecraft activities had written operations plans by April 1993, and these were shared with the Science Team. Members of the operations team who had been routinely interacting with the Science Team were deeply involved with operations after launch to carry out the plans they'd made over many months of collaboration with the Science Team.

12.1.6 The Flight

Figure 12.1-4 shows a timeline of the mission flown. A Titan IIG launched Clementine on January 25, 1994. The entire time in low-Earth orbit was extremely hectic as the operations team learned how to properly use the Kalman filter in the attitude-control subsystem. Initial, improper use resulted in loss of the spacecraft attitude needed to convert solar energy, so the spacecraft's battery discharged. Coping with these problems was complicated by the extremely short time for commanding during passes over ground stations. All hands were very happy to fire the solid-fueled rocket on February 3 in order to leave low-Earth orbit. The Apollo missions had large fuel reserves to support a direct trajectory to the Moon with arrival in about 4 days. Clementine used fuel-efficient and schedule-forgiving phasing loops to get to the Moon. Cislunar operations were much more relaxed than operations in low-Earth orbit. Commanding opportunities were much longer and the spacecraft responded well. The operations team rehearsed lunar operations several times during cislunar transit and collected some science data

On February 19, the spacecraft's 489 N thruster was fired to enter lunar orbit. On February 21, NCST maneuvered Clementine into the 5-hour period, polar orbit that it would use to take images of the entire lunar surface. After several days of testing to ensure proper exposure settings for the cameras and to perfect operations, they started systematic imaging on February 26 and continued until April 22. They used several more days in lunar orbit to make special observations, and the spacecraft propelled itself from lunar orbit on May 4.

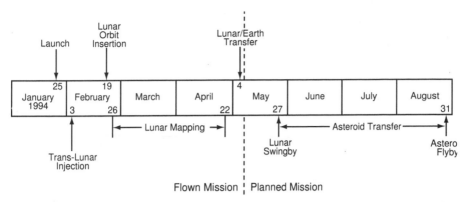

Fig. 12.1-4. Timeline of Clementine's Flight. Information to the left of the vertical dashed line describes the Clementine mission actually flown until May 7. Information to the right describes the rest of the planned mission and not the unplanned sequence of events leading to the present heliocentric orbit

The plan was to again use phasing loops around Earth and then leave the Earth-Moon system on May 27, with an assist from lunar gravity. However, an onboard malfunction on May 7 caused Clementine to use up the attitude-control fuel and left it in a spin at 80 rpm, which ultimately kept it from generating enough energy to keep operating. The flyby of the asteroid 1620 Geographos was no longer possible, and the mission essentially ended, although attempts to regain some control over the spacecraft continued for several weeks.

The fixed spin axis and high rate of spin made it impossible for the solar panels to track the Sun. By early June 1994, the spacecraft didn't have enough power to respond to commands or to modulate the RF carrier. However, the Deep Space Network was able to obtain intermittent pointing information when it could find the spacecraft's carrier. In late July, Clementine hurled into a heliocentric orbit as it passed the Moon. Based on the reasonably well known orientation of the spacecraft's spin axis and the orientation of the solar panels, there would be enough power generation again in late January or early February 1995 to operate critical spacecraft subsystems. Deep Space Network found some time to devote to a search for Clementine using its 70 m antennas. An attempt on January 27 found no signal. However, after a relatively easy search, they received a modulated signal during the next attempt on February 20. The signal was very weak because the spacecraft was about 30 million km away, but analysis of the telemetry fragments it contained verified that it was Clementine. On several occasions during March and April, NCST received solid telemetry and successfully commanded the satellite. The telemetry verified that there were no major failures of the spacecraft's subsystems. However, a thorough test wasn't possible because frequent loss of telemetry lock and the relatively long transit time made commanding laborious. By early May 1995, the spacecraft was over 45 million km distant, and Deep Space Network was no longer able to lock onto the signal. The spacecraft was also reaching a point in its heliocentric orbit where it didn't have enough power again. So Clementine is now really lost and gone forever—or at least until 2010, when Earth catches up with it again.

12.1.7 Clementine's Cost Reducers

Although NCST's main goal was to build and fly a spacecraft that could accomplish its mission, we believe several aspects of their approach to this task reduced cost and are worth discussing.

Incentive to Reduce Costs

Everyone knew Clementine had a limited budget. Although NCST wanted to ensure the program's success, they knew that if it grossly exceeded the estimate, the program would probably be canceled. Therefore, cost reduction hovered over all decisions. But the program leaders agreed they would rather have the program canceled than deliver a spacecraft that couldn't accomplish its mission.

Short Schedule

Somebody once said: "Time is money," and this is certainly true for spacecraft schedules. Personnel costs increase with schedule length. However, we can't truly shorten schedules by doing business conventionally. It's unconventional to order long-lead components before critical design review and highly unconventional to order before preliminary design review, as NCST did on Clementine. Yet, they saw no other way to get a delivery of flight items in time to support the schedule. NCST was well aware of the risks involved in creating fixed points in our design so early, but had enough confidence in the early designs to believe the risks were acceptable.

Short schedules demand flexibility. Schedules take on a life of their own, and we frequently forget they are just tools to create an orderly flow of activity. For a few days during integration and test, NCST changed the schedule almost hourly as bad news on deliveries rolled in. So, when an event is late, we must not slide the whole schedule and slip end dates. Instead, we must be willing to imaginatively juggle all intervening events to keep the same end date. Schedule slips increase mission costs because we have to pay the "marching army" for every day on the job, store and maintain the spacecraft or components, and pay mounting expenses for overhead. When NRL accepted the Clementine program in March 1992, a launch was planned for late January 1994. By January 1993 the launch date had been fixed as January 25, 1994. The launch did occur about 4 minutes late because of excessive wind velocities at high altitudes.

Short schedules demand preparedness, especially for government agencies. Laws regulating awarding of contracts by government agencies require months to pass before a large contract can be in place. Space missions do require large contracts. As a result of the LACE program, NRL had several suitable, large contracts already in place that could legally support Clementine. If they had not been in place, NCST couldn't have met the Clementine launch schedule. Note that, although we legitimately say NRL designed, developed, integrated, tested, and operated Clementine, 85% of the Clementine budget went to contracts with private industry to obtain the hardware, software, and people for the mission. Table 12.1-4 lists contractors and their areas of responsibility for Clementine.

Although a short schedule is a good cost-reducer, we believe Clementine's schedule was a little too short, especially when combined with the tight budget. For example, flight software for the later stages of the mission wasn't finished at launch. In late 1993 and January 1994, both flight-software development and operations

TABLE 12.1-4.   **Contractor Involvement in Clementine.** The left column identifies the project area receiving the products or services shown in the middle column. The right column lists the contractors who provided the products or services.

System	Engineering:	Aeronix; Allied Signal Technical Services; Barrios Technology; Fairchild; Midwest Engineering; Praxis
	Integration:	XEN
	QA:	Assurance Technology Corp.
Structure	Composite:	The Aerospace Corp.; McDonnell Douglas
Electrical power	Battery:	Johnson Controls
	Solar cells:	Applied Solar Energy Corp.
	Substrates:	McDonnell Douglas
	Drive assembly:	Schaeffer Magnetics
Attitude control	Reaction wheels:	Ball Aerospace
	Thrusters:	Kaiser-Marquardt; Rocket Research Corp.
	IMUs:	Honeywell; Litton Guidance and Control Systems
	Star trackers:	OCA; Thompson, CSF; Lawrence Livermore National Lab.
Reaction control	Tanks:	Atlantic Research
	Delta-V thruster:	Kaiser-Marquardt
	STAR 37FM motor:	Thiokol
Command, telemetry and data handling	Processors:	Honeywell; Integrated Device Technology; Telenetics
	Data-handling unit:	Innovative Concepts
	Data compressor:	Matra Marconi
	Data recorder:	SEAKR Engineering
Commu-nications	Transponders:	Loral Corporation
Payload	Cameras:	Lawrence Livermore National Laboratory
	IR Cameras:	Amber Engineering; Fermionics
	Intensifier:	General Atomic; Litton Electro-Devices Division
	CCDs:	Thompson, CSF
	Optics:	OCA; Optical Coating Laboratory
	Filter wheel motors:	Vernitron
	Laser:	McDonnell Aerospace Defense and Electronics Systems
	Integration support:	Research Support Instruments; Science Inquiries
	Charged-particle telescope:	The Aerospace Corp.
Software	Flight:	Interface Control Systems; Software Technology
	Ground:	Mnemonics; Protasis; Software Technology
	Mission planning:	Allied Signal Technical Services; Protasis
	Guidance control:	Space Applications Corp.
	System engineering:	SEAII
Data processing	Services:	Applied Coherent Technology; Innovative Concepts; Software Technology
Operations and mission planning	Services:	AlliedSignal Technical Services; Barrios Technology; Computer Sciences Corp.; Mnemonics; Protasis Research Support Instruments; Science Inquiries
	Hardware:	Silver Engineering

urgently needed the spacecraft simulator, but had to share it. The software developers were trying to make as much progress as possible with the flight software, and the operations team was trying to rehearse post-launch activity. Also, although all critical operations software was ready before launch, scheduling and mission-planning software that would have been extremely helpful to the operations team wasn't ready until several weeks after launch.

Small Spacecraft

Space mission costs are highly correlated with the spacecraft's size and mass. Larger, more massive spacecraft mean more components to procure or develop, and more complexity in the integration and testing. These requirements all add cost. A very significant element of cost is the launch vehicle. Launch vehicles in the Titan II class cost $20 to $30 million. The smaller Atlas variants cost $60 to $80 million, and a Titan IV costs $250 million or more. Both NCST and SDIO recognized that Clementine had to be kept small and light in order to be affordable. They achieved lighter weight by using lightweight sensors, electronics, solar arrays, battery, and composites. Mass savings on Clementine were magnified because it required 8 kg of fuel to carry 1 kg of spacecraft to the asteroid Geographos.

Using Readily Available Components

After NCST was first briefed on Clementine, one of the most important early tasks was to contact suppliers, especially of components using newer technologies, to assess the state of development and timely availability of flight hardware. Spacecraft design then strongly considered component availability and used as much readily available, space-qualified hardware as possible. NCST insisted on freezing the evolution of new-technology items at some acceptable configuration and passed up the opportunity to incorporate the very latest, but inadequately tested, improvements. Although NCST tried hard to use only space-qualified parts in all flight hardware, there were exceptions. They could accept the best available part consistent with the schedule. However, whenever space-qualified parts weren't available on time, their substitutes had to undergo special testing and failure analysis, which became more rigorous as the substitute parts became farther removed from space quality. Use of readily available components can reduce cost. They are often cheaper than specially developed components because they share development costs more widely, their interfaces are well defined and usually simpler, supporting tools and test materials are usually available, and they save the time and effort of technical people who have to interact with a supplier to get a special component ready on time. For Clementine, readily available components included items left from earlier programs. The LACE program previously used many of the computers, terminals, and other hardware used in Clementine's mission-operations center and ground station. Also, Clementine's commercial operating-system software made coding and testing of applications flight software easier, and that shortened the schedule and reduced cost.

Simple Decision Process

When a decision is pending, everyone waiting for that decision slows down, and the lost time translates into wasted money and missed deadlines. On Clementine, a few relatively accessible people were empowered to make system-wide decisions. They would consult among themselves and with others as necessary, make the decision, and immediately inform those directly affected. They would tell the rest of the team at the next status meeting. The goal was to make decisions within minutes to hours, but some of the nastier decisions obviously took longer. People who made the system-wide decisions also decided when management levels above the Clementine program should review a decision. This review was rarely necessary because higher management at NCST typically allows program management a fairly long leash.

Defining and modifying requirements were also relatively simple on Clementine because management treated a request to change requirements like other decisions. Where necessary, the person making the decision would try to eliminate or reduce systemic effects by negotiating a trade to compensate for the changed requirement. They tried to indicate clearly where margins were tight and where they could relax requirements, if necessary. They had to avoid impossible requirements and avoid including unnecessary requirements simply for completeness. They tried to specify the essential requirements quickly, so subsystem designers could get started earlier, and then deal with requests for definition of lesser requirements later. There were relatively few requests to define additional requirements.

Continuity of Staff

On Clementine, they had continuity of responsible people from beginning to end of the program. This means program management was responsible for all phases of the mission from design through operations. It also means the same engineers who designed a subsystem were deeply involved with its operation after launch. The same engineers who integrated and tested Clementine pulled 12-hour shifts, four days on and four days off, to operate the spacecraft after launch. The same mission planners who carried out the science operations after launch had been attending meetings of NASA's Science Team and otherwise working with the Science Team for almost two years before launch. When the Clementine mission abruptly ended, they were relieving the spacecraft engineers from shift duty because there had been ample opportunity for them to train their replacements.

Minimize Documentation

Documents require people and time to write them. However, it doesn't end there. They also require people and time to read them. Documentation can waste a lot of time, effort, and money, but some documentation is essential. They had very few documents on Clementine, but NCST never produces a lot of documents for a program. They also didn't spend much time modifying the documents they had. The top-level document for the program, the Mission Requirements Document, was only formally revised twice, even though it showed many changes. A formal revision was issued only when the pen and ink changes to the master copy became numerous enough to confuse readers.

A prime purpose of documentation is to transfer information clearly from one group to another in a disciplined way; this can happen and it can be useful. A documentation incident from Clementine is interesting. An external group was to provide flight hardware for Clementine, and it was commonly believed an interface-control document was essential. Groups from the external supplier and NCST met frequently for several months to create a useful document. The external group then submitted the document for review by their higher management, which insisted on so many changes and deletions that the negotiated document became useless. So groups from the external supplier and NCST met frequently for several months to create another useful document. Again their higher management modified and deleted until the document was useless. This cycle continued, but fortunately, engineers transferred the necessary information directly and efficiently. They integrated the hardware without the interface-control document that wasted so much time and effort.

Documents are useful, but they're expensive. We must balance these two facts to decide how much documentation a program really needs.

Integration and Testing in One Place

NCST integrated Clementine and did all testing, including the major environmental tests, in the same building at NRL. We've been involved in programs in which the spacecraft had to be carefully packaged and transported with a truckload of support equipment to another location for testing. The logistics of such a move, even if only a few miles, typically consumed the better part of a week. Clearly, testing at several different sites can significantly stretch a schedule.

Skunk Works Environment

People claim that NCST built Clementine in a Skunk Works environment. There's some truth to that assessment. A single manager was in charge, and he was involved enough in the day-to-day activity to know what was going on. They did all designing, developing, testing, and integration at the NRL site. A small, highly involved management team was empowered to make decisions quickly. There was little, but adequate, documentation. The NCST facilities aren't pretty, but functionally are of high quality. NCST is a relatively stable organization, and the Clementine team had worked together for some time.

12.1.8 Organizational and Political Factors

The management and technical approaches used to reduce cost on Clementine aren't new, but they're not universally used. Therefore, an important question remains: Why were SDIO and NCST able to adopt these approaches? This may be the most important question of all because there is little to gain if we know how to make space missions less expensive but can't do so. In retrospect, we can identify several factors that worked to the advantage of SDIO and NCST.

SDIO was a new organization and therefore was bureaucratically small and flexible. Organizations, as they mature, tend to develop established procedures and design philosophies. Many of the techniques for building less expensive spacecraft on faster schedules run counter to the procedures and design philosophies of NASA and the U.S. Air Force. To be fair, however, NASA and the Air Force often must take high-cost approaches to reduce risks in situations where a failure could be extraordinarily costly, such as the loss of a manned mission or the failure of a satellite vital to national security. Also, these organizations matured at a time when budgets weren't as constrained as they are today; although expensive, their approaches were affordable. SDIO lacked this "heritage" and thus had the chance to sponsor space missions carried out differently.

NCST has several organizational advantages that helped achieve a low-cost Clementine mission. It doesn't have a great burden of bureaucratic procedures. Its organization isn't bound to traditional approaches. It has a great incentive to innovate and keep cost low.

NCST is small, approximately 350 people, and thus naturally has fewer bureaucratic procedures. This small size not only reduces personnel costs and enables NCST to minimize the time and expense that bureaucratic coordination requires, but also frees NCST from the institutional constraints that can prevent cost-effective approaches. A new program is allowed to select procedures and approaches which appear useful, and reject those that do not.

Another NCST advantage is that NRL doesn't have an ongoing space program of its own. Once an organization is focused on a particular mission and technological solution, it often locks onto a single path of research and development. However, NCST builds spacecraft for other U.S. government entities. As a result, the NCST space programs have been highly diverse. Some of the early programs developed and tested basic space technology, such as gravity-gradient stabilization or time transfer, which formed a basis for the Global Positioning System. Some supported space-science programs and included several launches over several years, such as for research on solar radiation. Some supported specific DoD development, such as LACE and Clementine for SDIO. Because of this mission diversity, NCST has rarely been able to keep modifying a subsystem design for use on a sequence of new programs. Thus, NCST has avoided spending great sums of money to squeeze marginal improvements from a traditional technical approach instead of jumping to a newer and better approach.

NCST operates in much the same mode as a contractor. It is constantly having to find work from other government organizations that may need space systems. NCST is largely organized into groups which focus on specific space subsystems or skill areas. These groups have to market their specialties to sponsoring organizations within the Navy, DoD, or other government entities to obtain funding to pay salaries and advance their technologies. Because NCST is always competing for available funds in these specialty areas, there is great incentive to reduce costs. When a mission such as Clementine arises, these groups provide the system and subsystem engineering, and frequently the subsystems themselves, to support the space mission. Ideally, new technology and technology improvements which are mature enough will be incorporated into the subsystems so they can be tested in space. Of NCST's 350 people, 80 to 90 are administrative and clerical staff. Only a few of NCST's people would be assigned full-time to a program like Clementine. The rest would work on other space programs or continue technology development in their specialty areas.

12.1.9 Spacecraft Failure and Low Budget

Although Clementine achieved most of its mission objectives and is generally considered to be successful, we can't ignore the spacecraft's premature failure. The failure is especially significant here because much of the cost of present space missions goes to inspections, tests, and documents that we justify on the basis of enhancing reliability. Did Clementine fail because it was cheap? Would more money have made a difference? Would it have been cost effective?

Clementine failed because of the synergistic effects of multiple subtle errors. After the failure, with knowledge that errors existed and evidence limiting their probable locations, a team of very competent software engineers needed 5 weeks to identify the sequence of events that led to Clementine's failure. With the spacecraft simulator in the same configuration as the spacecraft at the time of failure, it was extremely difficult to cause the simulator to reproduce the failure. Each of the software errors could produce its effect without harm to the spacecraft. Only when they occurred simultaneously under specific conditions did the failure result.

Additional inspections and tests could have detected the errors. But we can't predict how much more testing they'd have needed to provide a high probability that they would find enough of the errors to ensure the spacecraft's survival. More inspections to reveal these errors would have significantly extended the testing time and caused a launch slip because they weren't obvious errors. The resulting cost increase might have jeopardized the program. Of course, other subtle errors may have been present whose unlikely combination could cause spacecraft failure. Suppose they omitted from an extended test plan the test that could have revealed one of these errors. In this case, they'd have been in the lose-lose situation in which they'd have spent additional scarce funding to ensure reliability, but the failure occurred anyway.

NCST tried to design the spacecraft and mission plan so recovery from reasonable failures would be possible. Clementine experienced several problems while in low-Earth orbit. In each case, the spacecraft shut down autonomously and put itself into a configuration that ensured survival over an extended period. While the spacecraft remained in this dormant state, ground personnel analyzed the problem and developed corrections.

Similarly, the long, indirect trajectory to the Moon allowed considerable margin for error. The opportunities for transfer between trajectory phases occurred frequently, so if they had to delay a transfer—as they did for the departure from low-Earth orbit—another opportunity soon followed.

12.1.10 Lessons Learned

The first category of lessons learned consists of the design and operational innovations that produced a low-cost mission which could survive significant deviations from planned or expected sequences and still achieve its main objectives. We can easily apply these design features and operational approaches to future programs.

The second category of lessons learned—the organizational traits that permit NCST to adopt these technological and managerial innovations—may be more difficult to achieve. NASA has expressed its intention to adopt Clementine-like approaches and has been able to start new programs that have many of the efficiencies Clementine demonstrated. Indeed, Lewis and Clark, two satellites planned under NASA's Small Satellite Technology Initiative, resemble Clementine in many respects. Each adds collecting scientific data onto a mission primarily intended to demonstrate new technologies in space; each costs about $50 million; and each is to be completed in about a year.

Possibly the most important thing to recognize from Clementine is that reducing the cost of space missions will require more than innovations in technology and management. It will also require opportunities and incentives for such innovations to be generated, offered, and adopted.

Pluto Express Summary

Pluto Express, to be built by JPL, is a 2-spacecraft flyby of Pluto. Final mission definition is still evolving. The projected cost data here are for the Pluto Express FY95 configuration.

         Spacecraft dry mass: 73.5 kg each   Average power: 71.5 W
         Propulsion: Hydrazine   TT&C: 150 to 450 bps

Launch: Molniya with Star 48

Orbit: Interplanetary with 10-year Jupiter gravity assist flight to Pluto

Operations: NASA's Deep Space Network's 70 m station net

Status: Mission cancelled for budgetary reasons.
                New Pluto/Kuiper Belt mission being studied.

Cost Model for 2 spacecraft (FY95$M): See page 348 for explanation.

	Initial Expected Cost*	Small Spacecraft Model**	Projected Cost‡
Spacecraft Bus	$1,140M	$12M	$226M
Payload†	$50M	$1M	$9M
Launch	$565M	$332M	$43M
Ground Segment††	N/A	—	Incl. spacecraft cost
Ops. + Main. (annual)	$40M	—	$6M
***Total** (through launch + 1 yr)*	**$1,795M**	—	**$284M**

*   Initial expected cost is based on the buy of 2 additional, unmodified CRAF/Cassini spacecraft.
** The Small Spacecraft Model does not include interplanetary spacecraft & is not intended for this application.
†   Does not include cost of the Russian "Drop Zond" to be deployed to the surface of Pluto.
†† Ground segment development cost was incorporated into spacecraft cost to facilitate cost trades during mission definition.
‡   An inflation factor of 1.000 has been used to inflate to FY1995$ [SMAD, Table 20-1].

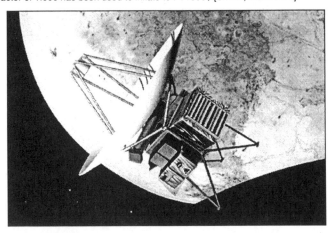

Fig. 12.2-1.   Pluto Fast Flyby '94. The satellite is shown here in its 1994 configuration. The truss structure is used for connection to the injection stages and the mounting of the high-gain antenna. An artist's rendition of Pluto's surface is shown behind the spacecraft.

12.2   Pluto Express

Richard S. Caputo, Robert L. Staehle
Jet Propulsion Laboratory, California Institute of Technology

12.2.1   Background
12.2.2   1992 Baseline Mission and Flight System
12.2.3   FY93 Mission Requirements
12.2.4   FY94 Mission Requirements
12.2.5   FY95 Pluto Express

Over the last 4 years, the Pluto Express pre-project's[*] mission requirements have changed rapidly due to budgetary and other societal pressures. Now, the cost goal for the entire mission is about 10 times less than it would have been if we had used carbon copies of the Cassini flight system (Saturn mission) to fly to Pluto. After adjusting Phase C/D costs for two spacecraft, the Pluto cost goal is actually less than the cost cap for new, low-cost Discovery-class missions, which is $150 million (FY92) without launch system or mission operations.

In each of the last 4 years, the Pluto team met in succession increasingly difficult cost ceilings and other top-level requirements (see Table 12.2-1): (1) performing acceptable science with a cost limit of $400 million (FY92) for mission development (project start to launch plus 30 days), a short project (launch by FY98), and less than ten years transit to Pluto; (2) adding many new technologies within this cost cap, science, and schedule; (3) halving NASA's life-cycle cost, now considering launch and mission operations; and, most recently (4) roughly halving the cost from project start to launch plus 30 days. During this period, other requirements were introduced and strengthened, such as strong educational outreach.

In this chapter, we want to stimulate discussion, but we are not committing to work. We summarize how we've adapted to a world in which a science mission to the last known planet is valued less and less each year by a political process that is still responding to the end of the Cold War. The techniques to keep up with these rapidly changing requirements cover a wide range: using advanced technologies, involving foreign partners, completely revising JPL's concept of a project from sending a space-craft with instruments to sending a "sciencecraft," directly integrating cost into the design process, completely rethinking the flow of the project—starting with system-level design and testing and evolving internal requirements as the design matured—rethinking the way people carry out a project, and other management techniques.

In a sense, Pluto is our "gateway to the stars." Our first mission offers an opportunity to set a new approach for exploration in the next millennium. So despite constricting fiscal resources, NASA Administrator, Mr. Goldin, has challenged us to lead low-cost exploration of the outer solar system and beyond.

---

[*] Formerly named "Pluto Fast Flyby." At JPL, *pre-project* means an activity expected in the future to be a line item in the Federal budget. For the rest of this chapter, we use the more familiar term *project*.

Much of the work described here was performed at the California Institute of Technology's Jet Propulsion Laboratory under contract to the National Aeronautics and Space Administration, sponsored by the Office of Space Science and the Office of Space Access and Technology. The authors wish to acknowledge Stephen Brewster, Linda Lievense, Nancy Livermore, Arnold Ruskin, and Giulio Varsi for providing considerable assistance.

TABLE 12.2-1. Evolution of the Pluto Mission.

Time	Mission	Cost Objectives	Constraints	Results and New Mandates
FY92	Pluto Fast Flyby 92 Baseline	$400M (FY92) for mission development	• Launch quickly • Get data back from Pluto quickly • Meet science as defined by science community	• Met cost goal within constraints • Few new technologies required
FY93	Pluto Fast Flyby 93 Baseline	$400M (FY92) for mission development	• Introduce many new technologies • Rest same as FY92	• Introduced nearly 20 new technologies • Met mission development cost goal • Reduce mission ops cost by half
FY94	Pluto Fast Flyby 94 Baseline	Life-cycle costs about $550M	• Lower project funding profile • Consider U.S. & Russian launch vehicles • Rest same as FY93	• Reduced life-cycle cost to $580M (FY93) • Extended project lengthened from 3.5 to 4.5 years • Reduced life-cycle cost by half
FY95	Pluto Express	Mission development cost of $250M, including launch system & excluding advanced power system	• Consider non-Radio-isotope Power Sources (RPS) • Rest same as FY94	• Sciencecraft approach • Advanced technology • Pathfinder for low-cost exploration of outer solar system

12.2.1 Background

The present mission to Pluto started in early FY92 with the realization that a satellite could fly by and collect science data from the last known, unvisited planet, after a cruise time of only 6 years, compared to Voyager's 12 years to reach the same distance. The mission would require combining a very light flight system (35 kg) with a powerful launch system (Titan IV/Centaur plus additional upper stages). This approach has annual launch opportunities with a direct trajectory, avoiding the need for gravitational assists from bodies such as Jupiter (with its high-radiation field). The U.S. Postal Service spurred this concept by issuing a stamp series on planetary exploration in October 1991. One stamp referred to Pluto as "Not yet explored." Clyde Tombaugh discovered Pluto in 1930 from the Lowell Observatory. Before proceeding with this work at JPL, we talked to Dr. Tombaugh, who gave his "permission" in 1992 to send the first exploratory mission there.

The initial spacecraft concept was largely a single-string design (no redundancy), with an optical camera to do bare-minimum imaging and a radio to perform radio science of the Pluto/Charon system [Salvo, 1993]. This initial concept did not yet rigorously consider overall mission reliability, costs, or project schedule. Nor had we conferred with the appropriate scientific group to determine acceptable science at Pluto/Charon. We believed that two spacecraft could cost <$200M in FY92 dollars.

Based on this initial concept, NASA was interested enough to start more detailed studies because people expected such a mission to cruise much longer (15 to 20 years) and to cost several billion dollars. Also, Pluto is in an attractive position just past perihelion and moving away from the Sun in its 248-year orbit. Pluto's tenuous atmosphere is expected to condense as it becomes colder sometime early next century (2005–2025), after which 200 years will pass before anyone can study its atmosphere again. Thus, the suggested short cruise time was a key to NASA's initial interest.

Voyager's 1989 encounter with Neptune's moon Triton boosted this interest because Triton is nearly Pluto's twin in size and present solar distance and has revealed a complex geology and atmosphere. Finally, some theories about the solar system's evolution indicate that thousands of planets[*] similar to Pluto may have been in the early solar system. Greater knowledge of Pluto might reveal the dominant forces that shaped our solar system's evolution [Stern 1992, 1995].

NASA placed several requirements on the potential mission to Pluto. The cost to produce the flight systems ready for launch (from project start at the beginning of Phase C/D to launch plus 30 days) was to be less than $400M (FY92). (This figure would not include the launch system, mission operations, tracking data analysis, and pre-start costs in Phases A and B). The spacecraft was to launch as soon as practical, with travel time to Pluto as short as possible (less than ten years). The Outer Planets Science Working Group determined "mandatory" science objectives without which the mission would not be worth funding [Staehle, et al., 1992a]. Table 12.2-2 lists them as Category 1A objectives. Along with these science objectives were more detailed specifications of spectral and spatial resolution, signal-to-noise ratio, and coverage for Pluto, Charon, and Pluto's atmosphere. Numerous lower-priority objectives (Categories 1B and 1C in Table 12.2-2) were made optional to avoid driving up costs.

12.2.2 1992 Baseline Mission and Flight System

By the end of FY92, our initial design effort produced a mission, flight system, and project design that met all the requirements. We did not use a micro spacecraft because it was inconsistent with the mission goals of low cost, acceptable reliability, early launch, and minimum acceptable science. We briefly considered an orbiter mission but dropped it because of the extremely long cruise time (>16 years)—even with a 35 kg mass that we couldn't achieve with technology credible at that time [Salvo, 1993]. The development was to take 3+ years, starting in October 1994; launch would be in January or February 1998; cruise on a direct trajectory would take 7.5 years; and encounter with Pluto would occur in August 2005. We were to achieve Earth occulation of both Pluto and its moon Charon with a velocity of about 16.5 km/s relative to Pluto. Four instruments were proposed to meet the minimum science objectives: a visible camera to map for surface geology and morphology, an infrared imaging spectrometer to map the surface composition, an ultraviolet spectrometer to measure atmospheric composition, and an ultrastable oscillator and signal processor to measure atmospheric temperature and pressure as a function of altitude during uplink radio occulation. Ambitious goals for the payload comprising all the instruments were established: cost <$30M (without reserves), mass <7 kg, and power <6 W (without contingency). (Note: In July 1994, we completed working prototypes of the most difficult instrument components so we were confident that we could better these goals for flight hardware.)

---

[*] In this context, a planet is a body orbiting the Sun that is big enough to assume a nearly spherical shape due to gravitational forces.

TABLE 12.2-2. **Science Objectives Covered by "Strawman" Payload.** These three priority classes of objectives (1A, 1B, and 1C) were agreed in 1992 and have remained stable. No priorities have been set within each class. As it turns out, while the Pluto Express mission is designed only to cover 1A objectives, it can also meet some of the lower-priority objectives using the same instruments, with minimal cost impact.

Category	Science Objectives	Pluto Express
1A	• Characterize global geology and morphology	Yes
	• Surface composition mapping	Yes
	• Neutral atmosphere composition and structure	Yes
1B	• Surface and atmosphere time variability	Some
	• Stereo imaging	Some
	• High-resolution terminator mapping	Yes
	• Selected high-resolution surface composition mapping	Yes
	• Characterization of Pluto's ionosphere and solar wind interactions	Partial
	• Search for neutral species including H, H_2, HCN, C_xH_y, and other hydrocarbons and nitriles in Pluto's upper atmosphere. Isotopic discrimination where possible	Yes
	• Search for Charon's atmosphere	Yes
	• Determination of bolometric bond albedos	Yes
	• Surface temperature mapping	No
1C	• Characterization of the energetic particle environment	No
	• Refinement of the bulk parameters (radii, masses, densities)	Yes
	• Magnetic field search	No
	• Additional satellite and ring search	Yes

Optical resolution using the spacecraft's camera will be better several months before closest approach than the Hubble Space Telescope. Within 100,000 km (reached ~1.7 hours before closest approach), imaging pixel size is 1 km on the surface directly below the spacecraft. Because the approach is almost precisely from the direction of the Sun, views of the terminator and at middle phase angles are possible for only a few minutes around closest approach. Thus, the camera's CCD detector must have a 2.5-sec readout rate. With all instruments fixed to the body of the 3-axis-stabilized spacecraft, the spacecraft must reorient and settle quickly. In the dim light at 31 AU the camera optics are sized to provide adequate exposures of about one second [Staehle, et al., 1992b].

The mission used two dual-string spacecraft, each with block-redundant and cross-strapped components. This is necessary to have a better than 90% chance of at least one working spacecraft after the long cruise. One of the spacecraft would be delayed in flight to arrive at Pluto about six months later so that the first spacecraft's results could be used to guide reprogramming of the second encounter. The second spacecraft would view the "other" side of Pluto and Charon. The spacecraft's velocity plus the planet's slow spin (one revolution in 6.4 Earth days) limits the high-resolution view of the sunlit surface to about half the area of Pluto and Charon during a single encounter.

During early approach, the science camera takes optical navigation images, which are transferred to the ground for processing. Within a day before closest approach, navigation must be done onboard because of the long two-way light times (about

8.6 hours). At near encounter, science data is stored in solid-state memory for post-encounter playback at 40 bps. The data is downlinked at X-band from the spacecraft's 1.5 m parabolic, high-gain antenna, using one 8-hour pass per day, to the Deep Space Network's 34 m antennas. The 400 MB of data are returned in less than a year.

We kept the cruise at 7.5 years by using a Titan IV with a Centaur as the first upper stage plus a Thiokol Star 48 and Star 27 as second and third upper stages. Hydrazine propellant onboard allows us to correct for solid rocket motor thrusting as well as to do trajectory corrections and retargeting.

The mission-development project is "hardware rich" because we were breadboarding and brassboarding key components and testing a brassboard system of most of the spacecraft. Two identical spacecraft would be built and environmentally tested at the system level to verify integrity. We also planned a prototype that we could qualification test and refurbish to serve as a flyable spare.

We estimated that a small radioisotope thermoelectric generator (RTG) augmented with capacitors for short-peak loads would be the most cost-effective, lowest-mass, most reliable approach to a mission at the edge of the solar system, although we left alternative power options (solar, battery, and fuel cell) open for further study. The power needed ten years after launch is 63.8 W electric (including 30% contingency) at 14 V, using five general-purpose heat source (GPHS) modules.

The spacecraft is 3-axis stabilized. The inertial sensor has a miniature star camera with a wide field of view and three solid-state rate sensors to provide an attitude reference during propulsion maneuvers. Control is through cold-gas (N_2) thrusters along all three spacecraft axes; nitrogen pressurant comes from a monopropellant hydrazine tank. Pointing knowledge is 1.5 mrad and stability is 10 microrad over one second. Fast slews of 90 deg require 2.7 minutes (zero rate to zero rate) plus settling time.

A central computer handles all commanding, sequencing, and computations. It is a 1.5 MIPS, single-board computer with rad-hard parts (25 krad), and several candidate processors are available. Very-large-scale integration, application-specific integrated circuits, and surface-mount packaging reduce mass. Power strobing minimizes power consumption. Direct lines connect to serial interfaces and a high-rate (5 Mbps) science interface. A 400+ MB solid-state memory uses high density packaging and includes error detection and correction. A data-compression chip provides lossless compression of the science data before storing it in memory.

The main structure is a hexagonal aluminum bus which also holds the propellant tank. Truss structures are used for the adapter to the injection stages and to mount the antenna to the bus. Avoiding articulated or deployed mechanisms lowers cost and increases reliability. A blowdown monopropellant system, using only off-the-shelf components, provides propulsion. The thermal-control subsystem uses the excess heat from the RTG to keep the propellant tank and the spacecraft bus warm. Radioisotope heater units were baselined to heat the exposed thrusters. The high-gain antenna and RTG both shadow the bus from the Sun in nominal Earth-point attitude.

The resulting flight system's dry mass is 140 kg including 29.4 kg mass contingency. The total wet mass (including monopropellant hydrazine to perform 350 m/s delta-V) is 164 kg. Table 12.2-3 summarizes results from the '92 baseline mission and flight-system design. The estimated mission development cost is $327M (FY92). When $37M to be spent in FY94 (Phase B) is included, the total mission development cost estimate is $363 million (FY92). Table 12.2-4 shows the top-level cost breakdown for the '92 baseline approach. This approach met all of NASA's initial

TABLE 12.2-3. Results of Pluto Mission Design and Project Definition.

Mission Name (year)	Results

Pluto Fast Flyby '92 Baseline	• Mission development cost = $363M (FY92)
	• Project (Phase C/D) lasts 3.5 years (preproject is 2 years)
	• 7.5-year cruise to Pluto
	• One year of data playback after encounter
	• Science is acceptable minimum (Category 1A)
	  —Global geology & morphology
	  —Surface composition mapping
	  —Neutral atmosphere composition & structure

Pluto Fast Flyby '93 Baseline	• Mission development cost = $310M (FY93) ($383M when Phase B included) • About 20 new technologies introduced • Flight system mass is reduced from 165 kg to 119 kg (wet) • 8.2-year cruise • 3-month data playback • Science is acceptable minimum (Category 1A)
Pluto Fast Flyby '94 Baseline	• NASA's life-cycle cost is $620 M (FY94) (reduced from $1100 M) (FY93) • Mission-development cost = $306M (FY94) ($340M with Phase B) • 4.5 year project duration due to funding-profile limits • 9.3-year cruise • 3-month data playback • Science is more than acceptable minimum (Category 1A + atmospheric probe)
Pluto Express (FY1995)	• Costs <<'94 baseline • 3+ years for Phase C/D • ~10-year cruise • U.S. or Russian launch vehicles • <10 JPL people staff the cruise mission • Sciencecraft development

requirements for the mission, including early launch (February 98), short cruise (< 10 years), and the 1A science objectives.

Lower-Cost Features in the '92 Baseline

The main way we reduced development cost was to reduce the science objectives to only about 25% of those of recent outer planet missions. These are the Category 1A objectives mentioned above that were defined by the Outer Planets Science Working Group (OPSWG) as minimally acceptable. If we had not worked hard with the OPSWG to arrive at this acceptable minimum, the objectives and their corresponding instrumentation requirements would have tripled or quadrupled. Thus, scoping the science was key to reducing the mission's complexity, mass, power requirements, and cost.

Interestingly, the resulting suite of four instruments is able to achieve about 75% of the total Category 1 science (Table 12.2-2).

The next most important way we kept costs low was to keep the rest of the spacecraft simple. For example, it has neither an instrument scan platform, unfolding

TABLE 12.2-4. Breakdown of Estimated Top-Level Costs for Pluto Project for Phase C/D and the Last Year of Phase B. Illustration examples only.

WBS	Area	Year of Baseline Design			
		FY92	FY93	FY94	FY95
A0000	Payload system	$27M	$28M	$32M	$7M
B0000	Spacecraft system	$115M	$128M	$137M	$94M
C0000	End-to-end mission ops system	$13M	$14M	$17M	$13M
D1000	Launch vehicle integration	($2M)	$1M	$2M	$1M
E1000	RPS[1] Integration	N/A	$1M	$1M	$1M
10000	Project science	$9M	$8M	$5M	$5M
20000	Project engineering	($3M)	$10M	$5M	$3M
30000	Mission design engineering	$6M	$5M	$4M	$2M
40000	Testbed integration & test	N/A	$2M	$2M	$5M
50000	Flight integration & test	$11M	$8M	$9M	$7M
60000	Launch approval engineering	$2M	$2M	$4M	$2M
90000	Project management	$18M	$12M	$13M	$7M
99000	Mission development reserves	$89M	$79M	$68M	$36M
	Mission development subtotal	$290M	$298M	$299M	$183M
Exxxx	RPS	$37M	$51M	$30M	$45M
	APA[2] and SBIR[3] Tax	$36M	$35M	$12M	$7M
	Mission development total	$363M	$384M	$341M	$235M
D0000	Launch system	N/A	$515M	$132M	$43M
	NASA program total (through launch + 30 days)	N/A	$899M	$473M	$278M
70000	Mission operations & data analysis	N/A	$203M	$148M	N/A
	NASA program totals[4], all years	N/A	$1102M	$621M	N/A

1) RPS: Radioisotope Power Source; 2) APA: Allowance for Program Adjustment; 3) SBIR: Small Business Innovative Research; 4) Some totals vary from text because of roundoff

antennas, nor panels that open. Cold-gas thrusters control its attitude, rather than the more complex reaction wheels or moving secondary mirrors. There is no low-gain antenna.

The third major way we lowered costs was to use little new technology and to use existing parts (spares from previous missions), with high inheritance from previous designs and technology. In general, the spacecraft used components that would have been qualified within 2 years (by 1994), with minor exceptions. The RTG used well-proven silicon-germanium thermocouples, as well as standard, general-purpose heat-source modules already safety tested for the Galileo Jupiter mission. The propulsion subsystem is entirely off-the-shelf, and the high-gain antenna is residual Viking hardware. Many components are inherited from the Cassini mission. The solid-state power amplifier is based on commercially available parts in a new component design. The telemetry-control unit is a reduced-function device using Cassini pieces repackaged in a smaller form. Exceptions to inherited hardware are relatively low-risk developments.

Using known U.S. launch systems with a good database (e.g., a safety data book already developed for Cassini) simplifies launch approval and review under the National Environmental Policy Act. This reduces schedule and cost risks in these areas.

Although not part of the cost goal for mission development, approaches to mission operations and DSN tracking were defined to reduce overall project cost. Prior missions often ignored mission operations costs during early design phases. Pluto Express planned from the beginning instead to use a low-cost operations center patterned on the successful cooperative JPL-University of Colorado operations center for the Solar Mesosphere Explorer mission.

Key to lower mission-operations costs are mission and spacecraft-design features, such as long periods of unattended operation during cruise and single weekly passes of only four hours to track the probes and collect data. Onboard data processing reduces downlinking and analysis of engineering data.

We also began to develop spacecraft capabilities that allow uplinking of cruise commands without elaborate simulation and constraint checking. We will pre-plan and test the encounter command sequence during cruise and adjust it slightly just before closest approach. The computer has a large onboard memory to capture all science data and return it over a limited downlink through routine daily DSN passes for up to a year following the encounter. We use a progressive development philosophy: develop the basic mission-operations system at the start of the project and use it to support subsystem tests, spacecraft tests, calibration and post-launch operations.

The resulting mission-development cost of $363M (FY92) is much lower than we'd expect from previous experience. For example, even a duplicate set of spacecraft based on the CRAF/Cassini program with a repetitive procurement would cost two to three times more. A Pluto mission derived from the *Cassini* orbiter would have involved more than a dozen instruments. It would have been more capable, but its abilities would far exceed what the Outer Planets Science Working Group mandated for the first Pluto mission. Table 12.2-5 summarizes these cost-reduction techniques.

12.2.3 FY93 Mission Requirements

At the end of FY92, as Table 12.2-6 shows, the NASA Administrator required the mission to use a lot of new technology to support national interests. Future space missions would use these technologies, and some were likely to spin-off to commercial uses. In addition, public outreach and student involvement were to be emphasized as important aspects of the project.

We had to introduce advanced technologies without relaxing any other requirements, such as the cap of $400 million (FY92) on mission-development cost. Because introducing many new technologies usually increases cost and schedule risk, this was a severe challenge.

The Pluto team responded in early FY93 by beginning to acquire new technologies for the Pluto project from sources in industry, academia, and federally funded research and development centers (FFRDCs). While introducing new technologies may lengthen development time, they are also expected to reduce spacecraft mass and thus reduce cruise time in compensation. The '92 baseline contains our fallback position for all subsystems to control development risk.

TABLE 12.2-5. Main Techniques to Meet the Pluto Mission's at Requirements at Lower Cost.

Mission Name	Principal Techniques

Pluto Fast Flyby '92 Baseline	• Get science community's support to limit required science to highest priority (Category 1A)
	• Have only three instrument functions plus radio science needed to meet science requirements
	• Use available technology and, in some cases, use devices left over from previous missions (e.g., 1.47 m Viking antenna)
	• Keep the design simple (no articulations or deployments, cold-gas thrusters, no low-gain antenna)
	• Limit new technology to that with greatest system benefit, e.g., solid-state power amplifier borrowed from commercial parts
	• Use energetic, U.S. launch system (Titan IV/Centaur) with:
	  –Known database for easier review under the National Environmental Policy Act
	  –Shorter cruise time
	  –Direct trajectory without Jupiter Flyby with minimum need for radiation hardened parts

Pluto Fast Flyby '93 Baseline	• Maintain science community support for Category 1A science objectives and four instrument functions • Vastly expand use of new technologies (15 to 20 items) with advanced technology program • Keep the design simple • Reduce use of breadboard and brassboards • Don't build a full-system prototype for use as flight spare • Use test bed for development of flight and ground software • Use common command architecture for flight and ground • Use spare RTG from Cassini project instead of new power system fuel • Operate the mission with combined JPL and university team • Teaming and organizational changes (see Table 12.2-7) • New design approach (see Table 12.2-8) • Supportive NASA role (see Table 12.2-9) • New information system (see Table 12.2-10) • Improved project control (see Table 12.2-10) • Improved procurement practice (see Table 12.2-11)

Pluto Fast Flyby '94 Baseline	• Team with Russia to:
	  –Use Proton launch vehicle
	  –Carry Drop Zond to directly study Pluto's atmosphere
	• Use project design center for design-to-cost and directly connect to test bed for "system"-level prototyping
	• Directly connect to test bed for:
	  –System-level prototyping using breadboard, brassboard, and engineering models
	  –Software development and verification
	• Use a wide range of design and technology features (see Table 12.2-12)
	• All '93 baseline techniques for reducing cost and risk

Pluto Express	• "Sciencecraft" versus spacecraft plus payload • Concurrent design and cost • System-level testbed • Real and virtual co-location of project • New work breakdown structure & fully developed management & control techniques • Close alignment with New Millennium program

TABLE 12.2-6. **Key Advanced Technologies for FY93 Pluto Mission.** New technologies were added in response to NASA Administrators' call for higher technology content, coupled with added FY93-94 "Advanced Technology Insertion" funding to build and evaluate new hardware and software prototypes.

System	FY92 (kg)	ATI Goal (kg)	FY93 (kg)	Technology*
Spacecraft system				• Electronic packaging; MMIC, MCM, ASIC
Tele-communication	25.20	16.80	12.75	• Micro-packaged digital receiver (MMIC, MCM) • High-gain antenna with composite structure • High-efficiency SSPA (MMIC)
Electrical power	23.2	12.5	19.4	• High-efficiency DC/DC conversion
Attitude control	2.70	2.10	6.65	• Miniature star camera • Inertial-reference unit containing a ring-laser gyro
Spacecraft data	7.0	4.5	6.5	• High-density data storage, ASIC, MCM packaging
Structure	20.0	14.6	14.6	• Composite and thermal zoning bus
Propulsion	20.1	13.1	9.9	• Tiny cold-gas thrusters with low-leakage, miniaturized components (valves, regulators)
Thermal control	4.0	3.5	3.7	• Light-weight louvers, multilayer insulation
Science	9.0	7.0	7.0	• Integrated electronics, ASIC, MCM packaging • Light-weight, thermally stable materials
Mission operations and data analysis	$TBD		$150M	• Concurrent development of test beds for flight & ground systems • Common command architecture and language for flight and ground • Smart editing of engineering data • Spacecraft tolerant to onboard & ground faults
Total	111.2	74.1	80.5	
Contingency	29.5	20.1	31.3	(FY92: 26.5%, FY93: 38.9%)
Total Dry S/C	140.7	94.2	111.8	
Propellant	24.6	16.1	6.9	(FY92: & ATI goal: ΔV=350 m/s, FY93: ΔV=130 m/s)
Total Wet S/C	165.3	110.3	118.7	

* MMIC=Monolithic Microwave Integrated Circuit; MCM=Multi-Chip Module; ASIC=Application-Specific Integrated Circuit; ATI=Advanced Technology Insertion; SSPA=Solid-State Power Amplifier

After establishing a list of potential sources of new spacecraft hardware and software technology, we held workshops describing project needs for new technology and sent more than 1,200 requests for information. After evaluating the requested information, we wrote 16 procurements or agreements for prototype hardware and software. Also, a NASA research announcement went out early in 1993 to find and demonstrate promising instrument technologies. The key technologies developed through these initiatives are listed in Table 12.2-6 [Staehle, et al., 1993].

A key indicator of improved performance is reduced subsystem mass. If the launch system is the same, this translates into reduced cruise time for direct trajectories to Pluto. The spacecraft mass (including payload) decreased from 165 kg in the '92 baseline to 119 kg in the '93 baseline. Improvements were greatest in the telecommunications and propulsion subsystems and significant in structure and electric power.

The project schedule was delayed one year from the '92 baseline, with project start set for FY96, a Phase C/D of 3.5 years, and a dual launch in January and February 1999. The estimated cost for mission development (project start to launch + 30 days) is $322M (FY93). Because of the large number of new technologies, we needed a lot of funding before starting the project. Adding FY95 (Phase B) gives us a total of $383M (FY93) which is essentially the same as estimated at the end of FY92. Table 12.2-4 shows the breakdown of '93 baseline costs for Phases B and C/D.

To keep the cost under the cap of $400M (FY92) while using much more new technology, we increased project risk by dropping the building and system testing of a flight prototype. Hardware development would go directly from engineering model at the assembly level to flight units. Besides staying within the cost cap, we maintained the objective to launch as soon as practical (just 5+ years for Phases A, B, and C/D). This required much more money after the first year, and the Phase B budget was $61M (FY93) to complete its work within one year. Estimated project life-cycle cost was $1,100M (FY93), from Phase B to one year after encounter at Pluto for data analysis, and including the launch system and mission operations for 10 years.

The baseline launch system was still the Titan IV-Centaur plus Star 48 and Star 27 solid upper stages. The cruise time to Pluto is 8.2 years, and one spacecraft will be delayed a year during cruise so we can adjust the second encounter based on the first encounter's results.

Figure 12.2-2 shows a layout of the flight-system design for the '93 baseline and its improvements over the '92 baseline. Table 12.2-6 shows the resulting system mass, which is about 28% (nearly 50 kg) lower. The four instruments (visible camera, infrared imaging spectrometer, ultraviolet spectrometer, and the ultra-stable oscillator for the radio receiver) were the same as in the '92 baseline. The flight system still used two identical spacecraft, each dual-string with block-redundant and cross-strapped components to have acceptable reliability after the long cruise to Pluto.

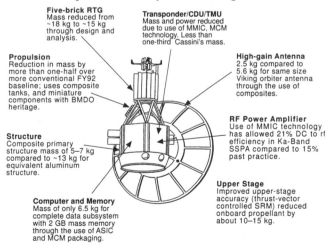

Five-brick RTG
Mass reduced from ~18 kg to ~15 kg through design and analysis.

Transponder/CDU/TMU
Mass and power reduced due to use of MMIC, MCM technology. Less than one-third Cassini's mass.

Propulsion
Reduction in mass by more than one-half over more conventional FY92 baseline; uses composite tanks, and miniature components with BMDO heritage.

High-gain Antenna
2.5 kg compared to 5.6 kg for same size Viking orbiter antenna through the use of composites.

RF Power Amplifier
Use of MMIC technology has allowed 21% DC to rf efficiency in Ka-Band SSPA compared to 15% past practice.

Structure
Composite primary structure mass of 5–7 kg compared to ~13 kg for equivalent aluminum structure.

Computer and Memory
Mass of only 6.5 kg for complete data subsystem with 2 GB mass memory through the use of ASIC and MCM packaging.

Upper Stage
Improved upper-stage accuracy (thrust-vector controlled SRM) reduced onboard propellant by about 10–15 kg.

Fig. 12.2-2.   Flight-System Design for '93 Baseline

Introducing new technology allowed us to reduce the mass by nearly 50 kg. For example, the telecommunications subsystem uses a lighter composite-structure antenna, high-density electronics packaging, and more efficient RF amplifiers. The mass of the subsystem was halved to about 13 kg and it uses only 22 W while transmitting. The key to reducing the transponder's mass was using monolithic microwave integrated circuits and packages of multi-chip modules. We used an all-digital receiver (12 W Pluto vs. 15 W best practice) and increased conversion efficiency for DC to radio frequencies (21% Pluto vs. 15% best practice). These changes reduced the overall power requirement and power system's cost. A 70 m ground antenna quadrupled the downlink data rate to 160 bps from the 40 bps we previously got using X-band (8 GHz) on the spacecraft's 1.5 m high-gain antenna. Advanced electronics packaging in the spacecraft computer increased the storage volume for science data from 400 MB to 2 GB while slightly reducing mass and staying within the power goal of 6 W.

The '92 baseline design had a mass of 2.7 kg for the attitude-control subsystem based on a star-tracker camera weighing less than 0.5 kg, featherweight valves and regulators, and milliNewton thrusters. As a reserve against the possibility that a micro-sized star camera could prove inadequate or difficult to qualify for the Pluto mission, the mass estimate for this subsystem was raised to 6.6 kg.

The propulsion subsystem's mass went from 20 kg to 10 kg by using miniature pressure regulators and valves, a pressurant/propellant tank overwrapped with composite, and a propellant-management device based on surface tension. We improved the injection accuracy by using 3-axis stabilization of the upper stages, which allowed us to reduce the mass of the hydrazine monopropellant from 24.6 kg to 7 kg.

The '92 design using an all-aluminum structure changed to a mix of aluminum and graphite-epoxy composite. We also reduced the mass of cabling and connectors to lower the mass of the structure and cabling subsystem from 20 kg to 14.6 kg. Improved thermal zoning with the RTG eliminated the need for small, separate, radioisotope heater units and for controllable electrical heaters in the thermal subsystem. The small size and short cable runs enabled central power conversion to several voltages.

We considered advanced power-conversion technologies such as alkali metal thermoelectric converters and thermophotovoltaic converters to see whether they could reduce the power subsystem's mass. But they were too immature to meet the Pluto mission's schedule (FY99 launch). A change in the RTG's support structure did, however, reduce mass from 23.2 kg to 19.4 kg for the '93 baseline.

Lower-Cost Features in '93 Baseline

Reducing the mass by nearly 50 kg also reduced mission operations costs on the direct trajectory because the launch system could impart a higher Earth-escape velocity, resulting in a shorter cruise.

To reduce power costs, we planned to use the spare fuel elements from the Cassini RTG rather than new fuel. Although old nuclear fuel generates less heat than new fuel, a small RTG could be built with sufficient power by using six of the general purpose heat source modules. Two flight units and a fueled spare were possible, saving about $25M.

To further reduce mission operations costs, we planned to use a combined team of JPL and university personnel. JPL personnel would perform operations at critical times or for anomalous events and university students and professionals at a remote site would perform routine tracking and data-collecting operations 8 hours per week.

Another step to reduce costs was dropping the full-up prototype (which acted as a flight-spare spacecraft) even though this increased the risk of completing the project within schedule and budget.

A number of other compensating risk-reduction and cost-saving features were introduced into the '93 baseline project in widely ranging areas such as teaming and organization (Table 12.2-7), design approach (Table 12.2-8), NASA role (Table 12.2-9), information system and project control (Table 12.2-10), and procurement (Table 12.2-11). Table 12.2-5 summarizes these cost-reduction techniques.

A project's organization and teaming arrangements have much to do with its productivity. Although technology receives most of the emphasis in describing cost reduction, the system staffing also powerfully drives cost. Table 12.2-7 summarizes 18 organizational changes that increased the effectiveness of people on the Pluto project. It is difficult to quantify the cost reduction for each of these suggested changes and it is almost impossible to estimate their combined effect. They are part and parcel of a basic re-engineering of the way JPL does projects.

TABLE 12.2-7. New Approach to Teaming and Organization in '93 Baseline.

- Skunk Works™-style co-located team involved throughout the project
- Mixed team: youth plus experienced seniors who are open to new approaches
- Interdisciplinary team with full range of skills needed to conceptualize the mission and bring the project home, (creative and innovative types, plus those who can follow through and prefer more detailed work)
- Workforce cap with team members working nearly full-time or consultants working small amounts of time as needed
- Flat rather than hierarchical organizational structure
- Team members accountable as individuals (not line organization) and responsible for deliverables, reliability, and cost through end of mission
- Rewards (salary) based on job done (not longevity or number of people supervised)
- Other government labs, private contractors, and scientists providing instruments as full team members
- Teams concerned with launch approval and compliance with the National Environment Policy Act involved early in trades on the spacecraft and launch-vehicle designs
- A focus on customer needs within team as well as with sponsors and constituency
- A credible threat of "no mission" used to maintain cost cap
- Frequent, informal, honest communication among concurrent team, top management, and sponsors to foster willingness for team members to set their own way of doing things (many paradigm shifts)
- Project manager accountable to assistant director of the laboratory and able to hire and fire team members
- Separate science teams for instrument development and for encounter to avoid cost of maintaining large science team throughout the cruise
- Hands-on student involvement to meet cost and educational goals
- Product-based breakdown for work: the delivered product determines the work organization's structure versus working in skill areas usually based on organizational structure
- Work organization structured so developers of mission operations and ground data systems are in the same work element as the flight command and control system
- Groupware to allow online sharing of up-to-date information and files, with the person responsible for that work element responsible for maintaining the information
- All project meetings available by telecom or in person

Skunk Works™ trademark property of Lockheed Advanced Development Co.

These changes to the way JPL carries out projects refocuses us from an institutional structure based on skills that, at one time, reflected the way a project was organized. The old structure supports the organization more than the project needs, except when these two needs overlap. As shown in Table 12.2-7, we need to support projects more directly. Whatever the changes are called (creating "soft" projects, re-engineering), they need to focus the talent and energies of JPL's staff directly at the current project. This re-engineering is under way at JPL; if successful, it will reduce costs in the ways we've outlined.

Table 12.2-8 summarizes our suggested approach to system design for the '93 baseline. It overlaps with some of the organizational aspects in Table 12.2-7, but Table 12.2-8 emphasizes the design process. Most of its suggestions aim to change how the team works together: concurrent versus serial design, concurrent development with system interaction versus more independent and separate development, and so on. Some of the planned approach, however, is purely technological, such as inserting advanced technology via breadboards and brassboards.

Clearly, conditions external to a project directly affect project cost. Table 12.2-9 shows how NASA headquarters influences these costs, especially in their decisions on launch approval and compliance with the National Environmental Policy Act as well as on budget support for the project. Some of the items in this table are beyond NASA's control; they reflect the project's larger political context.

TABLE 12.2-8.   New Approach to Flight-System Design.

• Design all systems concurrently while keeping track of life-cycle cost, mass, power, performance, and schedule
• Hold people (not organizations) responsible for deliverables
• Integrate software for flight and ground data systems; design a single end-to-end information system
• Use breadboards to insert advanced technology; brassboard hardware for almost all subsystems; system test the brassboard equipment
• Establish length of development schedule and don't change it
• Use common high-order language and same operating system and software in ground tests, flight operations, and the onboard RISC processor
• Simplify approaches to flight operations, the role of requirements in design, and project documentation
• Base reliability on Class C mission approach with some tailoring for critical items
• Limit spares to assembly-level (integrated) items that have procurement times greater than six weeks
• Eliminate radioisotope heating units by using RTG "waste heat" via a thermal zone configuration design*

* This applies to the baseline concept only. Other power source-options may require RHUs. A final decision on the power source will not occur until after mission-options trade-offs during Phase A/B are complete.

Table 12.2-10 recognizes the increasingly important role of information systems and project-control techniques and summarizes those included in the '93 baseline. Some of the suggestions try to simplify a system that has become overly complex and counterproductive (one-page requirements documents, no document unless requested and needed, and so on). The rest focus on improved planning and control, such as using integrated schedules and earned value, and on reducing the burden of external influences such as planning to keep disruptions low.

TABLE 12.2-9. NASA's Role for a Cost-Effective Project.

• Early decision on project elements that drive launch approval and compliance with the National Environmental Policy Act, such as new project start dates and launch dates

• Early inter-agency agreement on how to generate the environmental impact statement, book of safety data for the launch system, and safety analysis report

• Timely release of notice of intent, which is part of producing the environmental impact statement

• Timely choice of launch vehicle

• Larger early investments to reduce life-cycle costs

TABLE 12.2-10. Improved Information System and Project Control in '93 Baseline.

• Use simplified, flexible drawings and system for releasing them

• Thoroughly record important work

• Keep requirements documents at one page or less

• Don't generate a document unless the team carrying out the project needs it

• Don't generate a document until a user requests it

• Document new designs by "as-builts" (earlier generation plus changes and deltas)

• Manage risk throughout mission development by identifying, tracking, and reducing risks for technologies, cost, schedule, and the overall program

• Establish reserves and allocate margins for cost, schedule and system design

• Use integrated schedules and a simplified earned-value system, with frequent statusing to maintain better control of the project

• Plan to reduce potential disruptions; take a pessimistic (low-budget) approach to detailed planning for next fiscal year

Procurement processes directly and indirectly affect a project's cost. Table 12.2-11 lists how we hope to use them advantageously.

12.2.4 FY94 Mission Requirements

The '93 baseline Pluto mission met or exceeded all of NASA's requirements, including development cost and substantial new technology. The cost for mission development in Phase C/D is $369M (FY92). When the cost of the last year of Phase B is added and the total is inflated to FY93 dollars, the cost is $384M (FY93).

This development cost was part of NASA's life-cycle cost of $1,100M (FY93), which also includes the launch system, mission operations, and data analysis through one year after the encounter with Pluto. At the end of FY93, however, NASA sought to reduce the mission's life-cycle cost, not just its development cost, and once again changed the mission's requirements. Thus in FY94, the mission had to meet life-cycle costs that were about half of the $1,100M (FY93) (Table 12.2-1).

In addition, funding for Phases A and B was less than that needed to best develop the new technologies and to complete the project quickly at lowest overall cost. All other previous requirements remained, such as substantial new technology, early launch, short trip-time, public outreach, and student involvement.

We responded to these new requirements by extensively redesigning the mission. We considered new launch systems and power subsystems while examining new

TABLE 12.2-11. Improved Approach Procurement in '93 Baseline.

- Establish performance specifications and requirements before committing to contracting
- Base "make or buy" decisions on cost and most qualified source
- Buy hardware and software, not designs and studies
- Set difficult specifications where they count (to achieve goals for cost, mass, power usage) and relax elsewhere
- Cap costs in contracts
  - Communicate importance of cost performance to contractor CEOs
  - Don't allow contract modifications without primary accountable team member's approval (the primary accountable is the technical lead reporting to the project manager)
- Reduce agenda in the RFP
- Streamline university contracting (two-page contract plus cover letter)
- Use contractor procedures
  - Reporting format
  - Product assurance
  - Delegate contractor responsibility to get good bids for subcontracted work
- Use frequent, informal discussions of cost and schedule status between JPL and contractors' representatives
- Use simple tools to monitor and manage cost and schedule
- Solicit contractors' ideas on:
  - Commercial applications
  - Competitiveness
  - Educational benefits
  - Ways for government to amplify benefits
- Contract for current item with option to produce next step in development
- Include contractors in project team and avoid adversarial relationships to improve communications, avoid misunderstanding, and reduce costs
- Provide contractors with timely funding unless agreed otherwise
- Use "just in time" approach

spacecraft designs and scenarios for mission operations. This effort resulted in the '94 baseline project and mission design. The 1994 Technology Challenge Team reviewed the technical approach, as well as most of the project approach and cost estimate. Many review-board members were from small, quick, low-cost projects such as the Naval Research Laboratory's Clementine project (Sec. 12.1) and the NEAR project at Johns Hopkins' Applied Physics Laboratory project. They strongly endorsed the technical and programmatic approach in the '94 Pluto baseline.

To substantially reduce NASA's life-cycle cost, we changed the proposed baseline launch system from a Titan IV/Centaur to a Russia-provided Proton, in response to Russia's expressed interest in participating in the mission. The Star 48 and 27 for the final launch stages remained the same. Table 12.2-3 shows the overall results of the '94 baseline effort and Table 12.2-4 shows the cost estimates. The cost for the launch system alone went from $515M (FY93) to $133M (FY94) a reduction of nearly $400M. (Although the Protons are to be provided at no charge to the United States, the U.S. will still pay for U.S.-provided upper stages and integration.) In return for the Russians' Protons, they proposed to have the U.S. flyby spacecraft carry Russian atmospheric probes, called Drop Zonds, which would be dropped on impact trajectories prior to closest approach. The Zonds transmit scientific data to the flyby spacecraft for later relay to Earth.

NASA's total life-cycle cost for the project is $621M (FY94) from Phase B to end of mission, including the launch system, mission operations for 10 years and one year for data analysis after encounter at Pluto. This is about half the equivalent cost of the '93 baseline mission.

The mission is still designed to meet the Category 1A science objectives and to get to Pluto in fewer than 10 years. The project start is delayed one year from the '93 baseline and set for FY97. Adopting a more gradual budget build-up increased the length of Phases A, B and C/D to 6.5 years, with Phase C/D at 4.5 years. A dual launch was scheduled for January and February 2001 with direct trajectories to Pluto.

The estimated cost for mission development (project start to launch plus 30 days) to deliver the flight systems is $306M (FY94) including a 30% reserve held at JPL. This is 30% lower than the original cost cap of $400M (FY92) for Phase C/D. If we add FY96 (Phase B), the total is $340M (FY94), which is 15% less than the cost estimate for the '93 baseline.

To lower the mission-development cost, while still introducing new technology and stretching the project to match a lower budget profile, we won't build brassboard models, thus increasing risk. Now we will generally go from breadboard to engineering model before building flight equipment and build brassboard models only by exception.

To minimize this increased risk, we changed how we'll handle technical systems. Two of the more important changes are (1) design-to-cost techniques supported by the Project Design Center, which allow quick, simultaneous subsystem and system trades while considering technical and cost factors, and (2) a flight-system testbed used to prototype software and hardware early at essentially the system level. Thus, we'll develop the project faster and cheaper by:

- Building early to work out interfaces and find and solve problems
- Using a rapid-prototyping testbed to allow parallel interaction among the instruments, spacecraft, and information system—including the flight and ground data systems
- Limiting floating requirements by understanding capabilities through prototyping
- Going directly from simulating hardware and software elements, to prototyping in the testbed, to flight hardware and software

Both the Project Design Center and the Flight System Testbed started at JPL during FY94.

The FY94 baseline emphasized more rigorous management techniques, such as using more extensive cost and schedule planning than JPL typically uses. This includes networked schedules integrated for use at the project's high levels and tracking scheduled "events" (receivables, deliverables, start and finish of key activities, milestones, etc.) down to the lowest levels of the project. The cost plan will correspond to these "events," and we'll use a simplified earned-value system to relate work completed to actual costs and to compare work and cost to plans. Cost-to-complete, critical-path, and key-event tracking will continue over the life of the project. Finally, we will use concerted risk management to identify, assess, and reduce risks.

The biggest change in the technical approach is baselining Protons with PAM-D and Star 27 solid upper stages for the launch system. The cruise time to Pluto is 9.3 years, with one spacecraft delayed six months during cruise to allow for feedback from the first encounter. The spacecraft design will also fit a back-up launch system, a Delta II (7925), which uses an Earth-Jupiter, gravity-assist trajectory with a November 2001 launch. Using the backup Delta with gravity assist, the spacecraft will cruise 13.3 years and encounter Pluto/Charon in early 2015. Shorter flight times are possible with a closer approach to Jupiter, but that subjects the spacecraft to a higher radiation dose.

The '94 system's dry mass is 158 kg, including 26 kg of contingency and 15 kg for the Drop Zond and associated equipment. The wet mass is 182 kg. This is an increase of 63 kg over the '93 baseline design, mainly due to the Drop Zond and changes to the launch system.

The four instruments were the same as in the '93 baseline, but we added a science-data processor which is a clone of the spacecraft data computer.

As with prior baselines, the flight system uses two spacecraft, each dual string with a combination of block-redundant and cross-strapped components, for acceptable reliability after the long cruise to Pluto.

Lower-Cost Features in the '94 Baseline

Table 12.2-5 summarizes the techniques used to meet the '94 mission requirements. The single largest factor in lowering cost is using a Proton instead of a Titan IV in the launch-system. Also, to save upper-stage costs, we'll use a spin-stabilized stack instead of the 3-axis stabilized stack in the '93 baseline. This combination also saves about $55M compared to a Shuttle option, which we also considered. Solar electric propulsion was evaluated and found to give flight times comparable to that of the baseline launch system. But this stage hasn't been designed and the cost to develop, qualify, and operate such a stage is likely to be much greater than that of the baseline or backup launch system for a FY97 project start.

Essentially all of the advanced technologies included in the '93 baseline design listed in Table 12.2-6 are part of the '94 baseline. In addition to using advanced technologies, we also used other design or technology changes to reduce the project's life-cycle cost, as shown in Table 12.2-12. Table 12.2-13 shows various other features of the '94 baseline design.

12.2.5 FY95 Pluto Express

The FY94 Pluto Fast Flyby approach still didn't meet NASA's needs for advanced technology and total cost. To respond to these needs, the Pluto team developed a six-part strategy:

- *Sciencecraft*: Instead of soliciting individual scientific instruments after much of the spacecraft design has been cast, the project will call for an integrated set of science investigations from a single team early in Phase A. This science team and the engineering team will join to build the end-to-end mission. The combined team will use operations, encounter design, and instrumentation as starting points to select technology, design the flight system, and build software, and do everything else needed for a successful mission. Of course, cost and technology will influence these decisions.

TABLE 12.2-12. Features of Design and Technology that Reduced '94 Baseline Costs.

Telecommun-ications Area	• Double the downlink data rate by increasing the antenna's diameter from 1.5 m to 2 m, and increase the transmitter power from 3 W to 5 W. • Increase the X-Band antenna's gain 15% by using a center-fed approach without increasing cost or mass. • Include "hooks" for a Ka-band downlink to increase robustness of the design if 70 m stations of DSN are unavailable (Ka-band solid-state power amplifier must be provided from funding sources outside the project based on its value for other missions). • Incorporate radio science into the digital receiver to reduce mass and power of radio science hardware without affecting communications. • Jointly develop and buy a transponder with the Mars Surveyor Program to save cost to each program.
Power/Pyro Area	• Use centralized power conversion which increases efficiency, reduces cost through modularity, and adapts well to late changes in requirements.
Propulsion Area	• Use a blow-down system with a combined pressurant and propellant tank with replenishment nitrogen bottle.

- *Concurrent Design and Cost*: Instead of designing to meet technical require= ments and then checking the cost to see how the design team did against the cost requirement, we're developing design-to-cost tools so that we can understand and estimate the cost of any part of a design concurrently while moving toward the final design.

- *Early Use of System-Level Testbed*: In the past, projects have completed their paper designs from conceptual to preliminary to detailed levels. Then, they built from the part level to the assembly, subsystem, and system levels. Instead, we're giving the Flight System Testbed system-level functions starting with simulated subsystems for both hardware and software. Work then proceeds in the Flight System Testbed with increasing fidelity to bread-board and engineering models. Fully capable flight and ground system software will be developed and tested at the system-level. Eventually, we'll test much of the flight hardware in the Flight System Testbed before environmentally testing the whole system. This approach will flush out the more difficult system-level problems early, when they can be handled less expensively. This approach should allow a cheaper project and one that moves more quickly after the Phase C/D project start. This cheaper approach requires that a greater fraction of project resources be invested during Phases A and B but the absolute funding required is still much less than prior missions. Prototype hardware and software have already been developed and tested in the Flight System Testbed for some components.

- *Real and Virtual Co-Location through a "Paperless Project"*: To help carry out of the project's new approaches, we'll have all information within a few computer mouse clicks on each team member's workstation. This includes all current information, whether in draft, preliminary, or final form, and whether as text, tables, sketches, photos, or drawings. We'll start with a compatible work environment, with full connectivity to all kinds of comput-ers and every location. We'll take full advantage of hypertext links and soft-ware such as Netscape to make current information easily available. We'll generate hard copy much faster than usual, but only when it's requested.

TABLE 12.2-13.   Other Ways to Reduce '94 Baseline Costs.

- Use 70 m antenna as much as possible for encounter data downlink to transmit four times faster than the 34m antennas
- Have clear interface with science instruments and limits on amount of data return
- Design the flight system for mission operations (MO) to reduce or eliminate uncertainty in MO requirements
- Use "beacon cruise" operations
  - Listen versus track for 2-bit health status
  - Listen as opportunity permits
  - Track at will from non-DSN sites
- Combine JPL with university for multi-center operations to best take advantage of skill mix and cost
- Do no cruise science even if gravitational flyby
- Navigate onboard autonomously
  - Hardware capability onboard
  - Improve base software during cruise
  - If onboard not ready by encounter minus one year, use ground as degraded backup
- Reduce cost impacts of mission reliability
  - No reliability requirement for UV, IR, or radio-science instruments except to protect spacecraft
- De-rate equipment (operate electronics at lower than rated temperatures)
- Monitor spacecraft health autonomously onboard
- Use redundancy only where risks are highest
- Establish functional requirements for the spacecraft and don't change them
- Establish architecture of spacecraft hardware and software and don't change it
- Establish mission design and don't change it
- Limit new technology to items that are significant contributors to mission success
  - Acceptable effects on and risks to cost and schedule
  - More attractive mission due to importance of technology transfer
- Up-to-date cost accounting
- Earned-value system to predict cost-to-complete by major work elements
- Up-to-date work accounting using integrated network schedules and receivable/deliverable system
- Generate a simplified or agreed-to cost model for total life-cycle cost, so NASA can see how externally driven project changes affect cost
- Push basic inspection back to vendors and don't duplicate so much inspection (don't pay for pieces that don't work)

- *New Work Breakdown Structure (WBS) and Use of Fully Developed Management and Control Techniques*: A simplified WBS will do away with JPL's traditional spacecraft subsystem structure. The approach we're considering will have all electronics integrated into one WBS area by taking full advantage of recent advances in packaging of multi-chip modules. The rest of the sciencecraft will be partitioned into four areas: sensors and devices such as the science sensors, star camera, telecommunications; etc.; structure and thermal; and integrated flight software. Thus, five WBS elements will now capture the entire flight system. The WBS's system-level elements will decrease in number leading to reduced management, improved and easier technical design and integration, and reduced cost.

In FY96, JPL will establish a new system for managing and controlling projects and Pluto Express will be one of the first projects to use it fully. This will allow better schedule and cost planning, as well as better project control. When we update our business system in FY97, actual, real-time costs will be more visible. Combined, these two new systems will allow a better planned and executed project at lower overall cost and risk.

- *Close Alignment with the New Millennium Technology Program*: The Pluto project and the New Millennium Program are closely linked, with dual appointments of key staff and frequent, informal coordination among many other team members. New Millennium will develop and demonstrate in flight many key technologies needed for Pluto Express. Pluto will not only use the demonstrated technologies, but will also try to use the actual devices without change in design by exercising a next-unit purchase option in the contracts opened by New Millennium. In some areas, this will allow Pluto to eliminate most development costs and to obtain flight units at recurring costs. In addition, The Pluto team will use flight system automation capabilities to be developed by New Millennium to reduce the cost of mission operations (Phase E).

Figure 12.2-3 shows one possible sciencecraft configuration that takes advantage of these advances. Using an advanced radioisotope power source, a flat antenna, integrated microelectronics, and other innovations, we estimate the sciencecraft's dry mass to be 74 kg, including 6.7 kg for the science payload and 9.6 kg for contingency. The Russian Zond, estimated to be 15 kg, and the hydrazine monopropellant at 14 kg are not included in the dry mass estimate.

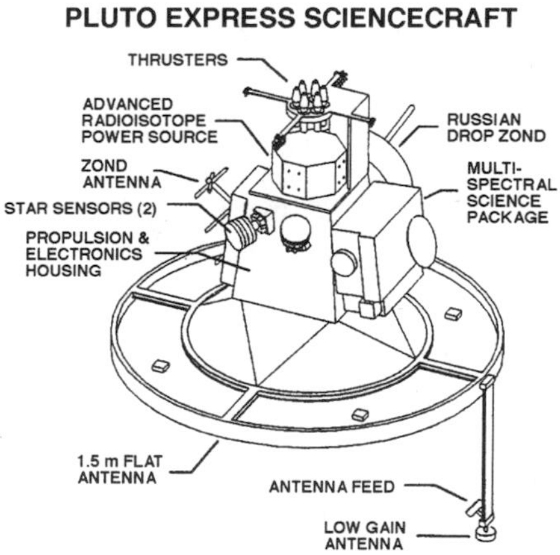

PLUTO EXPRESS SCIENCECRAFT

Fig. 12.2-3. Pluto Express Sciencecraft.

The peak power used at Pluto encounter is 72 Watts, including a 15% contingency allowance, which is provided by a radiosotope power source (RPS). This power source is assumed to use alkali metal thermal-to-electric conversion, which is an advanced conversion technology. A thermal output of 500 W is used to warm critical elements inside the bus, an important feature in a mission 30+ AU from the Sun. Multi-foil insulation blankets and louvers are used to create several thermal zones, and no electric heaters appear to be needed for temperature control even though the solar intensity varies by a factor of 1,000 over mission lifetime.

A 1.5 m diameter high gain antenna is a predominant feature in Fig. 12.2-3, and a flat reflectarray is indicated. This design uses thousands of printed elements to simulate a parabolic reflector. This design allows configurational flexibility and has minimal mass. The antenna's main structural support ring carries launch loads and also supports a hydrazine monopropellant tank through a graphite structure. We will use graphite composite material to form the modular bus structure surrounding the propellant tank.

We will use multi-chip modules in an integrated three-dimensional stack for most of the electronics, greatly reducing volume, mass, power, cabling, and the amount of radiation shielding required. The module stack is made up of block-redundant flight computers (one for engineering and one for science), solid-state mass memory (2 GB DRAM), power control electronics for instruments, electronics, and pyrotechnic initiators. The two computers and their large data storage memories are cross-strapped together so that one processor can perform all functions and access both memory blocks in the event of a failure. Other electronics, e.g., for valve drivers, propulsive thrusters, and telecommunication, will be housed in separate packages to reduce interference problems. Attitude control, pointing, and propulsion are accomplished using equipment similar to the FY94 design.

A small deep space transponder communicates with Earth, using a monolithic microwave integrated circuit in the mulit-chip module. The output from this transponder will feed a Ka-band solid state power amplifier with a downlink at 32 GHz with a data rate between 150 and 450 bits/sec, depending upon the Deep Space Network ground antenna configuration. The uplink will use X-band at 7.1 GHz. A low gain antenna is used for initial communication after launch and for certain emergencies.

We are developing software in a "plug and play" testbed and we will prove out software blocks as they are written. As the design matures, the testbed will evolve from breadboard to prototvpe to flight hardware. We expect to run a full set of flight software in the testbed before all the hardware is delivered. Also, we are developing a large amount of autonomy, self-monitoring, and internal fault protection for both software and hardware.

We are planning our approach to mission operations from the outset and anticipate a ground team of less than ten 40-hour per week JPL employees during cruise between planets. This level of staffing depends upon using a "beacon cruise" style of monitoring, where onboard software determines when the sciencecraft needs to communicate with the ground. The sciencecraft will transmit an uncoded carrier that indicates three sciencecraft states: okay and no attention is needed, data are ready for downlinking, and a serious problem exists. If a serious problem exists, then we will assemble an emergency response team to resolve the problem. Some months before the planetary encounter, we will form and train a larger ground team to plan and implement the science encounter portion of the mission.

In summary, we are using advanced technologies, advanced hardware and software design and integration techniques, and advanced organization and management practices to meet the challenges of substantial new technology, early launch, short trip-time, public outreach, and student involvement. As a result of these advances, we can complete the mission for a fraction of the FY94 baseline cost. Cost estimates for the FY95 baseline sciencecraft shown in Fig. 12.2-3 are given in Table 12.2-4. The estimated cost of mission development (Phase B till 30 days after launch) is $235 million (FY95) for two sciencecraft, including a 25% reserve held at JPL. This is more than a $100 million reduction in this area from the FY94 estimate. The cost for the launch system was reduced from $133 million (FY94) to $43 million (FY95).* We also expect considerable cost reductions in the area of mission operations, but we have not quantified them yet.

We now expect Pluto Team to meet the current set of requirements for this mission to the edge of the solar system. But in today's world of chancy funding and varying priorities, nothing is sure. Certainly the variations encountered through the Pluto Fast Flyby/Express history to date bear this out.

References

Cochran, A. L., H. F. Levison, S. A. Stern, and M. J. Duncan. 1995. *Astrophysical Journal*, December 10.

Horan, D. M. and P. A. Regeon. May, 1995. "Clementine—A Mission to the Moon (and Beyond)," *1995 NRL Review*, NRL/PU/5230-95-274.

Nozette, S. 1995. "The Clementine Mission: Past, Present, and Future," *Acta Astronautica*, vol. 35 supplement, p. 161.

Nozette, S., P. Rustan, L. P. Pleasance, D. M. Horan, P. Regeon, E. M. Shoemaker, P. D. Spudis, C. H. Acton, D. N. Baker, J. E. Blamont, B. J. Buratti, M. P. Corson, M. E. Davies, T. C. Duxbury, E. M. Eliason, B. M. Jakosky, J. F. Kordas, I. T. Lewis, C. L. Lichtenberg, P. G. Lucey, E. Malaret, M. A. Massie, J. H. Resnick, C. J. Rollins, H. S. Park, M. S. Robinson, R. A. Simpson, D. E. Smith, T. C. Sorensen, R. W. Vorder Bruegge, and M. T. Zuber. 1994. "The Clementine Mission to the Moon: Scientific Overview," *Science*, vol. 266, p. 1835.

Regeon, P. A., R. J. Chapman, and R. Baugh. 1995. "Clementine 'The Deep Space Program Science Experiment'," *Acta Astronautica*, vol. 35 supplement, p. 307.

Salvo, C. G. 1993. "Small Spacecraft Conceptual Design for a Pluto Fast Flyby Mission," paper AIAA-93-1003, AIAA/AHS/ASEE Aerospace Design Conference, Irvine, CA, February.

---

*The Russian Proton considered for the FY94 baseline is replaced in the FY95 baseline by a Russian Molniya (with Star 48 upper stage), again at no charge to NASA. As in the earlier case of the Proton, each sciencecraft will carry a Russian atmospheric probe, called a Drop Zond, to be dropped on an impact trajectory prior to closest approach. The Zonds will transmit scientific data to their flyby sciencecraft for later relay to Earth.

Staehle, R. L., D. S. Abraham, J. B. Carraway, P. J. Esposito, E. Hansen, C. G. Salvo, R. J. Terrile, R. A. Wallace, S. S. Weinstein. 1992a. "Exploration of Pluto," paper IAF-92-0558, 43rd Congress of the International Astronautical Federation, Washington, D.C., September 1992.

Staehle, R. L., S. Brewster, D. Caldwell, J. Carraway, E. Hansen, P. Henry, M. Herman, G. Kissel, S. Peak, C. Salvo, L. Strand, R. Terrile, M. Underwood, B. Wahl, and S. Weinstein. 1993. "Pluto Mission Progress Report: Lower Mass and Flight Time Through Advanced Technology Insertion," 44th Congress of the International Astronautical Federation, Graz, Austria, October 16–22, 1993.

Staehle, R. L., J. B. Carraway, E. Hansen, C. G. Salvo, R. J. Terrile, and S. S. Weinstein. 1992b. "Exploration of Pluto: Search for Applicable Satellite Technology," 6th AIAA Utah State University Conference on Small Satellites, September 1992.

Stern, S. A. 1992. "The Pluto-Charon System," in G. Burbidge, D. Layzer and J. G. Phillips, eds, *The Annual Review of Astronomy and Astrophysics (1992)*, Annual Reviews, Inc., Palo Alto, CA.

Stern, S. A. 1995. "Pluto and the Kuiper Disk," in *Ices in the Solar System* (C. deBergh, M. Festou, and B. Schmitt, eds), Kluwer, in press.

Chapter 13

Communications, Test, and Applications Missions

13.1 RADCAL

13.2 ORBCOMM

13.3 AMSAT

13.4 PoSAT-1

Historically, the amateur radio satellite organization, AMSAT, has been the long-term world leader in creating low-cost communications satellites. These have advanced over time from unsophisticated, simple designs to much more complex spacecraft that include sophisticated payloads, 3-axis stabilization, and large ΔV maneuvers. Although the work has been done with volunteer labor which dramatically reduces cost, the technical lessons from this series should not be ignored. The AMSAT experience has been the starting point for a number of other low cost satellites, such as the very successful UoSAT series.

Test and applications satellites are also very amenable to application of many cost reduction techniques. Typically, there is a willingness to trade on requirements to reduce cost. Testing can often be done with short-duration, low-cost missions in which a higher level of risk is both acceptable and appropriate. Commercial applications, such as materials processing, Earth observations, or low-Earth orbit communications systems frequently require substantially reduced cost to be economically viable. The new generation of communications constellations, such as ORBCOMM, may prove to be a principal driver of low-cost and reduced-cost spacecraft in low Earth orbit.

Among the large number of reduced cost spacecraft in this category are:

- The AMSAT series beginning with OSCAR-1 in 1961
- The UoSAT series launched by Surrey Satellite Technology beginning in 1981
- TUBSAT-1 and -2, German experimental satellites launched in 1991 and 1994
- MSTI-1, which was 53 weeks from conception to launch in 1992
- HealthSat, a good example of specialized communications applications launched in 1990 (HealthSat-1) and 1993 (HealthSat-2)
- Bitsy, a 1-kg communications and remote sensing demonstration satellite being developed for the Air Force

The case studies presented here are particularly good examples of the process of reducing space mission cost. Many additional examples can be found in the list of small spacecraft in the Appendix. The Industrial Space Facility, designed by Westinghouse but not built, is a good example of applying cost reduction techniques to larger spacecraft.

RADCAL Summary

The *Radar Calibration Satellite* was built by DSI for the U.S. Air Force. It has four major payloads and is gravity-gradient stabilized.

Spacecraft dry mass:	92 kg	No propulsion system	
Average power:	25 W	TT&C:	19.2 kbps

Launch: Scout launch from Vandenberg Air Force Base, June 25, 1993

Orbit: 823 km circular at 90 deg inclination

Operations: Operated by Air Force personnel (not included in cost)

Status: Still operational as of January 1996

Cost Model (FY95$M): See page 348 for explanation.

	Expected Cost	Small Spacecraft Model	Actual Cost*
Spacecraft Bus	$35.8M	$7.8M	$4.4M†
Payload	$18.0M	$0.7M	Incl. in spacecraft cost
Launch	$16.6M	$7.1M	$12.2M‡
Ground Segment	$37.8M	—	Incl. in spacecraft cost
Ops. + Main. (annual)	$4.5M	—	N/A
Total *(through launch + 1 yr)*	**$112.7M**	—	**$16.6M + O&M**

*An inflation factor of 1.106 has been used to inflate to FY1995$ [SMAD, Table 20-1].
†Includes cost of 2 ground stations.
‡Estimate based on Scout launch vehicle cost.

Fig. 13.1-1. RADCAL. The zenith (top) of the satellite shows 4 omnidirectional and 4 GPS antennas.The gravity-gradient boom is not deployed. Four 150 MHz beacon antennas are shown deployed from the lower edge. The UHF helix and 2 C-band transponder antennas are telescoped out from the bottom. Solar cells cover the side panels and top.

13.1 RADCAL

George Sebestyen, *Defense Systems, Inc.*

RADCAL is a sophisticated satellite that Defense Systems, Inc. (DSI) built for the Air Force and launched in 5 days less than a year for $4 million.* This chapter describes the satellite and the key factors that contributed to its low cost. I believe these factors are equally applicable to other satellites and hold the key to significant, achievable cost reduction for all satellites.

RADCAL was launched on June 25, 1993 aboard a dedicated Scout launch vehicle from Vandenberg Air Force Base, CA into a 823 km polar orbit. This was the nineteenth satellite built and launched by DSI. The mission objectives of this 92 kg satellite are to:

- Calibrate 70 C-band Air Force range radars

- Provide military UHF store-and-forward and transponder ("bent-pipe") tactical communications

- Test a "peak-power tracker" to extract maximum power from the solar arrays

- Test spacecraft attitude determination by interferometry with the onboard GPS receiver

The principal requirement of the mission is to provide C-band transponder service for the tracking radars while maintaining 5 m knowledge of instantaneous spacecraft position during radar calibrations. Two C-band transponders and redundant antennas provide the radar calibration. Satellite position determination is accomplished by two independent systems. One, redundant GPS receivers with ground processing, provide better than 5 m position accuracy; the other, 150 MHz and 400 MHz beacon transmitters in conjunction with the Navy Doppler Tracking Network provide about 3 m tracking accuracy. The GPS data is sampled every 6 seconds and the data stored onboard is dumped to the master ground station at 19.2 kbps. Telemetry and command and control are also accomplished by sharing the store-and-forward UHF communications payload. For UHF communications with small, handheld ground terminals a 65 W transmitter and a 6 dB gain helix antenna are provided. All transmitters and receivers have synthesized frequency control. Satellite scheduling is accomplished by stored schedules uplinked from the ground.

Satellite operations are conducted by the Air Force. The times when each of the 70 range radars are to be calibrated are determined by the ground control station from

---

* This schedule and cost performance merited special Air Force citation and award.

the geographic locations of the radars and the satellite orbital parameters. From this data the following times are determined:

- When the satellite C-band transponder must be operated

- When the satellite is over the geographic areas where communications service is to be provided

- When it passes over the ground control stations

From this and from special experiment or engineering requirements, the detailed satellite operating schedule is determined for about a week in advance. The schedule is uploaded by the ground control station during a pass and is replenished or changed when necessary. During the rest of the satellite passes over the ground control station, satellite telemetry is monitored only.

13.1.1 Satellite Technical Description

As shown in Fig. 13.1-1, RADCAL is an octagonal, gravity-gradient stabilized spacecraft (the gravity-gradient boom is not deployed in Fig. 13.1-1). The technical characteristics are listed in Table 13.1-1. The structure is a 76.2 cm diagonal, 41.9 cm high eight-sided cylindrical prism comprised of heavily milled and pocketed aluminum top and bottom plates interconnected by eight "stringers" and covered by side panels. The gravity-gradient boom at the center of the spacecraft also serves as a structural member interconnecting the top and bottom plates. The spacecraft was designed for 15 g's axial and 9 g's lateral acceleration and has a lowest resonant frequency in excess of 50 Hz. While not used in this mission, the structure was designed to be a general purpose 3-axis stabilized spacecraft bus capable of supporting several hundred pounds of mission payload. In this mission, the side panels and the top are covered by solar cells. Electronics modules are mounted on the bottom (nadir) plate and on the interior of the top plate. In addition to solar cells, the top also contains the gravity-gradient boom, hysteresis rod libration dampers, four GPS antennas, and four elements of the command and telemetry antenna system as well as the pyrotechnic release mechanism for the 150 MHz tracking beacon antennas.

Mounted on the bottom of the spacecraft are the telescoping UHF helix antenna, telescoping redundant C-band transponder antennas, pyrotechnic releases for these antennas, the four elements of the 150 MHz beacon antenna and the upper half of the 50 cm Marmon ring, V-band separation system.

DSI also built the launch vehicle interface and separation system. It consists of the lower half of the Marmon ring, the V-band restraint system, the redundant pyrotechnic bolt cutters to release the V-band, and the structure to attach the separation system to the E-25 Scout adapter section used to mount the spacecraft to the last stage solid rocket motor.

The eight sides and the top of the satellite contain 8,155 cm^2 of solar cells that provide a minimum of 18 W of orbit average power under worst case (Sun in orbit plane), end-of-life conditions. When the Sun is not in the orbit plane up to 25 W of orbit average power is provided. Two temperature-controlled, current-limiting battery charge regulators charge the redundant 150 W·hr NiCd batteries. The payload

TABLE 13.1-1. RADCAL Technical Characteristics.

Launch	• Dedicated Scout from Vandenberg Air Force Base, June 25, 1993
Orbit	• 823 km circular, polar (90 deg inclination)
Payloads	• Dual C-band transponders • GPS position and attitude determination • Peak-power tracker • UHF tactical store-and-forward and "bent-pipe" communications
Structure	• 8-sided, 76.2 cm diamter, 41.9 cm high • Milled and pocketed aluminum top, bottom, and longerons • Designed to 15 g's axial, 9 g's lateral acceleration
Electric Power	• 18 to 25 W orbit average power • Solar cells on sides and on the zenith face • 150 W peak +28 V power • +28 V, ±15 V, +5 V DC/DC converters • Redundant 150 W·hr NiCd batteries • Temperature-compensated, voltage-controlled charge regulators • Onboard computer-controlled pyrotechnic release mechanisms • Peak-power tracker (flown as an experiment)
Attitude Control	• Gravity-gradient stabilization • 6 m gravity-gradient boom with 2.27 kg tip mass • Hysteresis rod damping • 3-axis magnetometer sensor • 65 A·m² torque coil • GPS position and attitude data • 150 MHz and 400 MHz tracking beacon transmitter
Thermal Control	• Passive control with paints • Computer-controlled battery heaters
C&DH	• Spacecraft and payload scheduling • Telemetry data formatting • Command interpretation • Housekeeping • Communications system control • Mass memory for message storage • Payload and satellite control from the ground
Communications	• UHF telemetry and command link • Synthesized 10 W and 65 W transmitters • Synthesized UHF receiver • Frequency shift keying and binary phase shift keying modulation (ground commanded) • 4.8 kbps and 19.2 kbps programmable data rates • Store-and-forward and "bent-pipe" communications modes • Dynamically allocatable "mailboxes" for message storage • Earth coverage helix and omnidirectional acquisition mode antennas • 150 MHz deployable beacon antennas

and bus electronics are powered by 4 DC/DC converters supplying 150 W of 28 V, ± 15 V and 5 V. In addition, a separate 15 V regulator powers the high power amplifier to provide the 65 W of transmitter power.

The peak-power tracker, switched in or out as an **experiment**, uses a subset of the solar cells so that at no time is the mission jeopardized by the experiment.

The attitude control system consists of a 6 m long gravity-gradient boom with a 2.27 kg tip mass and two hysteresis rod dampers. Attitude knowledge is provided by a 3-axis magnetometer, while a coil that produces a magnetic field along the Z-axis of the satellite (the Z-coil) is used to invert the spacecraft should it stabilize in an inverted position. Two redundant GPS receivers and the 150 MHz and 400 MHz beacon transmitters (both with ground processing) are used to provide spacecraft position information with better than 5 m accuracy. The beacon transmitter signals are processed by the Navy Doppler Tracking Network which obtains satellite position data from the observed doppler shift as a function of time while the satellite is in view of the Navy Doppler Tracking Network tracking stations. GPS position with the required accuracy is obtained after ground processing of the downdumped GPS data. Spacecraft attitude as a function of time is also computed on the ground by processing the downlinked GPS phase data.

The RF subsystem consists of a 10 W UHF synthesized transmitter and receiver for telemetry and command and control. The same transmitter and receiver are used for store-and-forward and "bent-pipe" communications with ground users. The transmitter and receiver are capable of frequency shift keying and binary phase shift keying transmission and reception at data rates up to 19.2 kbps. A 65 W amplifier is used to transmit satellite data to ground terminals. Internal, ground-controlled switching of RF components permits using 10 W or 65 W transmitters, and omnidirectional or Earth coverage antennas. A diplexer enables "bent-pipe" transponder activities and analog voice communications.

The C-band transponder provides 500 W double pulse replies upon illumination by ground tracking radar pulses. Redundant antennas are provided. The 400 MHz beacon shares the communications system UHF antennas, while a separate deployable dipole antenna system is provided for the 150 MHz beacon transmitter. The four GPS antennas are mounted on the zenith face of the satellite, and they provide two orthogonal 62 cm interferometry baselines for attitude determination. Between the mission payloads, the position determination subsystem, and the UHF communications system, the spacecraft uses a total of 15 antennas.

The digital processor subsystem contains nine circuit boards. In addition to the CPU and memory I/O boards, the computer contains two status I/O boards for sampling and A/D converting 64 analog and 32 binary quantities (voltages, currents, temperatures), a computer power management board that also contains several hardware timers and counters, two boards for control of the transmitters and receivers and to perform carrier baud rate recovery. In addition, a 4 MB memory stores data and communications, while a special interface board collects GPS data.

The computer performs all telemetry data collection and formatting, command interpretation, power and redundancy control, and scheduling. It also controls deployables, manages the electric power system, processes attitude determination data and controls the magnetic coil of the attitude control system. The computer also manages packetized store-and-forward communications and allocates dedicated "mailboxes" to different ground users. All onboard software is written in C.

Two ground stations were constructed. One is deployed at the Air Force Space-craft Ground Control Facility in Sunnyvale, CA, the other is at DSI. The ground stations use essentially the same UHF communications suite as the spacecraft, except that they contain high gain UHF helix antennas that, using open loop tracking based on the satellite ephemeris, provide high signal strength, and a measure of immunity against local RF ground interference.

13.1.2 Construction, Testing, and On-orbit Performance

The contract to begin development of RADCAL was received on June 1, 1992. Essentially all of the components were designed and built in-house and were largely based on DSI components that had flown on previous spacecraft. Satellite construction was completed at Christmas of 1992 when power was first applied to the completed, assembled satellite. System integration, testing, and environmental testing were performed over the next 5 months. Then the spacecraft was transported to Vandenberg Air Force Base where preparations for launch took another 3 weeks.

A combined Preliminary Design Review and Critical Design Review was held 3 months after program start. A total of 19 formal deliverable documents and several versions of each documented the design. Numerous technical interchange meetings and point papers resolved different technical issues that arose during the development. A total of 54 test plans and procedures were used to document the test program. All parts were approved by the Defense Electronics Support Center and the Air Force. Parts meetings were held to resolve parts and quality issues every 2 months. The Air Force and The Aerospace Corporation personnel witnessed essentially every test. A collaborative relationship existed between DSI, the Air Force, and The Aerospace Corporation with the common objective to produce a quality spacecraft within a very challenging schedule.

The mechanical tests performed in the first 5 months on the spacecraft included static load test, modal survey, separation test, and pyro shock test. Antenna patterns were also measured early-on in a full-scale mockup of the satellite. Electronics modules were subjected to functional tests, thermal tests, and thermal cycling. Deployables were tested at various temperatures and under restraints to test excess deployment strengths. Harnessing was subjected to high pot testing, and the satellite underwent a multitude of performance qualification and verification tests during integration and test. When total system performance testing was completed at room temperature, the satellite was thermal cycled while operating to thoroughly exercise it and to find workmanship or design problems not uncovered during room tempera-ture testing. Our experience is that if we subject the spacecraft to eight thermal cycles with failure-free performance over the last five, we can successfully pass subsequent thermal vacuum testing.

After the spacecraft has passed thermal cycling, it is subjected to EMI/EMC, vibration, mass properties, full-up separation testing, static and spin balance, mag-netic grooming and system-level, end-to-end testing (with the ground station). At the same time the spacecraft is continuously being subjected to qualification and verifi-cation testing and running mission scenarios on the satellite.

At the launch site, verification testing was performed between each operation, mating the spacecraft with the last stage motor, spin balancing the combination, mating to the launch vehicle, and checkout of spacecraft functions from the block house.

The spacecraft computer was on during launch but the CPU was in "reset." Upon separation from the launch vehicle, the spacecraft sensed redundant contact closures and turned on the onboard computer. This initiated a set of pre-programmed activities—deployment of the multiple telemetry and command antennas, start of telemetry transmission and command reception operations, and self-test functions. First ground contact with the satellite was made on its first pass over the launch site on a 2 deg elevation pass. All DSI satellites are designed with excessive link margins enabling horizon-to-horizon communications.

The spacecraft was released from the launch vehicle spinning at 3 rpm. After de-spin through heat dissipation in the hysteresis rods, preprogrammed deployment of the gravity-gradient boom over the South Pacific resulted in an immediate stabilization of the spacecraft right side up. Subsequent commands deployed the remaining antennas.

While initialization activities took place, the rest of the spacecraft functions were checked out. Total checkout took 4 weeks. Control of the spacecraft was turned over to the Air Force on July 25, 1993. DSI continues to provide support to the Air Force for ground operations and anomaly resolution.

Since it was launched, the RADCAL satellite has been performing its mission under the control of the Air Force Spacecraft Ground Control Facility (Det-2). Initial checkout and testing was performed by the DSI RADCAL program engineers and software personnel from its own ground station during a 3 week period of time. All system functions are nominal. Gravity-gradient pointing accuracy varies between 4 deg and 7 deg, owing in large measure to the magnetic cleanliness of the satellite.

The GPS attitude determination system, used when radars are calibrated, achieves approximately 0.3 deg attitude knowledge accuracy. I believe this is the first spacecraft to fly GPS with attitude determination capability. In accordance with expectations, the peak-power tracker achieved a 17% increase in electric power availability. Position accuracy achieved is on the order of 3 to 5 m, with better performance by the Navy Doppler Tracking Network than by the GPS position determination systems.

13.1.3 Contract, Cost, and Schedule

The RADCAL contract is a fixed-price, incentive contract with a target cost, ceiling price and share ratio above and below target cost. The target fee is quite small. However, the contract contains provisions for a unilateral Air Force award fee, subject to certain performance objectives. The first half of the award fee was earned upon successful on-orbit acceptance on schedule. The second half of the award fee is earned based on satellite performance after one year of on-orbit operations. DSI has earned a high percentage of the award fee, making this program not only successful but quite profitable. We have previous experience with this type of contract with the Air Force with similar good results. I think it is a tough but fair contract type which permits the reward of good contractor performance. It also makes the contractor responsive to the customer while providing him with the contractual backing to resist unreasonable requests. Notwithstanding the fact that this is a good type of contract, it is the friendly and constructive relationship between the contractor and the customer that makes it possible to achieve success despite the rapid schedule.

The contract document is quite short and focuses on functional specifications rather than technical specifications. This has enabled DSI to make major trade-offs to minimize cost while adhering to the contract specifications. Despite significant additions to the capabilities of the spacecraft during the early period of contract performance, the broad wording of the contract functional specifications resulted in the Air Force deciding that these additions were within scope and did not warrant additional funding. This would have resulted in a contract overrun had it not been for the fact that DSI had the latitude to make major trade-offs which permitted it to include the added features at no increase in cost. The contract was completed exactly within budget.

Construction of the satellite and ground stations, including on-orbit initialization, were performed in a little less than one year for $4 million. The expenditure profile versus time of this program (as well as those of all other DSI satellite programs) was linear; it did not exhibit the usual S-shape resulting from a slow buildup, an accelerating middle, and a phase-down toward the end. There are two reasons for this. One is that DSI builds essentially everything except solar cells in-house. So, instead of ordering components we have to pay for when they are delivered (giving rise to a deferral of expenditures and the S-shape), we build the components in-house, expending labor as soon as a component is released to production. The second is the short schedule. There is no time for buildup. We have to begin with a running start.

13.1.4 Approach to Low Cost

There is no magic formula or trick to achieving low cost. The secret, if any, is to do what every good program manager has known how to do since the beginning of the space age:

- Assemble a capable, small team of people
- Set a short schedule
- Make major trade-offs and technical decisions rapidly and decisively
- Practice judicious concurrency between fabrication and design
- Do not procrastinate or analyze unnecessarily
- Do not let anyone slow you down
- Fire anyone who suggests running a program by matrix organization

The secret, if any, to producing a high-technology, quality product is to take bold and innovative approaches but implement them ultra-conservatively and test the finished product exhaustively under realistic operational conditions.

The key to having harmonious relations with your customer is to give him the keys to your factory (so he can have completely free use of information and access to your people), anticipate his concerns and questions (and have answers), but run faster than he can ask questions, so he cannot slow you down.

In the next few paragraphs, these and other practices and how they were applied are discussed in more detail.

Functional Rather than Technical Specifications

Most DSI contracts specify functional requirements and quantitative performance objectives of the satellites. Within broad boundaries we can make engineering tradeoffs to achieve these objectives, permitting trading off characteristics of one subsystem against another. This facilitates selecting an implementation that meets cost goals. It permits "design-to-price." Unduly detailed technical specifications tie the hands of designers. I believe this is one of the most leveraged factors responsible for low cost.

In RADCAL the cost per watt of solar cells could be reduced by using larger solar cells. To make use of larger cells, we increased the available planar areas by reducing the number of sides of the satellite from 16 to 8 and by increasing the height of the spacecraft. In addition, by reducing the bus voltage we reduced the number of solar cells in series. All of this permitted us to make a substantial decrease in the power system cost.

Experienced, Small Project Team

DSI is organized around project teams. A program starts by assigning an experienced core project team. Each member of the team is responsible for his or her subsystem from initial conception to the launch pad. We do not have separate systems engineers, preliminary designers, engineers to develop the hardware, integration, and test engineers, and another team for the launch campaign. Each of our team members is not only a subsystem specialist; he is also a systems engineer. Because of his experience of having taken other satellites "from cradle to grave," we assign less experienced staff members to the experienced team leaders to learn. On the next program the now experienced team member is assigned subsystem responsibility.

The small team is augmented by technicians, assembly, machine shop and other needed functions on an "as required" basis. At any given time, there are typically 3 times the number of people working on a program than the average number of full-time equivalent people. This practice keeps costs down. Instead of assigning a large full-time staff to the program, the average spending level is much lower. People are removed from the program and assigned elsewhere when no longer needed.

Vertical Integration

DSI builds almost everything in-house. This has two advantages. One is lower cost of components, the other is immediate availability. Since we do not have to make a living by building, for example, attitude control system components for sale, we do not have to support a whole company making these, and we do not have to maintain a full engineering, marketing, program management, and manufacturing organization for each subsystem (as would a subsystem vendor). Thus, our price is substantially lower for the same subsystem. The other advantage of vertical integration is the ready availability of the component. If we need the component in a hurry, we expedite it in our own shops.

Short Schedules and Concurrency of Development and Manufacturing

A short schedule is a blessing in disguise to keep costs low. One simply does not have enough time to spend much money. It also forces us to practice concurrency between engineering and manufacturing.

In the customary development process, very senior engineers start the program doing systems engineering and front end design. The same team then prepares for the preliminary design review, eliminates deficiencies identified during the review, and then continues to detailed design and the Critical Design Review. Only then does development and fabrication begin in earnest. This is an expensive way of doing business. By delaying start of fabrication until the design is completed, the heavy front-end design effort comes to an end and the people who did it have little to do after Critical Design Review. So they invent additional things to do, refine, study, or redesign. In any case, this effort costs money and delays the schedule.

DSI operates differently. We assess initially which subsystems or components will not change during detailed design; and we release those to production right away. In this way, while the engineers complete the system design or the design of new components, already mature components are in fabrication to be ready for checkout and subsystem integration when an engineer becomes available from system design. Knowing that whenever his front-end engineering is completed, work is waiting for him disincentivizes procrastination and eliminates prolonging unnecessary analyses. Thus we may have much of the hardware completed by preliminary design review while other parts of the system have not yet been designed.

This method of operation would not be possible if it were not for very experienced engineers that have, over several programs, developed a "feel" for knowing when the risk of proceeding with fabrication is low, and when more analysis is needed before it is prudent to proceed. Concurrency is not a **four letter word** if practiced by seasoned engineers. In RADCAL, the computer boards and much of the electric power system were released immediately, even though the exact method for interfacing GPS to the computer was not yet known. In the end a new circuit board had to be added to the computer. By proceeding with fabrication immediately, digital subsystem integration with the spacecraft could be completed early, missing only the GPS-related functions.

Make Major Technical and Cost Trade-offs Rapidly and Decisively

The program manager (or company management) should do a "cost at completion" estimate every 2 to 4 weeks and monitor cost performance personally. At the earliest sign of trouble (technical, cost or schedule), the program manager must immediately take the **long view.** Nearly all problems of cost and schedule are rooted in technical problems. In case of technical trouble, the manager must apply his best judgment to assess how long it is going to take to solve the problem, how much it is going to cost, and then add 30% to his estimate. If the grand total is not within his budget, he had better rapidly think of alternative technical approaches. These technical trade-offs must be made quickly, they must be sweeping in scope, and the decision to continue or to change must be made decisively. **Small changes do not result in large savings.**

Production Coordinator to Expedite Manufacturing

The key to low-cost manufacturing is the assignment of a production coordinator to each program. He assures that the schedule is kept, that nothing falls through the cracks and that problems are solved to meet production schedule requirements. The production coordinator participates in engineering meetings. He is thoroughly familiar with the product before it is built. He expedites releasing engineering drawings, reminds people of their schedule obligations, makes sure that all the parts needed have been ordered, tracks the status of each subassembly and moves the work to the next process step. The production coordinator participates in the weekly program status reviews where he accounts for each subsystem to the team leaders and identifies shortages and problems. This disseminates information among project team members, solves problems on the spot and permits each member of the team to speak up if a proposed change would adversely affect him. The production coordinator is the key to meeting a tight manufacturing schedule.

Do Not Try to Save Money in Testing

It is possible to hurry design and fabrication. It is foolish to hurry testing. DSI spends a great deal of time testing and exercising the system under simulated scenario conditions. It is only through use that problems become apparent that did not show up in normal checkout. These tend to be mostly conceptual problems. It is only through running mission scenarios that it is discovered that as designed is not necessarily as intended. It often takes 5 months from the time the spacecraft is fully assembled with all subsystems checked out to the time it is ready to be shipped to the launch site.

Holding Program Budget Responsibility Tightly

Budget management practices used on RADCAL and the views expressed here are nearly diametrically opposite those of customary management teachings and practice. It is customary to divide the program into work packages and to assign to each major work breakdown structure element a cognizant manager, give him a schedule and a budget, and then hold him responsible for both during the performance of the program. This traditional approach almost guarantees a cost overrun for two reasons:

- Some of the work packages may have been underbid by mistake and their budgets will have to be increased. Assigning budgets and cognizance for each work package ties the hands of the program manager; it is difficult for him to reduce the budget of a work package manager who has been doing a good job to augment another that is underfunded.

- Typically only a small percentage of work package managers are good money managers. Those who are not will overrun their budgets, some very badly. Those who are good will bring in their tasks within budgets, but not much under budget. The net result is that the overall program will overrun.

By contrast, at DSI the program manager keeps the entire budget to himself. He allocates this budget between the work packages and a reserve, but does not tell anyone what the budgets are. If, while discussing the technical approach, schedule, and cost of a work package with its team leader, the cost estimate is higher than anticipated, the team leader is asked to think of other ways of doing his task. This is done iteratively until the budgetary goal is reached or the program manager is convinced that the budget must be increased. If the estimate comes in lower than planned, then after assuring that it is valid, the estimate is accepted, adding to the management reserve.

In this way each work package manager can perform his task according to his own approach, while the program manger can build up a management reserve that will invariably be needed later. The management reserve must be large. Norm Augustine has said (based on some actual data) that "a program manager always underestimates his 'cost-to-complete' by 30%." He was correct.

13.1.5 Conclusion

There is no magic or technology advance needed to make dramatic reductions in the cost of any spacecraft. Tight management, small project core teams, short schedules, opportunity to make large cost-technical trade-offs, and concurrency between engineering and production are the keys to low cost. The keys to a quality high technology product are making bold technical decisions, implementing them ultra-conservatively, and subjecting the finished product to extensive testing.

ORBCOMM Summary

ORBCOMM was built by Orbital Sciences Corporation for commercial, two-way, global, wireless communication.

Spacecraft dry mass:	33.1 kg	Propulsion:	N_2 thrusters
Average power:	270 W BOL	TT&C:	19.2 kbps

Launch: "Piggyback" on Pegasus with Microlab-1. FM-01 and FM-02 were launched from Vandenberg Air Force Base on April 3, 1995

Orbit: 775 km circular at and 70 deg inclination

Operations: OSC operations center, Dulles, VA

Status: Still operational as of January 1996

Cost Model (FY95$M): See page 348 for explanation.

	Expected Cost	Small Spacecraft Model	Actual Cost*
Spacecraft Bus	$25.0M	$2.7M	$10.7 M
Payload	$18.0M	$0.7M	Incl. in spacecraft cost
Launch	$16.6M	$3.4M	$3.0M
Ground Segment	$294.8M	—	$0.7M†
Ops. + Main. (annual)	$21.5M	—	$1.3M
Total (through launch + 1 yr)	**$375.8M**	—	**$15.7M**

*An inflation factor of 1.000 has been used to inflate to FY1995$ [SMAD, Table 20-1].
†Ground Segment cost is for spacecraft control center only. Network Control Center and Gateway Earth Station costs are excluded.

Fig. 13.2-1. ORBCOMM. Shown here with one of the two solar panels deployed at right angles to the cylindrical "can" main body. Also deployed is the copper tape quadrifilar antenna array of three antennas. This is packed by literally squashing it into a shallow trough.

13.2 ORBCOMM

Gregg E. Burgess, *Orbital Sciences Corporation*

ORBCOMM intends to provide two-way communication of data packets through-out the world and to make this communication reliable and affordable. It is an entirely commercial enterprise from launch vehicle to satellite, so it doesn't rely on any government funding. It's also the first satellite system in which a single company—Orbital Sciences Corporation (OSC)—is responsible for constructing the satellite, launch vehicle, and ground station; marketing services; and handling satellite and launch operations. OSC is thus uniquely leveraged to control the cost of the entire system. ORBCOMM's constellation of satellites in low-Earth orbit will meet widely varying needs for ground communications:

- Tracking containers
- Locating trucks and monitoring shipment condition
- Relay of naval data
- Relaying data from remote scientific-monitoring stations
- Monitoring pipeline leaks
- Searching and helping to rescue backpackers
- Notifying emergency services of personal-car breakdowns
- Transmitting and receiving communications for personal alphanumeric beepers

ORBCOMM is the first of the *little LEO* satellite constellations to win FCC approval. Like other little LEO systems, ORBCOMM handles non-voice communications and provides global coverage in near real-time. Another moniker for these systems is NVNG, which stands for *Non-Voice Non-Geosynchronous*, to differentiate them from the filings of geosynchronous communication satellites that the FCC is accustomed to. Several *big LEO* systems are also in various stages of development, including Motorola's Iridium and TRW's ODYSSEY. These systems will provide real-time voice communications to special telephone handsets, with coverage ranging from global to specific geographic regions.

Table 13.2-1 shows the system's top-level goals. Cost is of paramount importance to the ORBCOMM system, and affects all aspects of the design. Development and recurring costs must be low to enable the privately funded venture to proceed without undue financial risk, and end-user costs must be low to allow this new communications market to develop.

TABLE 13.2-1. ORBCOMM System Goals. ORBCOMM's primary objective is two-way, global, wireless communication. The system architecture is designed to meet as many goals as possible within cost constraints.

System Parameter	Goal	Affected System Elements
Global coverage	100%	• Number of satellites • Number of orbit planes, inclination, altitude
Near real-time availability	10 min. max wait for pass, mid-latitudes	• Number of satellites • Number of orbit planes, inclination, altitude
Subscriber terminal capability, geolocation, size, and cost	2-way comm; Locate to 1 km; Handheld, $100–$200	• Satellite transmitter power • Power subsystem, attitude control; • Satellite receiver noise floor terminal
Message size	up to 256 bytes	• Number of simultaneous users
Message cost	<$1	• Operation costs • Cash flow for Return on Investment • Market
Launch cost	Minimize	• Number of satellites & launch vehicles • Satellite mass and size • Low-cost launch vehicle • Number of planes
Satellite's recurring cost	<$1.6 million	• Parts selection • Redundancy level
Satellite's development cost	Minimize, total <$170M for development and launch of first 26 sats	• Subcontract strategy • Team size • Process formality
Cost of ground-station operations	Minimize	• Software design • Autonomy of gateway Earth station
Design life	4 years	• Parts selection • Qualification levels • Redundancy
Worldwide system acceptance	Frequency usage Local telecomm control	• Location of ground station • Frequency

13.2.1 System Architecture

The minimum ORBCOMM system consists of the following elements shown in Fig. 13.2-2:

- 26 satellites
  - 2 in 70-deg, 775-km, circular orbits
  - 24 in 3 planes of 8 each; 45-deg, 775-km, circular orbits
- Four gateway Earth stations in the U.S. (AZ, WA, GA, and NY)

- At least one gateway Earth station in each member country

- Network-control center: one in each country; primary in Dulles, VA

- Spacecraft control center in Dulles, VA

- Subscriber terminals

A fifth plane of satellites with 45-deg inclination may be launched to provide additional coverage. The first plane of two satellites launches with a standard air-launched Pegasus rocket, having a co-manifested payload of the OSC-built Microlab satellite. The succeeding planes launch eight at a time using a Pegasus XL, which can carry more weight. Member countries will add gateway Earth stations as the system grows. With more than 40 countries signed up for the ORBCOMM system, satellites will be in view of a gateway Earth station almost all of the time.

Two-way communications for subscribers move between terminal and a satellite. The satellite must be in view of a gateway Earth station for normal communications. A store-and-forward capability is available for remote users. The subscriber terminal logs onto the satellite, and a channel is allocated based on background traffic. The terminal communicates its message to the satellite, which relays it to the network-control center (NCC). Depending on the destination indicated in the message, the NCC will route the message either over landlines to an end user's computer network or back up to a satellite in view of a destination terminal for relay. ORBCOMM complies with international telecommunications agreements by transmitting inter-national messages between NCCs over existing landlines or satellite systems such as INTELSAT.

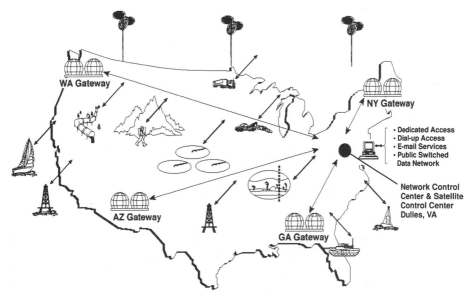

Fig. 13.2-2. United States ORBCOMM Network Locations. The ORBCOMM connects U.S. users through four gateway Earth stations and one network-control center located in Dulles, VA.

Unattended gateway Earth stations communicate in VHF, with the satellites using locally hosted software and ephemerides transmitted from the NCC. The NCC determines which satellite each gateway Earth station will track autonomously at a given time and sends the acquisition request to that station. All network layer commands to the satellite originate in the NCC. The spacecraft control center, located with the NCC, commands and monitors the telemetry of satellite subsystems. Command messages go through the NCC to the gateway Earth station and then up to the satellite. Although foreign NCCs control the network software onboard the satellite for local communications, all telemetry goes to the spacecraft control center in Virginia, and all spacecraft bus commands originate there. When foreign affiliates have gateway Earth stations, nearly constant monitoring of a satellite will be possible, depending on the orbit. Gateway Earth stations will hand off the satellite as it circles the globe.

A separate subsidiary of Orbital Sciences Corporation, ORBCOMM Development Corporation (ORBCOMM), designed the overall system and procured all major segments, including the satellite and gateway Earth Stations. ORBCOMM also secured outside financing by teaming with Teleglobe of Canada. Table 13.2-2 details the responsibilities of ORBCOMM's organizations. ORBCOMM reports directly to OSC's president and has its own stock issue. ORBCOMM is a market-driven enterprise, with intense schedule pressure to deliver the first system in this market. Cost containment is essential because we can't rely on the government to provide a safety net. Stock analysts closely track program progress, so key milestones and announcements of agreements with new countries affect the stock price of both OSC and ORBCOMM.

TABLE 13.2-2. ORBCOMM Organizations Responsibilities. Orbital Sciences Corporation's organizations are responsible for every aspect of the ORBCOMM system. Other electronics firms such as Tadiran, Torrey Sciences, and Panasonic manufacture subscriber terminals in addition to the Magellan subsidiary.

ORBCOMM Development Corporation (ORBCOMM)	
• Design system	• Procure launch vehicle
• Write software for and design NCC	• Procure subscriber terminal
• Write software for satellite payload	• Operate satellite after 90 days
• Procure satellite	• Obtain outside Investment
• Procure gateway Earth station	• Market

Orbital Sciences Corporation Space & Electronics (Virginia, Colorado, and Maryland)	
• Design and build satellites	• Write and design software for SCC
• Integrate payload software	• Conduct first 50 days' on-orbit operations

Orbital Sciences Corporation Launch Systems (Virginia and Arizona)	
• Fabricate launch vehicle	• Operate launch base
• Design and integrate Gateway Earth Stations (GES)	

Orbital Sciences Corporation Magellan Division & Other Electronics Firms
• Design, manufacture, & market subscriber terminals

Because of ORBCOMM's extraordinary cost and schedule pressures, we had to take a non-traditional approach to developing a satellite-based system. ORBCOMM levied a set of top-level requirements (listed in Table 13.2-1) that define **what** the satellite must do, not **how** to do it. This provides OSC the maximum flexibility to design the most cost-effective satellite possible. ORBCOMM retained end-to-end responsibility for network software, including the flight code that handles message traffic onboard the satellite. This enabled concurrent development of the satellite segment and the system communications software.

The program has two phases. The development phase includes design of all system elements, construction, and launch of five satellites. We built the first three satellites—ORBCOMM-X, OXP-1, and OXP-2—to monitor the background-noise spectrum in the VHF band to prove the concept of dynamic channel allocation. A small team in Boulder, CO built these simple satellites, which are essentially orbiting spectrum analyzers with passive attitude control. This team includes several people who had built amateur radio satellites. The satellites launched as piggyback payloads on the avionics shelf of the Pegasus upper stage. OXP-1 and OXP-2 were successful; they provided months of data on the characteristics of the VHF band. The data showed that enough "holes" existed in the spectrum where other users weren't using the channels to allow dynamic channel allocation to work.

The first two operational satellites are intended to be fully functional, revenue-producing elements of the ORBCOMM system. Because they will be used to transmit customer traffic, the recurring costs of the satellites are capitalized, which has significant tax advantages and makes this development affordable. These first two, ORBCOMM FM1 and FM2, launch into a 70 deg inclination orbit which fills the polar gaps in coverage that the lower-inclination planes don't cover. Teleglobe, OSC's partner in ORBCOMM, invested $10 million in Phase 1. Because Pegasus had additional capacity, we sought a rideshare to reduce launch costs. This effort resulted in the OSC-built Microlab, which is flying two experiments: the OTD, a lightning imager built by Marshall Space Flight Center, and GPS-Met, a high-atmosphere-monitoring system that uses GPS-signal occultation. Microlab uses the same bus components developed for ORBCOMM, plus an S-band communication system developed for the APEX and Seastar satellites.

The second phase is production and launch of the 24 lower-inclination satellites, plus two spares. These satellites will work the same as the FM1 and FM2, but will have some design upgrades. Based on the results of FM1 and FM2 testing, Teleglobe invested an additional $70 million to fund this phase.

Design of the Space-Segment's Mission

The primary goal of the mission design was to minimize the satellite's launch and recurring costs while maximizing coverage. Global coverage is essential because a significant part of the customer base requires remote operations, including polar regions. We quickly eliminated global real-time coverage as a requirement because it would more than double the number of satellites and launches. Instead, we established for the high-traffic, mid-latitude regions a near real-time objective: a wait of no more than 5 to 10 minutes for satellite acquisition. A constellation of three planes with seven satellites each, operating at an inclination of 45 deg and an altitude of 775 km meets this requirement. The original altitude planned was 785 km, but we lowered it to reduce the probability of collision with Motorola's Iridium system. We added an eighth satellite to each plane to provide continuing coverage if a single satellite fails in each plane. Failure tolerance is thus built in at the constellation level, rather than requiring hardware redundancy in each satellite. This requirement of eight satellites per Pegasus launch became the primary design driver for the satellite configuration. (See Fig. 13.2-3.) We added a near-polar plane at 70 deg to fill the polar gaps in coverage. This inclination was the minimum necessary to ensure 100% global coverage. Higher inclinations would expose the satellites to more radiation, reduce co-manifesting options, and make the active magnetic attitude control more challenging. Thus, the ORBCOMM system requires only four Pegasus launches. Figure 13.2-4 shows a plot of instantaneous coverage using this constellation. ORBCOMM doesn't require interplane stationkeeping because there are no satellite crosslinks. Small orbit-insertion maneuvers are required for the satellites at 45 deg inclination to ensure the semimajor axis is the same between planes. This reduces relative drift of the line of nodes of the planes to maintain distributed coverage. The resulting drift over the 4-year life of the satellites is small enough to require no further corrections. We use stop-drift maneuvers during the first 90 days to put the satellites on station, evenly spaced within each plane. The relative in-plane position is maintained without propulsion by actively managing the ballistic coefficient. Solar panels are flared or feathered during eclipse periods to achieve the desired relative motion.

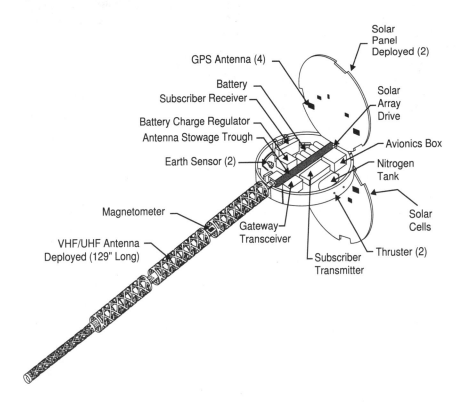

Fig. 13.2-3. ORBCOMM Satellite Configuration. The satellite features steerable solar panels, a compact disk shape that maximizes launch vehicle packing, and a 129-inch array of quadrifilar antennas.

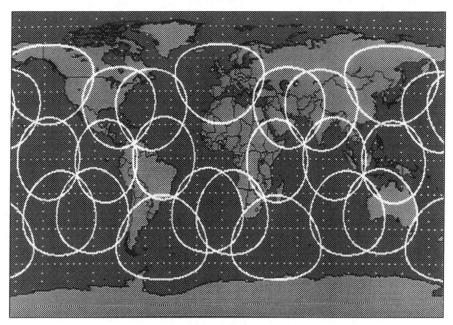

Fig. 13.2-4. Snapshot of ORBCOMM Global Coverage. The three planes of eight satellites at 45-deg inclination and single plane of two satellites at 70-deg inclination provide instantaneous nearly global coverage. Satellite motion provides near-continuous coverage at mid-latitudes.

13.2.2 Approach to Satellite Design for FM1 and FM2

Management and Engineering Processes

Engineering for ORBCOMM was very challenging on several fronts. The schedule was aggressive, cost was a significant driver, and the technical performance demanded of such a small satellite was exceptional. The performance, size, and especially recurring cost constraints dictated that much of the design would require new development. In order to reduce the risk of developing the program, we baselined at the outset a full engineering-development model and flight-quality qualification model for environmental testing.

The satellite-design effort was based on dedicating a small team of engineers to the program. The organization was project, not matrix, based. Team experience levels were mixed on the program. Much of the engineering team handling the bus system was young, with an average 5 years of experience in the industry. Several people transferred from the concurrent APEX satellite project, bringing institutional knowledge with them. People with 10 years of satellite-based experience, developed at OSC and other companies, managed the program and oversaw the system engineering. People with at least 20 years of satellite experience did the communications design. The small-team approach enables fast decision making, critical to keeping a highly developmental program on schedule track. The program manager had complete authority for design decisions. We used outside consulting support as needed in areas of particularly ground-breaking technology including attitude control and determination, digital-signal processing, and antenna design.

Subsystem engineers had wide responsibility for designing and buying subsystems. To keep cost down, some engineers managed more than one contract. The engineers basically stay with the program from "cradle to grave," being involved in procurement, design, development, and flight-level testing for both units and the system, as well as launch and on-orbit operations.

Initial system engineering was complete in about 6 months; it produced the basic, disk-shaped bus configuration and defined all subsystems. A very lean staff did this engineering; yet, the configuration held up well. System-engineering staffing was intermittent throughout the program, which resulted in some compromises. The most painful was underestimating the amount of difficulty in designing antennas and subscriber receivers, which caused the greatest schedule delays. System-engineering analysis wasn't complete for solar-panel usage and applying the GPS receiver in space. Late in the program, we discovered a problem with shadowing of the panels but solved it with rewiring that didn't affect the schedule. An incompatibility between the GPS receivers' satellite-search algorithm and mounting of the antennas on panels made it difficult to reliably fix positions for FM1 and FM2.

The customer didn't require formal design reviews. As a result, we held informal internal reviews as subsystems reached detailed hardware design or procurement. A system-interface review coordinated important interfaces and predictions of system performance. We kept formal design documentation to a minimum in order to focus on completing design work and testing. Traditional configuration management controls all production drawings and final software builds, as well as interface-control documents for the customer's communications system.

Computers were available in the engineering-development unit early in the program in order to speed software development. ORBCOMM used a proprietary real-time operating system, OSX. The software team mixed application engineers and dedicated software engineers. Specifications weren't developed in areas such as attitude control, for which application engineers wrote and tested their own code. We compiled world simulations for attitude control and flight code on PCs to verify algorithms. The flight code was recompiled for the target 68302 processors without any translation. This process greatly speeds up the transition to the real-time environment and eliminates translation errors. We developed payload software using the SDL CASE tool on UNIX workstations, which enabled us to simulate complete network functions without requiring target processors. It also allowed us, early on, to complete critical network tests and define subscriber software for the terminal manufacturers. The SDL-generated code required changes during conversion to flight code to meet limitations on the target processors' memory and speed.

Four satellites were built for ORBCOMM: an engineering-development unit, a qualification model, and the two flight spacecraft. Hardware and software development produced a full-up testbed for the engineering-development unit. All hardware was present, except the antenna and solar panels. We used the engineering development unit extensively to test system software while building flight hardware.

Unit, subsystem, and system testing used formal, controlled test plans. All units went through qualification testing, and we completed a full qualification program with a satellite built out of flight-quality hardware. Circuit-card assemblies went through full qualification testing. System tests included thermal vacuum cycling and vibration past nominal limits.

ORBCOMM was designed from the beginning to be a single-string system. The severe mass limits imposed by the requirement to place eight satellites on each Pegasus launch disallowed extra mass for redundant systems. Instead, the number of satellites made the system reliable. The cost of a full qualification program was justified to ensure the reliability of the single-string design was as high as possible and consistent with the 4-year design life.

Schedule, recurring cost, and performance and weight ratios kept us from considering space-rated (S-level) parts. We used Class B industrial parts and screened commercial parts throughout, mainly with surface mounts to achieve the needed packaging density. Wherever possible, such as in the 68302 microprocessor built on epitaxial layer, we selected components that exhibited some hardness to radiation. The higher performance available from using current, state-of-the-art, commercial parts versus lower-performance S-level parts allowed us to shift tasks from hardware to software, a key element of the ORBCOMM strategy. We tested selected parts suspected to be radiation "soft" in a particle accelerator. Early on-orbit operations have revealed only one soft part; unfortunately, these are the critical chips used to process digital signals in the radios. They work fine after reset, which occurs autonomously as part of the health and maintenance function.

ORBCOMM uses many subcontracted components. The ORBCOMM philosophy was to make only what we couldn't readily buy. The main way to control cost was using fixed-price contracts or purchase orders throughout, and using suppliers accustomed to fixed-price business. Several traditional aerospace firms participated and accepted fixed-price development contracts for the first time. Subcontractors were entrusted to complete design work with little oversight when they were building units that had significant heritage. Generally, this approach worked well, but several subcontractors had trouble with performance, cost, and schedule. The program either gave them cost and technical help or changed vendors. Table 13.2-3 lists the components of the ORBCOMM system and whether they were developmental (subcontract or built at OSC) or non-development items. OSC completed the integration, system testing, and all software (except for the COTS GPS receiver).

OSC's satellite team operated the satellite for the first 90 days on-orbit. The same people who designed and tested the satellites also operated them and worked through initial on-orbit anomalies. Training started several months before launch, and subsystem engineers helped draft the on-orbit operating procedures. The software for ground-station control also served for ground-system testing, which eliminated any possibility of incompatibility on orbit. We planned to operate 24 hours per day for the first month, then transition to "as-needed" when satellite operations became routine. ORBCOMM staff trained throughout the checkout period and assumed control after maneuvers were complete. Routine operations require at most one controller per satellite. Because the satellites are very autonomous, we'll need operators only to send occasional stationkeeping and transmitter-power commands. Most operators were hired after the satellites were launched, so they had no input into the satellite design.

TABLE 13.2-3. **ORBCOMM Satellite Development Content.** The ORBCOMM FM1 and FM2 satellites contained items developed at subcontractors (DevSub), Orbital Sciences Corporation (DevOSC), and Non-development Items (NDI).

Unit	Dev Sub	Dev OSC	NDI	Unit	Dev Sub	Dev OSC	NDI
NiH$_2$ batteries			X	Coarse Sun sensors	X		
Battery charge regulator		X		Antenna components	X		
Solar panel substrate	X			Antenna integration		X	
Solar cells, panel integ.	X			Subscriber receiver		X	
Solar array drive	X			Subscriber transmitter		X	
Heaters			X	Gateway transceiver		X	
Thermistors			X	UHF transmitter	X		
Blankets & radiators			X	Antenna base hinge		X	
GPS receiver & antennas			X	Antenna tape hinges		X	
Flight computer	X			Structure	X		
Att. control electronics	X			Non-explosive actuators	X		
Magnetometer			X	Brackets		X	
Magnetic torquers	X			Tip mass		X	
Earth sensors	X						

Satellite Characteristics

I've shown ORBCOMM's basic characteristics in Table 13.2-4 and discussed some of these characteristics below.

Packaging trades in the first phase of the program resulted in the best solution for fitting eight satellites in a Pegasus fairing. Using a stacked cylindrical can allowed us to combine the best packing in the cylindrical volume with the largest surface area for solar panels. Solar panels stowed on each side of the can-shaped bus structure provide maximum area. The satellites deploy from the booster one at a time from the top satellite down. Separation springs between the satellites ensure a sufficient relative drift rate to begin separation around the orbit. We determined the maximum diameter based on the top satellite's clearance from the fairing. A sequencer on the Pegasus flight computer initiates the separation of each satellite with a unique fire code.

The avionics architecture is based on a distributed microprocessor. We considered development issues for hardware and software when selecting this strategy. Processors communicate with each other through messages passed by the proprietary OSX operating system over serial communication lines. A central flight computer controls five buses:

- Subscriber at 150 kbps, synchronous
- Gateway at 600 kbps, synchronous
- Avionics (power and attitude) at 19.2 kbps, asynchronous
- GPS at 19.2 kbps, synchronous
- Pegasus (only used for launch)

TABLE 13.2-4. ORBCOMM Satellite Features. The ORBCOMM concept packages a complete communications system in a lightweight, low-profile satellite.

Unit	Features	Mass (kg)	Power (W)
Structure & mechanisms	• AlBeMet and Al honeycomb structure • Non-explosive actuators for separation & deployment • Liquid-damped hinge for main antenna • Tape hinges between boom segments	9.5	0
Power	• Single-axis drive for the solar array • Second axis from yaw steering • Twin silicon-cell panels 270 W BoL • Charge regulator that tracks peak power • Regulated 5 V and unregulated 14 V buses • 5 common pressure vessel NiH_2 batteries	13.9	4.2
Propulsion	• Blowdown N_2 system • Two thrusters	2.5	0
Attitude-control actuators & sensors	• Two 5 $A \cdot m^2$ torque rods • One 5 $A \cdot m^2$ coil • 2 staring Earth sensors • 3-axis magnetometer • 6 coarse Sun sensors • Integral solar sail with antenna • Tungsten gravity-gradient tip mass	4.2	0.3
Thermal	• Passive radiators • Heaters on hinges and batteries	1.2	20.0
Communications	• Subscriber & gateway uplink 148.00–149.90 MHz • Subscriber & gateway downlink 137.00–138.00 MHz • UHF position beacon 400.05–400.15 MHz • Gateway transceiver, OQPSK modulation at 10 W output RF power, 68302 processor • Dual-subscriber transmitters, SDPSK modulation at 20 W each, 68302 processor • 7-channel subscriber receiver; can dedicate one or more to assignment system for dynamic channel allocation; 68302 processor • VHF subscriber antenna, 2-element, quadrifilar • VHF gateway antenna, 1-element, quadrifilar • UHF beacon antenna, 1-element, quadrifilar	12.0	94.0
Avionics box	• 6-channel GPS receiver with four antennas • Flight computer with dual 68302 processors, handling TT&C functions, GPS interface, and communications-network software • Attitude-control electronics with 68302 processor, signal conditioning, and actuator outputs	4.2	7.7
TOTAL		47.5	126.2

The distributed architecture allows concurrent development of subsystem software and minimizes software integration. Each subsystem has its own processor, and all used the Motorola 68302 to create a common environment for software development. This was especially advantageous for the electrical-power (system) and attitude-control subsystems. Because these subsystems must communicate with many peripherals, we had to work out discrete hardware inputs and outputs before integrating the system. A single satellite computer would have made this integration a much greater schedule risk. Units with a processor are the gateway transceiver, subscriber transmitter (two transmitters with one processor board), subscriber receiver (seven receivers with one processor board), attitude control, electrical power, and the flight computer. The subscriber transmitter and receiver, as well as the flight computer, hosted network layer software. The flight computer hosts the functions for handling commands, gathering telemetry, resetting the health and maintenance watchdog, and interfacing with GPS data.

Communications between software tasks on one processor use the same messages as interprocessor communications, so the distinction of separate hardware is transparent to the software. All software developers used a common set of configuration-controlled message spreadsheets on a network to guarantee message interfaces work the first time. Files in the C language automatically generate from the spreadsheets and compile with all flight code. This living "interface-control document" guarantees interfaces work the first time and eliminates updating of paper documentation.

The primary structure members are made of AlBeMet skins bonded to a honeycomb inner core. AlBeMet provides high strength for launch loads comparable to steel, but with the weight of aluminum. AlBeMet enabled the weight goals to be met while tolerating the high lateral launch loads experienced in the eight satellite stack. We kept mechanisms simple to ensure reliability. The solar-array drive is a single-axis design using a stepper motor. The base antenna hinge is a fluid-damped coil spring. The separation system uses Non-Explosive Actuators with redundant electrical paths and firing circuits. Antenna and solar panels are also deployed with non-redundant, Non-Explosive Actuators.

The antenna was the main challenge in structural design and resulted in several patent applications. Packing three antennas into a small trough with enough gain was a very difficult task. The initial design called for folding copper tape elements that were simple to stow. However, the antenna pattern from this configuration proved inadequate. The final design chosen was copper-tape, quadrifilar antennas bonded to a composite substrate. The antenna folds up at the tape hinges, and the antenna elements are literally squashed to fit into the trough.

The power subsystem provides unregulated 14 VDC to high-power switched loads, and regulated 5 VDC for the microprocessors. A central, regulated power supply reduced satellite weight and cost over distributed regulators. Peak-power tracking maximizes the power out of the panels by using a software algorithm to track current and voltage and select the best point to extract power.

The thermal subsystem is a simple passive radiator with heaters. The heaters maintain warmth for the hinges during deployment, as well as for regular operation of the solar-array drive and batteries. A radiator is on the bottom of the spacecraft shelf, and thermal blankets cover the top of the shelf. The back of the panels are painted black.

The attitude determination and control subsystem achieves 5-deg CEP nadir pointing with limited hardware by best using the software's capabilities. Sun pointing is achieved with a 1-axis gimbal. The second axis is provided by yawing the entire spacecraft to save the weight and cost of a second mechanical gimbal axis. No momentum wheel was available at the start of development that was light or inexpensive enough to work within the design constraints. Therefore, we developed an approach based on combining active magnetic control, drag torque balancing, and a modest gravity-gradient assist. Initially, the hardware consisted of a 3-axis magnetometer, three torque rods, and a 2-axis Sun sensor that supported fixed yaw control.

However, as power requirements increased, the satellite was required to follow a sinusoidally varying yaw profile, which necessitated 3-axis control and attitude determination. We reworked the hardware, replacing the Sun sensor with two lightweight Earth sensors and adding a gravity-gradient tip mass to increase gravity-gradient stability for pitch and roll. These new Earth sensors, flown for the first time on ORBCOMM, enabled 3-axis attitude determination. An integral solar and drag sail was added to the boom antennas to balance solar and aerodynamic torques from the solar panels. The attitude estimator relies on a detailed magnetic field model and GPS navigation updates to propagate attitude solutions.

The propulsion system is a simple, blowdown, nitrogen-gas system with a lightweight composite tank. It propels the spacecraft in only one direction, along the velocity vector, which is all it needs to acquire the initial orbit. Yawing the spacecraft gives it ΔVs along the positive and negative velocity vectors. Two thrusters with commandable pulse duration allow for on-orbit adjustment of residual torque.

The satellites must stay within a ± 5 deg box in the orbit plane at their stations 45 deg apart. Recognizing that the primary perturbation is atmospheric drag, resulting from ballistic coefficient variation, we developed a scheme to alter the ballistic coefficient by steering the solar arrays in eclipse and manage phasing. This technique reduced satellite weight by eliminating cold-gas mass.

The flight computer and GPS receiver handle telemetry tracking and control. The flight computer processes commands received through the gateway transceiver and distributes them to the destination unit. The command-handler software includes the capability for stored commands, which prove very useful for contingencies and long out-of-contact periods. Likewise, the flight computer polls units for telemetry through software messages and passes them down through the gateway link. Telemetry is also stored in a backorbit buffer, so the ground station can request data from the out-of-contact period, which can last up to 10 hours. GPS data used for ephemeris determination is part of the telemetry messages. The post-launch and system-reboot telemetry mode is a robust broadcast system. This mode continuously sends data down at a low rate to ensure that the ground gets data even if the command link isn't working. Normal operations disable this mode to communicate telemetry only during a pass.

A single VHF transceiver handles gateway communications for both uplink and downlink; it also encodes the data using offset quadrature phase-shift keying. TT&C and subscriber messages are simply two different types of network message traffic. We used DSP software to decode reset commands. The gateway communications concept accomplishes both TT&C and payload communications in one system. The network-control center assigns specific slots for TDMA to TT&C messages and subscriber messages depending on traffic loading. Additional slots are allocated to satellite telemetry during anomalies or important telemetry dumps from out-of-contact periods (*backorbit*). Subscriber messages go through the gateway and out the subscriber transmitter to destination subscriber terminals. Similarly, the subscriber receiver accepts terminal transmissions and sends them out the gateway. Network-layer communications software hosted on the flight computer intelligently allocates bandwidth to different messages and also provides 1 MB of store-and-forward message capacity.

The subscriber transmitter includes two VHF transmitters to handle expected traffic. They each can operate at 20 W output power. Subscriber messages are encoded using differential phase-shift keying. If one transmitter fails, the other covers all functions but covers fewer simultaneous terminals. A single 68302 processor controls the transmitters.

The VHF receiver's seven channels provide terminal communications and frequency scanning. One or two receivers are dedicated to the assignment system for dynamic channel allocation, which scans for frequencies with low background traffic. The Dynamic Channel Allocation Assignment System software decides which frequencies are usable, and this data passes through the subscriber transmitter to the terminal as part of the acquisition process. The data then immediately bursts up on the assigned channel. The other channels communicate with subscriber terminals.

The UHF beacon provides a continuous wave signal that terminals can use to coarsely determine position at a resolution of up to 500 m. Terminals include software that computes position using Doppler shift from the beacon and the satellite's ephemeris downlinked to the terminal.

Launch Vehicle Options

The ORBCOMM concept preceded the Pegasus program for air launched rockets. In fact, the impetus for Pegasus came from the fact that small satellites, such as ORBCOMM, were too expensive to launch. Therefore, OSC developed launch-technology before starting on ORBCOMM. Pegasus effectively opened up opportunities for many companies and customers to afford small satellites in low-Earth orbits. It continues as an essential part of the small-satellite revolution of the past few years. The Pegasus XL, which can boost heavier payloads, developed mainly because of ORBCOMM's missions for eight-satellite constellations. Thus, we had to select Pegasus as our launch vehicle. As a customer, ORBCOMM Development Corporation doesn't receive special pricing.

Spacecraft Control Center

The spacecraft control center uses a COTS, real-time, command and control package called OASIS. It runs on Sun workstations to provide all ground commanding, telemetry display, and archiving. The spacecraft control center developed in parallel with the satellite software to serve as the test station for system test. We saved a lot of money by not developing separate system-test and ground-station tools. In addition, test engineers who later became satellite operators were already familiar with the system; they didn't to need to climb a new learning curve to support operations. A conversion utility converts the flight-code software message spreadsheets into OASIS databases. This guarantees 100% compatibility between flight code and the ground station, which eliminates much schedule and technical risk.

13.2.3 Schedule and Cost

The ORBCOMM project was supposed to take two years from the start of satellite design to launch. Although the satellite's program management considered this schedule unrealistic, the market drove our team to push for a short schedule. The program ended up requiring 3.5 years, which is a remarkable achievement considering the high amount of new development, inclusion of a full qualification program, construction and test of two flight satellites, and the drain on resources at a small company with several parallel new development projects.

Shortcuts in system engineering at the beginning of the program meant to accelerate the schedule, particularly in designing communications hardware, drove up program cost and schedule later on. The subscriber receiver and antenna proved to be much more complex than anticipated due to the severe packaging constraints on ORBCOMM and the lack of technology for VHF space communications. Although these two items caused much of the schedule delay, many other areas of the program—such as attitude control, power, and software—greatly benefitted from the additional time. Interestingly, if the schedule had originally allowed a more realistic three years, we may have finished the program earlier by allowing more time up front for comprehensive design, analysis, and review.

ORBCOMM was a fixed-price contract between OSC and the ORBCOMM subsidiary, valued at $21.4 million for the development of the two flight satellites. The fixed-price contract was especially ambitious due to the high development content. The final price charged was fixed, but overruns caused by the schedule delays eliminated most of the profit for OSC. Most contractors did an excellent job of maintaining their fixed-priced quotes, even though some experienced internal delays in their schedules.

The recurring-cost goal in the original contract for the third through the twenty-sixth satellites, including labor, was $1.6 million apiece for satellites with a 4-year design life. The current estimate for the remaining satellites is approximately $2 million per unit. These satellites will have a 5-year design life and gallium-arsenide solar panels, and a small momentum wheel. The growth in recurring cost has been well controlled: the additional unit cost is going mainly toward increased performance and additional testing. The subcontracts to vendors and the contract between OSC and ORBCOMM will be fixed price for the remaining project.

13.2.4 Summary of Lessons Learned

Many of these lessons are being applied on the next phase of the ORBCOMM satellite production. For instance, ORBCOMM has a significant engineering staff, and we are porting the satellite southwest to a COTS operating system.

Use a Right-sized Team to Maintain High Motivation

The small-team approach is essential to maintaining project focus and creating the motivation to achieve aggressive schedules. But, it must not be small just for the sake of keeping labor costs low. If key positions are left unstaffed to reduce start-up cost, the project will pay dearly later.

Staff with Experts in Key Technical Areas Right at the Start

ORBCOMM suffered from a lack of coverage in all areas for the first year of the program. This turned into significant schedule delays in the long run, which runs up cost significantly. Spending an extra month at the beginning by focusing on hiring will pay off many times over later. A particularly lean area was system engineering throughout the program; system engineers quickly got involved in solving subsystem problems and actually creating flight code, rather than focusing on the "big picture." A project must maintain at least some system engineers whose only responsibilities are the system's overall functionality and interface management

Empower Team Members with Responsibility and Needed Management Tools

The most successful push in the program occurred late in the project when senior engineers were given the authority to construct schedules and provide bonuses for their teams based on meeting milestones. By having compensation authority, technical leads were able to give their people more effective incentives.

Limit Low-level Technical Requirements from the Customer

Not directing the "how" of an approach is indispensable to flexibility, but it demands great self discipline from the development organization. They must self-police to flowdown adequate requirements and functional documentation.

Use a Realistic Schedule and Keep the Team Updated Regularly

Although a demanding, top-down schedule can inspire a team to work hard to meet challenges, an unrealistic schedule is ultimately detrimental and can force decisions which later work against you and delay the schedule. In a highly developmental program, such as ORBCOMM, some conservation is particularly important to proper scheduling. Allow the program manager to craft a realistic schedule at the beginning of the program and then hold him or her to it. Realistically update the schedule as often as once a month to keep the team informed of progress and the end date they are striving for.

Use Distributed Processing to Shorten Schedules

This was very successful in allowing parallel software development. But we paid an overhead penalty for processing and software complexity on some units that would have run more efficiently without an operating system.

Provide a Strong Technical Challenge to the Team

A tough technical challenge such as ORBCOMM's packaging can be very inspiring to an engineering team. ORBCOMM is not unlike the Sony Walkman story, in which the CEO of SONY gave the engineers a wooden block and said "make it this size." The packaging challenge of fitting eight satellites on a Pegasus drove the invention of many creative solutions to the small-satellite problem, which will benefit the company on other programs for years to come.

Don't Use a Custom Operating System

It's a much lower schedule risk in the long run to use an existing, well-supported product and modify the system to work with it, rather than create a tailored operating system.

Provide Enough Technical Oversight for Contract Management

Technical oversight was lacking in a few cases, which did lead to some surprises during the development program. Individual engineers should limit oversight to one contract-best case, and two at most.

Understand the Financial Strength of Your Vendors, Especially with Fixed-price Contracts

In general, the fixed-price approach worked well. I can't overstate the importance of working with excellent vendors who have a track record of solid business relationships and technical ability. A company in trouble can drag down the entire project.

Trade Hardware Weight and Cost for Increased Software Capability

This approach paid off handsomely in the attitude-control subsystem and with the peak-power tracking algorithm in the electrical-power subsystem. However, it requires early recognition of the critical nature of software development and demands hardware and software experts early in the program.

Make Sure the Program Manager Controls Responsibility for Flight Code

ORBCOMM Development Corporation was responsible for the payload flight code that was hosted on several processors, but the lowest person that both the satellite developers and ORBCOMM reported to was OSC's president. This sometimes made resolving of schedule and technical issues difficult. Place at least the engineering parts of organizations which share complex technical interfaces and development under one authority.

Make Sure Application Engineers Write Software to Ensure Technical Quality and Advance the Schedule

But make sure you have software experts consulting on the project.

Use the Ground-Station Software for System Test to Ensure Quality

This approach provided many benefits in the areas of cost, schedule, and technical utility. The ground-station command and telemetry interface was guaranteed to work, and there were no surprises in this area on orbit.

Use a Common Database to Generate Flight Code and Ground-Station Code

We used the common Microsoft Excel® spreadsheets to define intertask command and telemetry messages, as well as to electronically generate the ground-station databases. This saved an immense amount of time and cost. It also guaranteed the ground-station and system-test tools worked the first time when tested with new code.

Test the System Extensively and Realistically in its Initial On-orbit Configuration

The first week of ORBCOMM operations was very difficult, because the two satellites operating in broadcast-telemetry mode together were only tested for five hours before launch. We exhaustively tested each satellite, but two birds together caused severe operational headaches. The dropouts in a real communications link weren't tested end-to-end before launch. This turned out to place a much more stressing load on the flight computer than expected due to frequent reacquisitions. A simple test set up would have revealed this problem on the ground. Much larger programs have also missed this lesson about realistic software conditions, with far more serious consequences than ORBCOMM's operational headaches.

Allow Enough Time to Train Operators

Pegasus' extremely rapid integration capability can prove to be both a blessing and a curse. Because we can integrate to Pegasus a few days before launch, the engineering staff may focus on integration to the last minute, leaving little time to prepare for operations. Because OSC used the satellite engineers to operate the satellite, we allowed little rehearsal time. This lack of rehearsal caused trouble during contingencies. This is not a problem on other launch vehicles, such as the shuttle, for which months can go by after the satellites are handed over to the launch-vehicle engineers. This problem is endemic to small programs.

Don't Allow Last-Minute Changes to Functions

A key, fail-safe function on ORBCOMM was disabled intentionally 2 weeks before launch to reduce processor resets on orbit. Unfortunately, this left the satellite unable to recover autonomously from a condition which developed soon after launch. The satellite thus launched in a software mode that had received little testing on the ground.

13.2.5 Conclusion

Initial on-orbit operations proved the validity of the ORBCOMM design concept. Both of the satellites experienced early problems with processor hangups that we later determined to be caused by higher-than-usual radiation events. Single-event upsets corrupted processor and DSP memory. Using backup software provided workarounds to these problems. One satellite wouldn't accept commands for six weeks, probably due to a DSP memory upset. This satellite operated autonomously the entire time, and the ground station was fully informed of its health through backup broadcast mode telemetry. Spotty GPS performance caused a bad enough ephemeris to cause the satellite to go into an attitude where power was lost and the system fully reset. Fortunately, it came back up fully functional. Although the satellite is basically single string in hardware, software backups proved extremely important. The thorough qualification program for hardware proved the validity of using commercial parts in the space environment.

Subscriber-terminal messages are transmitting regularly and beta testing with DoD and commercial customers commenced in the Summer of 1995. Commercial operation will commence in first quarter 1996. The ORBCOMM system went through rigorous U.S. Department of Defense beta testing during the annual Joint Warrior Interpretability Demonstration. Over 1500 messages, faxes, and GPS position messages were transmitted across six locations during the two-week demonstration. Data requests were sent from a simulated battlefield at Camp Pendleton, CA to Fort Gordon, CA, Fort Bragg, NC, and Fort Huachuca, AZ. The system was also used to track two vehicles: one an ambulance roaming Fort Gordon, the other, a vehicle travelling along the East coast. Orbital Sciences Corporation is vigorously pursuing design refinements while preparing to construct the rest of the constellation.

AMSAT Summary

AO-13 is an amateur radio spacecraft carrying 4 linear transponders. *AO-16* is the first amateur radio "microsat" acting as a digital store-and-forward file server.

AO-13: Spacecraft dry mass: 84 kg   Propulsion: 400-N N_2O_4/UDMH thruster
         Average power: 50 W BOL   TT&C: 400 bps

AO-16: Spacecraft dry mass: 9 kg   No propulsion system
         Average power: 5.8 W   TT&C: 1.2/4.8 bps

Launch: AO-13—Ariane 44LP from Kourou on June 15, 1988
          AO-16—Ariane 40 from Kourou on January 22, 1990

Orbit: AO-13—2,200 km × 36,000 km at 57 deg inclination
          AO-16—805 km × 789 km at 98.7 deg inclination

Operations: AO-16 is operated by AMSAT NA. AO-13 has command stations in
            England, Germany, New Zealand. Both are used worldwide.

Status: Both satellites are still operational as of January, 2003.

Cost Model (FY95$M): See page 348 for explanation.

	AO-13			A0-16		
	Expected Costs	Sm. Spc. Model	Actual Costs*	Expected Costs	Sm. Spc. Model	Actual Costs*
Spacecraft Bus	$34.4M	$7.0M	$0.96M	$20.2M	$1.1M	$0.18M
Payload	$18.0M	$0.7M	Incl. spc	$18.0M	$0.7M	Incl. spc
Launch	$66.4M	$9.8M	$0.28M	$16.6M	$0.7M	$0.03M
Ground Segment	$7.6M	—	N/A	$5.1M	—	N/A
Ops. + Main. (annual)	$1.1M	—	N/A	$1.0M	—	N/A
Total *(thru launch + 1 yr)*	**$127.4M**	—	**$1.24M**	**$60.8M**	—	**$0.20M**

*Inflation factors of 1.377 (AO-13) and 1.246 (AO-16) have been used to inflate to FY95$ [SMAD, Table 20-1].

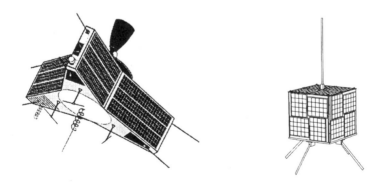

Fig. 13.3-1. AO-13 (left) is oriented with nadir-facing surface at the bottom. The sides are covered with solar cells. The 2 m omnidirectional antennas are on the arms. The antennas on the bottom are for 2 m, 70 cm, 23 cm, and 13 cm bands. **A0-16 (right)** has solar cells on all surfaces. The 2 m uplink antenna is on the center top surface and the 70 cm downlink elements are on the bottom edges.

13.3 AMSAT

Robert J. Diersing, *Radio Amateur Satellite Corporation*

The amateur radio community has a long and productive record of small satellite development and operation. The idea for the first amateur radio relay satellite is attributed to Don Stoner [1959]. In an article in the April 1959 amateur radio publication *CQ*, he suggested building such a satellite. A long series of events resulted in the formation of the Project OSCAR Association in California and the eventual launch of the first amateur radio satellite, OSCAR I, on December 12, 1961. The acronym "OSCAR," which has since been attached to almost all amateur radio satellite designations worldwide, stands for *Orbiting Satellite Carrying Amateur Radio*.

Since the launch of OSCAR I, many satellites have been designed, constructed, and operated by amateur radio satellite organizations. The complexity of the satellites has ranged from OSCAR I, carrying only a battery-operated beacon transmitter, to Phase 3-D, now under construction, which is a 3-axis stabilized satellite with multiple analog and digital communications transponders. Table 13.3-1 shows the evolution in costs of some of the satellites.

Because almost all amateur radio satellites are communications satellites, Table 13.3-1 includes the frequency bands and transmitter power as major identifying characteristics. I've given a single frequency band for beacons and an uplink and downlink band pair for transponders (e.g., 2–10 m for OSCAR 7). Besides an increase in the number of communications transponders and their RF power output, the other major cost and complexity factor is the presence of an attitude control system. Note that AO-10, AO-13, and Phase 3-D have active propulsion and attitude control systems, whereas all the other satellites use passive systems. The much larger 3-axis stabilization system, two propulsion systems, large number of transponders, and higher launch costs will make Phase 3-D the most expensive amateur radio satellite project ever attempted. I haven't used Phase 3-D as a case study because it is still under construction. It's scheduled for launch in late 1996.

13.3.1 History

Project OSCAR was instrumental in organizing the construction and launch of the first four amateur radio satellites—OSCARs I, II, III, and IV. OSCAR III was the first satellite to support communications relay as envisioned by Stoner, and about 1,000 amateurs in 22 countries used its relay capabilities [Davidoff, 1990]. OSCAR IV never achieved the planned orbit. Even so, it supported some amateur radio contacts including the first two-way satellite communication between the United States and the former Soviet Union. (See Table 13.3-2 for more information about OSCARs I–IV.)

TABLE 13.3-1.   Costs and Technical Details for Various Amateur Radio Satellite Projects.

Satellite	Launch Year	Cost	Transmitters/ Transponders	Tx Power (W)	Other Characteristics
OSCAR I	1961	$26	2 m	0.1	No solar generator, propulsion, or attitude control sys.
OSCAR 5	1970	$6,000	2 m 10 m	0.05 0.25	No solar generator, propulsion, or attitude control sys.
OSCAR 6	1972	$15,000	2 m–10 m 70 cm	1.0	Solar arrays, 4.5-yr lifetime, passive stabilization, no propulsion system.
OSCAR 7	1974	$38,000	2 m–10 m 70 cm–2 m 13 cm	8.0	6.5-yr lifetime, passive stabilization, no propulsion system.
OSCAR 8	1979	$50,000	2 m–10 m 2 m–70 cm	1.5	5.3-yr lifetime, passive stabilization, no propulsion sys.
Phase 3-A	1980	$217,000	70 cm–2 m	1.5	Launch vehicle failure, spin stabilized, solid-fuel rocket motor
OSCAR 10	1983	$576,000	70 cm–2 m 23 cm–70 cm	50	Spin stabilized, liquid-fuel motor, intermittent operation when Sun angles favorable.
OSCAR 13	1988	$925,000	70 cm–2 m 23 cm–70 cm 2 m–70 cm 70 cm–13 cm	50	Spin stabilized, liquid-fuel motor, still in operation, will re-enter in late 1996.
OSCAR 16	1990	$163,000	2 m–70 cm 2 m–13 cm	4.0	Digital store-and-forward, passive stabilization, no propulsion system.
Phase 3-D	1996	$4,500,000 (estimated)	70 cm–2 m 2 m–70 cm 70 cm–13 cm 70 cm–3 cm 70 cm–1.25 cm	200 250 50 50 1	3-axis stabilized, liquid-fuel motor and arc-jet thruster. Other uplink and downlink combinations possible through switching matrix.

SOURCE: Costs from [Jansson, 1987] except for last two projects.

While Project OSCAR was operating on the West Coast, a group with similar interests was developing on the East Coast. In 1969, the Radio Amateur Satellite Corporation (AMSAT) incorporated in Washington, DC. The first task of the newly formed organization was to refurbish the Australian-built AO-5 satellite and arrange for its launch. As seen in Table 13.3-3, AMSAT has participated in many international projects for amateur radio satellites, beginning with the Australia-OSCAR-5 project. Many countries now have their own AMSAT organizations and, although each operates independently, they often cooperate on large projects and other items of interest to the world community. Because of the widespread use of the "AMSAT" designation, the U.S. AMSAT organization is now called AMSAT-NA. Table 13.3-4 shows other projects sponsored by AMSAT and similar organizations outside the U.S.

The amateur radio community follows the usual convention of having one designation for a satellite before launch and another after launch. Thus, OSCAR 13 was known as Phase 3-C before launch. The AMSAT designator may be added to the name—for example, AMSAT-OSCAR-13, or just AO-13 for short.

TABLE 13.3-2. Amateur Radio Satellite Projects of the Project OSCAR Association.

Name	Launch Date	Lifetime	Notes
OSCAR I	Dec. 12, 1961	21 days	100 mW beacon transmitter. Battery depletion. No solar generator.
OSCAR II	June 2, 1962	19 days	Same design and failure mode as OSCAR I.
OSCAR III	March 9, 1965	18 days	Battery powering communications transponder had no solar generator. Transponder ceased operation at battery depletion.
		105 days	Beacon transmitter battery did have solar generator. Beacon operated until power system failed.
OSCAR IV	Dec. 21, 1965	85 days	Second satellite with a communications transponder. Didn't achieve the desired orbit because the launch vehicle failed. Satellite failure cause unknown. No telemetry system.

TABLE 13.3-3. Satellite Projects of the Radio Amateur Satellite Corporation in Cooperation With Other International AMSAT Organizations.

Name	Launch Date	Life or Status	Notes
OSCAR 5	Jan. 23, 1970	52 days	Built by students at Melbourne U., Australia. First satellite to have engineering and launch support from AMSAT-NA. No solar generator.
OSCAR 6	Oct. 15, 1972	4.5 years	First long-lifetime satellite. In service for over 4 years. Battery failure.
OSCAR 7	Nov. 15, 1974	6.5 years	First satellite to carry two linear transponders. 6-year lifetime. Battery failure.
OSCAR 8	March 5, 1978	5.3 years	Two linear transponders. 6-year lifetime. Battery failure.
PHASE 3-A	May 23, 1980	0.0 years	Launch-vehicle failure.
OSCAR 10	June 16, 1983	Limited operations	First OSCAR in high-altitude orbit. Two transponders. Operational when Sun angle is favorable. Radiation-induced computer RAM failure.
OSCAR 13	June 15, 1988	In operation[1]	OSCAR in high-altitude orbit carrying four linear transponders. Re-enters December 1996.
OSCAR 16	Jan. 22, 1990	In operation[1]	First amateur radio "microsat." Digital store-and-forward file server.
OSCAR 17	Jan. 22, 1990	In operation[1]	Educational microsat transmitting packet radio telemetry and digitized speech.
OSCAR 18	Jan. 22, 1990	In operation[1]	Educational microsat built by Weber State U. Primary experiment is Earth-imaging system.
OSCAR 19	Jan. 22, 1990	In operation[1]	Digital store-and-forward file server like OSCAR-16.
Phase 3-D	1996 (estimated)		Now under construction by international AMSAT team.

[1] As of January 1996

TABLE 13.3-4. Additional Amateur Radio Satellite Projects of AMSAT and Similar Groups Outside the United States.

Name	Launch Date	Life or Status	Notes
RS 1 & 2	Oct. 26, 1978	Not operational	Launched from the Soviet Union. Short lifetimes
RS 3–8	Dec. 17, 1981	Not operational	Launched simultaneously from a single vehicle. Some had communication transponders and some had Morse code "robots."
Iskra 2 *Iskra 3*	May 17, 1982 Nov. 18, 1992	~2 months ~1 month	Launched from the Salyut 7 space station. Built by Moscow Aviation Inst. Re-entered.
OSCAR 12	Aug. 12, 1986	3.2 years	Japanese-built satellite with both analog and digital communication facilities. Power system failed.
RS 10/11	June 23, 1987	In operation[1]	Satellite with communication transponders attached to a COSMOS navigation satellite.
RS 12/13	Feb. 5, 1991	In operation[1]	Satellite with communication transponders attached to a COSMOS navigation satellite.
OSCAR 20	Feb. 7, 1990	In operation[1]	Japanese-built replacement for OSCAR 12.
OSCAR 21	Jan. 29, 1991	~3 years	Joint German-Russian project attached to a GEOS research satellite. Loss or termination of control of main platform.
OSCAR 24	May 11, 1993	~4 months	French-built packet-relay satellite. Both of its transmitters failed shortly after launch.
OSCAR 26	Sept. 25, 1993	In operation[1]	File-server satellite using store-and-forward technology. Built by AMSAT Italy and based on the AMSAT Microsat design.
RS 15	Dec. 26, 1994	In operation[1]	Built by amateur group under supervision of Aleksander Papkov.

[1] As of January 1996

To provide wider coverage areas for longer time periods, design of the high-altitude Phase 3 series was initiated in the late 1970s. After losing the first satellite of the Phase 3 series to a launch vehicle failure in 1980, AMSAT-OSCAR-10 was launched and became operational in 1983. AMSAT-OSCAR-13, the follow-up to the AO-10 mission, was launched in 1988. The successor to AO-13, Phase 3-D, is already under construction and is scheduled for launch in 1996 [Harwood, et al., 1992; Jansson, 1991, 1993a, 1993b, 1993c; and Meinzer, 1993].

With the availability of the long access time and wide coverage of satellites like AO-10 and AO-13, it may seem that satellites with lower altitude orbits and shorter access times might become obsolete. This certainly might be true were it not for the incorporation of digital store-and-forward technology resulting in satellites gener-ically called *Pacsats*.

The first satellite with a digital store-and-forward feature was UoSAT-OSCAR-11. UO-11's Digital Communications Experiment was not open to the general amateur radio community, although designated "gateway" stations did use it. The first satellite with store-and-forward capability open to all amateurs was the Japanese Fuji-OSCAR-12 launched in 1986. By far the most popular store-and-forward satellites are

TABLE 13.3-5. Satellites with Digital Communications Facilities Operating in the Amateur Satellite Service.

Name	Launch Date	Life or Status	Notes
OSCAR 9	Oct. 6, 1981	8.0 years	Scientific and educational satellite built at the U. of Surrey in England. First amateur radio satellite to transmit plain ASCII text telemetry. In operation at re-entry.
OSCAR 11	March 1, 1984	In operation[1]	Built at U. of Surrey and is similar to OSCAR 9. Used to show that store-and-forward packet communications can work from low-Earth orbit.
OSCAR 14	Jan. 22, 1990	In operation[1]	Built by U. of Surrey and has several experiment payloads and a digital store-and-forward file server. Not operating in the Amateur Satellite Service.
OSCAR 22	July 17, 1991	In operation[1]	Built by U. of Surrey and has a store-and-forward file server and CCD camera. Operating exclusively in the Amateur Satellite Service.
OSCAR 23	Aug. 10, 1992	In operation[1]	Built at U. of Surrey by members of the Korean Institute of Technology (KAIST). Very similar to OSCAR 22.
OSCAR 25	Sept. 26, 1993	In operation[1]	Built at U. of Surrey by members of KAIST. Very similar to OSCAR 22.
OSCAR 27	Sept. 26, 1993	In operation[1]	Amateur radio payload on Interferometrics EyeSat-A based on AMSAT Microsat design.
OSCAR 28	Sept. 26, 1993	In operation[1]	Amateur radio payload on the Portuguese Industrial Consortium PoSAT based on U. of Surrey UoSAT design. Not operating in the Amateur Satellite Service.

[1] As of January 1996

the Pacsats using the *Pacsat Broadcast Protocol* [Price and Ward, 1990a, 1990b, 1990c; Ward and Price 1990a, and 1990b]. These Pacsats fall into two general categories—the *Microsats*, based on technology developed by AMSAT-NA, and the *UoSATs*, based on technology developed by the University of Surrey.

Amateur radio satellites have evolved to provide two main types of communication services—analog transponders for real-time communications and digital store-and-forward for non-real-time communications. An evolution has also occurred among groups building satellites that provide amateur radio communications. For many satellite projects most of the design, construction, and operations tasks are handled by radio amateurs who are members of their respective national AMSAT organizations. More recently, however, non-AMSAT groups interested in developing small-satellite technology have designed and built satellites that provide communications services to radio amateurs. In some cases these groups may receive partial funding for a project from an AMSAT or other amateur radio group. [See Table 13.3-5 for examples of satellites supporting amateur radio digital communications which have been designed and built by groups separate from any AMSAT organization.]

TABLE 13.3-6. AO-13 Technical Description.

System	Characteristics	Details
General	Mission type Launch vehicle Date of launch Satellite dry & wet mass Design life & actual life Orbital parameters	Real-time analog and digital communications Ariane V-22 June 15, 1988 84 kg; 140 kg 72 months; still operating 2,200 km × 36,000 km × 57 deg
Payload	Description	70 cm–2 m, 2 m–70 cm, 23 cm–70 cm, 70 cm–13 cm transponders
Electrical power	Power BOL & EOL average Battery type Total battery capacity	50 W; 35 W NiCd 10 A·hr primary, 6 A·hr secondary
Structures	Mass Material Dimensions & shape	7 kg Sheet aluminum 0.43 m height; 1.26 m span; Tri-star 1.35 m × 2.0 m with antennas
Attitude control	Stabilization type Attitude knowledge Pointing accuracy Sensors	Spin-stabilized; reorientation by electromagnetic torquing 5 deg (3 σ) 10 deg in-plane (3 σ) 5 deg perpendicular to plane (3 σ) Sun sensor (2 slits); Earth sensor
Propulsion	Thrusters Propellant & oxidizer Total Impulse	One 400 N liquid fuel Aerozine 50 / N_2O_4 293 sec
Telemetry, tracking & command	Uplink band & downlink band Uplink & downlink data rate Main processor	70 cm or 23 cm up; 2 m or 13 cm down 400 bps (uplink or downlink) SANDIA rad-hard 1802 CPU (RCA CDP 1802 architecture)
Data management & processors	Flight software lines Flight software language Ground software lines Ground software language Memory Processor	750 IPS 12,000* BASIC* 32 KB Same as TT&C
Thermal control	Type	Thermal blankets; coatings

*These entries refer to the software currently in use as described in the section "On-Orbit Operations" (page 545). But the initial version of the ground software was also IPS, about the same number of lines as the on-board software, and ran on Atari 400, 800, and 800XL machines. The original IPS software was used during the motor burns for reorientation.

13.3.2 Case Studies

I've chosen two of AMSAT's satellite projects as case studies—AMSAT-OSCAR-13 (AO-13) and AMSAT-OSCAR-16 (AO-16 or Pacsat)—for several reasons. First, they represent the development of analog transponders (AO-13) and digital store-and-forward technologies (AO-16) in the Amateur Satellite Service. Each has a different design heritage. AO-13 is based on the designs of AO-10 and Phase 3-A. AO-16, one of four AMSAT-NA Microsats, represents a new, innovative design with no heritage whatsoever, and yet, the elapsed time from program start to launch was 25 months. The orbits of the two satellites are radically different—

TABLE 13.3-7. AO-16 Technical Description.

System	Characteristics	Details
General	Mission type Launch vehicle Date of launch Satellite dry & wet mass Design life and actual life Orbital parameters	Digital store-and-forward communications Ariane V-36 on ASAP January 22, 1990 9 kg; NA 60 months; still operating 800 km × 800 km × 98 deg
Payload	Description	Digital store-and-forward communications. Four 2 m uplinks, two 70 cm downlinks, one 13 cm downlink
Electrical power	Power BOL & EOL average Solar cell type & area Battery type; no. of cells Total battery capacity	BOL orbit average 5.8 W BSFR; 0.176 m GE; Gates commercial grade NiCd; eight 6 A·hr
Structures	Dimensions; shape	0.23 m × 0.23 m × 0.21 m; cubical excluding antennas
Attitude control	Stabilization type	Passive; 4 permanent magnets; 7 hysteresis damping rods; different color paints on antenna surfaces to induce Z-axis spin.
Telemetry, tracking & command	Uplink band; downlink band Uplink; downlink data rate Main processor	2 m up; 70 cm or 13 cm down 1200 bps or 4800 bps NEC V40
Data management & processors	Flight software lines Flight software language Ground software lines Ground software language Memory Processor	4,800 operating system; 2,100 application C 4,800 C 8 MB NEC V40 (same as TT&C)
Thermal control	Type	Passive

AO-13 being in a high-altitude, Molniya-type orbit and AO-16 being in a nearly circular, Sun-synchronous, low-Earth orbit (800 km). Finally, AO-13 and AO-16 involved differing levels of international participation. For AO-13, AMSAT-DL (Germany) and AMSAT-NA shared most of the funding, design, and construction. But other international AMSAT groups participated. In contrast, AMSAT-NA people handled nearly all of the work on AO-16.

The AO-13 and AO-16 projects represent the spectrum of satellite projects in which AMSAT-NA and its members participate. AMSAT-NA contributed about $220,000 toward the cost of AO-13 and about $163,000 toward AO-16. Technical descriptions of AO-13 and AO-16 are in Tables 13.3-6 and 13.3-7. After approximately 7 and 5 years in orbit, for AO-13 and AO-16 respectively, both satellites remain in continuous operation.

Figure 13.3-2 shows some of AO-13's internal and external components. Figure 13.3-3 has the launch configuration for the satellites on Ariane V-22 mission which launched AO-13 (AMSAT P3C).

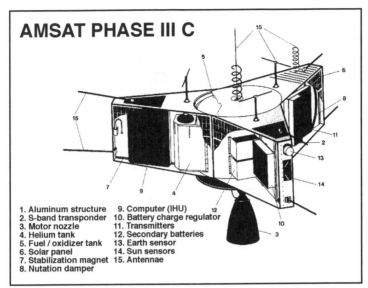

AMSAT PHASE III C

1. Aluminum structure
2. S-band transponder
3. Motor nozzle
4. Helium tank
5. Fuel / oxidizer tank
6. Solar panel
7. Stabilization magnet
8. Nutation damper
9. Computer (IHU)
10. Battery charge regulator
11. Transmitters
12. Secondary batteries
13. Earth sensor
14. Sun sensors
15. Antennae

Fig. 13.3-2. AO-13 External and Internal Components. (Drawing courtesy of AMSAT-DL)

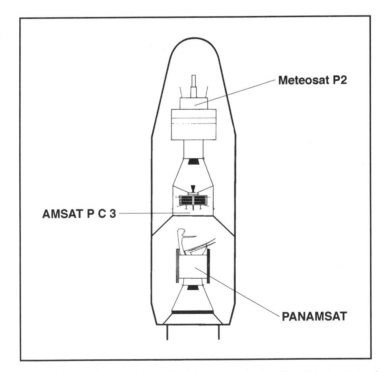

Meteosat P2

AMSAT P C 3

PANAMSAT

Fig. 13.3-3. AO-13 position on Ariane V-22 launch vehicle. (Drawing courtesy of AMSAT-DL).

AMSAT-OSCAR-16

More than 5 years have passed since the launch of four Microsat spacecraft developed by AMSAT-NA and other cooperating AMSAT groups. The Microsats were launched January 22, 1990, on Ariane mission V-35, the first mission to use the Ariane Structure for Auxiliary Payloads. (See Sec. 4.2.) All of the Microsats were placed in nearly circular, Sun-synchronous, low-Earth orbits (800 km). The four satellites, their primary missions, and owner/operators are: AMSAT-OSCAR-16 (AO-16 or Pacsat), store-and-forward file server, funded and operated by AMSAT-NA; DOVE-OSCAR-17 (DO-17 or DOVE), space-science education and promotion of international peace, funded by the Brazilian AMSAT organization BRAMSAT; WEBER-OSCAR-18 (WO-18 or Webersat), space-science education, funded and operated by Weber State University; and LUSAT-OSCAR-19 (LO-19 or LUsat), store-and-forward file server, owned and operated by Argentina's amateur satellite organization, AMSAT-LU.

While the Microsats were largely developed by AMSAT-NA, other organizations participated. For example, an engineer from AMSAT-LU did much of the spacecraft integration, and a Slovenian student studying in the U.S. helped with the transmitter design.

*Microsat Design Objectives**

The intended application of all four Microsats was digital store-and-forward communications. Even though all four satellites have this capability, only two—AO-16 and LO-19—provide store-and-forward messages to the amateur radio community as their primary mission. The other two Microsats—DO-17 and WO-18—use their store-and-forward ability to support their primary missions as educational tools. The objectives for the Microsat design were based on the experience gained from other AMSAT projects and the intended mission of supporting store-and-forward communications:

1. Eliminate wiring harnesses whenever possible: they're time consuming to make and are significant sources of failure.

2. Create a mechanical structure that could be completely assembled and disassembled in less than 30 minutes.

3. Create a solar-array design that minimizes the possibility of damage during handling and yet can be rapidly installed on the spacecraft body.

4. Use load-side power management to dynamically adjust the transmitter's power output in order to maintain an orbit-average power balance. The technique should be modifiable in orbit and should deliver every possible mW of RF power to the downlink transmitter.

5. Create a microsatellite design that can serve user data terminals employing omni-directional antennas.

6. Develop a suitable computer that will provide serial data communications over multiple channels, store at least 4 MB of data, and require less than 1.0 W average power.

7. Aim for a spacecraft mass of 10 kg.

---

* The following is adapted from [King, et al., 1990].

Design for the Structure and Thermal Control

The overall structure, known as the *Frame Stack Assembly,* is a stack of machined boxes (modules), each containing a major element of the spacecraft electronics. Usually, five modules comprise the stack but it can be extended, as it was for WO-18 which is the equivalent of seven modules. Each module frame has a useful volume of 200 mm × 184 mm × 40 mm. The center module, intended for the power subsystem, has a slightly larger useful height of 42.8 mm. Each module has a recessed area to allow a volume in the assembly for electrical connections. The circuit boards can be mounted within the module using small Delrin plastic blocks designed for this purpose, but other methods are possible.

Four solar panels mount in recessed areas formed by the module stack on each side of the spacecraft. These honeycomb panels provide significant support against shear loads once installed on the structure. A single standard tip jack centered on each panel provides quick, safe electrical connection. The frame itself is the electrical ground. The modules interconnect electrically through a 25-wire bus made with standard printed circuit material. This bus constitutes the entire wiring harness for the spacecraft and can be installed or removed in 5 minutes (including the locking hardware required for flight).

The top panel contains another solar panel and the VHF receive antenna. On AO-16 and DO-17 a small, bifilar-helix antenna for S-band is also mounted on the top panel. The spacecraft's most complex surface is the bottom. It contains another segment with four solar arrays (producing half the power of the other faces), the separation system, a microswitch for turning on the transmitters, and the transmitting turnstile antenna. The bottom is a major load-carrying surface and is machined from a single piece of aluminum. Gusset plates fasten its corners to the bottom module frame which contains the transmitters.

A single compression spring separates the spacecraft from its launcher plate. Concentric to the compression spring is a bolt which passes through the spring, the launcher plate, and a bolt cutter. Four locator pins on the spacecraft side of the interface mate to four locator pads on the launcher plate. These devices also counteract shear loads from the spacecraft. The bolt is tensioned from the underside of the launcher plate.

One of the advantages of a microsatellite structure is that we can expect the natural resonant frequencies of the tiny structure to be very high. Because the launcher's resonant frequencies are usually quite low (5 to 25 Hz), there is no resonant coupling between the spacecraft and the launch vehicle. Both versions of the Microsat module's stack configuration were formally tested at qualification and acceptance levels according to Arianespace documentation. The highest random level achieved during qualification-level testing was 14.6 g rms for 90 seconds. No resonances existed below 100 Hz, and the primary spacecraft resonances are from 200 to 400 Hz. The test structure was sufficiently well behaved at 14.6 g rms random so that it likely could have sustained at least 20 g rms.

The design uses completely passive thermal control. The external absorptance-to-emittance ratio of the solar arrays dominates the thermal balance. This value, which is near unity, was a given in the design. The design requires the largest possible area for solar cells, so it leaves only a limited surface for biasing the overall temperature. Of the overall surface area of 2,916 cm^2, only 591 cm^2 is available for thermal control.

Design of the Power Subsystem

One of the most difficult challenges in designing a microsatellite is maximizing power and developing a strategy for not wasting what is generated. The efficiencies of power regulators and RF generating equipment are critical to the design. Micropower consumption of computers and receivers is also a major challenge.

Even though solar cells with efficiencies of 18 to 23% were available, cost and availability drove this particular design to use *Back Surface Field Reflector* cells. The particular cells selected were 2 × 2 cm in size and were manufactured by Solarex of Rockville, MD. The fundamental unit of power for the Microsats is a solar-cell clip, which contains 20 series-connected cells. On each of the sides and top of the spacecraft, two sets of series-connected clips are wired in parallel to yield a maximum per-panel current of 0.35 A at 20.5 V when the panels are cold. The bottom surface contains four half-clips wired in series to produce a single 40-cell string, which gives half the power of the other surfaces. Some shading caused by the turnstile antenna blades also reduces the power delivered by the bottom surface.

Regulating the battery charge requires two steps. Maximum array power always transfers to the battery through manipulation of the solar array's operating point. Then, a duty-factor switch controlled by the flight computer dynamically adjusts the *Battery Charge Regulator's* load impedance to keep the voltage point of the array at the knee of the I versus V curve. This switch's overall efficiency is about 90%. Corrections to the operating point correspond to the measured temperature of the array. Bias adjustments may also be made as changes in the knee voltage occur resulting from radiation damage or other effects. In principle, the battery could become overcharged if loads don't demand all the power produced. Although this situation isn't likely to do serious damage to the batteries in such a small spacecraft, it does suggest that power is being wasted.

The second step in regulating the Microsat's power is to use a load-side management scheme. The power output of the primary transmitters varies continuously, under computer control, from a few milliwatts to about 4 W. When the battery is fully charged, the software increases the active transmitter's power until it reaches a break-even power condition. If the onboard computer later determines the power budget is negative, the software decreases the transmitter's power until it again reaches the break-even point. The power-management scheme ensures that every milliwatt of excess available power is used to generate RF for the user downlink.

A single, 8-cell, NiCd battery is used in each Microsat. The cells are GE goldtop series, size F. They're rated to work at temperatures as high as 70 °C. To match cells from commercial manufacturing lots for capacity and voltage, AMSAT has developed a proprietary program for testing and selecting these cells. The test program has very successfully identified flight-quality cells from commercial manufacturing lots and therefore saved a lot of money over using cells manufactured for space applications. Because the Microsat cells are operating near 0 °C, the typical operating voltage is slightly above 11.0 V in sunlight and decreases to about 10.4 V at the end of eclipse. Typical depth-of-discharge for each satellite during eclipse is 4% to 6% which, combined with the low temperature, should allow the batteries to last almost indefinitely. The batteries and power-regulation electronics are in the third module frame in the frame stack assembly.

Design of the RF Subsystem

The Microsat's RF subsystem consists of two redundant data transmitters and a 5-channel receiver. A quarter-wave, linearly polarized antenna is used with the receiver and a 45 deg canted turnstile antenna is used on the downlink. The latter is circularly polarized.

The packet-communications system employs data standards common to the Amateur Radio and Amateur Satellite Services. Packet-communications techniques developed by these communities use the Aloha form of *Carrier Sense Multiple Access*. Although throughput of this technique isn't high (18.4% maximum), it's simple to implement, and ground-station equipment is readily available. Each spacecraft has a single UHF downlink with a data rate of 1200 or 4800 bps. With a maximum throughput approaching only 20%, the total offered traffic should be about five times that of the downlink's data rate. We've achieved this level by using a VHF receiver with 5 channels, where each uplink channel's data rate can be adjusted by ground command for 1200 or 4800 bps operation. Uplink modulation is frequency shift keying.

A common front end serves all five receiver channels. Unlike most space applications, receiver G/T performance is relatively unimportant because user uplink signals have EIRPs in the range from 20 to 40 dBW. At these uplink power levels, and due to the level of carriers in adjacent frequency bands, it's more important that the receiver have excellent overload characteristics than a good noise figure. In the Microsat's receiver design, a high-Q filter doesn't precede the low noise amplifier (LNA). Instead, the receiver uses a GaAs FET transistor in the LNA. This transistor has a particularly high (large amplitude) front-end overload. In fact, the third-order intercept of the device used is above 0 dBm. Following the LNA is a helical resonator bandpass filter. The input signal at 145.9 MHz is downconverted to approximately 50 MHz, amplified, and then split into five separate channels. Each channel uses a Motorola 3362 FM receiver chip, and the signal within each channel is down converted two more times. The final intermediate frequency at 1.8 MHz passes to the discriminator and then to a slicer and data filter. Before passing raw data from the receiver to the flight computer, the system removes any DC component resulting from Doppler shift or error in a user's uplink frequency. A very steep-skirted bandpass filter (15 KHz wide) is placed after the second IF. It allows for a user's 1200 bps data on the nominal uplink frequency to pass through the filter, even at maximum Doppler shift, without distortion. For 4800 bps data, however, the user must compensate for Doppler so the modulation spectrum will pass through the filter. The flight computer monitors the signal strength and offset of each of the five receiver channels and provides the values as telemetry. Receiver sensitivity varies slightly among the channels and ranges from −117 dBm to −110 dBm for 10^{-5} BER. The total power consumed by the receiver is less than 0.25 W. The receiver occupies the top module in the stack.

The bottom module of each spacecraft contains two transmitters. Power output can be adjusted in 16 steps by controlling the voltage to the final two RF stages. It's approximately proportional to the square of the selected step. As I've already described, the power setting is under computer control and is used in a closed loop to manage the spacecraft's overall power budget.

The two UHF transmitters used in AO-16, WO-18, and LO-19 aren't identical. One transmitter emits standard PSK (±90 deg phase shift), and the other emits raised-cosine PSK. The efficiency of the standard PSK transmitter is about 63%, whereas the efficiency of the raised-cosine PSK transmitter is approximately 55%. The two VHF transmitters used in DO-17 use narrow-band frequency modulation; the flight units have an efficiency of about 76%.

In all spacecraft the two transmitter outputs feed into the two isolation ports of a 90 deg hybrid. The two remaining hybrid ports feed a turnstile antenna to produce circular polarization. Selecting one transmitter produces right-hand circular polarization; selecting the other produces left-hand circular polarization

Design of the Attitude-Control Subsystem

AMSAT has been using a passive magnetic technique to stabilize the attitude of small spacecraft in low-Earth orbit since 1974. The technique has worked well, so there has been no particular reason to increase the complexity of the Microsats by employing an active system.

Let's assume a particular spacecraft has good omnidirectional antennas and link margins are large (both of which are valid for the Microsats). If so, we barely need a system for attitude control. But we want to minimize polarization "tip null" fades that would result in some data loss, particularly on the downlink. Moreover, to avoid thermal gradients the satellite must rotate slowly.

The Microsat's attitude-control system uses four Alnico-5 bar magnets mounted to the outside of the spacecraft on the four edges of the cubical structure parallel to the Z axis. Seven hysteresis rods oriented in the X-Y plane are normal to the magnets. The rods are embedded in the battery-support plate, near the center of gravity, and run parallel to the X axis. Finally, the four blades of the turnstile antenna act as solar photon vanes. The blades are approximately 10 mm wide and are painted white on one side and black on the other. Because the blades are mounted in succession, the Sun always "sees" at least one black surface and one white surface.

The effect of using the permanent magnets and hysteresis rods is to quickly align the spacecraft Z axis with the local Earth-field vector at any point around the planet. The hysteresis rods damp any motion about the field lines. The four solar vanes impart a torque about the Z axis and reduce the thermal gradient across the spacecraft body. The solar torque is counterbalanced by both the hysteresis-rod damping and eddy current that we can't eliminate in parts of the spacecraft. For the hysteresis rods to damp this rotation, we need attitude deviations of the Z axis from the local magnetic-field vector on the order of 10 deg. The two damping torques place an upper bound on the rotation rate about the Z axis in response to solar torque. As a result, the stabilizing system causes a rotation of the Z axis, twice per orbit, in response to the Earth's dipole and then a rotation about the Z axis. Thermal considerations set the target value of the rotation period about the Z axis at 2.5 minutes per rotation. A large tolerance on this value is quite acceptable, and rotation rates from 0.25 minutes to about 20 minutes per rotation will still allow magnetic lock and result in acceptable thermal behavior. The thermal time constant of the Microsats is much longer than originally anticipated.

Design of the Flight Computer and Data-Handling Subsystem

The flight computer is used in a multitasking, real-time environment and, as such, it's part of all the spacecraft's other electronic subsystems. Data to and from the receiver and transmitter is handled through fairly high-speed serial links (designed for up to 100 kbps) using formal serial data control. Data moves between other subsystems in the spacecraft and the flight computer on a single serial-interface line, which is common to all modules, and a single return line. A specialized board known as the *Addressable Asynchronous Receiver/Transmitter* is used within each hardware module to provide serial communications with the flight computer. Telemetry signals are also handled through the AARTs, and multiplexed analog signals are routed to a single A/D converter within the computer through the 25-pin bus. Two wires forming a differential pair are employed for this function.

The Microsat's flight computer is state-of-the-art (for 1990) in terms of its weight, size, fabrication technology, and performance. Weighing in at 1,025 g and consuming an average power of 0.45 W, the computer is optimized for serial communications. The computer's clock speed is not outstanding at 9.830 MHz, but by using its DMA functions, it can support six simultaneous serial inputs at rates as high as 100 kbps each. The Microsats use only a fraction of this capability, operating at a maximum of 4800 bps per serial input. Three serial controller devices (NEC 72001) handle the incoming uplink data which is high-level data link control (HDLC) formatted and compatible with a variant of CCITT X.25 known as AX.25 [Fox, 1984]. The 72001s also provide up to six transmit outputs, although the current design only provides for a single line to the common transmitter modulation inputs. The data path to and from the 72001s is 8-bit parallel. The microprocessor used is an NEC V40, which is equivalent to an Intel 80C186 with a slightly modified instruction set. A serial input and output pair directly from the V40 is used to communicate with the four AART boards distributed on the 25-pin bus. In addition to the A/D converter mentioned above, the computer provides a utility-latched port, as well as headers for direct access to the processor's address and data lines. These headers are for experiments that may require maximum speed.

The flight computer uses four classes of memory. Two KB of Harris HM6617 ROM contain the boot loader. The main program memory uses 256 KB of error-detecting-and-correcting (EDAC) RAM comprised of Harris HM6207 devices. This memory is configured as 12 bits per byte and can detect two and correct one bit error per byte anywhere in memory. If a single output bus line were to fail out of the 12, the memory could carry on without a problem, although with a degraded ability to protect against errors. The computer counts any single-event error and software incorporates this value into the telemetry. Two MB of bank-switched RAM, segmented into 512 KB blocks, provides high-speed memory for general processing. We can enable, or disable, each block and can access this memory in 90 ns. The bank-switched RAM has no hardware-error protection but software-error protection is employed. In addition to the 2 MB of general-purpose RAM another 6 MB of memory is configured as a RAM disk. Access to this memory is slow but more than fast enough for packet-data communications. We can switch the 6 MB RAM off and on in 512 KB blocks just as we can the 2 Mbyte memory. Both mass memories are implemented with Hitachi's 62256L surface-mount RAM chips which are $32{,}768 \times 8$ bit memories. A total of 256 chips make up the two memories.

The computer module consists of three multilayered boards. The CPU and the RAM-disk mass-memory board are eight-layer, printed-circuit cards, whereas the 2 MB, bank-switched RAM is six-layer technology. The mass-memory board is populated on both sides with surface-mount chips. In fact, most integrated circuits in the flight computer are surface-mount devices. RAM chips have 0.021" lead spacing, whereas the other surface-mount devices are 0.050" between traces. Board interconnects use lead cables made of Kapton ribbon. All boards were conformally coated after final checkout. The flight computer contains 453 integrated circuits. Except for the boot-loader ROMs, none of the components used was high-rel or rad-hard. The HN.6617 boot-loader ROMs we used were qualified to MIL-STD-833B. The boards described above were complex enough to contract out to two firms who specialized in surface-mount technology. They laid out, made, and populated the boards. This was the first time AMSAT used an outside organization to make flight hardware. AMSAT people assembled the boards into the module frames and debugged the computers as required.

A unique identifying feature of Microsat is its use of serial communications to handle all data flow among modules within the spacecraft. Within each module except the flight computer, there is an inter-module interface board using an MC14469 Addressable Asynchronous Receiver/Transmitter. The AART boards comprise a small local area network (LAN) for communicating among the spacecraft modules.

Each AART receives 4800 bps data on a data line common to all units and has a unique single-byte address that each module must recognize before accepting the next byte as serial data. Depending on the value of the data, the unit may latch as many as three bits in a field of 24 available bits. Or, it may set two different analog multiplexers on the AART board to place a particular sensor value on the analog bus line. The A/D converter in the flight-computer module will read this value. By changing the most significant bit of the address byte, the AART unit can also return an 8-bit value to the flight computer on a second serial line common to all AARTs. If we use this feature, the flight computer must know to poll the particular module periodically for the expected data. The AART unit has no means of signalling to the computer that it has data ready. Each AART board has signal conditioning for up to 4 thermistors and up to 32 telemetry voltages. The board also contains its own precision 2.55 V reference. This level of telemetry capability per module has proved to be more than adequate, although the power module in each of the Microsats uses most of it.

Design of the Flight Software

The Microsat bus is a minimal architectural design. The single CPU carries out all software functions in the spacecraft:

1. *Telemetry generation*—Requires commanding the AARTs in each module and sampling all analog telemetry points, generating the real-time telemetry, and storing whole-orbit data if required.

2. *Ground command*—Includes adjusting targets for power management and switching on and off various hardware modules.

3. *Onboard autonomous control*—Includes controlling the load-side power management, scheduling experiments, and setting the software fuses to protect against high and low voltage conditions.

4. *Cleaning up single-event upsets*—Single bit errors are purged from memory with a process called "memory wash."

5. *Handling data-communications protocols*—Implements the Amateur Radio Service's standard data-link-layer protocol AX.25, which is a variant of LAPB, the X.25 link layer protocol.

6. *Applications*—Such as the Pacsat and Lusat, which store and forward messages.

The software design for the Microsat CPU had goals which were similar to those for the hardware design. The design was to be based on standard, proven, cost-effective components that could be integrated in interesting ways. Designers could then concentrate on the actual applications rather than on the nuts and bolts of the operating system and compiler.

The main goal of the design was to allow AMSAT to use more software engineers. Spacecraft control, protocol handling, and several large data-processing applications for the typical Microsat would require more software than all previous AMSAT missions combined. A Microsat would also be an on-going experiment requiring continuing software development over many years. Thus, the project needed more software engineers. Also, each sponsoring organization would be responsible for its own application software to control its hardware modules. The development system used would have to be accessible to many people in several countries and would, as always, need to be inexpensive.

Subsidiary goals followed from this main goal:

1. Employ a widely available development environment. This was a factor in the choice of the NEC V40 CPU for the Microsat computer. The V40 is an 80186; an Intel 808x style chip enhanced for embedded controller applications. As this chip is software compatible with the 8088 in IBM PCs, designers could use standard IBM PC development tools. Also, because the instruction set is the "native" PC set, we could use mainstream compilers rather than cross compilers. We chose the Microsoft C compiler for software development.

2. Bring together programs written at different times by different programmers in various locations and run them on a single Microsat computer. AMSAT can't afford the time or money consumed by numerous procedures, meetings, documentation, CASE tools, and other standard trappings, that allow binding of disparate software elements. Thus, we elected to permit each major function to be a separate program running in a multi-tasking environment. Peaceful co-existence among tasks is relatively easier to obtain when it's enforced by an operating system that manages shared resources such as memory, the telemetry system, and the data protocols. This is the software equivalent of the fast "rack and stack" mechanical structure and of the 25-wire harness; it's desirable for the same reasons.

3. Keep application programs for Microsat the same as a regular PC application. By allowing separate tasks, we can independently debug each one using standard tools such as Microsoft Codeview.

4. Make available all standard C language functions, including floating point.

5. Make sure the operating system provides communications between tasks.

6. Place the operating system in RAM, not ROM, so we can maintain and extend it on orbit. This is in keeping with the experimental nature of all AMSAT spacecraft. Only a small (though mission-critical) bootload routine is in ROM.

The operating system chosen was "qCF," developed by Quadron Service Corporation of Santa Barbara, CA. This system is designed for communications co-processor cards based on the 80186. The qCF system supports both IBM's Artic card and Emulex's DCP286i card. qCF allows execution of programs compiled by the Microsoft C compiler and also provides preemptive multitasking, timers, intertask communications, and interrupt- and DMA-driven HDLC I/O drivers. Quadron, three of whose founders are radio amateurs, ported qCF to Microsat and supplied I/O drivers for the 72001 communications chips. As the software was then compatible with the Microsat CPU and with the IBM Artic card we could use the Artic card to emulate the spacecraft's CPU.

Using Artic, qCF, and Microsoft C reduced the cost of a full-up development system and spacecraft simulator to that of an IBM PC clone, a $1,200 adapter card, Microsoft C compiler, and the software Quadron donated. With commercial-quality components available off the shelf software engineers could concentrate on developing the applications and not the development environment itself.

The University of Surrey also selected the qCF system for one of its spacecraft launched on the V-35 ASAP with the Microsats. Although its CPU is a very different design the high-level application programming interface is the same allowing AMSAT and UoSAT to share some applications. AMSAT provided the AX.25 communications handler and the I/O driver; UoSAT, in return, provided the file-system task and parts of the message and file-server system.

Let's briefly summarize the types of tasks running on a Microsat flight computer. The operating system kernel supplies the basic multitasking services. It manages the hardware timers, sets up memory, and loads and unloads tasks. The 8 MB data-storage area is managed as a RAM-based disk by the file-support task. The low-level C read and write subroutines, in the standard C library used by applications, are replaced by routines that format an I/O request and send it as an inter-task message stream to the file-support task. Acting much like an IBM PC's RAM disk driver, the file-support task provides blocking and de-blocking services as well as error correction for single-bit errors.

The most visible program to users on the ground is the file-server task. The major goal of the Pacsat and Lusat Microsats is to provide a bulletin board and file service. This interface is optimized for computer-to-computer transfers; the user's software at the ground station provides the human interface. Software and procedures have been developed to integrate the Microsats into the amateur radio service's packet radio network.

The AX.25 handler implements the LAPB-style communications protocol. It permits point-to-point connections between the file server task running on a Microsat and the many ground stations visible in the range circle.

The HDLC driver passes frames between the AX.25 handler and the uplinks and downlink. The hardware design supplies several DMA channels. Even so, there are more I/O channels than DMA channels so the HDLC driver must do both DMA- and straight interrupt-driven I/O. To get the most out of the available processor power, and to enable later Microsat missions to use even higher baud rates, the HDLC driver is written in assembly language.

A housekeeping task carries out algorithms for managing the spacecraft's power as discussed previously.

The telemetry software module periodically gathers data from sensors throughout the spacecraft by using the AART network. The data goes to the downlink for real-time monitoring and can also stay in the RAM disk for later transmission if required. The "whole-orbit" data format, where the values for telemetry channels are stored over several hours and are later downlinked, is an invaluable tool for analyzing the performance of low-orbit spacecraft.

Memory used for program and data storage is protected with hardware error-detecting-and-correcting (EDAC) circuitry and by the memory-wash task. Whenever an energetic particle normally filtered out by the atmosphere induces an error in memory, the EDAC will correct the error when a read occurs and place the correct data on the bus. But the hardware doesn't automatically write the corrected byte back into memory. Thus, if an error is allowed to linger then a second bit in the same byte may be corrupted, resulting in a multiple-bit error that can't be corrected. In a process called *memory washing*, a task periodically runs through the EDAC memory reading and writing every byte so the corrected byte is written back into memory over a damaged one.

Memory that stores users' files is not protected by hardware EDAC. The reason is economics; 12 bits are needed to store each 8-bit byte in hardware-protected memory. Bit errors in the RAM not protected by EDAC are corrected by a software memory-wash procedure. One 1024-byte RAM disk cluster at a time is read, corrected (if necessary), and re-written. The memory-wash cycle occurs at a rate high enough to wash the entire 8 MB of RAM in less time than it takes for the spacecraft to pass through the South Atlantic Anomaly twice—where the probability of a radiation-induced bit error is the most likely.

Performance After Five Years On Orbit[*]

The AMSAT-NA Microsats have compiled an excellent record of performance on orbit. This is true for the spacecraft and the onboard computer software. There have been a few subsystem and component failures, but none has caused the loss of a mission.

The power-generation, conditioning, and storage subsystems have had no problems. None of the Microsat onboard microprocessors have experienced any type of failure including single-event latchups. There have been no permanent bit failures in the EDAC-protected RAMs. One of the 16 AART communication paths has failed—the path from the DO-17 speech module to the computer module. Outside of this one AART path failure, more than one trillion AART commands have been issued successfully by the flight computers and acted upon by the receiving modules—none has been lost or interpreted incorrectly. Problems have developed with the 70 cm transmitter PSK modulators on AO-16 and WO-18 and the 13 cm transmitter PSK modulator on DO-17. In all three cases, they lost carrier suppression (which is equivalent to a reduction in modulation index). We believe a small change in value of

---

[*] Adapted from [Diersing, 1994a].

a piece part (capacitor) caused this problem, which is much more serious on DO-17. None of the transmitter modulator problems permanently affected the respective missions. For WO-18 and AO-16 operations were switched to the RC PSK transmitters. The near failure of the DO-17 13 cm transmitter modulator had significant impact on software uploading capability. Other aspects of the mission haven't been affected, however, because normal operations require only the 2 m transmitter.

The primary mission of AO-16 and LO-19 is to provide store-and-forward communications in low-Earth orbit. During the first 2.5 years in orbit, their application software evolved through several distinct stages. For about the first year of operation, AO-16 and LO-19 provided what is called digipeater service. With this mode of operation, two stations within the satellite's footprint could connect to each other using the satellite as a relay.

In late 1990 we began testing the first version of the file server which supported two modes of operation—connected mode and broadcast mode. With connected mode data transmitted on the downlink can be used only by the station establishing the connection even though all stations in the satellite's footprint can hear this data. On the other hand, in a broadcast mode, any station in the footprint needing information can use the downlink data. So, if several stations in the footprint need a particular file stored in the satellite then one broadcast request could satisfy all of them.

After nearly a year of uninterrupted operation, AO-16 suffered an onboard software crash on July 26, 1992. The crash was caused by the interaction between the spacecraft's software and user-written ground-station software. Of course, if there was a single, "factory-supplied" ground-station program, these types of software failures would be much less likely. However, a unique aspect of the Amateur Satellite Service is to allow users, who are so inclined, to write their own ground-station software.

AO-16 returned to operation quickly, but the file server didn't operate again until October 16, 1992. The intervening time was used to run engineering tests and ready a new version of the file-server software with enhanced abilities for broadcast mode. The most important of these new features was transmission of directory information in broadcast mode and cooperation between the satellite's and ground station's software to automatically fill holes in broadcast files and directories. The software providing these new features operated continuously from the time of its installation in October 1992 until September 1994, when we had to do a software reload. Since the September 1994 reload, up to the time of this writing, AO-16 has been in continuous service. Although the timeline has been slightly different, a similar progression of software installation has occurred on LO-19.

Details of typical configurations for ground-station equipment used to access AO-16 and LO-19 have been published [Diersing and Jones, 1992; Diersing, 1993a], as have descriptions of the software required to access the satellite's file system [Diersing, 1993b, 1993c]. I've included Fig. 13.3-4 to show a typical display on a ground-station computer for AO-16.

The screen in Fig. 13.3-4 consists of five major parts. Progressing from left to right, the bottom line shows the number of holes in the current ground station computer directory, that file 0x3e90d is being automatically downloaded, downlink data capture rate is 84% of the maximum possible, the total number of bytes received, the number of directory bytes received, and the number of file bytes received. The group of lines beginning with "PB:" shows the callsigns of stations that have made broadcast requests. Station callsigns followed by "\D" have made requests for directory entries while others have made file requests. The upper left part of the screen gives information about files being downloaded and the upper right part tells what type of broadcast information is being received. In this example, file 0x3e9e6 has 23 holes and is 42% complete after having received 57,543 bytes. The middle part of the screen (with the white background) shows various other informational messages that have been transmitted but are not parts of files or directory entries.

Although Fig. 13.3-4 shows a Microsoft Windows™ version of a user ground station program, an MS-DOS version was the first to be made available. There is now a version for Linux X-Windows available and a version for IBM OS/2 is under development.

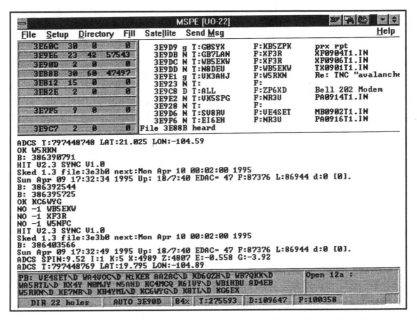

Fig. 13.3-4. Display from a Ground-Station Computer While Receiving Data from the Satellite Downlink.

Activity log files are generated by the file server daily. These activity logs can be downloaded and processed to extract usage statistics of interest. The data presented here from AO-16's activity log was collected over 3 years. It clearly shows the effects of progressing from the connected mode of operation to the broadcast mode.

Table 13.3-8 gives statistics for the broadcast mode starting when we installed the file-server software with enhanced broadcast mode. Note that the total transmitted byte count for 1993 is about 650 MB. At 1200 bps, about 4.75 GB could be transmitted in a year. Consequently, 650 MB uses only about 15% of the downlink, excluding HDLC overhead, telemetry transmissions, and other types of downlink data. Of course, much of the time AO-16's footprint doesn't include any populated areas, so we can't use it 100%. On the other hand, effective use would be higher than 15% if we could estimate how many stations in the footprint are using the same

TABLE 13.3-8. Broadcast-Mode Statistics for AO-16.

	1992 from 10–17	1993 complete	1994 through 5–31
Average Transactions (per day)			
Start + Fill Requests	794	646	727
Directory Requests	259	268	380
Average Byte Count (per day)			
File Bytes	1,860,244	1,260,069	1,154,442
Directory Bytes	419,080	501,573	653,423
Average Transactions (per month)			
Start+Fill Requests	20,104	19,649	21,942
Directory Requests	6,551	8,142	11,463
Average Byte Count (per month)			
File Bytes	47,126,172	38,327,111	34,864,153
Directory Bytes	10,616,681	15,256,183	19,733,372
Total Transactions[1]			
Start+Fill Requests	60,313	235,792	109,709
Directory Requests	19,653	97,698	57,313
Total Byte Count[1]			
File Bytes	141,378,516	459,925,333	174,320,760
Directory Bytes	31,850,043	183,074,192	98,666,858

[1] Note that 1992 and 1994 totals are not for an entire year

broad-cast data. Remember that many stations can be using the broadcast-mode data as a result of another station's request for a needed file or directory.

Table 13.3-9 shows the cash expenditures of AMSAT-NA for building and launching AO-16. It is interesting to note that one of the largest expenditures was for travel because of unplanned delays between delivery of the satellites to the launch facility and the time of launch. Another item of note is the cost to rent facilities. For small satellites and few people we don't need large integration facilities. In the case of the Microsats the laboratory was nothing more than a rented 2,000 ft^2 of office space in a multi-unit, single-story office building. Work space was a single room with a clean room built at one end.

TABLE 13.3-9. Itemized Project Costs for AMSAT-NA expenditures on AO-16 (FY89$).

Components	$14,883	Telephone	$13,822
Subcontracts	$16,995	Electronic mail	$13,531
Non-recurring engineering	$21,422	Travel	$38,028
Salaries	$3,070	Printing	$1,530
Equipment rental	$123	Postage	$4,847
Facilities rental	$3,206	Supplies	$1,366
Share of launch costs	$20,352	Photography	$658
License fees	$1,023	Advertising	$300
Liability insurance	$1,253	Accounting	$917
Other insurance	$262	Miscellaneous	$2,685
Documentation	$2,675	TOTAL	$162,948

When reviewing the costs shown in Table 13.3-9 remember that a project like AO-16 receives many different kinds of donations. The operating-system software was donated because the project was non-profit and some of its developers are amateur radio satellite enthusiasts. The application software was designed, written, and donated by the radio amateur software team supporting the project. Five different companies donated about $35,000 worth of equipment—including computers, radios, modems, and *Terminal Node Controllers*—that was first used for testing the spacecraft after integration and later used for command and control.

AMSAT-OSCAR-13

While people from AMSAT organizations outside the U.S. did some of the work on AO-16 and the other Microsats, AMSAT-NA members did most of the design, construction, and software development. In contrast, the largest part of the work on AO-13 was done by AMSAT-DL (Germany) and AMSAT-NA personnel with contributions from members of AMSAT groups in still other countries including Canada and Hungary.

Karl Meinzer of AMSAT-DL was head of the AO-13 project and Jan King of AMSAT-NA was project manager. AMSAT-DL provided all of the communications transponders and beacons except the Mode-S transponder. They also supplied the command receivers, demodulator, telecommand system, flight software, the fuel tank and its mounting fittings, the mounting for the 400 N motor, Sun and Earth sensors, separation timer, and documentation. AMSAT-DL was also responsible for spin balance and vibration testing, pressure and leakage testing of the fuel tank, as well as launch contract negotiations and management of the launch campaign. The Sensor Electronics Unit (SEU), separation interface, and the solar arrays were spares left from the AO-10 project. The cash expenditures by AMSAT-DL were approximately 1,300,000 DM (~1987) which includes the launch fee of 378,000 DM.

AMSAT-NA was responsible for the Mode-S transponder, thermal design, liquid propulsion system and supporting electronics, batteries, nutation dampers, mechanical and electrical integration, integration facilities and management, mechanical parts, refurbishing the spaceframe which was a spare from the Phase 3-A project, and thermal vacuum testing. The cash expenditures by AMSAT-NA are given in Table 13.3-10. The expenses listed reflect the fact that AMSAT-NA uses salaried personnel during critical phases of large projects such as AO-13.

TABLE 13.3-10. Itemized Project Costs for AO-13. AMSAT-NA Share Only (FY89 $).

Components	$12,617	Printing	$101
Subcontracts	$1,710	Postage	$8,523
Salaries	$103,334	Supplies	$2,436
Equipment rental	$587	Photography	$829
Launch Insurance	$14,000	Capital equipment	$4,556
Telephone	$14,547	Command stations	$8,490
Electronic mail	$232	Miscellaneous	$1,853
Travel	$42,093	TOTAL	$215,908[1]

[1]The AO-13 mission saved a lot of money by using spares developed for AO-10 and other previous missions. For example, the spaceframe itself was a spare, as were some of the transponders.

We first used the tri-star shaped structure on Phase 3-A and built two spaceframes. When Phase 3-B (AO-10) replaced Phase 3-A, we used essentially the same structure design and built a new spaceframe. The spare Phase 3-A spaceframe was used for Phase 3-C (AO-13). The cylindrical center section was sized to house the spherical solid rocket motor that was available for the first mission. The center section became the fuel tank for Phase 3-B and Phase 3-C. The spacecraft is spin-stabilized with the spin axis coincident with the center line of the central cylinder. The tri-star shape has proven to be an excellent design both structurally and in optimizing the illumination of the six solar arrays.

Attitude control with accurate pointing capability was essential during the motor burn and, later, for pointing the high-gain antennas toward Earth at apogee. For attitude control we use magnetic torquing against the Earth's magnetic field and Sun sensors for reference. An onboard computer, the *Integrated Housekeeping Unit*, has among its principal responsibilities taking data from the sensors, comparing the readings with three navigational-reference systems resident in memory, determining the spacecraft's attitude, and activating the torquing magnets contained in the three spacecraft arms to move the spin axis to the desired orientation. While this system doesn't permit rapid change experience has shown that it can effectively spin up the spacecraft from essentially no rotation to the 40–60 rpm necessary for motor burns and point the spacecraft with the necessary precision.

In addition to controlling spin rate and attitude, this unit oversees a wide range of housekeeping duties that include sequencing the operating schedules for the communications transponders; storing ground commands for later implementation; controlling the charge rate from the solar arrays so as to keep the NiCd storage batteries at the proper charge level; and most importantly from the propulsion standpoint, receiving, verifying, storing, and issuing motor-firing commands at the specified time and for the proper duration.

The Phase 3-A Propulsion System*

The first Phase 3 satellite propulsion system was designed to use an available spherical solid rocket motor. Although the Phase 3-A satellite never achieved orbit, I'll discuss some details of its propulsion system design because it's the first step toward the successful system aboard AO-13 (Phase 3-C).

The Phase 3-A propulsion system was designed around the Thiokol TE-345, a 13-inch spherical, solid rocket motor. This motor, originally designed as a retro rocket for the Gemini spacecraft, was donated to AMSAT. The electronics to start the ignition sequence were in the *Motor Ignition Unit* module. The electronics in the module controlled the ignition sequence and generated the high-current, low-voltage pulse necessary to ensure igniter performance. The cabling associated with the ignition unit included a heavy-duty relay that normally remains open in the igniter-firing circuit, interconnection with the housekeeping computer, and a safe/arm plug that completely isolated firing circuits before launch. Except for the motor and its two igniters, all other components used in the Phase 3-A propulsion system—cabling, connectors, and small electronic components—were readily available from commercial sources. The

---

* I've adapted this discussion of the satellite propulsion systems for the Phase 3 series from Daniels [1987, 1988].

main consideration was ensuring absolute reliability of the design and components to make sure the motor would ignite when commanded and not before.

The Phase 3-B Propulsion System

After we lost the Phase 3-A spacecraft we immediately started designing and building a replacement spacecraft. We kept the Phase 3-A satellite's basic configuration and design but incorporated a number of improvements in the communications modules. Beyond these, the most significant change was in the propulsion system. With improvements adding weight to the spacecraft and a launch opportunity that would provide a lower initial orbit inclination, the solid motor couldn't provide enough impulse to reach the desired orbit.

A breakthrough occurred when the German team, AMSAT-DL, was able to arrange for the aerospace firm, Messerschmitt-Boelkow-Blohm, to donate a 400 N bipropellant motor with associated hardware and ground-handling support. Calculations clearly showed that we could greatly improve total impulse and that the capability for multiple motor burns added significant mission flexibility. On the other side of the ledger, the propulsion system would be more complex and potentially more hazardous than the solid motor used in the previous design.

The configuration of the Phase 3-B spacecraft is almost identical to the AO-13 (Phase 3-C) design. The redesign used the total volume of the central cylinder for a two-chambered propellant tank, which was pressure fed by helium through a plumbing system designated the *Propellant Flow Assembly*. Because the propellants were hypergolic, simply opening the motor valves ignited them.

The motor provided by Messerschmitt-Boelkow-Blohm had originally been developed as a vernier engine for the Europa European launch vehicle and was designed to give about 400 N (90 lb) of thrust. The original propellants for the motor were nitrogen tetroxide (N_2O_4) and aerozine 50 (AZ-50). Following cancellation of the Europa project, the motor was selected as the apogee motor for the Symphonie communications satellite—a joint German and French project—with modification for use of monomethyl hydrazine (MMH) instead of AZ-50. In discussions with the Messerschmitt-Boelkow-Blohm engineering staff, we decided to further modify the AMSAT engine to permit use of unsymmetrical dimethyl hydrazine (UDMH) instead of AZ-50 because this propellant was more readily available at the launch site.

The propellants used raised major problems of material compatibility. Not only is N_2O_4 extremely active and will readily attack most organic material but it is also easily contaminated. UDMH, while not as corrosive, is equally subject to contamination. Both are extremely toxic and can't be handled without using protective suits and air filtering. To compound the problem, the close tolerance on valves and pressure regulators meant that the systems for transporting both the gas and fluid flow were extremely vulnerable to contamination. Although Messerschmitt-Boelkow-Blohm had agreed to provide certain key hardware items, such as the pressure regulator, explosive valves, and an assembly of check valves, AMSAT had to buy the remaining components and design and assemble the propulsion system. Materials that were to contact the propellants were restricted to selected types of stainless steel and aluminum alloys with teflon for seal material.

At first, we planned to make the two-chambered propellant tank from a special alloy of stainless steel. When we couldn't form this material properly, we decided to design an aluminum tank. Most importantly, we had to provide an intermediate bulkhead that would absolutely assure us that the two propellants couldn't come into contact through cracks or poorly welded seams. The tank was constructed in three sections milled from thick billets of aluminum alloy so that the intermediate bulkhead was an integral part of the center section and the welded seams were isolated to the individual propellant tanks. Thus, we avoid seams that, upon failure, would allow propellants to mix. Drain points were located so the effects of the spacecraft spin and propulsion thrust would keep the tube mouths covered by propellants. The tank took advantage of the full volume of the central cylinder with only enough space at the top and bottom for tubing and cable access.

Helium pressure forced propellants to flow to the motor. In order to provide enough helium gas to ensure total displacement of the propellants, we had to store the helium at a very high pressure. Because a commercial bottle wasn't available in a usable configuration and at an affordable cost, we developed an inexpensive but effective alternative. The AMSAT-DL group located a metal bottle that was inexpensive and commercially produced as a fire-extinguisher bottle, but this bottle was rated to a pressure well below the requirement of 400 bar (6,000 psi) for the helium-storage bottle. The solution was to encase the length of the bottle with carbon/epoxy fiber windings and thereby increase the bottle's burst pressure to a demonstrated 1132 bar.

The electronics module for starting and controlling the motor burn for the Phase 3-B spacecraft was designated the *Liquid Ignition Unit* to differentiate it from the Phase 3-A MIU. Its purpose was similar to that of the MIU but with one important additional requirement—it had to control the motor's burn time. Thus, we designed the LIU to count in 50 ms intervals so it could meter the duration of an individual burn.

Except for the electronics package, tankage, and interconnecting tubing, the remaining components of the propulsion system were assembled on a single mounting plate. The propellant-flow assembly was a consolidated module containing all fill valves, pressure regulation, filtering, and check valves to prevent backflow in the pressurization lines. It also included three normally closed explosive valves to isolate the high-pressure-helium tank and the two sections of propellant tank until the initial firing sequence was commanded. Also included in the assembly design was a precision pressure transducer to measure the low-pressure side of the system and transmit its measurements by telemetry to the ground. We took a much more coarse measurement of the high-pressure-helium bottle using a strain gauge attached to the wall of the bottle.

In designing the propellant-flow assembly, we tried hard to reduce the number of wrench-tightened fittings and we used welded connections wherever possible to reduce the possibility of leaks. In addition, we pressure- and leak-tested all complete subassemblies. The ignition sequence, following receipt of validated keys and commands from the integrated housekeeping unit, involved actuating the explosive valves in the proper sequence, allowing the system to stabilize, and then open the motor propellant valves. Valve operation in the 400 N thruster begins electrically but continues through helium pressure. Following ignition the LIU counted the commanded time intervals and then started a shutdown sequence. The Phase 3-B mission profile included two major burns to achieve the desired orbit with enough residual fuel for trimming maneuvers, if required.

The Phase 3-B spacecraft was successfully launched on an Ariane rocket from the Kourou launch site on June 16, 1983. Unfortunately, an unplanned collision between the spacecraft and the third stage damaged an antenna, changed the spacecraft's orientation, reduced the spin rate, and reversed the spin direction. But we reoriented the spacecraft, spun it up to the proper rate using the magnetic torquing system, and commanded the propulsion system into its first burn. Telemetry following the successful burn indicated some deviation from the expected burn time, but otherwise the system operated normally. The ground stations then started reorienting the spacecraft in preparation for the second burn.

During the somewhat lengthy reorientation, telemetry showed a steady drop in helium pressure. By the time the spacecraft was reoriented for the second burn, the helium pressure had fallen below the level necessary to actuate the motor valves. Thus, what was originally planned to be an intermediate orbit became the final orbit for the mission. We believe the problem developed from a leak in the seal of the high-pressure-helium bottle triggered by the rapid cooling of the bottle during the first burn. The orbit achieved proved to be very useful if not as effective as the one desired. It provided communications worldwide for about 3 years until radiation-induced memory errors caused the onboard computer to stop. Even with the onboard computer dead, AO-10 is still providing seasonal communications when the solar illumination generates enough power for transponder operation.

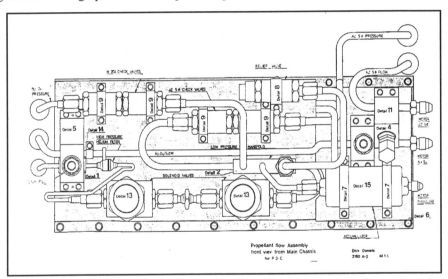

Fig. 13.3-5. Phase 3-C/AO-13's Propellant-Flow Assembly. (Drawing courtesy of AMSAT-DL.)

The Phase 3-C Propulsion System

Not long after AO-10 began orbital operations we started developing the Phase 3-C spacecraft. As with the Phase 3-B system, we improved the design of the communications subsystems. For the propulsion system we decided to continue with the 400 N thruster. Messerschmitt-Boelkow-Blohm (MBB) agreed to give us another motor, but they could provide only a few fill- and drain-valves because they had used other supporting equipment during the previous mission. The challenge, therefore, in developing the Phase 3-C propulsion system was to redesign the propellant-flow assembly so it could use available and affordable hardware.

Except for the 400 N thruster and three specialized fill- and drain-valves, we designed the entire system for commercially available and relatively inexpensive hardware. The design is a bare-bones approach emphasizing simplicity and economy. We reused the propellant- and helium-tank designs from Phase 3-B, except for modifying the propellant tank for AZ-50 rather than UDMH as the fuel and changing the helium tank to avoid the leak experienced with the Phase 3-B system.

Significant changes were made in the design of the assembly shown in Fig. 13.3-5. First, we replaced the normally closed explosive valve that had isolated the high-pressure-helium bottle with two electrically operated valves whose quality promised low leakage. Further, when the system is actuated, one of the valves stays open to start the flow of helium; the other operates in a feedback loop with the pressure transducer to regulate pressure in a "bang-bang" mode. In this operation the valve opens when pressure falls below a set level and closes when pressure is correct. Should the pressure rise above the expected level the role of the valves reverses; the suspect regulator valve now stays open and the other operates as a regulator.

We provided a further safety feature by adding a relief valve set to operate if the propellant-tank pressure rises significantly above the operating pressure. Anticipating undesirable pressure fluctuations from the pulsing of the regulating valve, we incorporated an accumulator. We also used redundant, commercial check valves to prevent backflow from the propellant tanks instead of the explosive valves and the specialized check-valve assembly used in the Phase 3-B design. The assembly looks very different but operates just like its predecessor. Figure 13.3-6 gives a schematic diagram of the entire propulsion system.

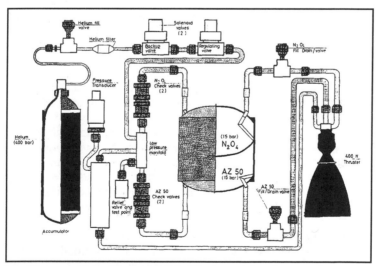

Fig. 13.3-6. Diagram of the Phase 3-C/AO-13's Propulsion System. (Drawing courtesy of AMSAT-DL.)

Redesigning the propellant-flow assembly meant we had to redesign the LIU. We kept the functions for verifying firing commands and timing the motor burn, but deleted the circuitry required to fire the explosive valves and incorporated new circuitry to control the pressure-regulating valves.

The launch and deployment of AMSAT Phase 3-C on Ariane mission V-22 occurred as planned on June 15, 1988. We launched at 11:19:04.33 UTC—6 minutes inside the first launch window. Twenty minutes into the mission the first of three satellites on the Ariane rocket, Meteosat, deployed. Phase 3-C followed 5 seconds later remaining within its canister. (See Fig. 13.3-3.) Panamsat was ejected from the SPELDA after the Phase 3-C deployment. Phase 3-C remained inside its carrying structure for some 60 minutes before it was released at 12:39:04 UTC. The beacon transmitter began sending telemetry at 14:03:38 UTC while the satellite was over the Indian Ocean. AMSAT Phase 3-C had officially become AMSAT-OSCAR-13 (AO-13) [Riportella, 1988b].

AO-13's initial orbit had an apogee of 36,000 km, perigee of 200 km, and inclination of 10 deg. The team discussed whether to use two or three motor burns to reach the final orbit but eventually used only two burns [Riportella, 1988a].

The first motor burn took place on June 22, 1988 at 18:57 UTC. Upon ejection from Ariane, AO-13's attitude was 269/-20 (Alon/Alat deg) in the Bahn (orbit plane) coordinate system, and the spin rate was about 5 rpm in the direction opposite to the one desired. Spin direction reversal, spin rate change, and the attitude adjustment operations were carried out simultaneously. The spin rate stabilized at 20 rpm after 8 perigees of the magnetic torquing while attitude adjustment to 75/-56 continued for five more perigees. Once we determined the attitude to be correct, we increased the spin rate to 26 rpm and started the first motor burn on orbit 16. The 50-second burn had a ΔV of 150.1 m/s. It raised perigee to 1,100 km and the inclination to 14 deg. The ΔV consumed was about 10 percent of capacity [Miller, 1995a, 1995b].

Before we fired the motor for the second (and last) time, we spun up AO-13 to 60 rpm. The gyroscopic effect from the higher spin helped stabilize the spacecraft during the longer burn and caused centrifugal force to empty the fuel tanks as completely as possible. We adjusted AO-13's attitude for the second motor burn to 302/-71 deg. The burn started at 21:03:31 UTC on July 6, 1988. It was a burn-to-depletion operation lasting 5m 41s with a ΔV of 1314 m/s. The result of the burn was a change in orbital inclination to 58 deg and a change in perigee to 2,500 km [Miller, 1995a]. AO-13's propulsion system had operated as planned to obtain the desired final orbit.

Radio ranging using one of AO-13's transponders and the engineering beacon determined the initial orbit, the intermediate orbit before the final motor burn, and the final orbit. AMSAT got some orbit-determination information from other tracking stations, but we had to use our own procedures because it wasn't easy for them to locate such a small satellite without knowing about where to look in the first place.

Communications Transponders

AO-13's primary mission is providing radio amateurs with access to analog communications transponders. There are a total of five transponders—four analog and one digital. One of the two operational transponders repeats a segment of the 70 cm band to the 2 m band (Mode B). The other transponder can repeat 70 cm signals in the 13 cm band (Mode S). A detailed list of transponder frequencies is in Table 13.3-11.

AO-13 also has two other analog transponders, which share a common downlink, but because a transmitter component failed after nearly 5 years of operation, they are inoperative. These were Mode J, the opposite translation of Mode B—2 m band to 70 cm band and Mode L—23 cm band to 70 cm band. For unknown reasons, the RUDAK digital transponder never worked.

The communications transponders in AMSAT's Phase 3 use a design technique that makes them convert DC to RF power more efficiently than other typical linear transponders. The technique used is known as HELAPS—*High Efficiency Linear Amplification through Parametric Synthesis*—and was developed by Karl Meinzer of AMSAT-DL [Meinzer, 1974]. The HELAPS design uses the *Envelope Elimination and Restoration* (EER) concept developed by L. Kahn [1952]. HELAPS was first used on the AMSAT-OSCAR-7 satellite launched in 1974. The design evolved further during the development of the Phase 3 series [King, 1993b] to include other techniques for increased efficiency [Doherty, 1936]. Although it's beyond the scope of this chapter to completely discuss the AO-13 transponder design, it's important to note that the typical transmitter power amplifier on an AO-13 transponder will exhibit an efficiency of at least 40 percent, with anywhere from 25 to 100% drive power. Such efficiencies are particularly important on a small satellite, which has limited space for batteries and limited surface area for solar arrays.

Because AO-13 provides only analog communication transponders, it doesn't keep activity logs like those on AO-16, so it's hard to estimate actual use. But even with the required seasonal reorientations of the satellite to get the most solar power, AO-13's operations are quite routine. Transponder schedules are published in the printed and electronic media, and many operators obtain the schedules directly from the AO-13's digital telemetry beacon at 400 bps.

TABLE 13.3-11.   **Communications Transponders on AO-13.**

Transponder No. 1	• Type: Linear, inverting • Uplink passband: 435.423-435.573 MHz • Downlink passband: 145.975-145.825 MHz • Bandwidth: 150 KHz • Output power: 50 WPEP, 12.5 W average
Transponder No. 2	Downlinks no longer operating probably due to a component failure in the 435 MHz transmitter at 5 years into the mission. • Type: Linear, inverting • Uplink No. 1 passband: 1269.641–1269.351 MHz • Uplink No. 2 passband 144.423–144.473 MHz • Downlink No. 1 passband: 435.715–436.005 MHz • Downlink No. 2 passband: 435.900–435.940 MHz • Uplink No. 1 bandwidth: 290 KHz • Output power: 50 WPEP, 12.5 W average
Transponder No. 3	• Type: Hard-limiting, non-inverting • Uplink passband: 435.602–435.638 MHz • Downlink passband: 2400.711–2400.747 MHz • Bandwidth: 36 KHz • Output power: 1.25 W continuous
Transponder No. 4	Never operated. • Type: Digital • Uplink: 1269.710 MHz, 2400 bps, BPSK • Downlink: 435.677 MHz, 400 bps, BPSK • Uplink No. 1 bandwidth: 290 KHz • Output power: 6 W

*On-Orbit Operations**

AO-13 requires a certain amount of attention from ground-control stations, mainly to adjust the satellite's attitude and transponder operating schedule in response to seasonal variations in the Sun angle and eclipse conditions. The ground-command network consists of three stations—one each in England, Germany, and Australia. AMSAT volunteers operate the stations, using less total time than one full-time equivalent. But note that this time estimate applies to current routine operations. Certain operations require much more time—immediately after launch, during preparation for motor firing, and while developing software to manage the satellite.

The volunteer's dedication and quality of the operations software have made it possible to manage AO-13's orbital attitude and schedule transponder operations. AO-13's management software, developed by James Miller [1993], consists of about 12,000 lines of BASIC code and runs on an Acorn RiscPC. We've recently converted the management software to run on PCs in preparation for the launch of AMSAT's Phase 3-D satellite in 1996. Table 13.3-12 summarizes the programs comprising AO-13's management system. It should be noted that prior to the development of the suite of programs described in this section, ground station software developed by Meinzer [1979] and written in IPS was used. This initial ground station software ran on the Atari 400, 800, and 800XL series machines and was utilized during the motor burns for orbit change maneuvers described previously.

TABLE 13.3-12.   Software Programs Supporting On-Orbit Operations and What They Do.

Program	Function
ILLPLAN	Determines Sun angle for various spin-axis orientations and time periods.
ATTHIST	Shows week-by-week summary of Sun angles and percent illumination given a certain initial orientation.
SQPLOT	Produces plot of squint angle versus mean anomaly to help schedule transponders.
MAGSIM	Simulates moving the spacecraft from one attitude to another. Generates the command sequence to accomplish the change.
ECLIPSE	Determines when the Earth blocks the Sun's illumination of the satellite.
MOONECL	Determines when the Moon blocks the Sun's illumination of the satellite.
SMOOTH13	Remove short-term variations from a group of Keplerian element sets supplied by NASA.
ATTPLOT	Produce graphical display of spacecraft attitude based on telemetered data from Sun and Earth sensors.
ATTFIX	Produce least-square numerical solution to the attitude-determination problem.

---

*This description of on-orbit operations is adapted from [Miller, 1993].

The first step in determining a new transponder schedule is to determine the Sun angle given various spacecraft attitudes within the orbit plane. The program *ILLPLAN* generates the required table of Sun angles versus spin-axis direction. We can scroll ILLPLAN back and forth in 7-day increments to observe long-term effects. The program considers a Sun angle greater than ±45 deg to be a "no go" condition. ILLPLAN also shows when the sun angle is greater than ±60 deg because, beyond these values, the sun would be out of the sensor's field of view. Once we've identified a candidate for spin-axis direction, the program *ATTHIST* can display a week-by-week summary of that attitude's effects.

Although we can use ILLPLAN and ATTHIST to arrive at an acceptable spin-axis orientation that will generate sufficient power, a usable transponder schedule must also take into account the squint angle—the angle between the spin axis (the direction in which the spacecraft's antennas point) and a line between the satellite and the user's Earth station. The program *SQPLOT* produces graphs of the squint angle versus mean anomaly as shown in Fig. 13.3-7. Given a typical ground station's latitude and knowing the beamwidths of various antennas, we can develop a transponder schedule. For example, transponders having antennas with the highest gain (smallest beamwidth) would activate when the squint angle is a minimum. Whether or not the minimum squint angle occurs at apogee depends, of course, on whether or not we've oriented the spin axis out of the orbit plane to achieve the best Sun angle.

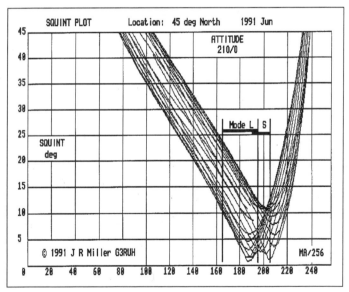

Fig. 13.3-7. Plot of Squint Angle as a Function of Mean Anomaly from Program SQPLOT.
Used to plan transponder schedule.

After we've identified a new spacecraft attitude, *MAGSIM* simulates moving from the current spin rate and attitude to the desired spin rate and attitude. (See Fig. 13.3-8 for a sample of MAGSIM's graphical output. The MAGSIM program also generates the command sequence for the required reorientation.) The spacecraft's attitude is altered by energizing torquing coils in each of its three arms. The magnetic torquing is carried out under control of the IHU. We typically do the torquing within ±10 deg of perigee and can change the orientation about 5 deg in one perigee pass.

Several other programs help us manage AO-13. *ECLIPSE* shows when the Earth will block the Sun's illumination. *MOONECL* does the same thing for cases in which the Moon blocks the Sun's illumination. Either of these eclipse conditions may require temporary changes to the transponder's operating schedule depending on the duration of the eclipse. *SMOOTH13* is used to process a group of Keplerian elements issued by NASA in order to remove short-term variations. It thus produces an averaged element set for use by AO-13's management programs.

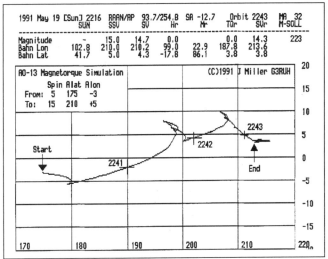

Fig. 13.3-8. Plot of Magnetorque Simulation from Program MAGSIM.

The programs mentioned thus far generate new satellite-attitude information and transponder schedules as required by seasonal changes in solar illumination. Once an attitude-change maneuver is over, AO-13's telemetry is monitored to verify the effects of the re-orientation. Specifically, *ATTPLOT* processes several days worth of date- and time-stamped data from Earth and Sun sensors. It then produces a graphical display of the spacecraft's attitude as in Fig. 13.3-9 and 13.3-10. *ATTFIX* uses the same input data as ATTPLOT, but instead of producing a graphical display, it numerically computes a least-squares solution to the attitude-determination problem

AO-13's suite of management software thus supports the complete cycle of operations from planning reorientation and changing transponder schedules, to generating the reorientation commands, to monitoring the effects of these commands.

The satellite's integrated housekeeping unit handles all command and control communications on the ground. The unit uses an RCA COSMAC 1802 microprocessor, 32 kbytes of random access memory and an instruction rate of 200 kips. The IHU runs the IPS (interpreter structure for processes) language designed by Karl Meinzer [1979]. Meinzer also developed the original version of the satellite-control software which the ground-control team later changed as required. Flight software with requirements for time-critical execution, such as clocks, analog multiplexer service, and telemetry buffering, is written in assembly language. Spacecraft-control

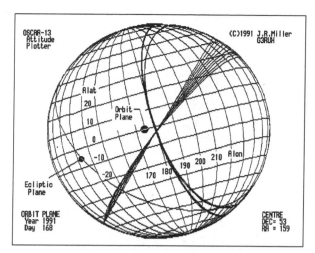

Fig. 13.3-9. Attitude-Determination Display Program ATTPLOT Produces after Processing Telemetry Data from Sun and Earth Sensors.

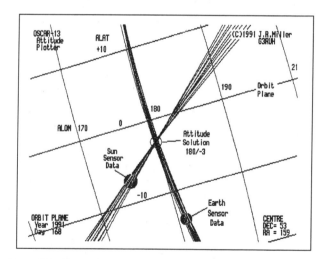

Fig. 13.3-10. Same Display as Fig. 12.1-11 Magnified Four Times.

routines are written in IPS and are compiled by the IPS interpreter as they are uploaded. The code that actually performs the spacecraft's various system services executes a chain of procedures: (0) Battery and watchdog time service; (1) Beacon sequencing; (2) Navigation; (3) Memory-error reporting; (4) Transponder scheduling; and (5) controlling the Mode-S transponder. Each process receives a certain time slice from the CPU before control passes to the next task in the chain.

The IHU and IPS software have proven to be very reliable. Since its 1988 launch, we've had to reload AO-13's onboard computer software only six times—four times because of operator error and twice for unexplained reasons (possibly operator errors or radiation-induced failures).

Commanding is usually done using the 24 cm (1269 MHz) uplink and the 13 cm (2401 MHz) downlink. In addition to the command stations many AO-13 users also monitor the beacon to decode telemetry and schedule information.

Due to the forces acting on its highly elliptical orbit, AO-13's perigee has been decreasing for some time. This was first noticed and studied by Kudielka [1990] and was later examined in more detail, using the best computer models available [Clark and Pavlis, 1991]. Current estimates are that AO-13 will reenter in late 1996 or early 1997 [Guelzow, 1992; Miller, 1992, 1994]. Figure 13.3-11 gives a graphical prediction of AO-13's reentry. Fortunately, AO-13's replacement—Phase 3-D—will be in orbit and operating before AO-13 meets its fate

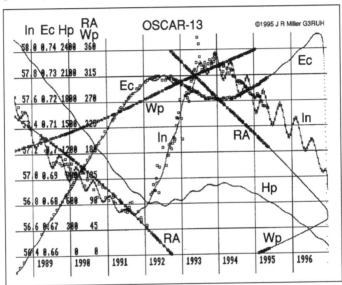

Fig. 13.3-11. Predicted Decay of AO-13's Orbit. IN=Inclination, EC=Eccentricity, Hp=Perigee Height, Wp=Argument of Perigee, RA= Right Ascension. Rectangles are NORAD 2-line Keplerian element values.

13.3.3 Ways to Reduce Cost

AMSAT-NA has built reliable, low-cost satellites in cooperation with other AMSAT organizations. I believe certain techniques can apply to projects in other sectors. I'll start with management structure and related personnel issues, but I'll include other issues, such as parts selection, because they're part of the overall project-management philosophy and are important to reducing cost.

One thing I can't discuss much is the nearly free labor AMSAT enjoys because of its volunteer, scientific, and educational nature. No commercial or governmental project can duplicate this labor savings so giving it a lot of attention here doesn't make sense! Of course AMSAT projects have used salaried people at critical phases, but they've kept such expenses to a minimum.

At the height of activity on a given project, we often see volunteers spending 8 to 10 hours per day on a project after 8 hours at their regular "day jobs." The hours-per-day logged on weekends can be even higher. We rarely record this labor, so we can't determine its actual dollar value. Of course, this stressful activity has burned out some volunteers. Some of them never participate in future projects, some reduce their level of involvement, and some return to their previous level of involvement after an appropriate rest period

Before I discuss specific cost-reduction issues, let's make clear what motivated development of the AMSAT philosophy in the first place. We're truly motivated to reduce costs when we have **no money** or much less than we'd spend if we did a similar project in the commercial sector. This lack of funds pushes us to try new, cost-effective techniques. The AMSAT philosophy continues to develop as we collect more information while applying and refining techniques. Refinement includes applying new technologies as soon as they are practical.

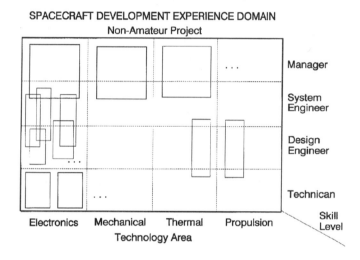

Fig. 13.3-12.   Mix of People on a Project for a Non-Amateur Satellite.

Management Structure

AMSAT has found it beneficial to use managers, engineers, and technicians from several disciplines in its satellite projects. Figures 13.3-12, 13.3-13 and 13.3-14 show mixtures of people in the context of skill level and technology area [King 1993a]. Smaller boxes indicate higher levels of specialization in a technology area and/or a skill level confined to a narrow range. Large boxes, on the other hand, indicate broader knowledge of technology areas and ability to function across a wide range of skills. Figure 13.3-12 illustrates a mix of people on a typical non-amateur project, for which the level of specialization is high. Moreover, people don't cross technology areas, so the project probably has some (probably expensive) procedure for interface control.

Figure 13.3-13 shows staffing for the AO-13 project; Fig. 13.3-14 shows the personnel structure on the Microsat project. The structure shown in Fig. 13.3-14, has one broadly experienced project manager with a couple of senior engineers covering multiple skill levels and technology domains. Technicians also cross technology areas

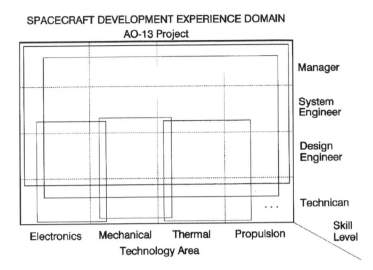

Fig. 13.3-13.   Mix of People for the AO-13 Project

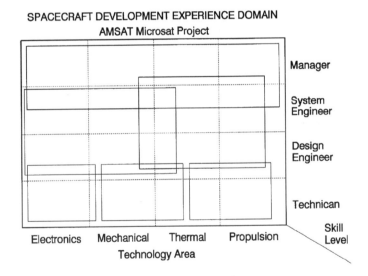

Fig. 13.3-14.   Best Mix of People for a Cost-Effective Satellite Project.

We have to consider two attributes of AMSAT people—motivation and skill level. Most people working on AMSAT projects aren't paid. Why, then, are they motivated to spend valuable time working on a satellite project? For the project manager it may be that a design concept could not come to fruition in any other way. For other participants there is a whole spectrum of possibilities. Perhaps the software designer wishes to take on the challenge of writing a reliable and fault-tolerant application for satellites. Maybe the technician has a strong philosophical attachment to one or more system design concepts or to applying the finished product. Possibly, a person derives satisfaction in working on something that will go into space. The point is that managers of non-amateur projects must choose a staff that is similarly motivated or create the motivation within the staff—probably some of both. When the staff isn't positively motivated system reliability and performance will suffer. Acceptable salary levels aren't always enough motivation to do quality work. This motivation comes about partly through training and partly through example. Management must first give the example and then choose people who can follow and propagate the example.

In amateur radio satellite projects, skill level of the participants means more than expertise in some required specialty. It means diversity of skill and the appropriate mixture of theory and practice. Many amateur radio operators, and not just those associated with satellite projects, started their hobby in grade school. So, by the time they reach the prime of their careers at age 40–50, they have 30–40 years of experience behind them. From these years of experience comes the ability to cross technology boundaries, to trade off cost and performance, to try innovative designs, to minimize failures, and to do what can't be done very easily—on a shoestring budget.

Parts Selection

The most important aspect of AMSAT's experience with parts is characterizing the in-orbit reliability of the lower MIL-HDBK-217F classes and unclassified parts.

Figure 13.3-15 shows a typical mixture of parts classification for non-amateur and the AMSAT Phase 3 and Microsat programs [King, 1993a]. I've already discussed the in-orbit problems and sub-system failures in the AO-13 and Microsat program, but recall that none of the failures has resulted in losing a mission. AMSAT's project management has observed certain "truths" [King, 1993a] about parts mixtures used in non-amateur programs

- The best parts available rarely fail
- Time-proven techniques command confidence
- Parts are not only reliable, but also have margin over the specified values

But,

- The best parts and techniques demand premium prices
- The schedule will always be long
- Using good parts can mask a poor design
- We don't know how lower-class and unclassified parts work in space

Having employed parts mixtures of the type shown for the Phase 3 and Microsat programs, AMSAT has found that:

- Good circuit design is more important than device technology
- We need a practical approach to reliability based on cost
- We can apply experience gained on much of the reliability-classification line to future projects

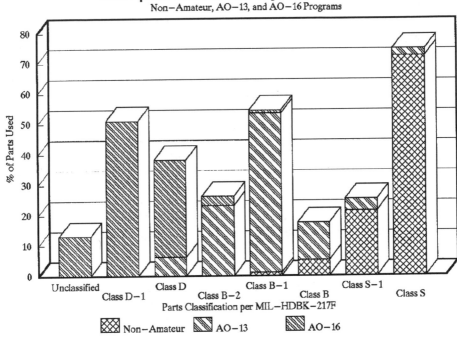

Component Reliability Classification
Non−Amateur, AO−13, and AO−16 Programs

Fig. 13.3-15. **Typical Mixture of Parts Classes for a Non-Amateur, AO-13, and AO-16 Satellite Projects.**

At the same time, the first use of components in flight poses risks, and the primary payload customer may be concerned with the parts choices made by the secondary payload customers. But appropriate testing before launch can largely mitigate these risks.

Radiation Issues

Parts reliability includes the question of radiation tolerance of components and systems, but AMSAT's experience differs from widely held opinion. We've found that radiation-hardness requirements are actually two to three times lower than industry practice. I'm not saying that AMSAT satellites haven't experienced radiation-induced failures. Indeed, AO-10's flight computer is inoperative because radiation destroyed its RAM. On the other hand, AO-13 has now been in a Molniya orbit for nearly seven years with no known radiation-induced failures.

AMSAT's experience with radiation issues [King, 1993a] has led to the following philosophy:

- Use radiation-hardened parts if they are available and affordable
- Use specially processed standard parts if they are available and affordable
- Try to use parts with gate geometries no smaller than 1.0 micron

- Use parts that are known to exhibit acceptable performance based on reliability experience
- Protect against memory problems by using EDAC and software memory wash
- Protect against corruption of the processor's setup table by using hardware-watchdog or fire-code methods
- Don't use more microprocessors than necessary
- Ignore the issue of single-event latchups

The AMSAT Microsats, which include AO-16 described earlier, clearly show that the AMSAT philosophy with respect to radiation issues is valid for low-cost spacecraft in low-Earth orbit. Each Microsat flight computer contains 453 integrated circuits and none are radiation-hardened. Only the boot ROM is MIL-STD-883. Considering the four Microsats together, more than 1,800 ICs are spread among four flight computers with more than 20 orbit-years (5 years per satellite) of operation and no identifiable radiation-induced failures. Perhaps one or more of the software crashes of unknown origin were radiation-induced, but such software failures have been so infrequent they have been hard to characterize. Single-event upsets have occurred in the various computer memories, but hardware EDAC and software memory wash have handled them as already described. When comparing the radiation environments of AO-13 and AO-16, remember that AO-16's orbit is beneath the Van Allen radiation belts whereas AO-13 passes through these radiation belts twice each orbit. Thus, AO-16 operates in a much more benign radiation environment than does AO-13.

Apart from the radiation-tolerance experience with ICs, AMSAT has found that solar arrays have degraded more slowly than predicted by industry-standard models.

Software Development

Software development costs have been virtually zero in all AMSAT projects to date, even though the software in the satellite and at the ground station has become increasingly more complex. As mentioned in the AO-16 case study, AMSAT-NA Microsats use Quadron's qCF multitasking operating system. Because several of the founders of Quadron were radio amateurs, they donated the license to use qCF. Even though qCF is a proprietary operating system, the tasks that run under its control aren't. Consequently, programs such as the housekeeping task and file-server system are available for other amateur radio (non-commercial) satellite projects. Software sharing occurs between AMSAT-NA and organizations with similar goals and projects. As an example, in exchange for the AX.25 communications handler and I/O drivers for the 72001 communications chips, the University of Surrey's UoSAT program provided the file system task and parts of the message and file-server systems [King, et al., 1990].

Another important aspect of software development illustrated by the AO-16 case study is that all the application software doesn't have be ready before launch. The evolution from digital repeater, to connected-mode file server, to broadcast-mode file server took place after AO-16 was in orbit [Diersing, 1994a]. Of course, the application will dictate whether software development can occur after launch.

The ability to develop code at distributed locations was very important during the Microsat program. For AMSAT projects, this ability will continue to keep software-development costs low. As long as a fairly inexpensive development environment allows software engineers to work independently someone who can write the needed code will volunteer for the task. With global telecommunications so much more available now than when the AO-16 project began, distributed software development and testing is all the more feasible. Even when someone must leave a software-development activity in favor of more pressing matters other AMSAT volunteers will continue the work in progress [White, 1993; Diersing, 1994b].

The case studies have also shown how ground-station software develops at little or no cost. In the case of AO-16, the University of Surrey charged nothing for the program suite consisting of PB, PG, PHS, and PFHADD. Because the complete specification for Pacsat protocols was published shortly after the satellites using it were in orbit people who wanted to write their own ground station software could do so. One example of user-developed software for accessing Pacsat is the WiSP program suite for Windows™. It's interesting to note that a Linux/X-Windows version of ground-station software for Pacsat actually appeared on the scene before the Windows™ version was available. For AO-13, I've already discussed Miller's comprehensive suite of programs to manage orbital attitude and schedule transponders.

It seems safe to conclude from the foregoing examples that people involved with amateur radio satellite projects are dedicated enough to ensure that software onboard satellites and at ground stations will continue to develop at little cost. Also, because the Amateur Radio Service operates in the public domain, things like protocol specifications are widely available at low cost. Of course software designers often have as much expertise on hardware as the hardware designers do. In fact, in many instances, software engineers feel just as much at home with oscilloscope probe in hand as they do sitting at the CRT and keyboard of the computer.

Cooperation with Educational Institutions

AMSAT-NA has sought to establish partnerships with educational institutions to assist in some of its satellite projects. For example, we have a productive relationship with the Center for AeroSpace Technology at Weber State University in Ogden, Utah [Jansson, 1989].

The concept of building low-cost satellites is not new at Weber State [Twiggs and Reister, 1992]. In April 1985, Nusat I was launched from a get-away-special canister on Challenger, the NASA orbiter. Nusat I operated nominally for 20 months until it burned up at re-entry. The cash outlay for Nusat I was less than $20,000 [Clapp, 1989]. In 1988, Weber State agreed to manufacture the major mechanical components for AMSAT-NA's Microsat project. Weber State owns and operates WO-18, one of the four satellites built as part of the Microsat project and a Weber State team designed and built its Earth-imaging experiment.

About the same time the Microsat project was under way, AMSAT-NA was investigating the feasibility of building a geostationary spacecraft called Phase IV. We enlisted Weber State's help for this project and completed a prototype structure in June, 1990. Additional work on antenna structures and deployment techniques was complete by spring 1991. Even though work on the Phase IV project ended because the amateur radio community didn't have enough money, the work on the Microsat and Phase IV projects has refined management interfaces and procedures between Weber and AMSAT-NA.

Weber State is strongly contributing to AMSAT's Phase 3-D satellite now under construction, by building the entire structure for the flight-model spacecraft, the electronics-module boxes, and the cylindrical section that will enclose and support the satellite on the launch vehicle [Phase 3-D, 1994].

Current Trends

While AMSAT has had to develop special philosophies and procedures to succeed, other organizations are seeking similar mixtures of fiscal, project, and personnel-management procedures. A recent article [Crabbs, 1995] states:

> As it was at the outset, the future of the U.S. space program—civil, military, and commercial—lies in the hands and minds of the current generation of under-graduate, graduate and post doctoral students. If these people are not trained correctly, do not have appropriate role models, and do not develop a passion for doing space research, the United State's program will fall in decline and we will become a second-rate space nation...
>
> Launching 20 small satellites a year at a total cost of $100 million, with four or five total failures, will still provide a huge science return for our money, and maybe even greater than if we had built one large spacecraft for the same $100 million. We will have trained more students, employed more people and generated a lot more ideas while solving a lot of problems...
>
> Passion is what makes it all work. Without passion, thousands of people merely go through the motions on a daily basis. With passion, real solutions to problems are developed, innovation is generated, excitement builds, fears are overcome and visions develop.

Without a doubt, the passion Crabbs talks about is a huge part of the amateur radio space program's success.

Until the past 5 years or so, universities have sponsored relatively few satellite projects, but this is rapidly changing. Some of the projects under way are SEDSAT at the University of Alabama at Huntsville [Wingo, et al., 1991a, 1991b, 1991c]; ASUSat at Arizona State University [Hewett and Reed, 1994; Rademacher and Reed, 1994]; and the SQuiRT microsatellite program at Stanford [Kitts, 1994; Kitts and Lu, 1994; Kitts and Twiggs, 1994]. Other projects are in progress abroad. Some of these projects originate in the amateur radio satellite program, either by virtue of AMSAT's leadership or through study of principles and practices already developed by AMSAT organizations throughout the world [Reeves, 1994]. More people are recognizing the value of small-satellite projects in the training of future engineers and scientists.

13.3.4 Summary

For more than 25 years, AMSAT-NA—in cooperation with other AMSAT organizations around the world—has been active in developing small communications satellites for use by radio amateurs around the world. During that time, ideas, techniques, and philosophies have evolved and matured to routinely construct reliable, long-lifetime satellites at a much lower cost than would be the case for a commercial satellite of similar size and capability. The AO-13 and AO-16 case studies reviewed earlier show that AMSAT can reduce costs for several reasons.

Having very little money in the first place makes cost reduction a **requirement**, not just something to talk about. Any individual AMSAT organization might be able to afford a project costing several hundred thousand dollars, whereas projects costing several million dollars require joint fund raising by multiple organizations. Although no AMSAT project has reached the $10 million mark, that cost would still be small by commercial standards. The requirement to drastically reduce costs leads to a change in **philosophy** about what should and shouldn't be tried and what can and can't be done.

The case studies show many examples of things done successfully that no one outside the amateur-satellite environment would have seriously considered. Parts selection and radiation-tolerance requirements are two examples of how AMSAT's philosophy differs from the mainstream. AO-13 has been operating in Molniya orbit for nearly 7 years despite most of its parts being MIL-HDBK-217F Class B-1 or lower. AO-16, and the other Microsats, have now been operating in low-Earth orbit at 800 km for over 5 years; yet, nearly all of its parts are MIL-HDBK-217F Class D or lower. Radiation-induced computer failures may have occurred on AO-13 and AO-16, but their onboard computers have failed so infrequently it's impossible to say with certainty a given failure was radiation induced. An extremely important aspect of a change in parts-selection philosophy is that, once it's put into practice, a new database of information accumulates which we can use in future design decisions.

The AO-13 and AO-16 projects have used project managers with **broad experience** in several technology areas. Many times the combined experience gained through vocation and avocation (amateur radio) extends to the systems engineers, design engineers, and technicians working on a project. Sometimes, a single person may cover multiple skill levels. The broader experience eliminates problems that occur when people operate in a single technology area or skill level. Recall that AMSAT-NA's use of broadly experienced managers, engineers, and technicians yielded four Microsats in slightly over 2 years from initial conception to delivery at the launch facility.

Finally, AMSAT projects use **almost no paid labor**. It is true that, in cases such as AO-13, salaried people have worked at critical times during the project but such expenditures have been minimal. While the commercial arena can't duplicate the "no paid labor" aspect of AMSAT projects, project managers must at least consider what motivates a person to carry out demanding, high-quality work for no pay.

The primary goal of AMSAT's satellite projects is enhancing amateur radio communications through facilities provided in the Amateur Satellite Service. The volunteer nature of the service, participating organizations, and people, dictates that we employ radically different satellite design, construction, and operations. Clearly, these techniques have resulted in many successful missions.

More important than any single cost-reduction strategy, what AMSAT hopes to offer is the encouragement to further develop and apply some of its philosophy. Multiple-satellite systems, by virtue of their redundancy, can afford to implement different design philosophies than have been used in large single satellites built to provide "all things for all people." If the time hasn't yet been right to adopt new ideas, I hope this chapter will bring that time closer.

PoSAT-1 Summary

PoSAT-1, a technology demonstration satellite developed by SSTL from the UoSAT series in cooperation with Portuguese academia and industry.

Spacecraft dry mass:	48.5 kg	No propulsion system
Average power:	20 W	TT&C: 9.6 & 38.4 kbps

Launch: Launched via Ariane from Kourou on September 26, 1993

Orbit: 788 km × 802 km at 98.7 deg inclination

Operations: The contractors (SSTL) installed ground station near Lisbon

Status: Operational as of January 1997

Cost Model (FY95$M): See page 348 for explanation.

	Expected Cost	Small Spacecraft Model	Actual Cost*
Spacecraft Bus	$26.4M	$3.2M	$1.15M
Payload	$20.1M	$1.1M	$0.38M
Launch	$16.6M	$3.7M	$0.27M
Ground Segment	$28.1M	—	$0.20M
Ops. + Main. (annual)	$2.5M	—	$0.10M
Total (through launch + 1 yr)	**$93.7M**	—	**$2.10M**

*An inflation factor of 1.066 has been used to inflate to FY1995$[SMAD, Table 20-1].

 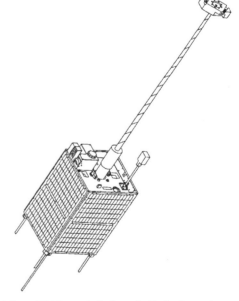

Fig. 13.4-1. PoSAT-1 shown on the left on the Ariane ASAP, ready for launch. Body dimensions are 33 cm × 33 cm × 53 cm. The top of the satellite contains Earth horizon sensors, Sun sensors, GPS antenna and magnetometer. The sides of the satellite hold GaAs solar arrays. As shown on the right, a 6 m gravity-gradient boom (not to scale) and navigation magnetometer are deployed after orbital injection.

13.4 PoSAT-1

Jeffrey W. Ward, *Surrey Satellite Technology Ltd.*
Harold E. Price, *Bektek, Inc.*

13.4.1 System Design
  PoSAT-1's Spacecraft Design;
  PoSAT-1's Operations and Ground Station
13.4.2 Creating a Low-Cost System
  Subsytem Reuse and Modularity;
  Designing for LEO; Mass Production
13.4.3 Operational Performance
  Operational Results; Failure Resilience
  in Practice
13.4.4 Conclusions

This book describes techniques which can be used to reduce space mission costs. These techniques will generally be applied after mission selection, in order to minimize the costs associated with achieving certain fixed program goals. In this chapter, we describe the PoSAT-1 technology demonstration microsatellite, which shows that extremely low-cost missions can be achieved by starting cost reduction before selection of mission goals. For many mission managers in inflexible programs, this will seem like an unattainable luxury; for managers faced with fixed budgets and more flexible orbital aspirations, however, our technique of selecting mission objectives to fit available resources will work very well.

Surrey Satellite Technology Limited (SSTL) has been investigating this low-cost end of the satellite market since the early 1980s and has applied the techniques described here to 12 missions. Each satellite has proven more capable than its predecessors, and each has fulfilled its owner's program goals. Unlike PoSAT-1, some of these missions had very tightly defined program goals. Because it was a technology demonstration and transfer program, PoSAT-1's program goals were broad, leaving great latitude when selecting mission objectives. Other spacecraft in the series, designed and constructed like PoSAT-1, have been developed on a more traditional basis—when the "mission defines the spacecraft."

The PoSAT-1 microsatellite shown in Fig. 13.4-1 was launched from Kourou, French Guyana on September 26, 1993 onboard Ariane Flight 59. It flew into an 800-km circular orbit with 98.6-deg inclination. PoSAT-1 is a 50-kg microsatellite carrying five payloads for technology demonstration and engineering research. Table 13.4-1 lists mass, power, and cost. SSTL built the satellite in the United Kingdom for the Portuguese PoSAT Consortium as part of a technology transfer program. SSTL managed PoSAT-1's design, construction, and in-orbit operation, so the PoSAT Consortium could quickly and inexpensively enter into satellite engineering. From initial contact between the two groups, to the end of the post-launch commissioning phase, this program lasted 21 months and cost $2.6 million. (See Table 13.4-2.)

Though PoSAT-1 was a rapid and inexpensive mission, we'll show that the satellite carries several complex payloads supported by a highly redundant, flexible bus.

The costs in Table 13.4-1 reflect the true commercial accounting of the PoSAT-1 Program, including overheads, profit margins, contingency planning and insurance, so this high ratio of capability to cost is not an illusion created by accounting techniques. On the other hand, PoSAT-1 was not an isolated mission. It was the eighth microsatellite built at Surrey, and the sixth that used a modular electronic and mechanical system developed in 1989 for the UoSAT-3 mission.

TABLE 13.4-1. **Mass, Power, and Cost of the PoSAT-1 Mission.** For this table, we've included the downlink transmitter (operating at 100% duty cycle) in the telemetry and command totals. Mass figures for each subsystem include the subsystem module box housing. Power figures in parentheses are peak power. Costs have been converted from Pounds Sterling to U.S. $K (FY95).

Item	Mass (kg)	Power	Cost ($K95)
Spacecraft Total	49.8	18.3 (36.7)	**$1,772K**
Mechanics and Structure	6.9	—	$189K
Power System	16.5	—	$355K
Telemetry and command system	5.6	7.4	$156K
Attitude Determination/Control System	10.5	1.4 (8.5)	$204K
Onboard Data Handling System	3.1	1.5	$137K
Thermal Control System	0	0	$13K
Wiring Harness	1.5	—	$25K
Payload Total	6.1	7.9 (19.3)	$226K
Radiation Monitoring Payload	1.0	0.5	$32K
Digital Signal Processing Payload	1.0	2.0	$35K
GPS Receiver Payload	1.7	2.0	$48K
Earth Imaging System Payload	1.3	1.7 (7.4)	$49K
Star Imaging Payload	1.1	1.7 (7.4)	$62K
Integration and Environmental Test	—	—	$150K
Ground Support Equipment	—	—	$40K
Launch Campaign	—	—	$72K
Post Launch Commissioning	—	—	$37K
Project Management	—	—	$166K
Ground Station	—	—	**$185K**
Training	—	—	**$159K**
Launch	—	—	**$266K**
Launch Insurance	—	—	$213K
3 Years Post Launch Operations	—	—	$239K
3 Years Software Development	—	—	$239K
PoSAT-1 Total Program Cost (launch plus 3 years)	—	—	$3,073K

TABLE 13.4-2. Principal Milestones in the PoSAT-1 Program. Intense activity on the project started in August 1992, after the launch of KITSAT-1, and continued for approximately 14 months until all onboard payloads had been commissioned.

Date	Elapsed Months	Milestone
January 1992	0	Initial contact between SSTL and the PoSAT Consortium
March 1992	2	Signing of launch-services contract
August 1992	7	Contract signing and project kick-off
August 1992	7	Launch of SSTL-built KITSAT-1
September 1992	8	Arrival of PoSAT Consortium Engineers at SSTL
November 1992	10	Decision on final payload complement
January–May 1993	12–16	Subsystem assembly and tests
June 1993	17	Spacecraft-integration tests at SSTL
June–July 1993	17–18	Environmental testing (thermal-vacuum, vibration, EMC)
July 1993	18	Shipment to launch site
August 1993	19	Initially scheduled launch date
September 1993	20	Actual launch date of PoSAT-1 and HealthSat-2
October 1993	21	First Earth images returned

The manufacturing infrastructure and technical legacy from these Surrey missions helped make PoSAT-1 cost effective. The clean rooms and laboratories used for PoSAT-1 existed from previous missions. Several members of the engineering team had already built five microsatellites, and some as many as seven. Existing infrastructure meant the mission had no significant start-up costs. The experienced team anticipated and avoided some problems, and they efficiently solved others. Later sections describe the effect of re-using equipment designs. Without these advantages, a team working on their first microsatellite could not replicate PoSAT-1's cost, schedule, and technical content. The cost-saving techniques used on PoSAT-1 are likely to apply best in organizations expecting to participate in many low-cost satellite missions.

The University of Surrey formed its commercial company, SSTL, in 1985, when they recognized they could build a self-sustaining program of low-cost space missions based upon the technology and techniques developed during the University-supported UoSAT-1 and UoSAT-2 missions. SSTL, operating as an "autonomous business unit" within the University, then built and launched two new UoSAT microsatellites (UoSAT-3 and UoSAT-4). They carried payloads from various paying and non-paying customers. These missions led to several commercial contracts for single-customer missions. SSTL's first fully commercial satellites, KITSAT-1 and S80/T (launched August 1992), illustrate the two distinct types of missions to which SSTL has applied the engineering and management techniques described in this chapter. S80/T was a turnkey microsatellite built by SSTL as subcontractor to Matra Espace. The satellite's single payload was provided by another subcontractor, with SSTL providing integration, test and commissioning support. The mission is thoroughly described by Allery [1994]. KITSAT-1 was a multi-payload technology demonstrator microsatellite constructed under a technology transfer program between SSTL and the Korean Advanced Institute of Science and Technology. The KITSAT-1 program was the first full technology transfer program undertaken by SSTL. It included academic courses and on-the job training for Korean engineers, constructing of KAIST's ground station in Korea, and the building and launching of the KITSAT-1 microsatellite. During the final stages of the KITSAT-1 mission, in early 1992, SSTL began negotiating with a group of Portuguese industrial and academic institutions for what became the PoSAT-1 program.

The momentum from these earlier missions was transferred to PoSAT-1. Although no parts or sub-assemblies were kept in stock, and many designs were altered substantially, an experienced team and tested infrastructure were in place for the new mission. Software, including development and test, ground station and flight, was also available for re-use. Start-up costs and the expensive early portion of the "learning curve" were avoided.

Another critical part of an inexpensive space mission is a suitable low-cost launch. There are many scenarios for low-cost launches (see Sec. 4.2), but with a total program budget around $3 million, only the least expensive piggyback options were sensible for PoSAT-1. Fortunately, the mission coincided reasonably well with the Arianespace launch of the SPOT-3 remote sensing satellite. This launch was the fourth Ariane launch to include the Ariane Structure for Auxiliary Payloads (ASAP)—see Sec. 4.2.2—a structure designed for launching multiple small, second-ary payloads. The PoSAT Consortium negotiated a launch on this ASAP in March 1992, at a cost of 2.5 million French francs (roughly $250,000). This event kicked off serious contract negotiations between The Consortium and SSTL. The availability of low-cost launches on the SPOT-3 ASAP also catalyzed five other low-cost micro-satellite missions: HealthSat-2, KITSAT-2, EYESAT, ITAMSAT, and STELLA.

In our experience, the somewhat peculiar pattern of the PoSAT-1 program is typical of technology-transfer missions. The launch opportunity (including a fixed launch date), mission prime contractor, satellite bus, and mission budget were deter-mined before any payload was identified. Of course, the PoSAT consortium set—and aimed to achieve—strong program goals, but many payloads and experiments could have fulfilled them. The program goals were

- To educate a core group of Portuguese engineers who could then lead subse-quent space activities—particularly industrial activities—in Portugal
- To transfer to Portugal satellite-engineering technology, so they could de-velop space activities rapidly and cost-effectively
- To establish in Portugal a national satellite program whose high profile would stimulate the Portuguese aerospace industry and their other highly technical industries by recognizing the nation's capabilities.

Political leaders of the PoSAT Consortium set these goals at the beginning of the program. PoSAT-1's technical team reported jointly to The Consortium and SSTL. Within budget and time restraints, they could select from a wide range of mission objectives and payloads. This arrangement saved money in two ways. First, the budget was fixed, and cost drove all engineering decisions. Second, we could select mission objectives that gave us the greatest "bang for the buck" rather than having to go as far in a fixed direction as the money would allow. This circumstance certainly allowed PoSAT-1 to incorporate a lot of new, complex technology. For example, we abandoned a laser-beacon experiment in favor of the more feasible and technically rewarding star imager. In the following paragraphs, we'll show how these constraints and freedoms tempered PoSAT-1's design.

13.4.1 System Design

A low-cost microsatellite must be deployed on a suitable mission. Technology-demonstration satellites, such as PoSAT-1, can serve the program's initial broad goals in many ways. Thus we must tailor these objectives to the satellite platform's capabilities and characteristics if we are to fulfill the mission with low-cost ground and space segments.

Traditionally, we enter system design with a known satellite bus and fixed budget. The PoSAT-1 technical team's first task was to select detailed objectives which met the program's goals within budget. These objectives also had to make the most of SSTL's microsatellite bus—specifically the KITSAT-1 version of the bus, which had just been verified in orbit. PoSAT-1's budget allowed us to modify the bus (in part by reinvesting SSTL's "profit") and to develop some new payloads. All changes and new developments had to be ready for integration within 11 months.

Because PoSAT-1 was to be Portugal's first national satellite, developed and launched under close media attention—all mission objectives were secondary to placing a working microsatellite built by Portuguese engineers into orbit. A "working" satellite was loosely defined as a controllable satellite fulfilling at least one of its experimental objectives. The team tightly controlled changes to the bus' basic subsystems (telemetry, telecommand, communications and power) to ensure they didn't threaten this key goal.

The next step was to choose payloads that would stimulate public interest, help train the Portuguese team and technologically stimulate Portuguese industry. Initial payload choices came from two sources—interests within Portugal and developments within SSTL. The leaders of the Consortium first proposed two new payloads for PoSAT-1: a laser beacon and a star sensor. Both of these proposals came from the Laboratório Nacional de Engenharia e Tecnologia Industrial, the Portuguese R&D center acting as the interface between SSTL and the PoSAT Consortium. They were seen as catalysts for R&D activity in Portugal and as the first step toward operational systems which the Consortium could sell or use on future satellites. At the same time, SSTL remained interested in the radiation, Earth-imaging, and digital-signal-processing payloads flown on KITSAT-1. We also wanted to fly a GPS receiver and certain bus upgrades. Early technical investigations revealed that mission constraints made a laser beacon or laser receiver impractical. The team determined to fly all of the remaining experimental payloads.

Equipped with a complete (if tentative) list of payloads and knowing the platform's and ground station's performance from previous missions, the team devised a mission scenario which would

- Exploit each payload independently of the others (in case of failure)

- Allow payload inter-operation for improved performance (nominal case)

- Form a coherent and justifiable mission from the group of payloads

- Not require expensive or risky increases in the bus' or ground station's performance

At first glance, this relatively luxurious approach to mission definition seems uniquely suited to technology-transfer missions. But we believe this approach will work for low-cost scientific and commercial missions whose budgets and schedules are tightly constrained. As in traditional space mission analysis and design, the process is iterative. We simply start it sooner, avoiding a mission-driven scenario and defining the mission to extract the most return from available technology, funding, and time.

TABLE 13.4-3. **PoSAT-1 Payloads and Their Mission Objectives.** Items marked † denote objectives that have been met, the others are in progress.

PoSAT-1 Payload	Heritage	Basic Objectives	Enhanced Objectives
Digital Signal Processing	KITSAT-1 hardware reused. New software developed.	Generate digital voice broadcasts for public relations exercises.†	Demonstrate minimum-shift keying in LEO.† Experiment with adaptive modulation techniques to improve the throughput in LEO communications environment
Global Positioning System	New development for PoSAT-1.	Demonstrate the use of modified commercial GPS receiver for determining the position of a LEO satellite.†	Demonstrate autonomous, onboard generation of Keplerian orbital elements. † Provide time and position services for other experiments, particularly Star and Earth Imaging Systems.
Earth Imaging System	KITSAT-1 hardware with improved noise performance.	Demonstrate low-cost panchromatic imaging with 100m resolution.† Generate public interest by returning understandable data.†	Do onboard cloud editing. † Do onboard image compression. † Illustrate potential applications of microsatellite remote sensing.
Star Imaging System	New development for PoSAT-1.	Investigate the use of uncooled, charge-coupled devices for star imaging.†	Provide enhanced, ground-processed, attitude determination for PoSAT-1.† Provide enhanced attitude determination with onboard processing for PoSAT-1.
Cosmic Particle Experiment	KITSAT-1 hardware and firmware.	Monitor particle energies and fluxes in PoSAT-1 orbit for correlation with the performance of onboard devices.†	
Store-and-Forward Communications Transponder	KITSAT-1 hardware and software re-used with modifications to reduce payload volume and increase data-storage capacity.	Provide all experimental data transfer between space and ground.† Provide a global system for e-mail messaging.†	Support ship-borne experiments in LEO communications.† Support demonstrations of military communications using LEO microsatellites.†

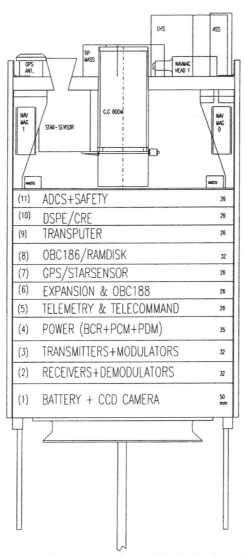

Fig. 13.4-2. PoSAT-1 Module Stack. Electronics modules in aluminum boxes form the primary structure of the spacecraft, with sensors in a bay at one end, and a nadir-pointing lens looking through the attach fitting at the other.

Each PoSAT-1 payload had basic and enhanced mission objectives (Table 13.4-3). The basic objectives usually require only the main payload hardware, a single computer onboard the spacecraft, and the communications links. We avoided objectives that required lengthy, continuous operation of the payload, bus, and ground station (particularly early in the mission). They demand more reliability from the software and hardware, as well as high-speed redundancy switching. (It's ill-advised to rapidly design a low-cost satellite and launch it as a secondary payload if its main mission is to monitor a one-of-a-kind event.) Clearly, relaxing the basic mission objectives increases the probability of success, **without** restricting the enhanced objectives, which become optional goals.

Each payload's enhanced objectives depend on successfully operating the payload hardware and developing more sophisticated software. Routinely improving payload operation through software is a common characteristic of modern microsatellite missions, and the PoSAT-1 bus and payloads are designed to support safe and **routine** software uploading. Enabling, indeed encouraging, post-launch software development has several benefits for small teams building satellites on short schedules. It means they can devote limited pre-launch engineering time to hardware and the low-level software in read-only memory. Furthermore, payload software is not required to be absolutely reliable. As long as it cannot threaten spacecraft integrity when it fails, software can simply be reloaded to fix faults or improve functionality. This relaxation of software criticality makes the software development process more rapid and less expensive than generally anticipated in a satellite mission. PoSAT-1's enhanced objectives were set knowing they'd be met well after launch, through continuing software development.

TABLE 13.4-4. SSTL Modular Bus Missions. These missions share a common heritage with PoSAT-1. In a "Mission derived from Spacecraft" specification, the mission was secondary to technical or developmental objectives. A "Spacecraft derived from Mission" specification is a more traditional approach, in which the spacecraft is designed to meet pre-existing mission requirements.

Spacecraft	Launched	Primary Mission	Type of Specification.
UoSAT-3	January 1990	Initially technology demonstrator, resold as commercial store-and-forward after launch.	Mission derived from spacecraft.
UoSAT-5	July 1991	ARS Store & Forward, Earth Imaging	Spacecraft derived from Mission, minimal specification.
KITSAT-1	August 1992	ARS Store & Forward, Earth Imaging	Technology Transfer, Mission derived from Spacecraft.
S80/T	August 1992	Commercial Radio Research	Spacecraft derived from Mission, detailed specification.
HealthSat-2	September 1993	Commercial Store & Forward	Spacecraft derived from Mission, minimal specification (heavily based on UoSAT3/5).
PoSAT-1	September 1993	Commercial research, Imaging	Technology Transfer, mission derived from spacecraft.
CERISE	July 1995	Military Radio Research	Spacecraft derived from Mission, Rigorous military-style specification.
FASat	August 1995	Commercial Research	Technology Transfer, Mission derived from Spacecraft.

The enhanced mission objectives also involve several payloads working together. On PoSAT-1, determining the spacecraft's attitude from star images requires orbital position information which the GPS receiver provides; then, we can use the attitude to accurately determine the Earth imager's orientation while interpreting images on the ground. Similarly, improved downlink throughput realized from advanced DSP modulation could improve all mission objectives by returning more data. Although each payload can operate on its own, the collection of payloads is coherent and powerful.

TABLE 13.4-5. PoSAT-1 Technical summary.

Launch	• Piggyback (ASAP) with SPOT-3 on Ariane 4 flight 59, September 26, 1993
Orbit	• 800 km, 98.6 deg inclination, Sun-synchronous
Structure	• 33 x 33 x 53 cm, L63/L192 Al, stacked module boxes.
Mass	• 50 kg
Payloads	• Wide-Angle Camera, 1500 km x 1050 km FOV, 2.2 km mean ground resolution, 810–890 m band, 568 x 560 CCD array. • Narrow-Angle Camera, 150 km x 100 km FOV, 200 m resolution, 610-690 nm band, 568 x 560 CCD array. • Star field sensor, 9.7 deg x 7.2 deg FOV, 350 μradians (approx. 1 arc-minute) per pixel, 568 x 560 CCD array, passively cooled to –5 °C, nominal attitude, magnitude 6 typical detection threshold. • GPS receiver. Trimble TANS-II receiver, single antenna. Average time to 3D fix from cold start, < 3min. Radar-confirmed accuracy, 130 m 3σ. • Charged-Particle Experiment (high energy particle flux detector), 512 channels from LET of 64 MeV cm^2/g^{-1} to 8360 MeV cm^2/g^{-1}. • Total-Dose Experiment. Measures long term accumulated ionizing radiation dose in a large gate-geometry device (RADFET). • Digital Signal Processor Experiment. TMS320C25 and TMS320C30 processors, software loadable through main onboard computers. Using a path to the main receivers and transmitters, can provide DSP modems for the main processors. • Store-and-Forward Communications. Provides file and message storage and retrieval using 16MB of onboard RAM.
Power	• Four 1344 cm^2 GaAs panels, 30 W BOL each panel. 20 W post conditioning available, 12 W typical load. Battery Charge regulator tracks peak power. • 10 NiCd cells, 6 Ah, 14 V capacity.
Attitude Control	• Control through Gravity Gradient, 6 m boom with 3 kg tip mass, active onboard attitude control for spin rate and spin axis, magnetic torquers in each axis. Typical accuracy ±2 to 3 deg.
Attitude Determination	• Attitude determination through optical sun and horizon sensors, and 3-axis magnetometer readings processed by an onboard computer. Typical accuracy of onboard determination 1 deg.
Thermal Control	• Passive through AL/Kapton structure coatings.
RF Links	• Downlink: UHF, 1 W, 5 W, 10 W switchable, CPFSK at 9600/38400 bps • Uplink: VHF, CPFSK/FM 1200/9600
Propulsion	• None

The team had selected PoSAT-1's payload and defined its mission three months after contract signing—11 months before launch. SSTL was to develop the new payloads (GPS and star sensor) while developing or enhancing existing payloads and bus subsystems. This kept design authority and detailed engineering of the mission, bus and payloads within a 20-person team operating in a single building under one mission manager. This arrangement maintains short and effective lines of communications, reduces or eliminates subcontract management overheads, and simplifies interface control and modification. These advantages contributed significantly to our ability to produce PoSAT-1 in the following nine months.

PoSAT-1's Spacecraft Design

The PoSAT-1 spacecraft is a member of a family of SSTL-designed modular spacecraft shown in Table 13.4-4. Much of our savings on a mission comes from reusing modules from previous missions. To keep cost and risk to acceptable levels, each mission must be a careful blend of "off-the-shelf" modules, upgrades, and new development. The modular bus easily captures this blend, which we explain further in Sec. 13.4.2.

Meerman [1994] describes PoSAT-1 in detail. Table 13.4-5 gives its technical characteristics. Among its more interesting features are the CCD imagers that appear in the wide angle camera, the narrow-angle camera, and the star-imaging sensor. At only a fraction of the price, the wide-angle camera produces images of a quality and detail comparable to the near IR spectral band of the NOAA AVHRR instrument. The images go into the store-and-forward system for file transfer, so images taken over one part of the globe are available at any other.

We overcome the limits of the low-cost, uncooled CCD for star sensing by taking four images at nearly the same time. Manipulating the images removes false stars caused by single-event upsets and sensor noise. This processing, as well as searching the star catalog, currently takes place on the ground in a post-processing environment. This algorithm has been enhanced to the point where it can be accomplished on board, and it is an ongoing activity to link the GPS data to the star sensor and provide autonomous fixes for more precisely determining where the Earth-imaging cameras are pointing.

The block diagram of PoSAT-1 in Fig. 13.4-3 shows one of the major tenets of SSTL's design. Low-cost modules, built in many cases from parts not rated for space, may fail. Accordingly, we give each processor that can support mission operations its own access to the telemetry and telecommand system, as well as the uplinks and downlinks. Because the processors may fail, the telemetry and telecommand systems have their own paths to the radio links and can be controlled directly from the ground. The latter feature is a holdover from the early days of UoSAT spacecraft, when processors weren't trusted at all. The PoSAT-1 spacecraft's main processors contain only rudimentary bootloaders in read-only memory. All operating software is uploaded after launch.

Even though the basic operating software is very reliable (UoSAT-3 has been running without a software outage or reload for 1250 days as of this writing), UoSATs still may need new software and complex, real-time operation of their payload. We assume software will need to be reloaded easily and quickly. The concept of continually upgrading the spacecraft's software was born of necessity—due to a quick launch schedule, SSTL developed most of the operational software for UoSAT-3 months after launch. Current spacecraft, including PoSAT-1, have all of the operating software available for load on the first pass, but the "field upgradable" concept has served well. Two of the UoSAT spacecraft have completed their original missions. UoSAT-3's owner has sold it to another company for commercial messaging experiments, and an educational consortium may take over the S80/T.

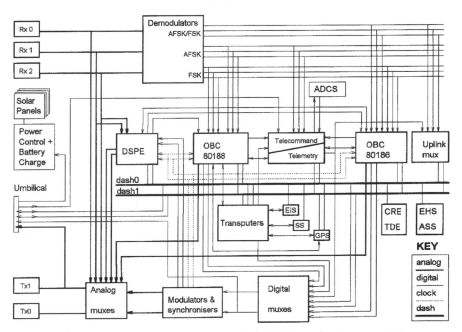

Fig. 13.4-3. PoSAT-1 Block Diagram. A typical UoSAT design contains several alternate data paths to maximize the likelihood that failed paths or components can be bypassed. Although myriad configurations are possible, a relatively small number are used during nominal operations.

Another interesting concept you can see in Fig. 13.4-3 is the DASH—a Local Area Network on a shared bus, with connections to all general-purpose computers and processors for experiments and payloads. The DASH is a multi-drop wire, and each connected device uses a collision detect multiple-access protocol. It moves data "peer-to-peer" between cooperating processes on each attached computer. Besides reducing the number of wires in the harness, this again provides redundancy by allowing any processor to control any experiment. Starting with the FASat-Alfa spacecraft (built 1995), we're supplementing PoSAT's 9600-baud DASH with a 1 MB Controller Area Network (CAN) bus.

PoSAT-1's Operations and Ground Station

The PoSAT-1's payloads and bus subsystems are configured to support low-cost operations from a single low-cost ground station. The ground station, developed from existing SSTL designs, was installed into a building for $185,000 not counting costs of minor changes to the building. The operations scenario requires, at most, one operator working full-time, without night or weekend shifts. This budget allows us to support normal and optional payload objectives, software development and fault recovery. Assuming a labor rate of $80,000 per staff year, the ground station and three years of operations cost $425,000—14% of the program cost. We designed the operations scenario and ground station to attain such low-costs.

The low-cost operations scenario for PoSAT-1 incorporates onboard autonomy with store-and-forward data collection techniques. We pioneered this combination with UoSAT-1 in 1981 and extensively developed it since then. Several of the PoSAT-1's experimental objectives require data return from arbitrary points in the satellite's orbit. Although data-relay satellites or large ground-station networks are suitable for some missions, the only practical low-cost way to return such data is to store it onboard the satellite and download it to the ground station for mission control. In general, large scientific missions do so by recording data on magnetic tape and playing it back when the satellite is in range of the ground station. Because is data serially accessed on a magnetic tape, it's heavily coded to prevent storage and transmission errors, then transmitted over a data link of consistently high quality. PoSAT-1 uses a different approach, which is both more flexible (allowing random data storage and retrieval) and more robust (permitting the use of smaller ground stations and poorer data links). PoSAT-1's data resides on board the satellite in a solid-state memory configured to act as a computer-file system. Whenever the satellite is in range of the control station, we can select and download any of the data files. This allows the ground station to determine what data is important to download and to change download priorities after the data is onboard. During data downloading, we use packet radio communications techniques to detect and correct errors while retransmitting packets lost due to communication errors. We first developed this system for store-and-forward messaging, as described by Ward [1990]. It permits engineers to use communications links to low-Earth orbit more efficiently than traditional methods, returning more data to smaller, simpler ground stations and using less energy onboard the satellite.

Using onboard autonomy on PoSAT-1 allows a single control station to exploit the payloads efficiently even though the ground station isn't continuously staffed and the satellite usually isn't within the ground station's communications coverage. The central onboard computer and payload-control computers in the major payloads do open- and closed-loop routines for autonomous operation. In routine operations, schedules for operation (e.g., Earth imaging) are prepared by experimenters, uploaded to the onboard file system, and then distributed to the payloads by the central onboard computer. In cases where an operation is particularly risky (e.g., operating high-power downlink amplifiers), closed-loop software in the onboard computer monitors critical telemetry points and ends the operation if it exceeds safe ranges. Whenever the satellite is within range of the ground station, automated software transfers any collected experimental data from the satellite to the ground station without operator intervention and without event-by-event scheduling. Although we can schedule arbitrarily long autonomy periods, weekday operations typically use 24 hours and weekends use 72 hours—so we don't have to staff the ground station during weekends.

Without ground-station interaction, PoSAT-1 would remain safe in orbit indefinitely. The attitude control would degrade slightly because onboard tracking algorithms wouldn't have updated orbital elements (presently provided by the ground station rather than the GPS system), but the satellite itself would suffer no damage. Furthermore, even if the onboard software were to fail completely, resulting in the loss of all high-level functions, PoSAT-1 would remain safe. It would be overly complex to say that the satellite would enter "safe mode." Instead, we designed the bus so virtually any likely mode is safe. Although this approach constrains the bus and payload design and adds cost, it makes operations, availability of the ground station, and faults in onboard software less critical. A failure-resilient bus and payloads are key to the PoSAT-1's operation.

As part of the PoSAT-1 program, a ground station was installed in Portugal to serve as primary mission control for post-launch commissioning and payload operations. It is maintained and operated by the Consortium with Portuguese engineers trained at SSTL, illustrating effective transfer of technical knowledge relating to PoSAT-1 and its experiments. Like the microsatellite, this ground station was a logical upgrade to SSTL's ground stations.

The ground station system design (shown in Figure 13.4-4) complements the mission design and operations scenario. Four IBM-compatible personal computers linked by a local area network connect through a communications switching unit to the uplink and downlink communications equipment. Although each computer is nominally allocated a specific task, they can run any IBM-compatible application. Ground-station software has simple, well-defined interfaces, which experimenters can extend with software that supports their payloads. These open interfaces, combined with the file-transfer model used in uplink and downlink communications allow the ground station to be easily enhanced or altered along with the PoSAT-1 payload software. This flexibility is in keeping with our system—rapidly developing software to support basic and optional mission objectives.

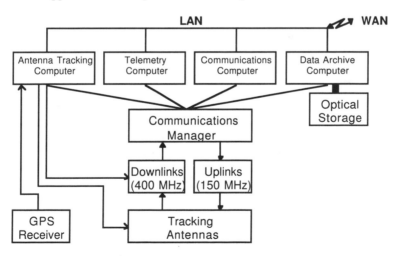

Fig. 13.4-4. PoSAT-1 Ground Station. The ground station is based on IBM-PC computers, commercial off-the-shelf radios (slightly modified), and a custom signal distribution multiplexor. The operation of the ground station is largely autonomous, and the operating software is an "open system" which experimenters can extend to meet payload requirements.

The ground station is largely autonomous. The Antenna Tracking Computer uses orbital predictions to determine when PoSAT-1 is within communications range. When the satellite rises above the ground station's horizon, the Tracking Computer uses the LAN to issue commands which configure the other computers for the necessary operations. The Communications Controller (essentially a computer-controlled patch panel for analog and digital signals) is also configured for the correct communications parameters. The Tracking Computer points the antennas and compensates for Doppler shift throughout the pass, while the other computers capture data, display telemetry, and store archival data. The Communications Computer manages data capture by downloading a list of newly created files from the satellite's onboard file system and then using configurable rules to determine the order in which files should be downloaded to the ground station. When the pass is over, the Tracking Computer shuts down the other applications and initiates post-pass data processing. All of this activity takes place on every satellite pass, whether or not an operator is present. Operators may override any automated activity (e.g., to upload new spacecraft schedules) or modify automation scripts within the ground station to achieve specific goals.

Experimenters throughout the PoSAT Consortium can access the ground station through a dial-up WAN connection. This connection, which supports electronic mail, allows the ground-station operators to coordinate operations with, and distribute data to, other users. The WAN link also provides a connection to the ground station and engineering team at SSTL. Data can also be transferred on IBM compatible media, including the archival optical disks.

Implementing a ground station of this complexity could prove expensive. We've adopted techniques for design and procurement to contain the cost while still providing an effective ground station. The most important is to **avoid custom items and specialized aerospace equipment**. PoSAT-1 operates in the UHF (400 MHz) and VHF (150 MHz) bands, which are next to communications bands used extensively by terrestrial and amateur satellite communications services. With minor modifications, radios and antennas from these services can be used to support satellite links. With data links between 10 and 50 kilobits per second, packet radio data equipment from the ground services—including data converters and modems—can also be used. This equipment brings a high level of sophistication (e.g., fully computerized radios) to the ground station at the low-costs typical of electronics for mass markets. The computers and computing equipment are IBM-compatible for two reasons. First, these machines are less expensive than the commonly used UNIX workstations. Second, they are typical of the computers used daily by the programmers, ground-station controllers, and engineers throughout SSTL and the PoSAT Consortium. In the PoSAT-1 ground station, custom electronics are used only for the Communications Manager and some experimental modulation systems.

Although these cost-reduction techniques may seem suitable only at relatively low data rates, we have recently investigated support for an S-Band data link at 1 Mb/sec. Even here, the standard satellite equipment from aerospace vendors is ten times more expensive than the low-cost commercial equipment available to the careful shopper. Personal computers now have more capacity than the UNIX workstations of the 1980s, and they'll be suitable for higher and higher satellite data rates—always at very competitive prices. The PoSAT-1 ground station indicates what is possible when the space segment is designed to be compatible with available low-cost ground-segment technology, and both segments are designed to strict budgets. The following sections address more cost-saving measures which we applied to the PoSAT-1 mission.

13.4.2 Creating a Low-Cost System

Subsystem Reuse and Modularity

Custom development should be minimized to keep costs low. Using off the shelf items, both in terms of actual hardware and in knowledge, has a cascade effect; design, parts procurement, documentation and certification, and in particular, software, have all been developed (and paid for) by previous missions. NRE for the current mission can be applied to those items that are sufficiently unique to require new development.

SSTL designed the UoSAT family to maximize modularity and reuse. Although many subsystem designs are slightly altered from generation to generation of the SSTL microsatellite bus, even partial design re-use provides significant cost savings.

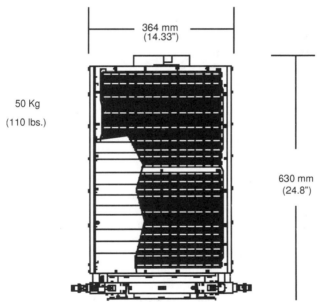

364 mm
(14.33")

50 Kg

(110 lbs.)

630 mm
(24.8")

Fig. 13.4-5. Standard UoSAT Modular Bus, Cutaway View. The inner structure is a stack of flat module boxes that support a skin of solar arrays.

Making a spacecraft modular and reusable must start at the beginning—in the structure. Many traditional spacecraft, including low cost spacecraft, consist first of a structure. Electronics modules are then attached at various places inside the structure, and interconnected with a three-dimensional wiring harness. SSTL's modular space-craft has no structure; it consists of a stack of module boxes. Solar arrays (including fold-out arrays) attach to the outside of these boxes. Each module's connectors are on the same facet, allowing the harness to be two-dimensional. We simply hang the har-ness on the outside of one surface of the stacked cuboid. (See Figs. 13.4-5 & 13.4-6.)

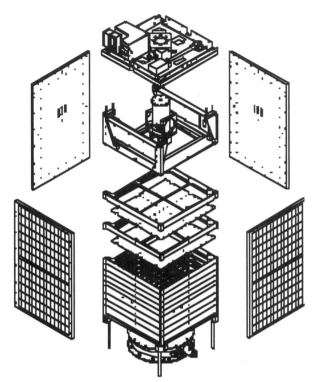

Fig. 13.4-6. UoSAT Modular Bus, Exploded View. An equipment bay at the top holds sensors and supports the gravity gradient boom assembly. Usually, eleven module trays make up the rest of the spacecraft.

Attitude determination sensors, CCD imager openings, and antennas, are placed on the nadir and zenith pointing "top and bottom" of the spacecraft. Solar arrays are hung on the other four surfaces. The equipment bay and array placement can be seen in Fig. 13.4-6. The facet that hosts the cable harness is at the back in this view, in a trough formed by the module box's outer wall, also shown in Fig. 13.4-7. Module boxes are milled from blocks of aluminum and can be configured to support a large PCB or contain chambers for receivers, high-power amplifiers, or other equipment.

We can't overemphasize the importance of a reusable, adaptable structure to a low-cost mission. Although we must balance, vibration test, and thermally analyze each spacecraft, these activities need not begin anew from first principles. That is, each satellite must go through acceptance tests but not qualification tests. The structure has qualified formally on several launchers (including Ariane and Tsyklon) and meets the specifications for many others. Testing and fault resolution are simpler because technicians can assemble, take apart, and reassemble the spacecraft in a few hours.

Another advantage of modularity is the ability to mix older modules with newer technology. Having a trusted fallback reduces risk. We test new modules, of course, but not to the level (and expense) of a "must-function" node containing a single-point failure.

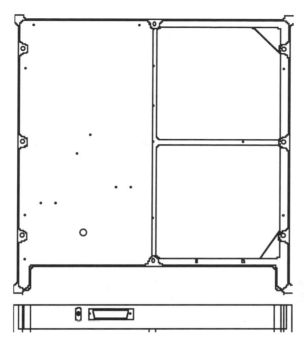

Fig. 13.4-7. Typical UoSAT Stackable Module Box. The U-shaped cutout forms a trough for the wiring harness. All intermodule connectors are in this area. Module boxes are manufactured using numerically controlled milling machines; each costs about $500.

PoSAT-1 mixes old and new technology, as all of SSTL's spacecraft do. (See Figs. 13.4-8 and 13.4-9.) As with other low-cost missions, we rely on parts that aren't rated for space. Radiation-tolerance data isn't available for many new parts. We must extrapolate from parts with similar technology that are rated, or on our own testing with high-rad sources. In general, we use old technology for the core support modules, so we can salvage part of the mission if a new technology fails.

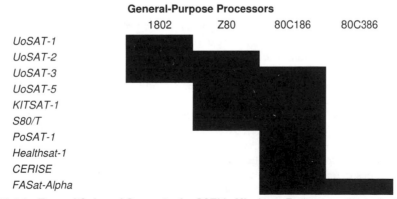

Fig. 13.4-8. Type of Onboard Computer for SSTL's Missions. Each generation typically flies the old standard, plus a new design, which in turn becomes the next standard.

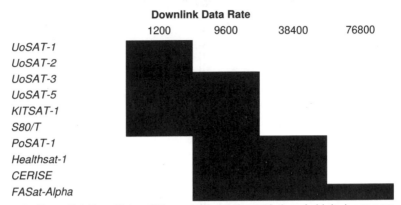

Fig. 13.4-9. Downlink Data Rates. This again shows the evolution of old design to new.

In moving from KITSAT-1 to PoSAT-1, SSTL

- Made the OBC186 and ramdisk modules smaller and combined them into a single module.
- Redesigned the Expansion/Z80 module to replace the Z80 part with an 80188 processor.
- Significantly redesigned the transmitter module to offer variable RF output and a switchable high-power amplifier.
- Used a new GPS receiver and star sensor imaging system.
- Slightly revised the transputer module to handle data from the GPS receiver and star sensor payloads, in addition to the wide- and narrow-angle cameras.
- Modified the modulators to operate the Digital Signal Processing Experiment as a modulator/demodulator.
- Mounted the GPS receiver's antenna and the star sensor's lens assembly in the zenith-pointing equipment bay.

Software modularity and reuse are also critical to a low-cost system. [Ward 1994] Thus, the main processors on SSTL's spacecraft use preemptive multi-tasking. The operating system allows each task to be a separate module, with all modules running on the same CPU. Examples are attitude determination and control, file transfer, power management, link management, imaging processing, or TT&C. The system we use (BekTek's "Spacecraft Operating System" product) allows us to develop tasks with standard C compiler and link them as a DOS-style.EXE, executable image. The image can then be uploaded "on the fly" to the spacecraft, where it is relocated and integrated into the program mix. Separate executables, and the preemptive dispatcher, allow modules to be developed, for the most part, without regard to other modules. Tasks interact through a message-passing service. For example, the TT&C module can and does offer a "telecommand service" using message passing, allowing other tasks on the processor to access the spacecraft hardware through a layer of abstraction (an Application Program Interface). Only the TT&C module needs to change whenever we develop a different hardware interface for TT&C.

The operating system that supports task execution is unaware of the spacecraft's particulars, so we can often reuse the operating system and support programs such as link control, file transfer, and program loading. The core of the operating system has remained unchanged since 1989 and is supporting 14 active missions (and 10 under construction).*

Designing for LEO

Low-Earth orbit (LEO) is the least demanding spacecraft environment, mainly because these altitudes don't encounter high radiation doses trapped in the Van Allen belts. The total dose of radiation is so low that most commercial electronic components will easily survive a 3-year mission. With careful attention to a few details, we can develop most of the hardware and software without much thought about space. We don't need heroic shielding efforts or a lot of money for radiation-hardened, S-rated parts. Careful, systematic system design can mitigate the effects of radiation and vacuum without exotic, expensive components.

One radiation effect we **must** counter in low-Earth orbit is the transient effect of charged particles, primarily manifested as Single Event Upsets. These occur regularly in low-Earth orbit, especially in the South Atlantic magnetic anomaly, altering the contents of semiconductor memories. SSTL's spacecraft use hardware-based error detection and correction for RAM used to store programs. PoSAT-1 provides 512 K bytes of program memory in each of its two main onboard computers, using 12 bits to store each 8-bit byte. Hardware error correction is essential to instantaneously correct program memory. Using this method to protect the main 16 MB file system memory in each processor would be prohibitively expensive in cost and volume. Fortunately, the computer's processing power can protect these memories because the computer software itself isn't run directly from them. A software protection algorithm, built into the file system software, is used for this memory. Using a Reed-Solomon scheme, 255 bytes are used to store 253, enabling two corrupted bytes to be corrected. A background task scans though memory, finding and correcting errors The scan rate is adjusted so that an uncorrectable number of errors does not occur before each memory block is visited. In most cases, this means a scan must be completed before a second trip though the South Atlantic Anomaly.

Other types of ram memory on board, such as the CPU's registers, can be corrupted, but this hasn't been a problem. In 25 orbit years of experience on SSTL's spacecraft, we've found no CPU crashes caused by a register upset.

Another concern to spacecraft designers is the thermal regime. A proper thermal design, particularly for spacecraft in Sun-synchronous polar orbits, will result in an environment far less demanding than many terrestrial applications. The PoSAT-1 main processor temperatures change less per orbit (8 °C) than the changes experienced by personal computers in a typical home office. Even the electronics mounted in the outer layers, such as the PoSAT-1 equipment bay, experience changes (80 °C) typical of an automotive engine controller.

While a benign radiation and thermal environment does not apply to all spacecraft, a low-cost low-Earth orbit mission must take the true (relatively low), risks into

---

* The operating system is now beginning to show its age; and we'll upgrade it to support 80386 protected mode in future missions.

Fig. 13.4-10. PoSAT-1 Photographs of Detroit and Cape Cod, Narrow-Angle Camera.

account and budget accordingly. Even when moving to orbits which encounter the Van Allen belts, designers must determine the expected environment and design to it, rather than simply expecting the worst and buying a satellite full of expensive components. S80/T and KITSAT-1 are in a radiation environment ten times worse than PoSAT-1; yet they use much the same technology. These satellites easily met their three-year mission design requirements.

Mass Production

PoSAT-1 was constructed at SSTL and launched on Ariane with the HealthSat-2 microsatellite. HealthSat-2 was designed for store-and-forward communications, and did not carry any of the same experimental payloads as PoSAT-1. Nevertheless, the two satellites still shared some bus subsystems, a common schedule, and many project team members. The savings of mass production—or at least reproduction—were shared by the two missions. Table 13.4-6 shows how PoSAT-1 and Healthsat-2 shared designs for the bus subsystems.

When building so few units, no volume discounts on parts are likely. Cost savings are largely in manpower and to a lesser extent facilities. For example, during parts procurement, the same number of outgoing orders and incoming shipments which would have been required for PoSAT-1 alone actually acquired parts for both PoSAT-1 and HealthSat-2. During technical development, the second and subsequent units benefit wherever there is a significant new process to be mastered. These savings appear during subsystem alignment and test, which might be done for several units in parallel or involve software and hardware bench test development which can be spread across the two missions. For PoSAT-1 and HealthSat-2, this was particularly apparent in the alignment of receivers (six units) and the testing of 80C186 onboard computers (three units). Development of standard software represents an even greater opportunity to divide and subdivide costs; as discussed in Sec. 13.4.3, the basic software for PoSAT-1 and HealthSat-2 is shared by 24[*] other satellites. Enhancements implemented for any SSTL onboard computer are immediately shared by the others.

TABLE 13.4-6. **Bus Subsystems for PoSAT-1 and HealthSat-2**. PoSAT-1 and Healthsat-2 were produced at the same time and had common subsystems. This allowed savings in procedures execution time, but not a bulk-buying discount.

Subsystem	Variations between HealthSat-2 and PoSAT-1
Power	None
Attitude Control	HealthSat-2 system simplified
Telemetry and Telecommand	None
80C186 Onboard computer and memory	None
80C188 Onboard computer and memory	None
Uplink receivers	None
Downlink Transmitters	HealthSat-2 provided with 10 W amplifier
Mechanical	Minor configuration changes
Basic Spacecraft Software	None

Parallel processing of the two missions continued during environmental testing. SSTL rented a single thermal-vacuum chamber which would hold both satellites, permitting the two missions to share these costs. Unfortunately, electromagnetic compatibility tests and vibration tests cannot be shared in this way, but there are some savings to be made in transportation and setup charges. Finally, during the launch campaign, logistics and management for two satellites need not be double those for one.

It is difficult to quantify the savings which resulted from the parallel processing of HealthSat-2 and PoSAT-1. Recently, SSTL has offered customers 10-20% savings on satellites which could be built in such a manner—truly identical satellites (which PoSAT-1 and HealthSat-2 were not) might save even more. For PoSAT-1 and Health-Sat-2, the mass-production savings were "distributed" in several ways. The missions' customers saw the effects in SSTL's ability to bargain—both missions saw more technical features and lower prices during initial negotiation. More importantly, because of increased efficiency and reduced staffing, more money was spent on improving the

---

[*] 14 on orbit, 2 launch or separation failures and 10 under construction in early 1996.

satellites. For example, the two missions shared development costs for a new backup onboard computer (based on the new 80C188 processor). Neither mission alone could have paid these costs. SSTL team members have a tongue-in-cheek saying that "three are as easy as one." Although not literally true, this saying captures the essence of the mass-production savings discussed here.

13.4.3 Operational Performance

Operational Results

During its first two years in orbit, PoSAT-1 met all of the customer's primary experimental objectives and many of the "optional" objectives (as marked in Table 13.4-3). In particular:

- **Store-and-Forward Communications.** The PoSAT-1 store-and-forward transponder has been used in several pilot projects for potential commercial and military uses. Experiments were done with ship-board ground stations involving the capture of specific images from the CCD camera which were then downloaded by the shipborne station. Portuguese UN forces in Angola have exchanged messages with stations in Portugal. Small, portable ground stations are being tested.

- **Digital Signal Processing.** The Digital Signal Processing Experiment relayed voice broadcasts to the Portuguese press during early satellite demonstrations. It has also been used to evaluate a practical MSK demodulator in low-Earth orbit. [Sun, 1995] Fig.13.4-10 shows some examples.

- **Earth Imaging**. Both the wide-angle and narrow-angle cameras are in use daily, having imaged 3400 scenes as of January 1996. [Sweeting, 1994] (See Fig. 13.4-10 for example photos.)

- **GPS**: The Global Positioning System receiver autonomously acquired position fixes which have been converted to smoothed orbital elements onboard. [Unwin, 1995]

- **Star Imaging System**: The Star Imaging System works and can determine the spacecraft's instantaneous attitude to within 0.05 deg. [Fouquet, 1994]

- **Charged Particle and Total Dose Experiments**: The Charged Particle and Total Dose Experiments have correlated radiation dosage with single-event upsets (in program and file-storage memory). [Underwood, 1995]

- **Spacecraft Health**. The spacecraft systems, batteries, arrays, transmitter, etc., have worked to specification, and show no signs of degradation.

Failure Resilience In Practice

Like many satellite missions, PoSAT-1 has been successful despite some small "anomalies." As we have described, the bus and payloads were designed to be failure resilient and to provide many operating modes controlled by software which can be uploaded after launch. It is informative to investigate one of the PoSAT-1 anomalies and show how it was resolved with the aid of new software.

The anomaly in question involves the CCD imaging system, shown in a much-simplified block diagram in Fig. 13.4-11. In normal operation, the Transputer sends an image request (including exposure settings) to the Microcontroller. The Microcontroller activates the CCD Imager through a special power switch, and captures the image into local random-access memory. The Transputer then reads the entire image into its own memory, where it can be processed and eventually prepared for downloading to the ground station. The system was operated in this nominal mode as soon as PoSAT-1 was stabilized, but it did not reliably produce images. The images it did produce had widely varying exposures and other more-extreme problems. Occasionally, a well-exposed image appeared. If the experiment system had been implemented entirely in hardware, or with software sequencing in read-only memory, the experiment would have been a failure. Fortunately, the PoSAT-1 Earth Imaging System is driven and sequenced by software which can be reloaded.

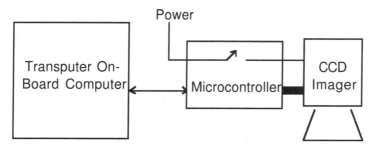

Fig. 13.4-11. PoSAT-1 Earth Imaging System.

The PoSAT-1 imaging team immediately decided to implement a closed-loop image capture program on the Transputer. This program exploits the fact that the Microcontroller which controls the CCD Imager also calculates image statistics, including an image histogram. Rather than transferring the entire image between the Microcontroller and the Transputer, the Transputer may request and receive only the histogram. This option was added to the Microcontroller software (in read-only memory) to support advanced automatic exposure loops. In the new closed-loop image capture mode, the Transputer commands the Microcontroller to activate the CCD Imager, and then to send the histogram. The software compares the histogram to certain good-image criteria, and repeats the capture process until a good image is indicated. When the acceptable image has been captured, it is finally transferred to the Transputer for downloading. The team implemented this closed loop software in less than a day, uploaded it, and tested it. The system immediately began to capture high quality images, and the closed-loop process is now the nominal PoSAT-1 image capture sequence. The Earth Imaging System was saved by software and by flexibility built into the system design.

The fault itself has been isolated to a DC-to-DC power converter first used on PoSAT-1. Poor documentation of this converter caused engineers to select a marginal capacitor value, and the converter does not start-up consistently at the imager's operating temperature (10 °C—well below room temperature but not a severe environment). This had not shown up during environmental testing because of the difficulty of conducting representative camera tests inside the thermal-vacuum chamber. It is not surprising to experience such difficulties during fast, low-cost missions. It is important to anticipate them and provide as many workarounds as practical. The PoSAT-1 Earth Imaging System functions routinely and successfully (with slight target timing degradation) despite this anomaly.

13.4.4 Conclusions

PoSAT-1 illustrates how a multi-mission technology demonstration microsatellite can be built rapidly and inexpensively. The translation of these lessons to operational scientific and commercial microsatellite missions must be done with care, but it certainly can be done. Using similar techniques, SSTL has built two operational radio monitoring microsatellites and two operational store-and-forward microsatellites. Each succeeding mission has had greater technical capabilities; yet, costs have remained the same. This increase in cost effectiveness comes partly from an improved ratio of semiconductor performance to cost and partly from SSTL's evolution of subsystem designs and management techniques.

PoSAT-1 results from a drive over seven previous missions to reduce space mission cost to the minimum level which will produce a meaningful mission. When available for $2 million to $5 million, microsatellite missions can fill new market niches—from educational aids such as KITSAT-1, PoSAT-1 and FASat-Alfa, to operational satellites such as HealthSat-1 & -2, S80/T, and CERISE. These missions are not the result of applying cost-reduction techniques to pre-existing mission objectives; they are the result of matching mission objectives to the capabilities of extremely low-cost space and ground segments.

Once SSTL has identified suitable mission objectives we develop a detailed design which maximizes performance within a fixed cost. We use state-of-the-art terrestrial microelectronics to build a computerized, configurable satellite. To achieve this safely, we apply the following rules of thumb:

- Radiation hardened electronics are not needed in a polar low-Earth orbit, but single event upset protection must be used
- Mission objectives should not require 100% system availability
- All instruments should be controlled by software which can be reloaded after launch
- The spacecraft bus should be resilient to software and hardware failures.

The points above will help you even if you are only building a single microsatellite as a one-off or first-time endeavor. If you intend to produce many microsatellites also consider these rules:

- Designing and building subsystems in-house increases control and reduces recurring costs
- A modular mechanical and electrical system design provides a basis for cost-effective product enhancement over several missions
- A policy of gradually introducing new technology can be exploited when missions are frequent.

The PoSAT-1 mission illustrates all of these rules of thumb for successful low-cost microsatellites.

References

Allery, M. N. and Castetbert H. 1994. "Microsatellite Operations in the first year of the S-80/T Mission," 2nd International Conference on Small Satellites Systems and Services, Biaritz, France, June 27–July 1, 1994.

Clapp, W. G. 1989. "Space Technology Education Through the Combined Efforts of Industry, Education, and Government." *The AMSAT Journal*, vol. 12, no. 3, November, pp. 35–37.

Clark, T. A. and E. Pavlis. 1991. "CHAOS: The Eccentricities of Eccentric Orbits." Presentation at the 3rd Phase 3-D design meeting, Orlando, FL, November.

Crabbs, R. F. 1995. "Where are the Space Visionaries." *Space News*, vol. 6, no. 1, January 9–15, p. 15.

Daniels, R. L. 1987. " 'Bare Bones' Propulsion for Small, Low Cost Satellites." *Proc. 1st Annual AIAA/USU Conference on Small Satellites*, Logan, UT, October 7–9.

———. 1988. "The Propulsion Systems of the Phase III Series Satellites." *AMSAT-NA Technical Journal*, vol. 1, no. 2, Winter, pp. 9–15.

Davidoff, M. 1990. *The Satellite Experimenter's Handbook*. Newington, CT: American Radio Relay League, Inc.

Diersing, R. 1993a. "The Development of Low-Earth-Orbit Store-and-Forward Satellites in the Amateur Radio Service." *Proc. IEEE 1993 International Phoenix Conference on Computers and Communications*, Tempe, AZ, March 23–26, pp. 378–386.

———. 1993b. "Characterization of the PACSAT File Broadcast System." *Proc. IEEE 1993 National Telesystems Conference*, Atlanta, GA, June 16–17, pp. 71–79.

———. 1993c. "Communications Services provided by the Pacsat-1, UoSAT-5, and Kitsat-1 Amateur Radio Satellites." *Proc. IEEE 1993 Global Telecommunications Conference*, Houston, TX, November 29–December 2, pp. 593–598.

———. 1994a. "The In-Orbit Operation of Four Microsat Spacecraft—Four Years Later." *Proc. 8th Annual AIAA/USU Conference on Small Satellites*, Logan, UT, August 29–September 1.

———. 1994b. "1993–94 Report on DOVE Recovery Activities." *Proc. AMSAT-NA 12th Space Symposium*, Orlando, FL, October 7–9, 1993, pp. 40–47.

Diersing, R. and G. Jones.1992. "Low-Earth-Orbit Store-and-Forward Satellites in the Amateur Radio Service." *Proc. 1992 IEEE National Telesystems Conference*, Washington, DC, May 19–20, pp. 8-7 to 8-14.

Doherty, W. H. 1936. "A New High-Efficiency Power Amplifier for Modulated Waves." *Proceedings of the IRE*, vol. 24, no. 9, September, pp. 1163–1182.

Fouquet M., et al. 1994. "PoSAT-1 Star Imaging System: In-flight Performance." *Proceedings of the European Symposium on Satellite Remote Sensing*, Rome, Italy, September 26–30, 1994.

Fox, T. L. 1984. *AX.25 Amateur Packet Radio Link-Layer Protocol*. Newington, CT: American Radio Relay League, Inc.

Guelzow, P. 1992. "The Date of OSCAR 13's Demise." *Satellite Operator*, no. 20, May, pp. 1–2, 12.

Harwood, K. A., M. Crookston, and R. Butler. 1992. "AMSAT P3D Satellite Project at Weber State University." *The AMSAT Journal*, vol. 15, no. 4, July–August, pp. 23–25.

Hewett, C. H. and H. L. Reed. 1994. "A Microparticle Recognition Experiment for Near-Earth Space On Board the Satellite ASUSat 1." *Proc. 8th Annual AIAA/USU Conference on Small Satellites*, Logan, UT, August 29–September 1.

Jansson, R. 1987. "Spacecraft Technology Trends in the Amateur Radio Service." *Proc. 1st Annual AIAA/USU Conference on Small Satellites*, Logan, UT, October 7–9 and *AMSAT-NA Technical Journal*, vol. 1, no. 2, Winter, pp. 3–8.

————. 1989. "AMSAT-NA Phase IV Project—Lessons in Distributed Engineering." *Proc. 3rd Annual AIAA/USU Conference on Small Satellites*, Logan, UT, ·September 26-28 and *The AMSAT Journal*, vol. 12, no. 3, November, pp. 11–15.

————. 1991. "The Shape of Things to Come." *Proc. AMSAT-NA Ninth Space Symposium*, Los Angeles, CA, November 8–10, pp. 44–52.

————. 1993a. "New Agreement with the European Space Agency on the Phase 3-D Design." *The AMSAT Journal*, vol. 16, no. 1, January/February, pp. 1, 6.

————. 1993b. "Phase 3-D Spacecraft Construction Activities." *The AMSAT Journal*, vol. 16, no. 4, p. 10.

————. 1993c. "The Phase 3-D Spacecraft." *Proc. AMSAT-NA Eleventh Space Symposium*, Arlington, TX, October 8–10, pp. 3–16 and *The AMSAT Journal*, vol. 16, no. 5, September/October, pp. 4–11.

Kahn, L. R. 1952. "Single-Sideband Transmission by Envelope Elimination and Restoration." *Proceedings of the IRE*, vol. 40, July, pp. 803–806.

King, J. A. 1993a. "Design Practices for Low Cost Spacecraft." Presentation given for AMSAT personnel, September 10.

————. 1993b. "Principles of HELAPS Transponder Amplifiers." *The AMSAT Journal*, vol. 16, no. 5, September/October, pp. 27–31.

King, J. A., R. McGwier, H. Price, and J. White. 1990. "The In-Orbit Performance of Four Microsat Spacecraft." *Proc. 4th Annual AIAA/USU Conference on Small Satellites*, Logan, UT, August 27–30.

Kitts, C. A. 1994. "Investigating the Integrated Control of Payloads with Amateur Satellites." *Proc. AMSAT-NA 12th Space Symposium*, Orlando, FL, October 7–9, pp. 91–97.

Kitts, C. A. and R. A. Lu. 1994. "The Stanford SQUIRT Micro Satellite Program." *Proc. AMSAT-NA 12th Space Symposium*, Orlando, FL, October 7–9, pp. 84–89.

Kitts, C. A. and R. J. Twiggs. 1994. "The Satellite Quick Research Testbed (SQUIRT) Program." *Proc. 8th Annual AIAA/USU Conference on Small Satellites*, Logan, UT, August 29–September 1.

Kudielka, V. 1990. "Langfristige Vorhersagen fur hoch-elliptische Satelliten-Bahnen."*AMSAT-DL Journal*, vol. 17, no. 2, June/August, pp. 5–7.

Meerman M. J. M., M. N. Sweeting, and M. G. Leitmann. 1994. "One Year in Space: Results from PoSAT-1." *Proceedings of the 8th Annual AIAA/Utah State University Conference on Small Satellites*, Logan, UT, August 29–September 1, 1994.

Meinzer, K. 1974. "A Linear Transponder for Amateur Radio Satellites." *UKW Berichte*, vol. 14, no. 2, 2nd Quarter, pp. 112–126.

———. 1979. "IPS, An Unorthodox High Level Language."*Byte*, January, pp. 146–159.

———. 1993. "The New Shape of P3-D—A Development Status Report." *The AMSAT Journal*, vol. 16, no. 3, May/June, pp. 9–12.

Miller, J. 1992. "May the Force be With You." *Satellite Operator*, No. 27, December, pp. 6–10.

———. 1993. "Managing OSCAR-13." *Proc. AMSAT-NA Eleventh Space Symposium*, Arlington, TX, October 8–10, pp. 46–55 and *Satellite Operator*, no. 37, October, pp. 1–9, 12.

———. 1994. "The Re-Entry of OSCAR-13." *Proc. AMSAT-NA 12th Space Symposium*, Orlando, FL, October 7–9, pp. 36–39 and *The AMSAT Journal*, vol. 18, no. 3, May/June 1995.

———1995a. Private communications. May 26.

———1995b. Private communications. September 11.

The Phase 3-D Design Team. 1994. "Phase 3-D, A New Era for Amateur Satellites." *Proc. AMSAT-NA 12th Space Symposium*, Orlando, FL, October 7–9, pp. 2–13.

Price, H. E. and J. W. Ward. 1990a. "PACSAT Protocol Suite—An Overview." *Proc. 9th ARRL Computer Networking Conference*, London, Ontario, Canada, pp. 203–206.

———. 1990b. "PACSAT Broadcast Protocol." *Proc. 9th ARRL Computer Networking Conference*, London, Ontario, Canada, pp. 232–244.

———. 1990c. "PACSAT Data Specification Standards." *Proc. 9th ARRL Computer Networking Conference*, London, Ontario, Canada, pp. 207–208.

Rademacher, J. and H. L. Reed. 1994. "Preliminary Design of ASUSat 1." *Proc. 8th Annual AIAA/USU Conference on Small Satellites*, Logan, UT, August 29–September 1.

Reeves, E. I. 1994. "A Cost Model for a University Satellite Program." *Proc. 8th Annual AIAA/USU Conference on Small Satellites*, Logan, UT, August 29–September 1.

Riportella, V. 1988a. "Ariane V-23 Launch Success Sets Stage for Phase 3-C Launch June 8." *Amateur Satellite Report*, no. 177, June 8, p. 1.

———. 1988b. "AMSAT-OSCAR-13 in Orbit, Under Control, and Working Perfectly." *Amateur Satellite Report*, no. 178, July 5, p. 1.

Stoner, D. 1959. "Semiconductors." *CQ*, April, p. 84.

Sun W., M. N. Sweeting, and M. S. Hodgart. 1995. "In-Orbit Communications Experiments Using PoSAT-1 DSPE for Optimizing LEO Microsatellite Communication Systems." *Proceedings of the International Conference on Telecommunications 1995*. Nusa Dua, Bali, Indonesia, pp. 157–160. April 3–5, 1995.

Sweeting M. N. and M. Fouquet. 1994. "Earth Observation by UoSAT Microsatellites: Six Years of Orbital Experience." *Proceedings of the European Symposium on Satellite Remote Sensing*. Rome, Italy, September 26–30, 1994.

Twiggs, R. J. and K. W. Reister. 1992. "Phase 3-D, A Student Manufacturing Engineering Challenge." *Proc. AMSAT-NA Tenth Space Symposium*, Washington, D.C., October 9–11, pp. 47–54.

Underwood C. I., et al. 1995. "Measurements of the SEE Environment from Sea Level to GEO using the CREAM & CREDO Experiments." *IEEE Transaction on Nuclear Science, No.6*. December 1995.

Unwin M. 1995. "A Practical Demonstration of Low-Cost Autonomous Orbit Determination using GPS." *Proceedings of the Institute of Navigation GPS-95*. Palm Springs, CA, September 13–15, 1995.

Ward J. W. 1990. "Store-and-Forward Message Relay Using Microsatellites: The UoSAT-3 PACSAT Communications Payload." *Proceedings of the 4th Annual AIAA/USU Conference on Small Satellites*, Logan, UT, August 27–30, 1990.

Ward, J. W. and H. E. Price. 1990a. "PACSAT Protocol: File Transfer Level 0." *Proc. 9th ARRL Computer Networking Conference*, London, Ontario, Canada, pp. 209-231.

Ward, J. W., H. E. Price, and BekTek, USA. 1994. "Re-usable Software for Microsatellite Onboard Computers." *Proceedings of the EUROSPACE Symposium on Onboard Data Management*. Rome, Italy, January 25–27, 1994.

———. 1990b. "PACSAT File Header Definition." *Proc. 9th ARRL Computer Networking Conference*, London, Ontario, Canada, pp. 245–252.

White, J. 1993. "Dove Progress Report and Future Operation." *Proc. AMSAT-NA Eleventh Space Symposium*, Arlington, TX, October 8–10, pp. 73–87

Wingo, D., Stluka, E., and C. Rupp. 1991a. "Amateur Satellite Communications and the SEDS AMSEP Project-Part I." *The AMSAT Journal*, vol. 14, no. 2, March, pp. 26–29.

———. 1991b. "Amateur Satellite Communications and the SEDS AMSEP Project—Part II." *The AMSAT Journal*, vol. 14, no. 3, May, pp. 6, 10.

Wingo, D., C. D. Bankston,, and J. Champa. 1991c. "SEDSAT 1 1991 Status Report." *Proc. AMSAT-NA Ninth Space Symposium*, Los Angeles, CA, November 8–10, pp. 186–190.

Appendix: Representative Small Spacecraft 1980–1995

	Name	Mass (kg)	Form	Function	Peri. (km)	Apo. (km)	Inc. (deg)	Launcher & Site	Launch Date	Prime Contractor
1	GFZ-1	20.6	Sphere	Geodetic laser reflector	400	400	51.4	Deployed from MIR airlock	4/19/95	K-T
2	Astrid (Freja-C)	27	Cubical	Magnetospheric research	966	1027	83	Cosmos; Plesetsk	1/24/95	SSC
3	SAFIR-R	38	Attached to RESURS	Communication	661	662	98.0	SL-16	11/4/94	DARA, OHB-System
4	APEX	208	Hexagonal cylinder	Experimental	360	2558	0	Pegasus	8/3/94	OSC
5	STRV 1A & 1B	53	Cube	Experimental	298	35624	99	Ariane 44 LP; Kourou	6/17/94	UK MoD DRA
6	MSTI-2	168	Cylinder	Strategic defense	342	455	96.9	Scout; WTR	5/9/94	BMDO & USAF Philips
7	DARPA Sat	204	Cube	Experiment	519	519	105	Taurus	3/13/94	Ball Corp.
8	BREM-SAT	63	Cylinder	In orbit	350	350	57	STS-60	2/9/94	ZARM, Univ. of Bremen
9	Clementine	233	Polygon	Space science	N/A	Helio-centric	66.7	Titan IIG; WTR	1/25/94	NRL
10	TUBSAT-B	50	Cube	Experimental	1120	1196	82.5	SL-14; Plesetsk	1/25/94	Tech. Univ. Berlin
11	Itamsat	12	Cube	Interferometric study	788	802	98.7	Ariane 40; Kourou	9/26/93	AMSAT-Italy
12	KITSAT-2	50	Cubical	Experimental	788	802	98.7	Ariane 40; Kourou	9/26/93	SSTL
13	HealthSat-II	50	Cubical	Communication	788	802	98.7	Ariane 40; Kourou	9/26/93	SSTL
14	Stella	48	Sphere	Geodetic studies	788	802	98.7	Ariane 40; Kourou	9/26/93	CNES
15	Eyesat 1 (OSCAR 27)	12.5	Cube	Interferometric study	788	802	98.7	Ariane 40; Kourou	9/26/93	Interferometrics AMRAD
16	PoSAT-1	50	Cubical	Experimental	788	802	98.7	Ariane 40; Kourou	9/26/93	SSTL
17	Temisat	42	Cube	Communication	932	967	82.5	SL-14; Plesetsk	8/31/93	Telespazio & KT
18	RADCAL	92	Octagonal	Radar calibration	750	884	89.6	Scout; WTR	6/25/93	DSI
19	ARSENE	150	Hexagonal prism	Radio amateur satellite	17699	37094	1.1	Ariane; Kourou	5/12/93	CNES
20	Kosmos	225		Military	1406	1417	82.6	SL-14; Plesetsk	5/1/93	
21	ALEXIS	115	Cylinder	Experimental	747	836	69.9	Pegasus; WTR	4/25/93	LANL
22	SEDS	39		Deployment system	20	20		Delta II; ETR	3/30/93	NASA Marshall
23	SCD-1	115	Octagonal prism	Environmental data relay	728	794	25	Pegasus; ETR	2/9/93	AEB
24	OXP-1	14.5		Communication	731	795	25	Pegasus; ETR	2/9/93	
25	MSTI-1	180	Cylinder	Strategic defense	292	378	91.2	Scout; WTR	11/21/92	BMDO; AF Philips Lab.
26	CTA	82		Canadian target assembly	164	243	47.5	STS	10/22/92	
27	Freja	214	Cylinder	Ionospheric studies	596	1759	63	CZ-2C; Shuang Cheng Tzu	10/6/92	SSC
28	Pion 5/6	50		Passive sub-satellite	224	233	82.5		8/19/92	
29	KITSAT-1	50	Cubical	Experiments	1304	1325	66	Ariane 42P; Kourou	8/10/92	SSTL
30	S80/T	50	Cubical	Experimental comsat	1303	1324	66	Ariane 42P; Kourou	8/10/92	SSTL
31	SAMPEX	158		Solar anomaly explorer	510	680	81.7	Scout; WTR	7/3/92	
32	SROSS 3	106		Gamma-ray & aeronomy studies	256	429	46	ASLV; Sriharikota	5/20/92	
33	Magion 3	52	Orbicular	Sub-satellite	435	3066	82.6	Interkosmos	12/18/91	ASGI
34	Kosmos	225		Military	1406	1417	82.6	SL-14; Plesetsk	9/1/91	
35	SARA	39	Cube	Jovian amateur radio astronomy	762	770	98.5	Ariane 4; Kourou	7/17/91	ESIEE Space / CNES
36	Tubsat-A	38	Cube	Comsat	763	773	98.5	Ariane 4; Kourou	7/17/91	Tech. Univ. Berlin
37	UoSAT-5	50	Cubical	Technology experiments	761	773	98.5	Ariane 4; Kourou	7/17/91	SSTL

See end of table for acronyms

Attitude Control	Sensors	Telemetry System	Data Rate	Transmit Power (W)	Avg. Power Gen. (W)	Power System	
See note 1	N/A	N/A	N/A	N/A	N/A	N/A	1
Simple spinner, 2 air coils & fluid nutation damper	Analog SS, two 2-axis flux-gate Mags.	Patch ants, turnstile ants	PCM bi-phase, 128	D: 2	42	Solar arrays NiCd batt	2
N/A	N/A	Linear polarized & circ. polarized ants	U: 0.3 / 2.4 D: 0.3 / 2.4	U: 5 D: 5		Powered from RESURS	3
3 magnetic coils (solid core), 1 wheel	1 SS, 1 Mags., 1 GPS	Dipole ants	U: 1 D: 128	D: 3.1	137	Si SA NiH$_2$ batts	4
Air coils / cold gas thruster, fluid nutation damper	V-slit & analog SS	Two dipole ants	U: 0.1 D: 1	U: 6 D: 1	30	GaAs SA, & NiCd batts	5
3-axis control, reaction wheels	ES & SS's	Nadir ant	U: 2 D: 1M		145	GaAs SA & NiH$_2$ batts	6
Active (spin stabilized); magnetic coils (solid core)	SS, ES, HCI & Mags (2-axis)	Micro-strip, Omni	U: 1 D: 320	U: 1k D: 2.5		GaAs SA, 2 NiCd batts	7
Momentum wheel with magnetic torquer	Mag, SS & Star	4 dipole ants	D: 9.6 U: 1.2	D: 2	30	Si-SA, 2 lead-acid batts	8
Attitude thrusters, 4 reaction wheels	Ring laser gyro, 2 Star, 2 SS's	High gain ant, Omni ant	D: 128			GaAS SA & NiH$_2$ batts	9
3 axis Mag, 3 reaction wheels	CCD-Star / rate gyro, CCD-SS	Single dipole ant (VHF)				NiCd batt	10
Passive magnetic stabilization		4 turnstile ants	U: 1.2 / 9.6 D: 9.6 / 38.4	D: 4	8-10	GaAs SA & NiCd batts	11
Gravity gradient, 12 air coils for active damping	HCI & SS's	Omni ants	U: 1.2 / 9.6 D: 9.6 / 38.4	U: 50 D: 10	20	GaAs SA & NiCd batts	12
Gravity gradient, 12 air coils for active damping	HCI & SS's	Omni ants	U: 1.2 / 9.6 D: 9.6 / 38.4	U: 50 D: 10	20	GaAs SA & NiCd batts	13
Spin stabilized							14
		Omni whip ants	300 / 9.6k baud	0.5-2.5			15
Gravity gradient & 12 air coils for active damping	HCI & SS's	Omni ants	U: 1.2 D: 9.6 / 38.4	U: 50 D: 10	20	GaAs SA & NiCd batts	16
Active stabilization with 2 magnetic air coils	Line CCD	TM / GPS / monopole ants (steel-band)	U: 1.2 D: 1.2 / 4.8	5		Si SA & batts	17
Gravity gradient, hysteresis rod dampers, torque coil	3-axis Mag & GPS position	Omni, GPS & beacon ants	U: 4.8 D: 192	U: 65 D: 10	18-25	SA & NiCd batts	18
Cold gas (N$_2$) thruster, 1 RPM for stability				U: 15 D: 2	60	GaAs SA & NiCd batts	19
							20
Air-coil & solid core coils	SS	RHCP crossed dipole ants	U: 9.6 D: 750	U: 20 D: 10	50	SA & batt	21
							22
Magnetic torquer		ants			60 (in Sun)	SA & NiCd batt	23
							24
3-axis control, cold gas thruster, reaction wheels	ES & SS's	Omni ant	U: 32 D: 1		132	GaAs SA & NiH$_2$ batt	25
							26
Spin stabilized, MTs, viscous ND, two 3-axis Mags	2 SS's, HCI	Whip ant	D: 524 / 1.2	D: 2	75	NiCd batts, solar array	27
							28
Gravity gradient 12 air coils for active damping	HCI & SS's	Omni ants	U: 1.2 / 9.6 D: 9.6 / 38.4	U: 50 D: 10	20	GaAs SA & NiCd batts	29
Gravity gradient 12 air coils for active damping	HCI & SS's	Omni ants	U: 1.2 / 9.6 D: 9.6 / 38.4	U: 50 D: 10	20	GaAs SA & NiCd batts	30
MW, 3 magnetic torquers, 3-axis Mag	2-axis SS, 4 coarse SS's	2 S-band transponders	D: 900	5	100	NiCd batt & GaAs SA	31
							32
							33
							34
Passive; permanent magnet		Orthogonal ants & quasi-omni ant		D: 0.5	5	Si SA, single batt	35
3-axis mag	CCD-Star / rate gyro, CCD-SS	Single dipole ant (VHF)				NiCd batt	36
Gravity gradient 12 air coils for active damping	HCI & SS's	Turnstile UHF & monopole VHF ants	U: 1.2 / 9.6 D: 9.6 / 38.4	U: 50 D: 10	20	GaAs SA & NiCd batt	37

Appendix: Representative Small Spacecraft 1980–1995

	Name	Mass (kg)	Form	Type	Peri. (km)	Apo. (km)	Inc. (deg)	Launcher & Site	Launch Date	Prime Contractor
38	ORBCOMM-X	22		Comsat	766	773	98.5	Ariane 4; Kourou	7/17/91	OSC
39	Microsat	22		Low-Earth orbit communications	359	457	82	Pegasus; Pacific	7/17/91	
40	Losat-X	74	Rect. parallelepiped	Low altitude experiment	402	416	40	Delta 2; ETR	7/4/91	DoD: USAF Space Sys.
41	ISES (REX)	85	Cylinder	Radiation experiment	770	871	89.6	Scout; WTR	6/29/91	
42	CRO	197		Chemical release observation	243	261	56.9	STS; WTR	5/2/91	
43	Kosmos	40		Military	1456	1470	74	SL-8; Plesetsk	2/12/91	
44	OSCAR 21	92	Tri-star	Radio satellites	953	1008	82.9	SL-8; Plesetsk	1/29/91	AMSAT-DL
45	PRC	4		Atmospheric balloon	775	804	99	CZ-4; Taiyuan	9/3/90	
46	Badr-1	52		Experiments	201	984	28.4	Xichang; CZ-2E	7/16/90	
47	Macsat	68	Cylinder	Military communication	605	765	89.9	Scout; WTR	5/9/9	DoD: USAF Space Sys.
48	Pegsat	178		Chemical release observation	453	645	94.1	Pegasus; Pacific	4/5/90	
49	USA 55	25		Deployed by Pegasus	487	661	94.1	Pegasus; Pacific	4/5/90	
50	Ofeq 2	160	Irreg. octoganol	Technology experiment	209	1577	143.2	Shavit; Yavne	4/3/90	Israel Space Agency
51	Hangoromo	12		Telemetry test				Muses A	3/19/90	
52	Debut	60	Orbicular	Boom/ Umbrella test	112	909	99	H-1; Tanegashima	2/7/90	
53	JAS 1B	50	Orbicular	Radio satellite	112	909	99	H-1; Tanegashima	2/7/90	
54	Muses A	185		Injected lunar orbit				M-3S-II; Kagoshima	1/24/90	
55	UoSAT 4 (OSCAR 15)	47.5	Cubical	Experiments	786	800	89.7	Ariane 4; Kourou	1/22/90	SSTL
56	UoSAT 3† (HealthSAT-1)	45.5	Cubical	Communications payload	785	800	98.7	Ariane 4; Kourou	1/22/90	SSTL
57	Microsats 1-4 (Oscar 16-19)	12	Cubical	Radio satellites AMSAT	784	800	89.7	Ariane 4; Kourou	1/22/90	AMSAT-NA & Weber State
58	Magion 2	50		Receives signals	504	2494	82.5	SL-14; Plesetsk	10/3/89	
59	Kosmos	100		Communications	1422	1440	82.5	SL-14; Plesetsk	9/14/89	
60	Pion	78		Atmospheric research sat.	255	272	82.5	SL-4; Resurs-F	6/1/89	
61	Ofeq 1 (Horizon 1)	155	Irregular octagonal	Experimental satellite	250	1150	142.9	Shavit; Negev	9/19/88	Israel Space Agency
62	SRS2 (Rohini)	150		Monocular stereo scanner				ASLV; Sriharikota	7/13/88	
63	Nova 2	165		Navigation	1150	1199	90	Scout; WTR	6/16/88	
64	Nova 3	174		Navigation	1150	1199	90	Scout; WTR	6/16/88	
65	AMSAT-Oscar 13 (AMSAT IIIC)	92	Tri-star	Amateur radio	2314	36495	57.4	Ariane 4; Kourou	6/15/88	AMSAT-DL
66	San Marco D (5)	236	Sphere	Upper atmosphere studies	263	615	3	Scout; San Marco	3/25/88	Uni. Rome NASA, FRG
67	Polar Bear	125		Polar experiments	960	1015	89.6	Scout; WTR	11/14/86	
68	OSCAR 12 (JAS-1)	50	Orbicular	Radio satellites	1480	1497	50	H-1; Tanegashima	8/12/86	
69	Glomr	52		Relayed ground signals	304	332	57	STS; ETR	11/1/85	
70	Navstar 11	163		Navigation	1988 7	20474	63.4	Atlas E; WSMC	10/9/85	
71	Planet A (Suisei)	141		Encountered Halley's Comet				M-3S-II; Kagoshima	8/18/85	
72	OSCAR	55		Navigation	1000	1260	89.9	Scout; WSMC	8/3/85	
73	Nusat	52		Air traffic control	301	339	57	STS	4/29/85	
74	Sakigake	141		Encountered Halley's Comet				M-3S-II; Kagoshima	1/7/85	

See end of table for acronyms

Attitude Control	Sensors	Telemetry System	Data Rate	Trans. Power (W)	Avg. Power Gen. (W)	Power System	
							38
							39
Active attitude control with 6 solid core torque rods	SS; GPS; Star; Mags; gyros	Ant; micro-strip patch	U: 1 D: 32	U: 1k D: 2.5	200 BOL	GaAs SA & NiCd batts	40
							41
							42
							43
Spin stabilized, re-orientation by electromag. torquing	ES & SS	Ants, 3 transponders	U: 0.4 D: 0.4	D: 50/50/1	50/35 B/EOL	NiCd batt & SA	44
							45
							46
Gravity gradient with boom							47
							48
							49
3-axis rate gyro, spin-stabilized	Mag, SS				53	NiCd batt, SA	50
							51
							52
							53
							54
Gravity gradient 12 air coils for active damping	HCl & SS's, Mag	Turnstile UHF & monopole VHF ants	U: 1.2/9.6 D: 9.6/38.4	U: 50 D: 10	10	GaAs SA & NiCd batts	55
Gravity gradient 12 air coils for active damping	HCl & SS's, Mag	Turnstile UHF & monopole VHF ants	U: 1.2/9.6 D: 9.6/38.4	U: 50 D: 10	10	GaAs SA & NiCd batts	56
Passive; 4 PM; 7 hysteresis damping rods	Webersat: CCD-camera, HCl		U: 1.2/4.8 D: 1.2/4.8	D: 4	5.8	NiCd batt & SA	57
							58
							59
							60
3-axis rate gyro, spin-stabilized	Mag, SS				53	NiCd batt & SA	61
							62
							63
							64
Spin stabilized, re-orientation by electro. torquing	ES & SS	Ants, 3 transponders	U: 0.4 D: 0.4	D: 50/50/1	50/35 B/EOL	NiCd batt & SA	65
							66
							67
PM & gravity gradient		2 turnstile ants	U: 1.2 D: 1.2	U: 100	8.5	Si SA & NiCd batt	68
							69
							70
							71
							72
							73
							74

Appendix: Representative Small Spacecraft 1980–1995

	Name	Mass (kg)	Form	Type	Peri. (km)	Apo. (km)	Inc. (deg)	Launcher & Site	Launch Date	Prime Contractor
75	NOVA 3	165		Navigation	1150	1198	90.1	Scout; WTR	10/12/84	
76	AMPTE 3 (UKS)	77		Magnetosphere measurement	1002	113417	26.9	Delta; ETR	8/16/84	
77	AMPTE 1 (CCE)	242		Magnetospheric studies	974	49817	3.8	Delta; ETR	8/16/84	
78	UoSAT 2 (OSCAR 11)	52	Cubical	Magnetospheric studies	674	693	98.2	Delta; ETR	5/1/84	SSTL
79	Ohzora	180		Middle atmosphere	317	503	74.6	M-3S-II; Kagoshima	2/14/84	
80	IRT	91		Rendezvous target	296	279	28.4	STS; ETR	2/5/84	
81	AMSAT-OSCAR 10	70	Tri-star	Radio satellite	3990	35461	27.1	Ariane; Kourou	6/16/83	
82	Rohini 3	42		Earth resources	384	785	46.6	SLV-3; Sriharikota	4/17/83	
83	Astro-B	185		X-ray observation	465	455	31.5	Mu-3s; Kagoshima	2/20/83	
84	Iskra	28	Hexagonal prism	Radio satellite	342	357	92.3	Salyut 7 (Progress)	5/17/82	ISKRA of MAI
85	CAT 4	217		Test equipment	241	34115	63.4	Ariane; Kourou	12/20/81	
86	UoSAT 1 (Oscar 9)	52	Cubical	Radio science	472	478	97.7	Thor-Delta; WTR	10/6/81	SSTL
87	PRC-10	28		Earth resources	235	1615	59.4	FB-1; Shuang Cheng Tzu	9/19/81	
88	Rohini 2	38		Earth resources	186	418	46.3	SLV-3; Sriharikota	5/31/81	
89	Astro (Hinotori)	185		Scientific	541	592	31.3	M-3S-II; Uchinora	2/21/81	
90	SROSS C2	113		Gamma-ray burst analyzer	433	917	46	ASLV; Sriharikota	5/4/91	

1. GFZ is a passive satellite.

KEY:

Prime contractors:

SSC	=	Swedish Space Corporation
OSC	=	Orbital Sciences Corporation
KT	=	Kayser-Threde
NRL	=	Naval Research Laboratory
DSI	=	Defense Systems, Inc.
BMDO	=	Ballistic Missile Defense Organization
SSTL	=	Surrey Satellite Technology Limited (University of Surrey)
UK MoD DRA	=	United Kingdom Ministry of Defense, Defense Research Agency
AEB	=	Agencia Espacial Brasileira
ASGI	=	Academy of Sciences Geophysical Institute, Russian Federation

Power system:

SA	=	Solar array

Sensors:

Star	=	Star sensor
SS	=	Sun sensor
ES	=	Earth sensor
Mag	=	Magnetometer
MT	=	Magnetic torquer
ND	=	Nutation damper
MW	=	Momentum wheel
PM	=	Permanent magnet
HCI	=	Horizon crossing indicator

Telemetry:

Ant	=	Antenna
Omni	=	Omni-directional antenna

Launch Sites:

ETR	=	Air Force Eastern Test Range
WTR	=	Air Force Western Test Range

Transmit Power:

U	=	Uplink
D	=	Downlink

† Originally known as Oscar 14.

See end of table for acronyms

Attitude Control	Sensors	Telemtry System	Data Rate	Trans. Power (W)	Avg. Power Gen. (W)	Power System	
							75
							76
							77
Gravity gradient 12 air coils for active damping	HCI & SS's, Mag	Turnstile UHF & monopole VHF ants	U: 1.2 / 9.6 D: 9.6 / 38.4	U: 50 D: 10	10	GaAs SA &NiCd batts	78
							79
							80
		Ants, transponder			50/35 B/EOL	NiCd batt & SA	81
							82
							83
No attitude control							84
							85
Gravity gradient 12 air coils for active damping	HCI & SS's, Mag	Turnstile UHF & monopole VHF ants	U: 1.2 / 9.6 D: 9.6 / 38.4	U: 50 D: 10	10	GaAs SA &NiCd batts	86
							87
							88
							89
							90

Index